Understanding and Conducting Research in the Health Sciences

Understanding and Conducting Research in the Health Sciences

Christopher J. L. Cunningham
Department of Psychology
The University of Tennessee at Chattanooga
Chattanooga, TN

Bart L. Weathington
Department of Psychology
The University of Tennessee at Chattanooga
Chattanooga, TN

David J. Pittenger
College of Liberal Arts
Marshall University
Huntington, WV

WILEY

Published by John Wiley & Sons, Inc., Hoboken, New Jersey
Published simultaneously in Canada

Library of Congress Cataloging-in-Publication Data:
Cunningham, Christopher J. L., author.
Understanding and conducting research in the health sciences / Christopher J.L. Cunningham, Bart L. Weathington, David J. Pittenger.
p. ; cm.
Includes bibliographical references and index.
ISBN 978-1-118-13540-2 (cloth) – ISBN 978-1-118-51494-8 (eBook) (print) – ISBN 978-1-118-51492-4 (ePDF) (print) – ISBN 978-1-118-51493-1 (ePub) (print) – ISBN 978-1-118-51490-0 (eMOBI) (print)
I. Weathington, Bart L., author. II. Pittenger, David J., author. III. Title.
[DNLM: 1. Biomedical Research–methods. 2. Research Design. W 20.5]

610.28'4–dc23

2013001307

10 9 8 7 6 5 4 3 2 1

We dedicate this book to the mentors, family members, students, and colleagues who have guided and shaped us personally and professionally.

Contents

Preface

> Understanding health requires understanding people and understanding people requires knowledge of behavioral and social science.
>
> —Christopher J. L. Cunningham, Bart L. Weathington, and David J. Pittenger

PHILOSOPHY FOR WRITING THIS BOOK

Professionals in the health sciences interact with people every day. It is increasingly the case that issues of cultural competence, humanism, interpersonal compassion, and patient perceptions of care quality are on the minds of those responsible for developing tomorrow's physicians. A shared characteristic of all these and many other issues is that they are based on perceptions, behaviors, and cognitions. These latter elements are core elements to human psychology and to the ways in which people interact with each other and their environments.

Given this reality, it is important to consider the role of behavioral and social science research within the health science professions. Unfortunately, many professionals-in-training in this area receive little to no training in basic social science research methods, focusing instead on biological or other forms of laboratory-based study. Although such highly controlled studies are important for the development and testing of new medications and treatments, the health sciences cannot simply ignore the more psychological and sociological factors that also impact the application of health science knowledge. This is our primary reason for writing this book.

The importance of social and behavioral processes within the health sciences is often overlooked. Because of this, we offer the present textbook to help teach core research concepts that can help health science professionals better understand and work with complex behavioral and social processes. For those of us who teach research methods, one of the most gratifying experiences is watching others develop a passion for conducting research and an understanding of the practical value of systematic information gathering and decision making. Because the vitality of any branch of science depends on a solid grounding on research methods and analytic strategies, it is our belief that understanding research methods is among the most important competency building blocks within the health sciences.

The present text represents our effort to provide a resource for health science professionals-in-training and for established professionals wishing to expand their expertise. The material we present may also be helpful to advanced undergraduate students who intend to pursue graduate work in the health sciences, especially given the increasing importance of this content in upcoming versions of the Medical College Admission Test and given the increasing emphasis placed on research within most graduate medical education programs. We have incorporated a variety of key features to make the material in this text accessible, interesting, and practically useful, without watering down the importance of systematically identifying questions and appropriate methods for answering those questions.

General Style

To keep this book accessible and useful, we have written this book in editorial style that is narrative in its tone and inviting for the reader. Although there is a clear conversational style to the text, there is no lack of rigor in the material that is presented. Throughout the text, we provide comprehensive accounts of behavioral and social scientists' best ideas and research methods.

Examples

A related strategy that we applied throughout this text was the selection of relevant and contemporary examples of behavioral and social science research within health-related fields. Our goal in selecting these examples was to illustrate important topics covered in the chapter and to show how health science professionals can use the research methods and analytical tools we present in this text to address complex and important questions.

Integration of Research Methods and Statistical Concepts

Those familiar with basic parametric statistics know that they come with many mathematical strings attached. If the researcher cannot ensure that the data and the design of the data collection methods meet these basic assumptions, then the inferences derived from eventual statistical analyses may be suspect. In some cases, a statistic is extremely robust to violations of its assumptions; other statistics fail to withstand even minor deviations from assumptions.

An assumption we made while writing this book is that that our readers have completed at least a general introduction to statistics course. Consequently, a certain level of statistical knowledge is assumed, especially regarding more basic concepts such as measures of central tendency, measures of dispersion, or standard scores. Nonetheless, many students have a tendency to forget or temporarily misplace much of their statistical knowledge following completion of their required statistics course. To address this, we review critical statistical concepts as they relate to specific methodological techniques. Additionally, we include a focused appendix that can serve as an initial statistics review.

Therefore, in writing this book, we wanted to ensure that students receive a comprehensive and detailed review of the best techniques for studying behavior and social phenomena within the health sciences. Consequently, where appropriate, our chapters provide a comprehensive review of research methods and the statistical concepts that support them. The review of the statistical principles is comprehensive yet conceptual and nontechnical. Students who have completed a course in statistics will find these sections to be a useful review of important topics. Students who have not studied statistics will find these sections a suitable and readable introduction to these topics.

For example, the review of sampling procedures examines the different methods researchers use to create representative samples and demonstrates how the central limit theorem allows one to make valid inferences using sample statistics. An additional recurring topic throughout the book is the issue of statistical power. Because of the importance of this concept, reviews of statistical power occur in many parts of the book. The goal is to show students that they can control the power of their research by adjusting several design parameters beyond sample size which is often limited in health science research.

ORDER OF CHAPTERS

We arranged the sequence of chapters to both match the steps in conducting research and aid readers in learning how to design and implement a research project. Consequently, the first few chapters present background information, ethics, and an overview of various research methods. The second section is focused on core research topics including bibliographic research and the writing of research reports, sampling, and identification/developments and data collection strategies. We present this information relatively early in the text as we have found that it is counterproductive to wait until the end of a research methods course to discuss these nuts and bolts elements (especially writing style requirements and proper reporting techniques). A concrete understanding of core research elements and what the finished project should look like aids student researchers in learning about and planning a research project.

The third and fourth sections of the present text discuss a variety of specific research designs, highlighting their utility, strengths, and limitations. The methods presented in this section include single-factor and multifactor designs, as well as single-subject and nonexperimental methods.

While writing this text, we worked to provide readers and instructors the flexibility to rearrange the order of the chapters without a loss of continuity. Ultimately, the goal of a structured research methods course is to produce informed consumers of existing research and informed producers of new or refined knowledge. We believe that covering the chapters in the order we have arranged them will aid in the achievement of this goal, but some may find that a different order of chapters better fits their style or current needs.

A text should challenge its readers beyond their current ability. Thus, while we labored to write as clearly as we could, we also labored to ensure that we challenge readers to extend beyond the bounds of their comfort and existing knowledge regarding research. It is not necessary that every chapter or topic must be covered in sequence or in total; this is a fairly dense book and we hope that both student and instructor will recognize that the book is a resource from which to draw information.

Pedagogy

Each chapter uses multiple methods to present the material, including clearly written text, familiar and interesting examples, and visual illustrations to help the reader understand complex and abstract concepts. Specific pedagogical features include

- **Integration of Research:** Each chapter includes example research studies and critical thinking exercises. The goal of these exercises is to help the reader apply critical concepts to a research scenario.
- **Knowledge Checks:** Each chapter contains several knowledge check questions, which consist of a series of questions that require the reader to apply the material to objective problems. These questions require more than rote memorization because they ask the reader to apply the material.
- **Multiple Presentations of Concepts:** Throughout the book, the reader will find a combination of text, pictures, and examples to illustrate various concepts.
- **Glossary:** Each chapter contains definitions of important terms.
- **Statistical Review and Integration:** The first appendix is a statistics review designed to help students remember and understand basic statistical concepts. Additionally, many chapters have a section that deals with the statistics underlying the topics covered in that chapter.
- **Statistical Tables:** An appendix contains a comprehensive list of commonly used statistical tables.

ACKNOWLEDGMENTS

Although we are the authors of the book, the present work was touched and improved by the involvement of others at various stages throughout the writing process. We sincerely appreciate the feedback we received from each other, external reviewers, and the editorial team at Wiley. The book you are about to read has been very much improved because of this revision process; any remaining errors, of course, are our own. We are especially grateful to the editorial staff at Wiley. In particular, we thank Tisha Rossi and Jackie Palmieri for helping us initiate this project. Countless other professionals working at John Wiley & Sons, Inc. also deserve our sincere thanks for helping us to transform our idea for this text into a reality.

CHRISTOPHER J. L. CUNNINGHAM
BART L. WEATHINGTON
DAVID J. PITTENGER

Part One

Overview of the Research Process

Chapter 1

Behavioral and Social Research in the Health Sciences

All the world is a laboratory to the inquiring mind.
—Martin H. Fischer

INTRODUCTION

The best potential solutions to health-related challenges are frequently changing, as promising new techniques, therapies, and medications are continually developed and tested. This never-ending process relies on science and also carries with it the requirement that professionals within the health sciences are able to understand basic research and its potential applications within their daily health-related work. Despite the fact that so much of the widely touted and shared health sciences or medical research is focused on relatively impersonal methods of treatment or aspects of the process of treatment (e.g., medication errors or the impact of new technologies), so much of what impacts the quality of life of individual patients,

Understanding and Conducting Research in the Health Sciences, First Edition. Christopher J. L. Cunningham, Bart L. Weathington, and David J. Pittenger.

their families, and entire communities is directly linked to behavioral and social factors operating within the person and the environment in which that person functions.

Understanding complex human behaviors and cognitions associated with health is no easy task. Without an empirical, scientific approach to gaining such insight about those who work in the health sciences and those who are served by the health sciences, it is likely that either or both sides of this relationship will be operating with potentially dangerous, incomplete information. Having a solid understanding and appreciation of basic behavioral and social research methods will help you to make higher-quality and better-informed decisions. This text is designed to provide a rich introduction to the basic behavioral and social science research methods that will help you generate new health-related knowledge and translate existing health science knowledge into practice.

WHY IS UNDERSTANDING RESEARCH METHODS SO IMPORTANT?

There are three core types of information that anyone seeking to understand human behavior must know: (1) how to conduct and interpret high-quality research, (2) how to develop and evaluate measures of human thought and behavior, and (3) how to use basic statistics to make sense of available and relevant data.

Why are the three core topics just highlighted so important? Think about it—researching, analyzing, and reporting are three of the most generalizable and valuable skills learned during higher education that can help you find a job, keep a job, and more generally, make a meaningful contribution to society. You can think big thoughts and theorize all day long, but without these three research-related skills, your great ideas will never translate into credible actions. We do not want your good ideas to be restricted by the boundaries of your mind. This is why we all sincerely hope that you are not dreading learning about research methodology or fearing something nonspecific about the process of research. There is nothing scary here, just a systematic approach to learning, understanding, and questioning that will benefit you, no matter what path you take over the course of your career.

The most efficient, credible, and ultimately useful techniques for studying and understanding human behavior apply the scientific method in some way, shape, or form. Proper test development and utilization help to ensure the best possible data are collected. Appropriate statistical techniques facilitate interpretation of these data. We firmly believe that the only way to become a legitimate consumer and producer of knowledge is to first learn how to work with the tools of the trade: the scientific method, tests and assessments, and statistics.

This book will help you learn how to conduct and understand research within the health sciences that addresses questions with behavioral and social science underpinnings. Such questions could involve interpersonal phenomena or person-environment linkages, among many other possibilities. We will cover how to develop and evaluate tests, surveys, and other measures of behavior. Throughout the chapters of this book we will also remind you of how basic statistics can help as you work to address specific research questions. If you feel you need a refresher in the most basic statistics, a review is included in Appendix A.

THE ROLE OF BEHAVIORAL AND SOCIAL SCIENCE

- *What are some examples of science in your life?*
- *Have you "researched" anything today?*
- *What are some big decisions or questions you are currently considering?*

H.G. Wells, a nineteenth-century author, predicted that "statistical thinking will one day be as necessary for effective citizenship as the ability to read and write" (as cited by Campbell, 1974). We strongly believe this prediction has come true. Although you may not plan to become a researcher yourself, working in the health sciences (and many other areas of life) will force you to confront issues that can be addressed only with the aid of scientific research. Consider the following example issues:

- What intervention technique is most likely to be effective at reducing alcohol consumption among teenagers in this community?
- What are the best ways to demonstrate empathy when interacting with patients?
- Are cultural competence training programs or interventions effective?
- Are there any side effects associated with this new medication that might negatively affect patients' quality of life?
- How can we improve patient safety and operational efficiency without reducing our level of compassionate care within this hospital?

These are just a few examples of relevant behavioral and social science-related questions that can commonly arise when working in the health sciences. Knowledge of basic behavioral and social science research methods can give you the confidence needed to ask these and other difficult questions and to actually find the important answers.

Take, for example, the classic legal case of *Daubert v. Merrell Dow Pharmaceuticals, Inc.* (1993). In this case, the Supreme Court ruled that judges, not jury members, must determine the merits and scientific validity of testimony given by expert witnesses. In response to the court's decision, the Federal Judicial Center (1994) developed the book *Reference Manual on Scientific Evidence* to help judges and lawyers understand the principles of research methods and statistics. As the authors of the book noted, "no longer can judges . . . rely on their common sense and experience in evaluating the testimony of many experts. . . . The challenge the justice system faces is to adapt its process to enable the participants to deal with this kind of evidence fairly and efficiently and to render informed decisions" (p. 1). As H.G. Wells predicted, the knowledge of the scientific method is now a vital part of our government and judicial system and, therefore, our everyday lives.

You are not alone if you fear research in general and perhaps statistics more specifically. Many otherwise very intelligent and confident people seem to freeze like a deer in the headlights when anything statistics-like appears because they do not understand the relevance or importance of these topics to their own lives. We hope that by the time you finish this text, you will understand the relevance of the scientific method, proper measurement, and appropriate statistics and have confidence in a newfound skill set that will serve you for the rest of your life.

THE SCIENTIFIC METHOD

When designing and conducting behavioral and social science research, perhaps the most important element is to ensure that you adhere to the scientific method. Knowing each of its steps and how they are interrelated will allow you to conduct the highest-quality research possible, in any domain. Perhaps the easiest way to remember the scientific method from start to finish is to learn the mnemonic *HOMER* (Lakin, Giesler, Morris, & Vosmik, 2007), which stands for

1. *H*ypothesize
2. *O*perationalize
3. *M*easure
4. *E*valuate
5. *R*eplicate, revise, report.

These are the core steps to the scientific method and they should sound at least vaguely familiar from various introductory science courses you completed in middle school, high school, and college. The rest of this text focuses on ensuring you will finish with a working knowledge of all five components.

For starters, however, it is important to note that good behavioral and social science research begins with the identification of hypotheses or informed expectations about the particular phenomenon you are about to study. Once these expectations are identified, you can then work on figuring out how you would collect data, which would allow you to test your expectations. This process of operationalization is the bridge that connects relatively abstract, high-level hypotheses to the actual measuring of data. When your data are then collected you have the challenge of evaluating them and either confirming or disconfirming your hypotheses. This is not the end of the research process, however, as good science is open to modification and improvement, hence the need to replicate, revise, and report (or share) your findings with others.

BRIEF HISTORY OF SCIENTIFIC REASONING

Science is a way of thinking about and explaining the world around us. The scientific method consists of the process used for collecting, analyzing, and drawing conclusions from data. Research methods and statistics are complementary techniques that we use to acquire information and to reach reasonable conclusions. When we speak of research methods, we refer to procedures for collecting information. When we speak of statistics, we refer to procedures for organizing, summarizing, and making inferences from the data. A little bit of history here can help you to understand how this scientific approach to understanding behavioral and social phenomena has developed.

Throughout human history, people have struggled to protect, maintain, and improve their health. It can be argued that one's health is the most valuable asset or resource one has, especially as time passes and as the end of one's life approaches. Despite this reality, as a science, the fields of behavioral, social, and now health sciences are relatively young. It is important to note that efforts to understand human behavior have long been a topic of interest. Ancient Greek philosophers wrote extensively about many familiar topics, including learning, language, memory, and dreams. Although many writers and great thinkers wrote about how they thought the mind works, none conducted anything that we would call an experiment. The problem is that internal mental events are necessarily difficult to observe and measure. Consequently, many philosophers believed that we could not observe or measure mental events in the same way that we observe or measure physical objects.

This perception exists even today and has resulted in the behavioral and social sciences being labeled as "soft" sciences, a term that suggests that other sciences such as chemistry and physics (the so-called hard sciences) are more accurate or empirically valid. Interestingly, essentially identical research methods are used across all of these scientific fields (Hedges, 1987). It is the subject matter that sets the sciences apart. Properly designed and implemented research in the social sciences can be as valid and replicable as any other research. Historically, though, before this research could be conducted, a profound shift in studying human social interaction and behavior had to occur.

Although Greek philosophers had a profound effect on the generations of scholars who followed them, it was not until the questioning of these ancient authorities that the scientific revolution occurred. During this revolution, seventeenth-century scientists decided that there was more to learn about nature than the ancient philosophers had described in their writings. One of the more articulate spokespersons for the new scientific revolution was Sir Francis Bacon. Much of the scientific method as we know it today evolved to overcome and protect us from several basic human biases or "idols" that Bacon (1620/1994) outlined in his seminal book on this topic.

Interestingly, Sir Francis Bacon (1561–1626) was not a scientist; he was a British politician who was interested in empirical science and became one of its strongest proponents. In 1620, he published a book on the scientific method titled *Novum Organum* ("the new instrument"). Bacon saw the scientific method as a better way of finding accurate answers to difficult questions. Like many of his contemporaries, Bacon distrusted the wholesale belief in everything that the ancient philosophers had to say. He (Bacon, 1620/1994) wrote, "For the ancients . . . out of a few examples and particulars, with the addition of common notions and perhaps some portion of the most popular received opinions, they flew to the most general conclusions or principles of the sciences . . . through intermediate propositions, they extracted and proved inferior conclusions" (p. 127). In essence, Bacon accused the earlier philosophers of making hasty generalizations that had little or no merit. He also argued that to comprehend the physical world, we must use the scientific method to ask and answer questions.

Bacon's most important and lasting contribution to the history of science may be his discussion of common human biases that can cause us to make irrational decisions or to ignore important information. According to Bacon, there are four main human biases that hinder our ability to think clearly. He referred to each of these biases as the **idols of the tribe**, **idols of the cave**, **idols of the marketplace**, and **idols of the theater**. Bacon's observations were as insightful then (early 1600s) as they are now. Indeed, we continue to rely on the scientific method, statistics, critical thinking, and analysis skills to overcome the obstacles to credible behavioral and social science research that each of these idols creates.

Idols of the Tribe

The first source of bias described by Bacon involves our human tendency to rely on intuition and common sense when trying to understand a complex phenomenon. Bacon (1620/1994) suggested that

> *The Idols of the Tribe lie deep in human nature itself and . . . it is wrongly asserted that the human sense is the measure of all things. It is rather the case that all our perceptions . . . are reflections of man [sic] not of the universe, and the human understanding is like an uneven mirror that cannot reflect truly the rays from objects, but distorts and corrupts the nature of things by mingling its own nature with it. (p. 56)*

Bacon recognized that many people have a tendency to believe that what they see and how they interpret events is accurate, and that their common sense is well informed and infallible. This tendency leads us to selectively perceive events around us, trust our first impressions, and then uncritically use those impressions to form "answers" or to make decisions.

A common example of the idols of the tribe is illustrated by self-fulfilling prophecies. A **self-fulfilling prophecy** occurs when we believe something is true and these preconceived beliefs then influence the way we perceive and react to specific events, ultimately confirming our initial beliefs (Baron, Graziano, & Stangor, 1991). In most cases, we are unaware of how our attitudes affect our behaviors and cognitions. Moreover, when we believe something to be true, we tend to remember events that align with our beliefs and forget or ignore events that conflict with our beliefs. This human tendency functions as a bias because preconceived ideas have considerable influence on how we interpret and react to cues in different situations.

Another example of the idols of the tribe is known as the **gambler's fallacy**. If a person bets on black three times in a row and red wins, most people believe that the next round *must* come up black instead of red. Some people will argue, "It makes *good common sense* that red cannot win four times in a row!" However, the probability of winning with black or red is 50% each time (unless the particular game is somehow rigged or weighted), making it entirely possible that the next round could come up red. Many people make this error because they trust their intuition and preconceived beliefs about probability; that is a sure way to lose a lot of money at the gambling tables.

Many researchers (e.g., Nisbett & Ross, 1980; Rosnow & Rosenthal, 1997) have examined the shortcomings of human decision making. The consensus among researchers is that humans tend to rely too much on intuition and common sense to make decisions. In summary, the idols of the tribe highlight our human tendency to depend too much on common sense and to the tendency to make consistent errors in logical reasoning.

How could this affect your ability to conduct credible research?

How can this risk be addressed by proper use of the scientific method?

Idols of the Cave

This second source of information processing bias develops from our exposure to culture, common practice, and education. According to Bacon (1620/1994), our life experiences shape how we look at things. Although our experiences are valuable, there are important sources of limitations. As Bacon (1620/1994) described them, "The *Idols of the Cave* arise from the individual's particular nature, both of mind and body, and come also from education, habits and by chance. Though there are many different kinds, we cite those which call for the greatest caution, and which do most to pollute clear understanding" (p. 61).

The problem with personal experience is that it is personal, unique to you. Chances are that your background and our backgrounds are very different. Who is to say which of us has a more *valid* or accurate worldview? Each of us has experienced different important or critical life events. These events shape our beliefs and perceptions and affect how we perceive things. Although these beliefs and perceptions help to make us unique, we also need to recognize their potential influence on our decision making and reasoning. Karl Popper (1902–1994), a famous philosopher, provided an interesting example of what can happen if we depend too much on personal experience. Early in his career, Popper worked with the psychotherapist Alfred Adler, who had developed a comprehensive theory of personality development from observations made within his clinical practice. Popper (1963) described the following episode:

> *Once . . . I reported to him [Adler] a case which to me did not seem particularly Adlerian, but he found no difficulty in analyzing in terms of his theory of inferiority feelings, although he had not even seen the child. Slightly shocked, I asked him how he could be so sure. "Because of my thousand fold experience," he replied; whereupon I could not help saying: "And with this new case, I suppose, your experience has become thousand-and-one fold." (p. 35)*

The problem relevant to our discussion is Adler's use of personal experience. Adler's status as a professional psychoanalyst in no way guarantees that his observations or conclusions are automatically valid. A moment's thought will reveal the limitation of personal experience in this situation. Adler was a therapist who treated people suffering from a wide variety of psychological ailments. His patients were hardly representative of the general population and, therefore, not the foundation for a comprehensive theory of personality development that describes all people. The idols of the cave, therefore, refers to our human tendency to depend on our personal experiences to explain and determine why things happen as they do. As we will soon see, we must do more than merely rely on personal experience to develop scientific explanations.

How could this affect your ability to conduct credible research?

How can this risk be addressed by proper use of the scientific method?

Idols of the Marketplace

The third bias that Bacon examined involves our use of language. Turning to Bacon (1620/1994), we read, "The *Idols of the Market-place* [*sic*] are the most troublesome of all; these are idols that have crept into the understanding through the alliance of words and names" (p. 64). Bacon recognized that our use of words shapes how we think about things. Consider an example regarding day care. Scarr, Phillips, and McCartney (1990) noted that during the 1950s and 1960s, developmental psychologists who studied the effect of child care examined the effects of *maternal absence* or *maternal deprivation*. Clearly, these emotionally charged phrases create a negative bias against women who choose to pursue a career while their children are infants and toddlers. Why use these phrases as if the mother deprived her children of food and water? What about the father's absence? If children suffer *maternal deprivation*, why don't they suffer *paternal deprivation* as well? Could it be that fathers are guilt-free because societal norms allow men to work outside the home? Furthermore, the words *absence* and *deprivation* evoke images of children warehoused in dangerous day-care centers. Scarr and her colleagues argued that these terms grew out of "fantasies about child development . . . mother-infant attachment . . . and the role of early experience for later development" (p. 255). These terms fell out of favor during the 1970s, when the rights of women to pursue a career became popular. Researchers then began to examine the benefits of day care. Thus, the idols of the marketplace reflect the power of language over our thought processes.

How could this affect your ability to conduct credible research?

How can this risk be addressed by proper use of the scientific method?

Idols of the Theater

The fourth and final idol in Bacon's list represents the potential biasing effects of education. Here we find Bacon (1620/1994) complaining that many of the things we learn may mislead us. "The *Idols of the Theatre*, on the other hand, are not innate, nor are they secretly insulated into the understanding, but are imposed and received entirely from the fictitious tales in theories, and from wrong-headed laws of demonstration" (p. 66). In other words, the idols of the theater are illustrated any time we accept an explanation without critically evaluating it first. In many cases, we automatically accept certain explanations because we learned them from someone we trust or see as an authority figure. Countless "scientific" theories have enjoyed this kind of dubious honor, including the now-debunked notions that the earth is the center of the universe and the world is flat. Apart from these seemingly ancient ideas, commonly accepted notions are all around us. Perhaps the best illustration of this is in Kohn's (1990) book on popular beliefs, in which he describes various common beliefs and their fallacy, including "No pain, no gain," "Competition builds character," "Like father, like son," and "Playing hard to get makes one more attractive."

The defining characteristic of the idols of the theater is our tendency to accept the truth of a statement without criticism. The best defense against this source of bias is simply to always think critically about what someone is asking you to believe.

How could this affect your ability to conduct credible research?

How can this risk be addressed by proper use of the scientific method?

BACON'S LEGACY

Bacon's primary legacy is that he clearly identified core obstacles to critical thinking as they apply to all branches of science to this day. Although the scientific method has been around for 400 years, the effects of his idols remain. Each of us can fall prey to the idols. Being aware of this can help you to understand why researchers use specific tactics when developing and conducting research. Researchers use research methods and statistics to overcome many forms of bias. Researchers also understand that we can never become complacent with the knowledge that exists today; tomorrow's research may change everything.

For these reasons, the primary lesson to take away from Bacon is that the idols of the tribe, cave, marketplace, and theater are always present, and we must work to guard against these biases whenever we utilize the scientific method to study and explain the behavior of people. Take some time to review Table 1.1 and think

Table 1.1 Summary of Bacon's Idols

Idols of the tribe	Biases due to overreliance on common sense and the tendency to make errors in logical reasoning
Idols of the cave	Biases due to dependence or reliance on personal experience to explain why things occur the way they do
Idols of the marketplace	Biases due to how we use specific words to describe phenomena in the world around us
Idols of the theater	Biases due to uncritical acceptance of explanations that people in authority tell us are true

of examples of Bacon's idols that are relevant in your own areas of interest or line of work.

OTHER IMPORTANT HISTORICAL FIGURES

The goal of this text is not to provide you with a history of science or a comprehensive listing of individuals who have influenced scientific thought. However, a brief review of two additional individuals (Gustav T. Fechner and John B. Watson), among many others, who had a significant impact on the current state of behavioral and social research methods follows.

On October 22, 1850, *Gustav T. Fechner* (1801–1887) discovered a way to measure mental events. All science relies on measurement, which is nothing more than assigning numbers to observations. All sciences have specific methods for measuring the phenomena they study. However, before Fechner's work, researchers had no objective method for measuring mental events. Fechner studied physics and human perception. In his research, he observed that there was not a one-to-one relation between the intensity of a stimulus and our perception of the stimulus. For example, imagine a friend asks you to hold out your hand and close your eyes. If your friend puts a pencil on your hand, you will notice its weight. Now imagine your friend putting this textbook on your hand. You will feel the weight of the book. What if your friend then places the same pencil on top of the book? You will probably not be able to detect the additional weight. Why are you able to feel the weight of the pencil in one situation but not the other?

Fechner reasoned that better understanding the relationship between changes in the intensity of a stimulus (a physical event) and changes in a person's perception (a mental event) of a stimulus could help us better understand how the mind generally functions. He then proceeded to conduct a series of famous experiments that we now recognize as the start of *psychophysics*. Even if Fechner's experiments are not the most exciting thing you have heard about today, his work is very important because it caused people to recognize that it is possible to study mental events using empirical techniques.

John B. Watson (1878–1958) is another important person in the history of behavioral and social science research methodology. In 1913, Watson wrote an influential paper titled "Psychology as the Behaviorist Views It." The paper began with the proclamation, "Psychology as the behaviorist views it is a purely objective experimental branch of natural science. Its theoretical goal is the prediction and control of behavior" (p. 158). This statement seems obvious now, but was written at a critical moment in the history of science (Murray, 1983).

The implications of developing a science of behavior extend well beyond psychology. At the start of the twentieth century, the scientific study of human behavior and cognition was a new phenomenon and scientists were searching for the best methods to conduct scientific research. At the time, many researchers used a procedure known as **introspection**. *Introspection* means to examine or look within. Whenever you think about your own thinking and mental events, you are using a form of introspection. Try this experiment in introspection: What reactions do you have when you read the word *health*? Although introspection can be revealing, it has several shortcomings. Take a moment to think of a few.

Perhaps the most troubling question is, *how do we know that the self-report is accurate?* When you are asked to introspect about something, will you report everything that occurs to you? Is it possible that thinking of work evokes a painful memory that you do not want to share? How complete is your report? Although you may report things of which you are aware, could there be reactions that you did not recognize as important and worthy to share with others? Is it possible that there are unconscious mental processes that you do not directly experience? The use of introspection troubled Watson because there is no way to verify the accuracy of an introspective report. The problem with introspection is that only one person can *experience or observe* your mental events—you. In science, researchers want to examine phenomena that others can see when they use the same procedures.

There are other problems with introspection. To what extent does your introspection influence the mental events you wish to study? Does thinking about your thinking affect your thinking? Are you confused? Try another thought experiment. Can you read and introspect about the process of reading at the same time? If you are like us, reading for content while introspecting is essentially impossible. As soon as we start examining the process of reading, we are no longer reading. When we read for content, we cannot introspect. Watson (1913) rejected introspection as a research tool and recommended that psychologists study behavior exclusively. He believed that by focusing on behavior, psychologists could engage in the objective study of all living creatures. For Watson, if you can observe the behavior, then you can conduct scientific research.

Watson's legacy to the study of people from a health sciences perspective is that he focused our attention on behavior. Watson has had a lasting impact on all research involving the study of behavior and social interaction. Many researchers today subscribe to the perspective of **methodological behaviorism**, a philosophical stance evolving from Watson's beliefs. Methodological behaviorism suggests that researchers should study overt and observable behaviors as the primary focus of their research. Researchers use observable behaviors to make inferences about the emotional, cognitive, and other mental processes that occur within a person. Behavior is the focal point of research dealing with human beings. As one general example, a great deal of emphasis in the health sciences now is placed on translating research-based knowledge into practice. What this really means, in most cases, is figuring out how to change human behaviors. Fechner's and Watson's work can help us to understand that the behaviors we can observe, and may seek to change, are intimately linked to underlying complex mental and cognitive events.

ASSUMPTIONS OF SCIENCE

Supporting everything we have discussed so far are two core assumptions, which operate behind the scenes of any good research study. All sciences make the same basic assumptions about their subject matter.

Behavior Is Determined

Our first assumption is quite possibly the most important. We believe that behaviors are caused or triggered by specific factors. This perspective is known as **determinism**. Someone who believes this (that all behaviors have a knowable set of causes) can be referred to as a *determinist*. You will learn that almost all researchers are determinists of one form or another. Sigmund Freud (1856–1939), for example, was a *psychical* determinist because he believed that human behavior reflected a series of unconscious drives and motivations. He believed that there are no accidents of behavior—everything we do reveals something about our character and unconscious drives.

In contrast, B.F. Skinner (1904–1990) was an *environmental* determinist because he believed that an individual's interaction with the environment produces changes in behavior. Other researchers are *biological* determinists because they believe that biological processes control many behaviors. Finally, some researchers are *sociocultural* determinists because they believe that cultural traditions, customs, and regulations control people's lives. When you examine different fields of study, such as human development, social behavior, marketing, or behavioral finance, you will find that researchers in each area conduct research to find the things that determine behavior. Regardless of their perspective, each type of determinist believes that by observing behavior and the surrounding conditions, we can infer the causes of the behavior.

Some people object to determinism and suggest that human behavior is only subject to **free will**. The principle of free will states that a person's soul or mind controls how he or she acts. Many religious faiths and philosophy theories suggest that humans are special because we have a spirit and self-awareness that guides us through life. These religions also teach us that we have the freedom to choose between the good and virtuous, or the evil and sinister. Thus, at first glance, it appears that there is quite a contrast between determinism and free will. Belief in determinism holds that we can explain observable behaviors by looking for and examining material causes. Belief in free will holds that each person is unique and that we cannot use the scientific method to understand human behavior.

It is not helpful to pit determinism versus free will. If you are willing to accept that people share some basic characteristics, then you will find that the scientific method does a good job of finding the causes of those common behaviors. Science does not have all the answers to important questions. Science, religion, philosophy, literature, and the arts are all different ways of knowing and experiencing our world. Each answers a unique set of questions using a different perspective. As Gould (1999) noted, science and religion are two ways of knowing. Both are equally important, yet both answer different questions. Taking a scientific perspective allows us to understand how things work, and when studying human behavior, this means trying to discover why people do what they do. Religion helps us to examine our values and to discover how we should behave. For many people, science and religion are not competing forces but rather complementary methods for addressing different issues of importance. In the same vein, determinism and free will can be viewed as complementary and not necessarily competing views.

We Can Measure the Critical Variables

A second assumption of science is that we can directly or indirectly observe (and therefore measure) the important causes of behavior. All sciences rest on a foundation of measurement. Fechner realized that we could use a person's behavior to make inferences about mental events. Physicists, chemists, and other scientists routinely use observable events to make inferences about the existence of things that they cannot directly observe. For example, no one has seen gravity, only its effects. Nevertheless, physicists can use the motion of the planets and stars to infer that there is gravity and to describe its effects. In business, we often study behavioral events and situations to make inferences about interpersonal and intrapersonal events that we do not fully understand and perhaps cannot directly observe.

REQUIREMENTS FOR SCIENTIFIC RESEARCH

Now it is time to focus on specific elements of research that, when combined, allow us to "be scientific" when doing research.

Empirical Analysis

Empirical analysis involves the gathering of data by observation and experimentation, with the goal of learning something. One important characteristic of empirical analysis is that it involves **measurement**, or the converting of observations into numbers. There are many different types of measurement, but just about all can be classified as either self- or other-observation, in which we use our own senses or someone else's senses to collect information.

Empirical methods are not the only way to gain insight into challenging questions. Within the health sciences, just about everything we "know" has come from scientists' efforts to observe and experience the phenomena of interest. Contrast this method with other ways of knowing. Mathematicians, for example, do not use empirical analysis but instead discover new ideas using deduction and formal proofs. Here is an example of the difference between the empirical method of knowing and the mathematical way of knowing. Imagine that you have 10 quarters in your hand and toss them in the air. What is the probability of obtaining 0, 1, 2, 3, . . . or 10 heads? There are two ways of finding the answer. The first method is empirical. You would toss the 10 coins, count the heads, and then repeat these steps several thousand times until you had enough samples to make a relatively accurate conclusion about the probability of each outcome. You will eventually come to the correct answer, if you are willing to spend the hours of drudgery tossing and counting coins.

The second method uses deductive logic and analytical techniques. If you know enough about probability theory and your way around mathematical proofs, you can derive an equation that gives you the correct answer. There is nothing wrong with either method, although most people find the mathematical solution more elegant and convenient. There are many times, however, when the analytical method does not work and the empirical method is the only alternative. We can use mathematics to solve the coin problem because we know several critical things to be true, such as the fact that each coin has a 50% chance of landing heads. From these facts, we can derive additional truths. Thus, the deductive method works well when we have the necessary information before us to solve a problem. *In many cases we do not have this information.* Consequently, we must go about gathering data so that we can answer the question. In other words, empirical and deductive methods both have strengths and weaknesses.

The following is an example that illustrates the potential weakness of sole reliance on deductive logic:

1. All students are human.
2. We are all students.
3. Therefore, we must all be human.

Although extremely simple, this example illustrates a categorical syllogism that contains two premises (statements 1 and 2) and a conclusion (statement 3). In deductive logic, if we accept the premises and use the appropriate rules of logic, then the conclusion is true. Now consider the deduction:

1. All unicorns are purple.
2. Annie is a unicorn.
3. Therefore, Annie is purple.

The conclusion about Annie's color is logically consistent if we accept the premises. This example illustrates a potential problem with finding answers by deductive logic or pure reason. If we accept the premises of an argument, then we must accept the truth of logically consistent conclusions. In the example of the unicorn, the conclusion is valid, although it has no bearing in truth—unless you can find a living purple unicorn. Sir Francis Bacon and many others recognized that deductive logic can lead to erroneous conclusions based on a false or unproven premise. Consequently, scientists who utilize empirical methods are working to verify the truth of specific premises with gathered data. With respect to this admittedly bizarre example, if we can obtain observable evidence that unicorns exist and are purple, *then* we can conclude that Annie is purple.

Public Verification

Public verification is another important feature of empirical research. Using the empirical method requires us to rely on our senses when gathering data. If we design our research so that it can be publicly verified, then we are measuring things in a way that others can replicate with similar results. Therefore, public verification implies that anyone who uses the same procedure should be able to observe the same general outcome. Watson (1913) emphasized this requirement of good science when he called for all researchers to drop introspection and to adopt the study of behavior. Studying your own mind is fine, but this pretty much ensures you will be the only researcher who can experience your thoughts and make your observations. In other words, your mental events would not be subject to public verification.

Your behavior and actions, however, are things that can be observed by anyone. Using a video camera, we can record your interactions with coworkers and team members, and any researcher can share those observations. We can also attach sensors to your body and monitor your heart rate, the sweat on your palms, and the electrical activity of your brain. We can give you a personality test as a way to measure how you perceive yourself. In each case, we have collected public information that others can verify. Public verification also means that anyone with the appropriate equipment can repeat an experiment. This facet of public verification is extremely important. Our ability to repeat or replicate experiments gives us greater confidence in the general applicability of our results. The more times we can repeat an experiment and obtain similar results, the more likely we are to agree that an effect we observed is real and not just a fluke, due to chance.

Systematic Observation

Systematic observation refers to the way we go about collecting information. Whenever we collect data, we want to make our observations under specific and controlled conditions. Doing so can help us to rule out alternative explanations for the outcomes we might be observing. Imagine that a pharmaceutical company claims to have a new medicine that can prevent obesity and lead to dramatic weight loss in those who are already overweight. Although this claim sounds great, we need to determine its truth. We can do this using systematic observation. For example, we should determine whether the medication, which has particularly noxious side effects, yields better results than more natural intervention options (e.g., diet and exercise). We could evaluate this within a controlled experimental framework by assigning patients to one of several conditions, for example, no treatment, **placebo** treatment, exercise and diet treatment, or new medication treatment.

In this example, the systematic observation comes into play as we measure differences in weight loss and side effects for the participants in each of the treatment conditions Yet another way to use systematic observation in this type of scenario might be to determine whether this medication works better for some people than others (e.g., perhaps only those over a certain BMI are likely to benefit to a degree that the noxious side effects are worth enduring).

The overarching goal of systematic observation is to examine a particular phenomenon under as many relevant situations as possible. We continue to repeat our observations and experiments to determine which conditions consistently produce the effect and what other possible factors aside from the training might influence the phenomenon. Unfortunately, many people do not recognize the necessity of systematic observation, tending instead to accept testimonials and/or personal opinions without question. **Testimonials** are not a form of systematic observation, although they are often treated as such. Testimonials are nothing more than an example of Bacon's idols of the theater. When people make a claim like this, we are supposed to believe what they say. Testimonials are also an example of the idols of the cave because they reflect personal experience. Watch any infomercial on television and you will hear many happy customers share their personal experiences with the product: "My life was really going nowhere fast until I learned about Bacon's Idols. Now look at me!" Good researchers shy away from putting too much emphasis or weight on testimonial claims that are neither systematic nor objective. *How do you think this can improve the quality of research?*

Control of the Environment

In all forms of research, we attempt to exercise **control of the environment** in some way. We do this to ensure that the conditions in which we make our observations are consistent and can be replicated by other researchers who might wish to verify our findings. Researchers have the greatest level of control when they conduct research in a laboratory setting because they can control many or all external environmental conditions. This control helps to reduce the number of possible factors that might influence a participant's behavior, thoughts, or feelings. There are many cases, however, in which direct control of the research environment is not possible. This is especially true when a **field study** is being conducted, but even here, a true researcher will try to ensure as much as possible that the environment is the same each time he or she collects data from that sample.

Rational Explanation

A **rational explanation** refers to the two basic assumptions of science: (1) Behavior is determined and (2) behavior follows a pattern that can be studied.

Rational explanations of behavior, therefore, include two essential components. The first is that the explanation refers only to causes that one can observe or confirm through public verification. The second is that the explanation makes a clear and logical link between the cause and effect. Explanations that are not rational are not scientific. Instead, these are typically called **pseudoexplanations** because although they may sound like sophisticated explanations of some phenomenon, they do not improve our understanding of the phenomenon in any way. A pseudoexplanation is also commonly referred to as a **nominal fallacy** or a **tautological (circular) explanation** in that this type of explanation uses the phenomenon to define itself. Thus, a pseudoexplanation is an example of the idols of the tribe, as it appeals to our desire for commonsense explanations.

For example, a typical early definition of a *reinforcer* was *a stimulus, produced by a behavior, that increases the probability that the individual will repeat the behavior*. This explanation is circular because there is no independent definition of the reinforcer. The definition uses the effect of reinforcement to define the property of reinforcement. Why is this technique a problem? Consider the following exchange.

Question: "*What is a reinforcer?*"
Answer: "*A reinforcer is anything that increases the probability of a behavior.*"
Question: "*How do we know that something is a reinforcer?*"
Answer: "*Because it increased the probability of a behavior.*"
Question: "*Why did the probability of the behavior increase?*"
Answer: "*Because we used a reinforcer.*"
Question: "*But what is a reinforcer?*"

The problem with this cycle is that we have no way of defining the reinforcer without referring to the behavior it affects. In other words, this type of definition tells us nothing about why a reinforcer works. Using the definition of reinforcement does not allow us to predict what things will serve as effective reinforcers. This definition also does not explain why a reinforcer will increase the probability of reinforcement.

Fortunately, David Premack (1959, 1965) discovered that high-frequency behaviors can reinforce low-frequency behaviors (the Premack principle). The advantage of this definition is that it breaks the circular definition, defining the cause as independent from the effect. More specifically, Premack's theory states that any high-frequency voluntary behavior will reinforce a low-frequency voluntary behavior. According to this definition of reinforcement, we can take several behaviors and categorically predict which will and will not be reinforcers. Consider this example: "For Alex, playing video games is a high-frequency behavior and studying math is a low-frequency behavior. Therefore, playing video games will serve as a reinforcer for studying math." We predict that video game playing is a reinforcer because it is a high-frequency behavior. We can then verify this hypothesis with an empirical test by allowing Alex to play video games only if he spends more time studying math. If there is an increase in the amount of time spent studying math (the effect), we can then say that the reinforcement (playing video games) caused the change.

Another feature of a rational explanation is that a researcher can empirically test and determine whether an explanation is correct. What if your professor told you that there is a special energy force that affects the brains of some people and causes them to be schizophrenic? The first question you should ask is, "Where's the empirical evidence?" What if the professor told you that no known apparatus can detect the radiation from this special energy force? At this point, you should realize that your professor is either losing his mind or offering you a classic pseudoexplanation. A better explanation is one that is objectively defined in a way that can be supported with observational data by you and other researchers who may wish to replicate your work. Indeed, many researchers have tested the accuracy of the Premack principle. Some have verified Premack's predictions, whereas others have not (Mazur, 1998). Using the results of these experiments, Timberlake and Allison (1974) were able to refine Premack's definition and offer a more comprehensive definition of reinforcement.

Parsimonious Explanation

In addition to being rational, scientists strive to make explanations *parsimonious*. Parsimony means simplicity. If you have difficulty remembering this concept, try to link it in your mind visually to a big fat kiss and remember that K.I.S.S. represents the "Keep it simple, stupid!" principle. In the present context, a scientific conclusion or explanation is parsimonious if it makes relatively few assumptions, does not refer to unobservable causes, and refers to specific causes. This requirement is also known as **Occam's razor**.

Please realize that we are *not* saying that simplicity automatically makes a theory correct. Instead, a parsimonious theory allows for specific predictions that researchers can directly test. Its value to science is its ability to generate many ideas for specific research projects. When possible and appropriate, simpler explanations often have more utility than more complex explanations.

Tentative Explanations

Whenever a researcher presents the results of a study, the explanation of the results is **tentative**. No single study can account for all the potential explanations of the results. You can think of any single study as a small step in a long journey. Although each step may take us closer to our goal, it may also take us in the wrong direction. Although the theory is useful, it is never fully complete.

As you read more about science, you will learn that researchers are continually revising their explanations for why things work the way they do. The change occurs because each study adds new information. Some new information may confirm what we already know and so we continue to use the theory to explain the phenomenon we study. Other new information, however, may indicate that the theory cannot account for specific events and must be revised or replaced. Therefore, it is the case that explanations of behavior are only as good as the data they have collected. Researchers recognize that as new data are collected, they may have to revise their explanations or develop new explanations.

CHAPTER SUMMARY

This chapter introduced you to research methods by briefly examining the history of science as it relates to research methods and by offering an overview of the meaning of scientific research. The goal of this chapter was to illustrate that understanding research methods and the basics of science is important whenever you are trying to understand how and why people behave in the ways they do. Researchers use the scientific method to conduct basic research to understand various behavioral phenomena. Research methods also have many practical applications. Regardless of your current or future career objectives, it is important to understand the foundations of science and research methods.

Sir Francis Bacon was an early advocate of empirical science. He believed that the scientific method would overcome several human tendencies that are obstacles to a better understanding of our world. He called these tendencies *idols* and identified four specific ones: *idols of the tribe* (common modes of thought that lead to irrational conclusions), *idols of the cave* (overreliance on personal experiences), *idols of the marketplace* (biases in beliefs based on the meaning and use of words), and *idols of the theater* (biased thought based on tradition, habit, or deference to authority).

Gustav T. Fechner recognized that researchers could indirectly observe or make inferences about mental events by observing reactions to physical stimuli. John Watson's contribution to research was his insistence that behavior is the proper target of research and that introspection is not a useful procedure for science. The objective study of behavior allows researchers to understand behavioral and cognitive phenomena. Therefore, many researchers in the behavioral and social sciences are methodological behaviorists.

Researchers believe that they can use the scientific method to study behavioral and cognitive phenomena. They base this belief on the assumptions that the behavior they study is determined by specific causes that can be measured. Scientific research, regardless of the discipline, incorporates the following seven general characteristics:

1. Empirical analysis is the process of learning through observation and experimentation and through quantifying observations.
2. Public verification requires that we conduct research that can be repeated by others and specifically that the variables we examine can be observed by everyone.
3. The systematic observation criterion requires us to make our observations under various conditions or settings.
4. Control of environment refers to our ability to conduct our research under consistent conditions. When researchers explain various phenomena, they also attempt to make their explanations rational, parsimonious, and tentative.
5. The rational explanation means the terms are clearly defined and can be independently assessed and defined.
6. Parsimonious explanations are specific, make few assumptions, and generate many testable ideas. Pseudoexplanations, by contrast, are circular in definition and cannot be directly or objectively assessed.
7. Explanations are tentative. Researchers recognize that their explanations must be revised in the face of additional research.

KNOWLEDGE CHECK

1. Think about a health-related issue that affects a large number of people in your community. How could the scientific method help researchers better understand this issue?
2. Many disciplines examine human behavior. The authors of many great novels write about the human condition and use their stories to describe

why people behave as they do. Describe the difference in perspective between a health sciences researcher and the author of a novel.

3. Many people believe that professional athletes have moments when they are "in the zone," during which their performance is greatly enhanced. There are also times when the athlete will be "in a slump." By contrast, statisticians argue that these phases do not exist and are nothing more than random events. Which of Bacon's four idols best describes the belief that athletes are in the zone or in a slump?
4. You want to buy a new car. A friend of yours, an auto mechanic, says, "Stay away from that car, my shop is always filled with them. I plan to send my kids through college on the work that model makes for me." How does this example relate to Bacon's idols of the cave?
5. Describe the meaning of introspection and why Watson objected to its use.

Use the following scenario to answer questions 5 and 6: Imagine that your friend believes that he has psychic powers. He claims that he can often guess what another person is thinking. Two of your other friends agree and claim that there have been several times when your friend has shown his psychic abilities. Given this information, respond to the following questions:

6. Why would you want to use empirical methods to confirm your friend's psychic abilities? Why not rely on the testimonials of your friends who are being honest when they say that your friend is psychic?
7. Your friend agrees to a test. You create a list of randomly selected common words. As you concentrate on the word, your friend tries to read your mind. He fails the test and is unable to guess any of the words. To explain the failure, he says, "Well you see, it only works when there is no doubt of my ability. You doubt my ability and that creates negative energy that blocks my ability to read minds." Based on what you read in this chapter, comment on your friend's reaction.
8. According to the text, what are the essential elements of scientific research? Describe how these are incorporated into business research.
9. Contentment is a mental phenomenon that we cannot directly observe; yet it is a common experience. Describe how a researcher might measure contentment and make it an observable phenomenon.
10. Why is public verification especially important for studying behavior?
11. Would science exist if there were no measurement? Defend your answer.

CHAPTER GLOSSARY FOR REVIEW

Control of the Environment A feature of empirical research. The researcher attempts to observe the phenomenon under identical conditions. Also implies that the researcher reduces the effects of distracting or nuisance conditions that will add confusion to the data.

Determinism A philosophical stance that natural events and human behavior are the result of an orderly sequence of preceding events that can be predicted using fundamental scientific laws.

Empirical Analysis Using observation and research methods involving the gathering of data to help with identifying answers to research questions.

Field Study Research conducted beyond the boundaries of a laboratory, in an environment in which the phenomenon under study tends to occur or exist.

Free Will A philosophical stance that human behavior is independent of external causes and that humans are free to choose how they will act.

Gambler's Fallacy An example of the idols of the tribe. The fallacy is a belief that random events follow a predetermined pattern. For example, many people believe that for six tosses of a fair coin, the pattern THHTHT is more likely than TTTHHH; both are equally likely based on laws of probability.

Idols of the Cave Bacon's phrase to describe the tendency to use one's personal experience as the foundation for truth or the measure of all things.

Idols of the Marketplace Bacon's phrase to describe how our use of words shapes our perception of and reaction to things.

Idols of the Theater Bacon's phrase to describe the tendency to accept a theory or statement as fact and fail to question its accuracy or generality.

Idols of the Tribe Bacon's concept to describe common errors in humans' thinking. These errors of thought are present, to varying extents, in all people and include overreliance on common sense and logical errors of reasoning.

Introspection A process by which one attempts to analyze his or her own conscious experiences.

Measurement The process of converting observations to numbers using a set of rules.

Methodological Behaviorism The belief that when studying human beings researchers should study observable behaviors. By observing the conditions under which behavior occurs, one can then infer the causes of the behavior or the presence of mental processes that cannot be directly observed.

Nominal Fallacy An example of a pseudoexplanation that makes the erroneous assumption that naming a phenomenon is the same as explaining the phenomenon.

Occam's Razor A version of parsimony that requires that we do not create more distinctions among things than is necessary.

Parsimonious Explanation A requirement in science that we offer explanations that make the fewest assumptions and require reference to few or no unobservable phenomena.

Placebo A false treatment condition in which participants are not exposed to any real stimulus but rather an imaginary placeholder such as a sugar pill or glass of water. Useful as a means of creating a control group without the participant knowing he or she is not getting the real treatment.

Pseudoexplanation An explanation of a phenomenon that does not really explain the phenomenon.

Public Verification The requirement that the subject matter of any empirical research must be observable to any person who uses the same procedures and equipment to examine the phenomenon.

Rational Explanation Offering a description or interpretation of a phenomenon that follows the rules of logic.

Self-Fulfilling Prophecy An example of the idols of the tribe. People will act in ways that bring about the result(s) they expected in the first place.

Systematic Observation A process in which the researcher varies the conditions under which he or she studies a particular phenomenon.

Tautological (Circular) Explanation A form of pseudoexplanation that involves circular definitions, which use the phenomenon to be described when trying to define its cause.

Tentative Explanation The recognition that all descriptions and explanations that arise from empirical research may be incomplete or inaccurate. Additional research may force us to revise our beliefs.

Testimonial A statement that a person makes about the truth of a fact or a claim based on personal experience.

REFERENCES

BACON, F. (1994). *Novum organum* (P. Urbach & J. Gibson, Trans.). Chicago: Open Court (original work published in 1620).

BARON, R.M., GRAZIANO, W., & STANGOR, C. (1991). Social perception and social cognition. In R.M. Baron & W. Graziano (Eds.), *Social psychology* (pp. 108–159). Fort Worth, TX: Holt, Rinehart and Winston.

CAMPBELL, S. (1974). *Flaws and fallacies in statistical thinking*. Englewood Cliffs, NJ: Prentice-Hall.

Daubert v. Merrell Dow Pharmaceuticals, Inc., 113 S. Ct. 2786 (1993).

Federal Judicial Center (1994). *Reference manual on scientific evidence*. Washington, DC: Author.

GOULD, S.J. (1999). *Rock of ages: Science and religion in the fullness of life*. New York: Ballantine.

HEDGES, L.V. (1987). How hard is hard science, how soft is soft science? The empirical cumulativeness of research. *American Psychologist*, *42*(5), 443–455.

KOHN, A. (1990). *You know what they say . . . : The truth about popular beliefs*. New York: HarperCollins.

LAKIN, J.L., GIESLER, R.B., MORRIS, K.A., & VOSMIK, J.R. (2007). HOMER as an acronym for the scientific method. *Teaching of Psychology*, *34*(2), 94–96.

MAZUR, J.E. (1998). *Learning and behavior*. Upper Saddle River, NJ: Prentice-Hall.

MURRAY, D.J. (1983). *A history of Western psychology*. Englewood Cliffs, NJ: Prentice-Hall.

NISBETT, R.E. & ROSS, L. (1980). *Human inference: Strategies and shortcomings of social judgment*. Englewood Cliffs, NJ: Prentice-Hall.

POPPER, K. (1963). *Science: Conjectures and refutations*. New York: Harper and Row.

PREMACK, D. (1959). Toward empirical behavioral laws: I. Positive reinforcement. *Psychological Review*, *66*, 219–233.

PREMACK, D. (1965). Reinforcement theory. In D. Levine (Ed.), *Nebraska symposia on motivation* (pp. 123–180). Lincoln: University of Nebraska Press.

ROSNOW, R.L. & ROSENTHAL, R. (1997). *People studying people: Artifacts and ethics in behavioral research*. New York: Freeman.

SCARR, S., PHILLIPS, D., & MCCARTNEY, K. (1990). Facts, fantasies, and the future of child care in the United States. *Psychological Science*, *1*, 255–264.

TIMBERLAKE, W. & ALLISON, J. (1974). Response deprivation: An empirical approach to instrumental performance. *Psychological Review*, *81*, 146–164.

WATSON, J.B. (1913). Psychology as the behaviorist views it. *Psychological Review*, *20*, 158–177.

Chapter 2

Ethics and Research

The essence of morality is a questioning about morality; and the decisive move of human life is to use ceaselessly all light to look for the origin of the opposition between good and evil.

—Georges Bataille

INTRODUCTION

From the moment we wake up in the morning until the moment we return to sleep in the evening, we confront situations that require us to act in one way or another. Typically, the choices that we make have little bearing on the lives of others. The decision whether to have a second cup of coffee or trudge off to the library to study may characterize the typical daily "dilemma" we face. There are, however, cases where our actions do directly affect the lives and well-being of others. These choices represent a special case of decision making because they involve moral behavior.

Anyone who conducts research or interprets and uses the results of research must be sensitive to and mindful of moral principles and ethical reasoning. As you will learn in this chapter, researchers are serious about their responsibility to act morally. All psychological and medical research affects the researcher, people who participate in the research, the sponsors of the research, and eventual "consumers"

Understanding and Conducting Research in the Health Sciences, First Edition. Christopher J. L. Cunningham, Bart L. Weathington, and David J. Pittenger.

of the research. Therefore, before we undertake any research, we must examine with considerable care the ethical implications of our work. In other words, we must consider the justification for and the consequences of our actions.

Researchers have a personal stake in examining the moral principles that guide their work. Within the health sciences especially, foremost among these principles is the belief that the researcher, as a healthcare provider, should do no harm. Although health sciences research is very important, the potential value of any given study does not grant an automatic license to the researcher to act without regard to the welfare of others.

Aside from this basic ethical tenet, there are also practical reasons for researchers to examine the ethical foundation of their work. Most professional organizations, such as the American Psychological Association and the American Medical Association, have developed clearly stated ethical guidelines.

Apart from professional organizations, the federal government requires all researchers receiving federal grants to conduct research on humans receive approval from an **institutional review board (IRB)**. An IRB is a group of researchers and other professionals who examine the researcher's proposed methodology to ensure that the researcher will protect the basic rights of the participants in the study. All large research universities and hospitals have an IRB and require all researchers, regardless of the funding source, to obtain IRB approval for any medical research.

In this chapter, we examine common ethical issues that arise when researchers go about the process of collecting data from humans. First, we examine several broad ethical principles and perspectives that are a part of any ethical decision. We then examine the ethical principles established by various professional organizations as they apply to research with people. Finally, we examine an array of case studies that illustrate common ethical dilemmas that researchers in the health sciences may face when conducting behavioral and social science research.

WHAT IS ETHICS?

What does it mean to be moral or to act morally? What is ethics? Morality and ethics represent a set of interconnected principles and ways for making choices. As we progress through this chapter, we will use ethics and morality to describe how researchers make decisions about their research design. For example, we will ask, "Is this procedure moral?" or "What is the ethical thing to do in this specific study?"

Although many people use the words *morals* and *ethics* interchangeably, there is an important difference between the two. In general, **morals** are the principles or rules that define what is right and wrong. For example, you may have learned that it is wrong to lie and that it is good to help others. By contrast, **ethics** is the process of studying moral standards and examining how we should interpret and apply them in various situations. An example of ethics is asking whether it is acceptable to lie to another person to spare them the shock and pain created by the truth or if it is acceptable to steal something to save the life of another person.

You should recognize that ethics is more than simply "doing the right thing." Although it is important that you behave correctly, it is also important that you understand *why* you make the decisions you do. One might even argue that your behavior is not ethical unless you can justify your actions as good to the core. Consequently, whenever you conduct research, you must ask yourself whether your actions are right or wrong and be able to justify your decision. Although there are many moral standards or principles, they all share four general characteristics. Understanding these characteristics will help set the stage for our review of specific moral principles and ethical deliberation.

Moral Principles Address the Well-Being of Others

At the heart of morality is the well-being of people who are affected by our behaviors. This concern for others is an integral component of our religions, laws, and social codes. Your parents, for example, hopefully taught you to believe that it is bad to hurt others intentionally. Consequently, you probably believe that assault, child abuse, theft, and murder are wrong because they harm others. Thus, the well-being of others is the central feature of all codes of conduct.

Moral Principles Transcend Other Standards, Including Self-Interest

Moral principles direct us to act without regard to our personal goals and interests. For example, during the 1960s, many people in the United States protested the segregation laws practiced in many states. These people believed that they had a moral obligation to oppose laws that systematically denied other Americans their civil liberties. By participating in marches and helping with voter registration programs, these people took considerable personal risk to stand for what they believed was right.

There is nothing wrong with pursuing self-interest; it is a reasonable goal. The question arises, however, when we must choose between a morally correct behavior and personal gratification. Would you cheat on an exam or submit a plagiarized paper to get an "A" in a course? If you knew you would not get caught, would this change your answer?

When faced with a conflict between self-interest and moral behavior, morality should lead the way. The application of this principle to research is clear. No matter how valuable the data may be to us, we cannot overlook the rights of the participants during the research process and the ethical implications of our findings to society in general.

Moral Principles Are Constants and Universals

The laws our governments make are often arbitrary and reflect many compromises among lawmakers. By contrast, moral principles are not arbitrary and do not reflect mere agreement among a few people. For example, there is no moral principle supporting the law that we drive on the right side of the road. That law represents an arbitrary and convenient standard for motorists to follow (i.e., we could change the law to require that people drive on the left side of the road). Another example of the arbitrary nature of laws is that each state has a different set of laws. In many cases, what is legal in one state may be illegal in a neighboring state. In contrast, we treat moral principles as if they are self-evident facts that apply to all people in all situations. Consequently, we follow moral principles because they reflect a universal virtue, not because some authority directs us to do so.

Moral Principles Are Impartial

When we examine moral problems, we attempt to solve them consistently and impartially. For example, some people oppose any research that uses animals because they believe that the research harms the animal's well-being. A person who opposes animal research may begin by endorsing the moral principle that one should act to maximize happiness and minimize suffering. They may also argue that animals, like humans, can feel pain and do suffer. These observations lead some people to believe that the dividing line between humans and animals is arbitrary and irrelevant. The conclusion from this line of reasoning is that the moral principle of maximizing happiness and minimizing suffering must be impartially and equally applied to humans and animals.

Ethics and Ethical Codes

Ethics is a process of investigation, criticism, and decision making. When we say that someone studies ethics, we mean that he or she examines how basic moral standards and the facts of a situation lead to a consistent moral conclusion. An **ethical code**, or **code of conduct**, is a set of rules established by and for a group of people. As we noted previously, the American Medical Association has created an ethical code of conduct that applies to all professional healthcare providers and arguably serves as a good guide for ethical conduct among any health science professionals.

APPROACHES TO ETHICAL ANALYSIS

Throughout history, people have devised different approaches to determine what constitutes ethical and unethical behaviors. We cannot examine all of these in this chapter. Instead, we will examine two of the more influential approaches and encourage you to enroll in a philosophy of ethics course if these concepts intrigue you. The two approaches that we will review are the principle of utilitarianism and the principle of rights.

The Principle of Utilitarianism

Jeremy Bentham (1748–1832) and John Stuart Mill (1808–1873) were two philosophers who wrote a great deal about the principles of utilitarianism. The fundamental perspective of **utilitarianism** is that ethical behaviors are those where the total positive outcomes are greater than the total negative outcomes produced by one's actions (i.e., when the positive consequences of the ends help to justify the means by which those ends were achieved). For utilitarianism, the primary focus of the ethical analysis is the consequence of our behavior. As Bentham often quipped, "the . . . truth is that it is the greatest happiness of the greatest number that is the measure of right and wrong." Therefore, a behavior is morally correct if the positive consequences outweigh the negative consequences for the greatest number of people. Mill, like Bentham, agreed that we must consider the consequences of our actions but was specific about the types of consequences we should examine. According to Mill, we should select behaviors that maximize happiness while minimizing suffering. Therefore, from Mill's perspective of utilitarianism, actions that create the greatest good or happiness for the most people are morally correct behaviors.

How would we apply this principle to business research? The goal of research, according to utilitarianism, should be to maximize happiness for the greatest number of beings. Consequently, from this perspective, researchers must justify a research project by satisfying two conditions. First, the researcher should be able to show that the results will produce a useful outcome that will benefit others. Therefore, if the research jeopardizes the well-being of the participants, the researcher must demonstrate that the benefits of the study outweigh the discomfort experienced by the participants. Second, the researcher must demonstrate that the specific methods employed are better than all the available options for collecting the data. In other words, the researcher must show that there is no other, more ethical method for collecting data of the same quality.

Here is an example of how we might apply utilitarian reasoning when gauging the ethicality of health sciences research. Imagine that a medical resident wished to examine a specific type of breast cancer stem cell that exists in patients' bone marrow. To conduct this research, the researcher decides to use a sampling procedure that identifies currently healthy women who are either at a low risk or high risk of developing breast cancer. The women participating in the study are then asked to provide tissue samples using bone marrow biopsy procedures, an extremely painful testing method that require considerable time for recuperation.

We could justify this research from a utilitarian perspective if we can successfully support two arguments. First, we must show that the results of the screening have the potential of bringing the greatest happiness to the greatest number of people; that is, we need to show that the results of the study will greatly advance the science of cancer detection and treatment. The women who are being asked to participate in this study may not realize a benefit from the study, although they will experience a procedure that does produce notable discomfort. Moreover, they are, by definition, healthy, although some are at risk for developing breast cancer. And it is not yet clear whether the results of the test would indicate the potential development of breast cancer. Second, we must show that this method is the only way to acquire the information; that is, what is it about a bone marrow biopsy that produces data not available from other tests including blood or tissue samples? Is there no other way to find evidence of identifying the presence of cancer stem cells in otherwise healthy women?

Disadvantages of Utilitarianism

Although intuitively appealing, a utilitarian perspective on ethics suffers from several disadvantages, most of which are due to our inability to firmly define benefits and costs. From a research angle, for example, how are we to know the true effect of any experiment? How do we determine the relative cost of a person's discomfort or his or her happiness as a result of participating in the research? Although we use these terms every day, how do we weigh or measure discomfort and happiness? What accounting system do we use to ensure that our experiments do not "go into debt" by creating more discomfort than happiness? What counts as a cost? What counts as a benefit? One person may think that the results are of considerable theoretical significance and thereby serve science. You, however, may think that the results are trivial and of no known practical value. Therefore, one of the essential problems with utilitarianism is that *value*, *discomfort*, and *happiness* are vague terms that are extremely difficult to quantify.

Advantages of Utilitarianism

Aside from these important disadvantages, utilitarianism does offer multiple advantages to a researcher attempting to demonstrate the ethics of his or her research. One advantage is that utilitarianism offers a clear rationale for conducting research that creates temporary discomfort in participants. The researcher can justify the temporary discomfort of a few people if he or she can show that the results have a recognized value and that there is no other reasonable way to collect the data.

Another advantage of utilitarianism is that it allows us to evaluate a research project that may involve conflicting ethical concerns. In the study described earlier, we acknowledge that the women will be asked to experience a painful treatment procedure with the hope that the results may in some way help future generations of women from suffering the consequences of breast cancer.

The Principle of Rights

The utilitarian perspective requires that we examine the consequences of our actions to determine whether a behavior is morally right or wrong. The perspective of rights is a completely different perspective on ethics because it requires us to recognize that the rights of other people must guide our actions at all times (regardless of the costs associated with doing so). When we speak of rights, we mean that every person is entitled to certain privileges regardless of who they are; that is, rights are universal principles applied to every person equally and impartially. The philosophical foundation for this concept of universal moral rights is most often attributed to the philosopher Immanuel Kant (1724–1804).

Kant formulated the principle of the **categorical imperative** as a means of determining fundamental moral principles. As you will see, the categorical imperative allows us to define moral principles and to identify our duty as moral agents. For Kant, doing the "right" thing was not enough. We often do the right thing because it serves our goals. Some people give money to charity because they can use the donation as a tax deduction, to win the admiration of others, and to express pity for the less fortunate. Although giving money to a charity may be a nice thing to do, it is not automatically the morally right thing to do. According to Kant, the individual's sense of moral duty must drive his or her actions. Consequently, a true charitable act would be to donate the money in such a way that no one knows who gave the money. The categorical imperative in this case is that giving should occur without the expectation of benefit to the donor. Following Kant's categorical imperative, it is not enough to accept blindly any code of conduct as a field guide of the right and wrong things to do in research. Rather, you should examine the moral principles that guide your work as a researcher.

An essential formulation of the categorical imperative is that we should "act in such a way that you always treat humanity, whether in your own person or in the person of any other, never simply as a means, but always at the same time as an end" (Kant, 1750/1964, p. 96). This formulation of the categorical imperative establishes that all people are equal, as are their moral rights. More specifically, this moral code requires that we respect the dignity of other people and refuse to treat them merely as a way to serve our self-interests. Operating within this ethical framework, we must ensure that our behavior as researchers does not deny others their basic rights. The principle also requires that we avoid actions that diminish the dignity or self-worth of other people.

The ethical perspective created by the principle of rights is different from the utilitarian perspective. As noted previously, the utilitarian perspective considers only the consequences of our behavior. By contrast, the principle of rights requires that we examine the intention of our actual behaviors throughout the research process (i.e., instead of justifying the means by focusing on the outcomes, now the focus is on the means themselves). In addition, the principle does not focus on the balance of positive and negative outcomes. Rather, the principle of rights examines the universal rights each participant has and the duty we have to respect those rights at all times.

Disadvantages of Principle of Rights

Several criticisms can be directed at a principle-of-rights approach to justifying the ethicality of research. The first is that the system has no way to balance the conflicting rights of all individuals involved in the research process. For example, the authors of this book are applied psychologists who believe that all people who are asked to participate in research should, provided that they have given informed consent and the personal choice not to participate. To us, helping expand the body of scientific knowledge is a moral duty. By contrast, another person may believe that scientific research is misguided and that no one should act to help researchers. Kant's categorical imperative treats these conflicting rights as equal and gives us no way to decide between them. Because this perspective does not consider the consequences of our behavior, we have no way of resolving the conflict.

A second problem with Kant's perspective is that it can sometimes be too absolute and can create conditions that we find intuitively unacceptable. For example, we may believe that confidentiality is a moral right that must be accorded to all research participants. According to Kant's categorical imperative, we must never divulge what someone did or said to us as a part of our research. Imagine that as a part of his research, a psychiatric nurse learns that a participant is sexually abusing children. Application of a categorical imperative involving the absolute right to confidentiality would lead us to not report what is clearly morally unacceptable behavior. According to Kant, the consequences of maintaining confidentiality are immaterial. If we believe that confidentiality is a universal rule, then we cannot violate the participant's confidentiality. If the nurse calls the child protection agency about the abuse, he will violate the universal rule and can no longer say that confidentiality is a universal moral right of all people.

Advantages of Principle of Rights

Kant's philosophical principles make clear that all people are equal and that we cannot treat them as if they were pieces of equipment. You will find that most professional codes of conduct within the health sciences endorse the perspective that all people receiving medical treatment and/or participating in research have specific fundamental rights. Specifically, these codes require health sciences researchers to recognize that we cannot conduct research merely to serve our self-interests. Instead, we must put the participant's interest ahead of our own research goals. In other words, our research procedures must respect the dignity of the people who participate in our research and recognize that they have the capacity to act freely.

MAKING ETHICAL DECISIONS

Life is complex and often unpredictable. Consequently, there is no way that we can create a single set of rules that prescribes what you should do in every situation that you will encounter. In a similar fashion, it is unacceptable for any researcher to attempt to justify the ethicality of research by using either the utilitarian or principle-of-rights criteria by itself. Instead, we try to formulate general

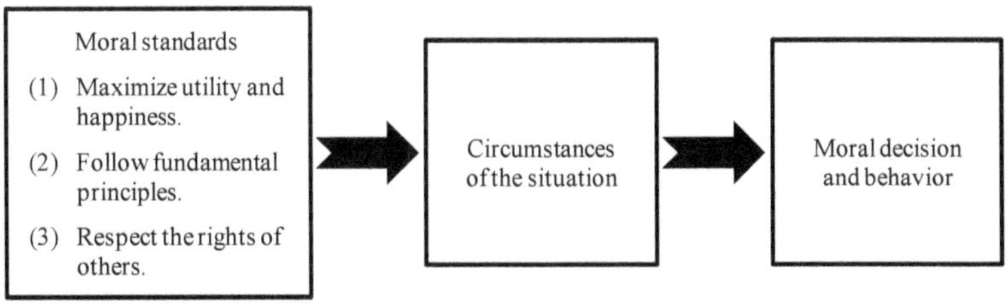

Figure 2.1 Ethical decision making.

standards or principles that can guide our decisions and behavior when conducting research.

Thus far, you have learned that the perspective of utilitarianism and the perspective of rights lead to different strategies for resolving ethical conundrums. In addition, each perspective has its relative advantages and disadvantages. These two perspectives can often lead to conflicting conclusions. Because research is a practical enterprise, we need to find some way to make practical decisions.

As we have already noted, ethical reasoning requires careful thought and deliberation. To examine an ethical issue, we must begin by examining the moral principles that we hold to be true. Next, we examine the specific circumstances of the situation. Using consistent and objective reasoning, we can determine how to behave.

Figure 2.1 represents the sequence of events that we use to come to an ethical decision. The moral standards are those principles that we hold to be true. This approach requires that we attempt to balance these moral standards within the circumstances of a research project. The result of this discourse is the moral decision and our course of action. Using fundamental moral principles and the facts of the situation, a person can come to a rational and logical decision regarding the ethicality of the research.

ETHICAL HEALTH SCIENCES RESEARCH

Research that uses humans as participants is a relatively new phenomenon (Resnik, 1998). Before the twentieth century, most scientists conducted research on topics unrelated to human behavior and the body. Toward the end of the nineteenth century, however, scientists began to study humans in detail as physicians began to use the scientific method to study various diseases and their cures. Behavioral and social phenomena also are relatively recent foci for research within the health and social sciences. Unfortunately, until very recently there was no clearly stated ethical code to guide researchers except their own personal beliefs and consciences.

The need for an ethical code for researchers studying humans became apparent after World War II. During that war, Nazi scientists conducted extremely cruel experiments on prisoners held in the concentration camps. These "experiments" were little more than protracted torture. As a response to these atrocities,

BOX 2.1 ***The Nuremberg Code of Principles Governing Research Using Humans***

1. **Informed Consent** The participant must understand what will happen to him or her in the research and then volunteer to participate.
2. **Social Value** The results of a study should benefit society.
3. **Scientific Validity** Only trained scientists who employ careful and well-designed studies should conduct research.
4. **No Malfeasance** Researchers must conduct studies that are safe and minimize risk of harm to the participants.
5. **Termination** The participant may withdraw from the study for any reason, and the researcher must stop the study if the participant is at risk of injury or death.
6. **Privacy** The researcher must preserve the privacy/confidentiality of the participant.
7. **Vulnerable Populations** Researchers need to use special caution to protect the rights of those who cannot act for themselves, including children, the developmentally delayed, or the mentally disabled.
8. **Fairness** Selection of the participants for the research and assignment to treatment conditions must be fair, consistent, and equitable.
9. **Monitoring** The researcher must continually monitor the study to ensure the safety of participants.

Note: The original Nuremberg Code included principles 1–5. Principles 6–9 are more recent additions to the code (Resnik, 1998).

researchers developed the Nuremberg Code (Resnik, 1998). The Nuremberg Code, summarized in Box 2.1, consists of nine fundamental principles that researchers must use when conducting research with humans.

As you should see, the Nuremberg Code is a reaction to horrible treatment of people who could not defend themselves. As you read the components of the Nuremberg Code, can you recognize the components that represent a utilitarian ethical perspective? How about the principle of rights? For example, how would you describe the "no malfeasance" requirement or the "social value" criterion? Why do you think principles 6–9 were added to the list?

Unfortunately, horrible and immoral experiments on humans did not only occur at the hands of Nazi scientists. Recently released documents from the U.S. Department of Energy reveal that during the 1940s and 1950s, U.S. government researchers conducted experiments on unsuspecting American citizens to determine the effects of nuclear radiation (Welsome, 1999). None of the people knew that they had been exposed to these deadly conditions. In another case, physicians at the Tuskegee Institute purposefully withheld treatment from 399 people who had contracted syphilis, for the sole purpose of studying the long-term effects of this dreadful disease (Jones, 1982). Unfortunately, these examples are not isolated cases (Resnik, 1998).

This bleak history of ethical abuses in research with human participants illustrates the need to create and enforce clear ethical guidelines for social and behavioral research. The **National Research Act**, which became law in 1974 (PL 93-348), established the IRB that we described earlier. The law requires that researchers receiving federal money to conduct research with humans and animals must demonstrate that their research methods meet minimal ethical standards and

BOX 2.2 ***Ethical Codes of Conduct Relevant to Health Science Professionals***

American Medical Association
Code of Medical Ethics
American Psychological Association
Ethical Principles of Psychologists and Code of Conduct
American Sociological Association
ASA Code of Ethics
American Statistical Association
Ethical Guidelines for Statistical Practice

that the IRB review the research proposal. As mentioned earlier, most research institutes and institutions of higher learning now have IRB committees that review any and all research involving humans and/or animals regardless of how the research is being funded. Box 2.2 highlights several of the professional associations and codes of conduct likely to be relevant to researchers working within the health sciences.

COMPONENTS OF AN ETHICAL RESEARCH PLAN

Ethical research does not happen on its own. It takes a conscious effort and careful planning by the researcher and ethical oversight committees (e.g., IRB) to ensure that the participant experience and impact on society more generally are carefully considered before any data are actually collected. The following subsections detail critical elements to an ethical research plan.

Competence, Accuracy, and Validity

When developing a research plan for behavioral and social research in the health sciences, researchers need to ensure that (1) the members of their team have the competence to gather and analyze the data to be collected, (2) the data will be collected objectively and accurately, and (3) the results will answer the questions that were the basis for the research. Simply put, it is a waste of time, money, and reputation to conduct a research project that fails to satisfy these basic criteria. Indeed, this book is dedicated to helping you learn how to conduct effective research. Thus, we will touch only briefly on these matters here.

Start by considering the matter of competence. Does the researcher or one or more members of the research team have the skills necessary to collect and analyze the data? Assume a physician approaches a pharmaceutical company with a proposal to test one of the company's drugs for a medical condition for which the drug has not yet been approved. In order to conduct the research, members of the research team will need the skills required diagnose the participants in the study, collect the appropriate data, and perform the correct statistical tests of the data.

The design itself is an ethical matter as the research team will need to ensure that the data can provide useful information regarding the effectiveness of the drug; that is, the research team could collect data that are meaningless or not valid considering the purpose of the research. Here are some simple examples. Assume that the researcher proposes to determine if the drug can be used to treat hypertension. Should the researcher compare the effectiveness of the drug relative to a placebo or should the researcher examine the effectiveness of the drug relative to other know medications for the treatment of hyper tension? Who should be included in the study? Hypertension has many causes. Would it be appropriate to test the drug on people with different levels of hypertension or whose hypertension is caused by specific conditions?

Voluntary Informed Consent

Voluntary **informed consent** is the hallmark of all research involving human participants. Ethically speaking, researchers cannot coerce or force people to participate and researchers must show that the people who participate in the study did so of their own free will. In addition, the participants must understand that they are free to withdraw from the study at any time and for any reason without penalty.

It is generally good practice when conducting a study to address the following points when communicating these rights and freedoms while recruiting potential study participants:

1. the purpose of the research, expected duration, and procedures
2. their right to decline to participate and to withdraw from the research once participation has begun
3. where appropriate, the foreseeable consequences of declining or withdrawing
4. where appropriate, reasonably foreseeable factors that may be expected to influence their willingness to participate, such as potential risks, discomfort, or adverse effects
5. where appropriate, any prospective research benefits
6. limits of confidentiality
7. incentives for participation
8. whom to contact for questions about the research and their rights as participants.

As you might imagine, informed consent contracts for health sciences research can be quite complicated and can span several pages. As an example, if a researcher is studying the effectiveness of a new drug, the informed consent would include a lengthy section explaining that the person may receive a placebo that has no effect on the underlying medical problem. The same document will also list all the currently known risks with taking the medication as well as the risk of not receiving any treatment—due to the placebo—the participant may experience. Also included in the document will be a statement regarding potential conflicts of interest; that is, the informed consent will reveal who is sponsoring the research—a pharmaceutical company, a private organization, or a federal agency—and the researcher's associate with the sponsoring agent.

A typical informed consent document will often require the participant to initial each page, parts of the document, or both, to illustrate that he or she understands specific components of the research. While the informed consent can be very long and take quite some time for the participant to read and understand, it is an essential step to ensure the person understands the risks and limitations for participating in the study.

Informed Consent and Minors

It is important to understand that in the United States and some other countries, minors cannot legally give informed consent. This means that you must obtain it from their parent or legal guardian. For example, if you wanted to study children's reaction to new medical treatment, you would first need to obtain the consent of the child's parents to conduct the research. You will then need to ask the child to participate in the study (to provide their assent, which is different than consent). As with all research, the parents and the child have the right to withdraw from the study at any time. The definition of what constitutes a "minor" for research purposes varies depending on the state and/or funding source (e.g., some federal sources of funding treat anyone under 21 years of age as a minor). Because these definitions vary, it is important for you to become familiar with the definitions relevant in the area in which you will be doing research.

Filming or Recording Behavior

If you plan to videotape or photograph the participant's behavior, or if you want to make an audio recording of an interview, you must first obtain informed consent from the participant. In addition, you will need to explain how you will use the recording. A researcher may need to record a participant completing neurological test of manual dexterity to track changes in symptoms over time. In such cases, you will videotape their test performance for research purposes but never present the tapes outside of the laboratory unless you have the written permission of the participant.

When No Informed Consent Is Needed

There are a few special conditions when you may not need informed consent from research participants before you begin to collect data from them as a part of a study. These conditions primarily occur when you need to passively and unobtrusively observe people's behavior in natural contexts. For example, you might go to a large pharmacy and observe shopping behaviors as people browse through over-the-counter medication. If you do not plan to interact with the patrons, you can collect data without asking for their permission. You could also use the stores' video campers to observe shoppers' behavior.

State law may control recording people's actions and conversations. For example, some states make it illegal to record a telephone conversation unless both people understand that the conversation is being recorded. Therefore, you may want to consult a legal authority concerning what you can and cannot record without consent.

Anonymous questionnaires and surveys are also typically exempt from the informed consent requirement. Therefore, if you wanted to send a questionnaire through campus mail to all students, you would not have to have the students' informed consent unless the data you collected allowed you to learn who completed the questionnaire. Even though this is often the case, we still recommend you include an informed consent form with this type of survey, but instead of requiring signed consent, you indicate that completion and return of a survey constitute that person's consent to participate.

Deception in Research

Deception in research means that you lie (or fail to tell the whole truth) to the participants about some important part of the research in which they are a participant. You may lead a person to believe that the other people sitting at the table volunteered to participate in the study. The truth may be that the other people are confederates or accomplices of the researcher and are following a script that you prepared.

There are two types of deception: **deception by omission** and **deception by commission**. Deception by omission means that you withhold important facts from the participant. Deception by commission means that you purposefully lie to or mislead the participant about an important component of the research.

Deception by omission occurs in all research to some extent because it is impossible and impractical to tell the participants everything about the purpose and details of the research. For example, we do not describe the hypothesis we want to test, nor do we describe every aspect of the research design to participants in advance of collecting data. If we did, the information could bias the participants' behavior. Therefore, we describe the general themes of the research that will allow the potential participants to know what may happen to them during the study. This type of deception can be addressed to some extent with a comprehensive debriefing statement that provides additional details about the research to participants once the data collection is complete.

Deception by commission raises serious ethical concerns, however, because the researcher is lying to the participants. Although deception by commission has many negative consequences, there are times when we must use it as a part of the research. For example, a researcher may need to create an elaborate cover story for the study. Previously we examined a researcher's use of participant observation where the researcher pretended to be a newly hired employee. The cover story hid the true purpose of the study and allowed the other employees to follow their normal work behavior.

The concept of debriefing requires researchers to reveal their use of the "cover story" and why they needed to use the deception in order to collect authentic information. The goal of debriefing is twofold. First, we want to be honest with the participants and let them know the rationale for the study and the value of the data. This should allow the participants to understand that they have contributed information that will help answer important questions. The second goal is to judge how people respond to the study. Clearly, if people begin to object to the methods of the study, the researcher may need to consider the consequences of the project and to determine if the risks are greater than the benefits.

Confidentiality

The health sciences researcher must be concerned about the confidentiality of all parties included in the research project, including the people being studied, the organization that has commissioned the research, and any third party that may be involved in the project.

The first group to consider is the people you are studying. Regardless of the purpose of your research, you have an obligation to ensure the information you collect remains confidential. Confidentiality means that the identity of your participants remains a secret. In most cases, confidentiality is relatively easy to preserve. One way to maintain confidentiality is to use a unique but meaningless number to identify each of the participants. For example, you might use the number "2–3–47" to record the data of a person. The number is meaningless in that it does not identify the person but does allow you to file the data. In addition, when we publish the results of the research, we typically present the data as averages across groups rather than highlighting the performance of individual people along with their names and home addresses.

There is a subtle difference between confidential and anonymous. In both cases, when the researcher reports the data, the reader will not be able to determine how individuals performed or responded in the study. Technically, confidentiality means that the researcher does not reveal how participants acted in the study. By contrast, anonymity means that the researcher does not link the participants' names with the data being collected. Here is an example: A researcher wants to study patients' use of illegal drugs and prepares a questionnaire asking patients questions like "I smoke marijuana" or "I use heroin at least ___ a day." These are pretty serious questions as people are being asked to verify that they are engaged in illicit behavior. If the researchers say the data are confidential, it means that they know how each person responded but will not release the information. If the researchers say the data are anonymous, it means that they collected the data in such a way that it is not possible to link answers to the questions to a single person.

In addition to protecting the confidentiality of participants, health sciences researchers often need to consider the confidentiality of the research sponsor. There are many reasons that a particular corporation or funding source may not want to let others know about its product development or concerns about the status of its development of a new drug or medical device. For these reasons, those conducting sponsored research may be required to sign nondisclosure agreements.

RESEARCH IN ACTION: ETHICAL DILEMMAS

What follows is a series of brief and ethically complex research scenarios. Read each with care and then consider the ethical issues raised by the scenario.

Informed Consent

A researcher wants to conduct a large-scale study examining cholesterol levels in a large segment of the population. To gather sufficient blood samples, the researcher will need to use many clinics that take blood samples. At these clinics, the technicians will be asked to draw the blood samples required by the patients' physician and one additional sample for the research project. For the purpose of the study, the researcher needs to know only the patients' age and sex.

KNOWLEDGE CHECK

1. Does the researcher need to have people give informed consent and why?
2. As an alternative to using multiple clinics, would it be appropriate for the researcher to advertise the people can have free blood tests—diabetes screening, cholesterol levels, and similar tests—for participating in the study?
3. Imagine that during the project, the researcher decides to conduct a new test on the blood sample that was not a part of the original research plan. Can the researcher conduct the new test without the participant's consent?
4. As a part of the research project, the researcher finds that several of the participants have a life-threatening disease. What obligation does the researcher have to notify the participants of this condition?

Long-Term Research Project

A researcher begins a 4-year clinical trial for a new drug to treat the symptoms of Parkinson's disease. The length of the project was set because the researcher wanted to compare the rates of the disease's progress for those administered the drug and those who were not given the new drug.

KNOWLEDGE CHECK

1. For this study, should the researcher use a placebo to ensure the effects of the new medication? What control conditions should the researcher use to examine the effectiveness of the drug as a treatment for Parkinson's?
2. The researcher decides to conduct a comprehensive evaluation of each participant's neurological condition every 6 months to track the progress of the disease. What should the researcher do if after the first test she finds that participants receiving the new drug appear to show less deterioration compared with the other participants?
3. During the research project the Food and Drug Administration (FDA) approves a new drug for the treatment of Parkinson's disease. Many of the participants in the study wish to be treated with this new medication. Should the researcher encourage the participants to remain in the study to test the effectiveness of the current drug being tested?

Historical Research

A medical researcher wishes to examine the use of specific surgical techniques practiced in different regions of the county. To conduct the study, the researcher plans to pick hospitals at random and to review the medical records of patients treated for a specific medical condition.

KNOWLEDGE CHECK

1. Can the researcher review the records without the patients' consent?
2. What limits should be placed on the researcher regarding the types of information that can be reviewed?

Trauma Interview

A medical school researcher wants to study women who receive medical treatment at an emergency department and who may have been the target of domestic violence. The interview is designed to cover a wide range of abusive behaviors the women may have experienced including the extent of physical assaults, sexual violence, and psychological abuse. For the study, whenever emergency department personnel suspect that a woman is seeking treatment due to domestic violence, a nurse or social worker will conduct the interview after treatment.

KNOWLEDGE CHECK

1. Imagine you are a member of this medical school's IRB. What are potential objections to conducting this study?
2. What should the researcher do if women in the study provide information that a crime has been committed?

CHAPTER SUMMARY

All researchers have a responsibility to act in a moral and ethical manner. This responsibility is especially critical for psychologists because their research directly affects the lives of the people and animals that are a part of their research. This chapter reviewed basic moral principles and the process of ethics. Moral principles are the beliefs we have that help us distinguish between right and wrong. Ethics is the process of examining how we react to specific situations and make decisions on how we should act.

The four moral principles at the heart of research are (1) the concern for the welfare of other beings, (2) the belief that moral principles stand before our interests and secular laws, (3) the belief that moral principles are universal truths that do not change across time or situations, and (4) the belief that moral principles are impartial and treat everyone equally.

Two of the major philosophical systems that guide ethical analysis are the principle of utility and the principle of rights. Utilitarianism directs us to seek ways to maximize happiness and minimize discomfort. Therefore, when a researcher examines a research design, he or she must consider the consequences of conducting the study. Specifically, the researcher must show that the results of the study justify the method by demonstrating that there is no other way to collect the data and that the importance of the data outweighs any discomfort the people or animals in the study may experience. By contrast, the principle of rights requires that the researcher focus on preserving the dignity and self-worth of the people participating in the research. Research that puts the interest of the researcher above the welfare of the participants is morally wrong.

CHAPTER GLOSSARY FOR REVIEW

Categorical Imperative An ethical perspective developed by Kant that requires that we always respect the humanity and dignity of other people and never to treat others as a means to an end.

Deception by Commission The intentional act of deceiving the participants by telling them something incorrect about the research.

Deception by Omission A form of deception in which the researcher withholds information about some conditions of the research from the participant.

Ethical Code or **Code of Conduct** A list of rules and guidelines that direct and guide the behaviors of members of an organization.

Ethics The process of examining one's moral principles and behavior.

Informed Consent A requirement for all human research that ensures that the participant understands the purpose of the research, his or her rights as a participant, and the potential hazards of participating in the research.

Institutional Review Board (IRB) A group of people who examine the researcher's proposed methodology to ensure that the researcher will protect the basic rights of the participants.

Morals Fundamental principles that distinguish right from wrong.

National Research Act A federal law (PL 93-348) that established the existence of the IRB. The law requires that any researcher receiving federal funds to conduct research on humans must first receive IRB approval of the research.

Utilitarianism A perspective of ethics that requires one to examine the consequences of one's behavior and to ensure that it maximizes happiness and minimizes suffering.

REFERENCES

JONES, J. (1982). *Bad blood.* New York: Free Press.

KANT, I. (1964). *Groundwork of the metaphysics of morals* (H.J. Paton, Trans.). New York: Harper & Row (original work published in 1750).

RESNIK, D.B. (1998). *The ethics of science: An introduction.* New York: Routledge.

WELSOME, E. (1999). *The plutonium files: America's secret medical experiments in the cold war.* New York: Random House.

Chapter 3

The Foundations of Research

Science knows only one commandment—contribute to science.

—Bertold Brecht

INTRODUCTION

Researchers look for patterns of behavior and then try to explain them using the scientific method. The techniques researchers use to collect and analyze data are methods that help answer specific questions. This chapter introduces you to many of the basic concepts that form the foundation of the research methods used in contemporary research. We begin with the research **hypothesis**.

The hypothesis is important in all research because it determines the type of data that the researcher will collect, the methods used to collect the data, and the statistical procedures used to analyze the data. There are as many different strategies for collecting and analyzing data as there are forms of research hypotheses. After discussing some of the most common forms of hypotheses you are likely to encounter, we then turn our attention to the process of data collection from a sample of participants. In this section, we delve into the issue of how samples of data can allow us to make inferences about a broader population.

Understanding and Conducting Research in the Health Sciences, First Edition. Christopher J. L. Cunningham, Bart L. Weathington, and David J. Pittenger.

THE HYPOTHESIS IN RESEARCH

What is a hypothesis? Most people believe that a hypothesis is an "educated guess." Although this is technically correct, this definition misses the full purpose and role of a research hypothesis. Therefore, we would recommend that you set that definition aside and opt instead for one that is a bit more informative: *A hypothesis is a specific prediction about the relation among two or more variables based on theory or previous research.* Yes, the first definition is shorter and easier to remember, but this second one is more accurate and actually gives you a clue about how researchers develop and use hypotheses.

Hypotheses Come from Theory or Previous Research

Most research hypotheses do not materialize out of thin air nor are they wild guesses. In general, hypotheses come from two general sources, existing theories and previous research. Hypotheses also often come from direct replication and extension of previous studies because our confidence in research findings dramatically increases as consistent findings emerge. This is especially true when replication-type studies use different methods, samples, and measurement approaches.

Consider a simple example about the relationship between cholesterol and blood pressure. As a generality, we can say that "All things being equal, people with high levels of cholesterol have high blood pressure." Therefore, if you want to study the effects of cholesterol, then your best starting point would be the large existing literature and knowledge base from the previous research in this area. You can use this information to help you generate a hypothesis, and when you do so, you are really saying, "If the results of previous research are consistent, then when I conduct my study, I should obtain similar results." This is why we can often deduce or derive a hypothesis from an existing theory.

What is a theory? A **theory** is a broad set of general statements and claims that allows us to explain and predict various events. This definition of theory differs from the common or lay use of the word *theory*. Many people often equate *theory* as interchangeable with *hypothesis* and treat both as a guess about why a situation or relationship exists. This view is incorrect as the theory represents a general statement, whereas the hypothesis is a specific prediction.

Here is an example of a simple theory from which we can derive several hypotheses: Excessive levels of cholesterol can collect on the interior of blood vessels, thus narrowing the vessels and increasing blood pressure. This theory offers a general explanation for the relation between cholesterol and blood pressure. We can generate several hypotheses from this theory:

- Changes to diet may reduce cholesterol levels and thereby reduce blood pressure.
- Medications that lower cholesterol may also lower blood pressure.

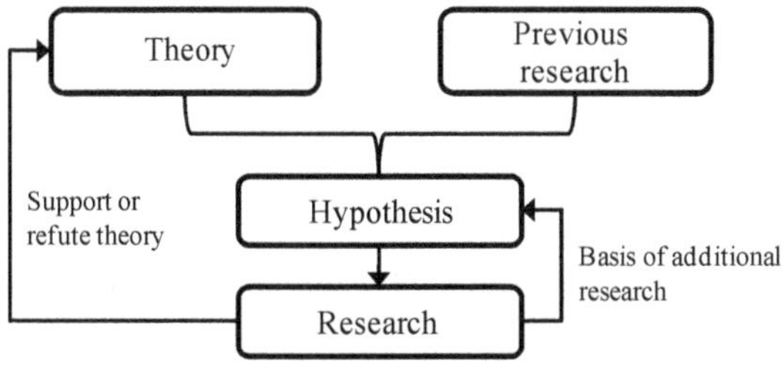

Figure 3.1 A diagram of the role of theory, hypothesis, and research.

- Removal of cholesterol from the lining of blood vessels may lower blood pressure.

Any hypotheses we form, however, should be based on as much existing empirical information as possible. Our ability to confirm or refute each hypothesis will help us evaluate the overall accuracy of the theory. Figure 3.1 represents the interrelations between previous research, theory, and the research hypothesis. The individual research hypothesis is the product of theory, previous research, or both. For confirmatory research, the data derived from a study can influence a theory as well as serve as a guide for additional research. For exploratory research, the results of a study may inspire additional research.

Figure 3.1 also illustrates that the results of a study can serve as the foundation for additional research as well as a more formal test of the theory itself. As a generality, research hypotheses designed to test the predictions of a theory represent **confirmatory research** because the researcher plans to use the data to confirm or refute the validity of an existing theory. It may also be the case (and often is) that existing theories are modified or changed based on the results of research. The process of research and theory formation is dynamic and cyclical. Existing theory should always be evaluated and, if necessary, modified based on current research.

One of the many virtues of a hypothesis is that it helps protect researchers and the research process from being influenced by the idols of the theater, the uncritical acceptance of information, including that which is included in theories. Bacon (1620/1994) warned us not to accept ideas or a theory merely because they sound correct or are popular, often repeated, or presented by an authority. Rather, we should determine whether we can empirically verify the hypotheses derived from the theory. Most researchers will support a theory as long as it generates interesting and empirically testable hypotheses. Researchers begin to doubt a theory when the data that are collected do not confirm its predictions or when someone offers a different theory that accounts for the same phenomenon while also incorporating any new and unexpected findings.

It is important to emphasize that all research is not confirmatory. Much research in the health sciences especially is exploratory in nature. **Exploratory research** is not necessarily guided by theory but rather by the need to answer specific questions. A researcher may, for example, want to study the effects of long-term exposure to a chemical or other product added to food to increase its shelf life. Bacon (1620/1994) advocated the collection of empirical data from many diverse sources and suggested that data collection should be free of preconceived notions in order to avoid the idols of the cave, the willingness to depend on personal experience to understand phenomena.

All this is not to say that exploratory research operates without hypotheses. The primary difference is that exploratory hypotheses merely postulate a tentative explanation or prediction rather than the more confident ones that are characteristic of confirmatory research. For example, an insurance company notices that some physicians are sued less frequently than others. Because the company wants to lower its risks, it begins to study the characteristics that distinguish higher levels of litigation from lower levels. How would you proceed; what variables would you examine? How about the physician's specialty (general practice vs. obstetrics); level of training and board certification of physician; age, gender, and education of patients? Clearly, there are a lot of variables in play. Indeed, one factor that appears to greatly influence the rates of litigation is the time the physician spends with the patient explaining the procedure and the risks. Doctors who spend a few minutes more talking with their patients about the patient's condition and treatment are less likely to be sued for malpractice. (Rodriguez et al., 2008). The hypotheses going into this type of study would be primarily exploratory, as the company researchers are not fully confident going into the study what the likely predictors of litigation will be.

Hypotheses Direct Our Observations

When we prepare a hypothesis, we focus our attention on the type of information that we will study. Specifically, the hypothesis identifies the variables that we will examine and the data we will collect. In this way, a hypothesis is like a brief summary of your research design. Good hypotheses help us to *operationalize* our constructs into concrete, measurable variables. A good hypothesis states how these measured variables are related. More details on "good" measurement are provided later in this chapter.

A **variable** is any characteristic of a person or research condition/environment that can take different values. A symptom of a disease is an example of a person-level variable because it exists in different amounts (i.e., it varies) across different people. Similarly, sex is also a variable, as some people are men and others are women. We can also examine the influence of different research conditions on participants. In an experimental context, any element of the experiment that can change is a variable. The opposite of a variable is a **constant**, a numerical value or characteristic that, by definition, cannot change. For example, a minute always contains 60 seconds and $\pi = 3.1416...$

An empirical hypothesis will generally state the relation between two or more variables. Consider the hypotheses listed earlier and illustrated in Figure 3.2. In each case, the hypothesis states that changes in one variable correspond with changes in another variable.

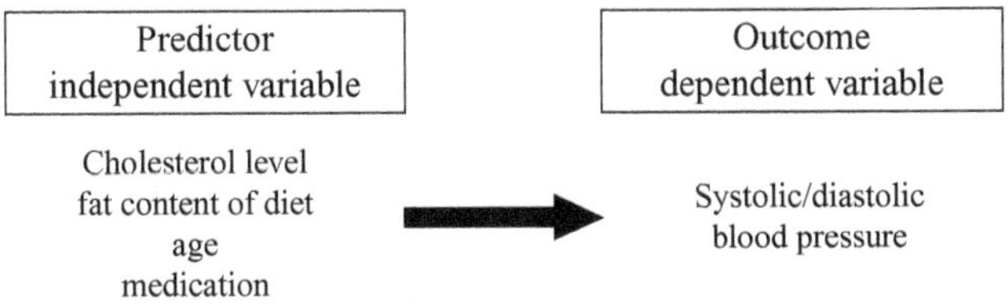

Figure 3.2 The relation between variables described in a hypothesis.

Included in this example are four possible variables that have been hypothesized to predict or account for a second variable. In this example, we want to find the variables that allow us to explain or predict a patient's blood pressure. As in most research, there are two types of variables in this example: the dependent variable and the independent variable.

The **dependent variable** is the focus of the study and is the condition that the researcher wants to explain, describe, change, and/or possibly predict. You will sometimes see the term *response* variable or *outcome* variable used synonymously with *dependent* variable. As an example, the first hypothesis in Figure 3.2 states that blood pressure may be affected by a number of variables. In this example, blood pressure is the dependent variable because we see it as the result of such things as age and diet.

The other variable in a hypothesis is the explanatory or **independent variable**, the variable we expect to cause or account for a change in the dependent variable. A hypothesis explicitly describes how we expect the independent variable(s) to explain, describe, and predict the dependent variable(s). In this example, we note that things like age and diet affect blood pressure.

As you will soon learn, different types of research have different types of independent variables. In some cases, the researcher can control or manipulate the levels of the independent variable. We call these **manipulated independent variables** because the researcher modifies or changes the amount or level of the independent variable and because the variable is independent of (i.e., not affected by) other variables in the study.

Another type of independent variable is the **subject variable**. A subject variable is a characteristic or condition that the researcher cannot manipulate or control but that may play an important role as an independent variable. As a rule, if a characteristic exists prior to the start of a research project, it is likely to be a subject variable. Sex, height, weight, and age are examples of these types of subject variables. Whenever people participate in a research project, they bring unique characteristics to the study. Although researchers cannot change the characteristics of the participants, they often use these characteristics as subject variables to predict meaningful dependent variables or outcomes.

Imagine a researcher conducting a survey using the variables presented in Figure 3.2. What are the subject and manipulated variables? In this case, age is the subject variable as it is an attribute of the participant. By contrast, the researcher can manipulate the medication a person receives when conducting the study.

How would we classify some of the other variables such as diet? The answer depends on how the researcher used the variable. If the purpose of the study is to compare the blood pressure of people who have a diet high in sodium with that of people who have a diet that is low in sodium, the diet is a subject variable in that the researcher determines the person's sodium intake and then measures blood pressure. In contrast, the researcher may want people to follow different diet plans. Because the researcher now asks people to, for example, eat a low-sodium diet and others to eat a normal diet, the diet becomes a manipulated variable.

In summary, a good hypothesis identifies an expected relationship between at least two variables. One of these variables is the dependent variable that the researcher hopes to explain, describe, and predict. The other variable, the independent variable, is the one that the researcher uses to explain the results of the study. The independent variable will typically be either a manipulated variable or a subject variable.

Hypotheses Describe the Relationships among Variables

Finding the orderly relationships among variables is the goal of any science. Thus, one of the most important components of a hypothesis is that it describes the relationships that we expect to observe between the variables. When we prepare a hypothesis, we use different verb phrases to describe the relation between independent and dependent variables. For example, we can state that when one variable increases, the other variable will also increase: *As the sodium intake increases, blood pressure increases.* We can also predict that an increase in one variable will lead to a decrease in the other variable: *Taking drug X causes a decrease in blood pressure.*

Hypotheses Connect Research to Populations

Most hypotheses are general statements about relationships among variables at the level of a population. The hypothesis *People with high levels of cholesterol are more likely to have high blood pressure* implies that the relationship between cholesterol and blood pressure is universal, applying to all people in the target population. If this hypothesis is correct, we should observe an association between cholesterol and blood pressure in that population and in a representative sample of people from that population.

Populations and samples are familiar terms. We often talk about the population of the United States or the population of a city. Similarly, we often hear the word *sample* in day-to-day language. Even walking through a grocery store can present you with tasty samples of new foods and products. The companies that market these products hope that your experience with the little samples will be enough to compel you to buy their product in the future. The meanings of population and samples are similar in research, but there are a few additional details that we need to present.

In research terminology, a **population** is a group of people or things that share one or more characteristics. In some cases, the definition of a population will be broad and inclusive. In other cases, the population will be much more specific. Consider a researcher who is interested in blood pressure. The focus of this researcher's study could be broad and could include all adults, or be extremely narrow and limited to adult men over 65 who have a family history of stroke. The definition of the population will then define the nature of the sample. In the health sciences, it is common to refer to these definitional details as inclusion and exclusion criteria. These criteria clearly define which individuals are to be included as participants and which people are not to be included as participants in a given study.

A **sample** refers to a subset of the population. In most cases, when researchers speak of samples, they are referring to a group of individuals or things that represent a broader population. If we assume that a sample is representative of the population, then we can assume that any conclusions we make based on data gathered from this sample will generalize to the population. Researchers define populations using empirical characteristics or publicly verifiable characteristics. These definitions can be general (e.g., all children) or specific (e.g., males between 8 and 12 years old). In general, we improve our ability to describe, explain, and generalize our findings regarding behaviors when we specifically define the population of interest beforehand and succeed in representatively sampling from that population when we collect data. You will read more about sampling in Chapter 7.

TYPES OF HYPOTHESES

Hypotheses differ in terms of the specific types of predictions that researchers are making. In some cases, researchers will examine the correlation between two variables; in other cases, the researcher will determine whether there is a meaningful difference between two groups. The type of hypothesis that a researcher uses determines the research methods that he or she will use as well as the types of conclusions the data allow. There are four general types of research hypotheses that you are likely to see in research from any scientific discipline, including the health sciences: (1) estimation of population characteristics, (2) correlation among variables, (3) differences among two or more populations, and (4) cause and effect.

Estimation of Population Characteristics

The goal of this form of hypothesis is to estimate the characteristics, or parameters, of a population using information or data gathered from a sample. We use special terms when we estimate and describe population parameters, including **data, descriptive statistics**, and **parameters**. Populations are generally too large to use for our research. Therefore, we use a representative sample of the population to estimate the population parameters. In all research, we collect data. In most health-care research, these data are quantitative, though it is becoming increasingly common across most business research to also gather qualitative, or nonnumerical, data for a richer understanding of the phenomenon under study. We will discuss qualitative research methods in Chapter 17. Regardless of the type of data collected, the assumption guiding research is that the data adequately reflect the actual variables of interest.

Once the numerical data are collected, we typically calculate descriptive statistics. A descriptive statistic is typically a number of some form that helps us to organize, summarize, and describe the data. In addition to describing the sample, we can use descriptive statistics to estimate the value of population parameters. A parameter, therefore, is a number that summarizes and describes a characteristic of a population. As a quick summary, **statistics** *are numbers based on direct observation and measurement and refer to the sample.* **Parameters** *are values that refer to the population. In most cases, we use statistics to estimate parameters.* If the sample is representative of the population, then we can infer that what is true of the sample is also true of the population. A common practice in research is to represent statistics using Roman letters (e.g., *A*, *B*, *C*, *D* . . .) and to represent parameters using Greek letters (e.g., α, β, γ, δ, . . .). In most contemporary research, it is common to use the letter *M* to represent the sample mean and the symbol μ (mu) to represent the population mean, the letters SD to represent the sample standard deviation, and the symbol σ (sigma) to represent the population standard deviation. Although statistics are variable in that they will change each time we collect data from a different sample, we consider population parameters to be constants.

What would a hypothesis for this type of study look like? An example of a hypothesis pertaining to the estimation of population parameters is *What is the typical systolic and diastolic blood pressure of men between 45 and 55 years of age*? In this hypothesis, we want to know the relation between a subject variable (men between the ages of 45 and 65) and the dependent variable (blood pressure measures). Another way that we can say the same thing is to write $M \approx \mu$. This expression states that the sample mean we obtain from our collected data will be approximately equivalent to the value of the overall population mean. As stated earlier, if the sample is representative of the population, then our best estimate of a population parameter is the sample statistic that we estimate from the data we collect.

Correlation between Two Variables

Another useful form of hypothesis involves stating that there is a correlation between two variables. A correlation between variables means that they are linked such that as one variable changes, so does the other. This does not mean, however, that one variable necessarily causes the other. To be clear, correlation does not always indicate causation.

As an example, there is a correlation between years of education and overall health. If we collect a sample of adults and ask them to tell us their level of education (e.g., high school, associate's degree, college, graduate school) and then assess their general health, we will find that there is a positive correlation between education and health. People with less than a high school education tend to have poorer health as adults. By contrast, people with greater levels of education tend to be healthier. The presence of this correlation, however, does not explain why exactly it is true that adults with more education tend to be healthier.

When we form correlational hypotheses, we can predict that there will be a positive, a negative, or no correlation between the two variables. A **positive correlation** means that an increase in one variable corresponds with an increase in another variable. The correlation between years of education and overall health represents a positive correlation—an increase in years of education corresponds with an increase in health. By contrast, a **negative correlation** means that increases in one variable correspond with decreases in the other variable. Trick question: *What is your prediction about the correlation between one's golf score and the time spent practicing?* There is a negative correlation between the two variables. Remember that in golf, the lower the score, the better.

When there is **no correlation**, there is no systematic relation between the two variables. There is no correlation, for example, between shoe size and intelligence. People with small feet are just as likely to be bright as people with big feet. Similarly, dull-witted people are equally likely to have small or large feet. Figure 3.3 illustrates three types of correlation: (*a*) a positive correlation, (*b*) no correlation, and (*c*) a negative correlation.

When we speak of a correlation at the level of a population, we often use the parameter ρ (rho) to express the size of the correlation between the two variables. When we predict that there is a positive correlation between the variables, we write $\rho > 0$. When we predict that there is a negative correlation between the variables, we write $\rho < 0$. If there is no correlation, we write $\rho = 0$. When writing or speaking of an observed correlation as a sample statistic, we typically use the Pearson product moment coefficient, r.

Difference among Two or More Populations

There are many cases when we want to know whether members of separate populations are, on average, different from each other. In this situation, the independent variable may be either a manipulated independent variable or a subject variable that we use to classify people. The second variable is a common dependent variable that we measure for all participants in the study. Once we have the data, we can determine whether there are meaningful and statistically significant differences among the groups.

In both of the following examples, we want to determine whether the typical behavior of members of one group is different from members of another group. As with the other hypotheses, we use sample statistics to make inferences about

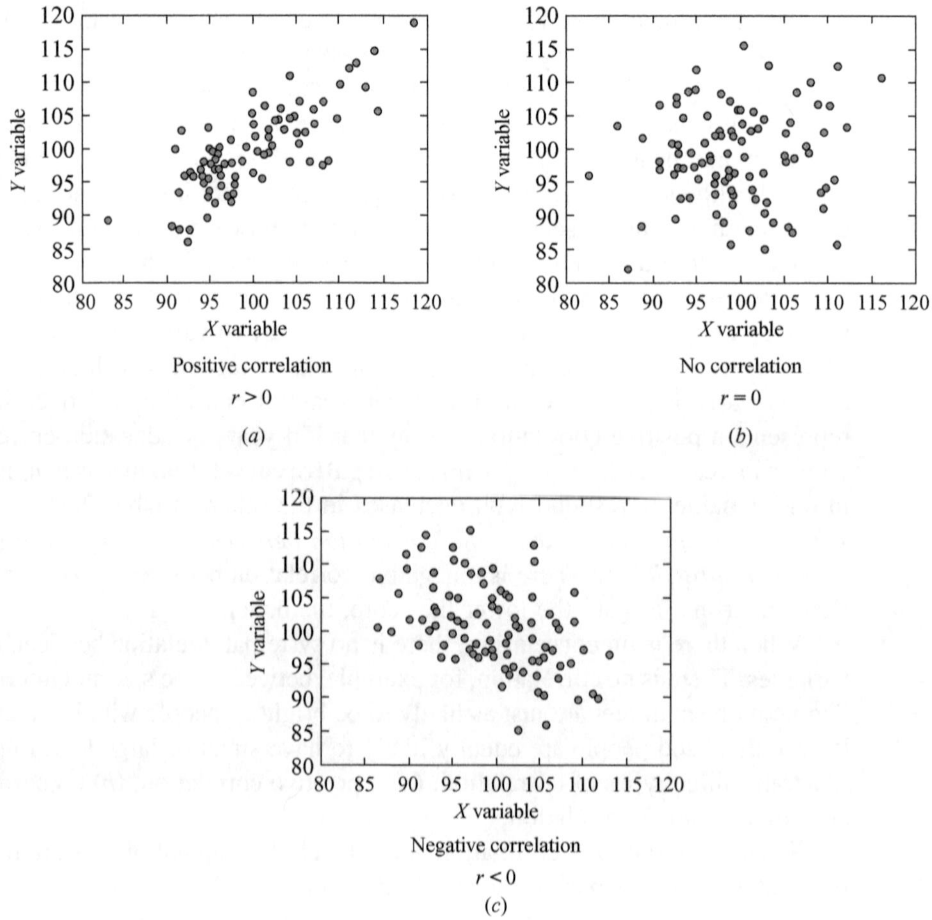

Figure 3.3 Three correlations: (*a*) a positive correlation, (*b*) no correlation, and (*c*) a negative correlation.

the population parameters. Therefore, when we compare two groups, we compare the difference between the sample means. If the difference between the sample means ($M_1 - M_2$) is sufficiently large in a statistical sense, we then infer that the population means are different from each other. As an example, if you wanted to predict that the mean of one group is greater than another, you would use the statement $\mu_1 > \mu_2$. In this example, μ_1 represents the mean of the first population and μ_2 represents the mean of the second population.

EXAMPLE 3.1 *Payment Method Influences Purchases*

Are people who use a credit card more likely to buy "junk food" than people who use a debit card? Recently, Thomas, Desai, and Seenivasan (2011) conducted an experiment that asked people to make purchases using either a credit card or a debit card. In this case, the dependent variable was the amount spent on junk food, whereas the independent variable, a manipulated variable, was the method of payment.

EXAMPLE 3.2 *Is There a Sex Difference for Health Care?*

Do men and women differ in their use of general health care? In this case, we can use sex, a subject variable, to create our comparison groups. The sex of the participant, male versus female, is a subject variable because it is a condition that the researcher can measure but not control. The common dependent variable is the number of times people see a physician for a routine checkup.

Cause and Effect

The last general form of hypothesis is the most demanding hypothesis to test. For this hypothesis, we want to prove that changes in the independent variable cause changes in the dependent variable. Discussions of cause and effect can quickly become complicated and have been the subject of considerable philosophical analysis and debate for quite some time. It has become common in contemporary research to base determinations of cause and effect on the criteria established by John Stuart Mill (1986). Mill argued that one must show three things to infer a cause-and-effect relationship:

1. **The Cause Must Precede the Effect, Temporally (*X* Must Come before *Y* in Time Order):** This requirement recognizes that time moves in one direction, from the past to the future. Consequently, events that occur in the past can influence future events, and it is impossible for events in the future to influence events in the past.
2. **The Two Events in Question Must Be Correlated (*X* and *Y* Must Covary with Each Other):** If we believe that the independent variable causes the dependent variable, then we should observe a consistent relation between the two variables. We can demonstrate the relation between the independent and dependent variables by examining the correlation between the two variables or by comparing the differences among group means. The difference between the groups is evidence that the independent variable is correlated with the dependent variable.
3. **All Other Possible Explanations for the Relationship between *X* and *Y* Should Be Ruled Out or Explained:** This third criterion is difficult to achieve because it is often difficult to account for all possible explanations except for the one implied in the hypothesis.

Recognize that all the other hypotheses describe the population parameter and the general relationship among variables; they do not allow us to infer cause and effect. When we conduct research to examine cause and effect, we still examine the correlation between the variables and compare the differences among groups. Therefore, we will still need to determine whether $\mu_1 > \mu_2$. If we find that this statement is correct and the experiment meets the criteria for determining cause and effect, we can then infer that the independent variable caused the dependent variable, or IV $\rightarrow$ DV. The arrow indicates that changes in the independent variable caused changes in the dependent variable. Table 3.1 presents a summary of

Table 3.1 Features of the Four General Hypotheses Used in Empirical Research

Type of hypothesis	Purpose of research	Mathematical hypothesis	Written hypothesis
Estimate parameter	Use sample statistics to estimate population parameters.	$M \approx \mu$	What is the average score of fifth-grade students taking a reading comprehension test?
Correlation between variables	Measure two variables to determine their correlation.	$\rho > 0$ or $\rho < 0$	There is a positive correlation between time spent studying and exam grade.
Difference among populations	Compare two or more sample groups to determine if they are different.	$\mu_1 > \mu_1$ or $\mu_1 < \mu_1$	Men, on average, are taller than women.
Cause and effect	Determine if the independent variable causes differences in the dependent variables.	IV $\rightarrow$ DV	High levels of dopamine cause schizophrenia.

the four types of hypotheses. Be sure that you understand the similarities and differences among the four.

KNOWLEDGE CHECK

1. Describe the differences between a theory and a hypothesis.
2. Consider each of the following statements and determine whether they meet the criteria for a hypothesis.
   a. Corporations should promote healthy lifestyles for their employees.
   b. Patients with a particular disease should join a social network (e.g., Facebook®) of people with the disease.
   c. Perceived health is a subjective appraisal.
   d. Patients will be more compliant with a medical treatment if the physician spends more time talking about the side effects of the treatment.
   e. Members of a family are likely to have similar diets even when living as independent adults.
   f. People with low self-esteem are less likely to seek medical help.
3. Jackie is interested in patients' perceptions of their physician. Her hospital uses a standard evaluation form that asks patients a series of questions regarding their health and their evaluation of the physician's effectiveness. Jackie asks the Director of Nursing if there is a relation between the patient's health and evaluation of their physician.
   a. Rewrite Jackie's question as a hypothesis.
   b. What is the independent variable?
   c. Is the independent variable a subject or a manipulated independent variable?
   d. What is the dependent variable?

4. Tim wants to examine the effect of having a celebrity promote flu vaccinations. He creates two videos that are identical in every way except for the person reading the script. In one video, the woman reading the script is a popular movie star. For the other video, the woman is an unknown professional actor who has similar attributes as the movie star. Tim has people watch one of the two commercials and then asks as series of questions that ask the individual to rate their willingness to purchase a flu shot.
   a. What hypothesis or hypotheses do you believe that Tim can test with this research? State these in your words.
   b. What are the independent variables?
   c. For each independent variable, determine whether it is a manipulated independent variable or a subject variable.
   d. What is the dependent variable?
5. Carlos administers two tests to 500 adults (250 men and 250 women). The first test is a standard level of routine exercise. The second test is a measure of willingness to participate in a medical research project.

   What hypothesis or hypotheses do you believe that Carlos can test with this research? State these in your words.
   a. What are the independent variables?
   b. For each independent variable, determine whether it is a manipulated independent variable or a subject variable.
   c. What is the dependent variable?
6. Rebecca conducted a study that examined how people react to different types of antismoking commercials. Sixty people volunteered to participate in the study. Half the participants read an advertisement that listed all the negative consequences of smoking. The other half read an advertisement telling people how tobacco is an addictive drug. Rebecca then gave all the participants a survey that asked them to rate the belief that the tax on tobacco products should be increased.
   a. What hypothesis do you believe that Rebecca tested with this research? State the hypothesis in your words.
   b. What is the independent variable?
   c. Is the independent variable a subject or a manipulated independent variable?
   d. What are the dependent variables?

MEASUREMENT

At the heart of all research is measurement, a special form of empirical observation. Measurement allows us to record and quantify our observations in an objective and consistent manner. Therefore, measurement is a means to an end in that it allows us to answer the question, "What do we have?" When done well, measurement can also improve the objectivity of our research efforts. When preparing to start a research project, our priority is objectivity. Proper measurement can reduce various forms of bias from entering into our observations, helping us to address Bacon's idols.

Attempting to define objectively our variables allows other researchers to choose to accept or reject our definitions. This also makes it easier for researchers to accurately replicate studies and to repeat measurements. One criterion of all science is that it depends on public verification. Good measurement can help to ensure that any competent person is able to observe the same phenomenon that another researcher has observed. Can you see how public verification protects us from the idols of the theater? If any person has the ability to verify a theory or to replicate an existing finding, this means we do not have to rely on any singular authority to tell us what is true or false. All of us have the opportunity to collect the data and to evaluate the results for ourselves.

The remainder of this chapter and section of this textbook discusses more basic elements of good measurement. To begin with, it is important to note that all forms of measurement share four essential features: (1) operational definition, (2) measurement scale, (3) reliability, and (4) validity.

Operational Definition

Measurement is the process of assigning numbers to objects, situations, or observations using a rule. An **operational definition** is the rule we use to convert our observations into numbers and our theoretical constructs into measurable variables. Consider the following questions:

1. What is the distance between Cleveland and New York?
2. How many Americans are diagnosed with multiple sclerosis?
3. Do patients perceive their physician as an expert in health care?
4. What is the volume of the typical stomach?

What does each of the questions have in common? First, each question refers to a specific construct. In the order of the questions, the constructs are

1. distance
2. number or frequency
3. perception
4. capacity.

Second, we can answer each question using some form of measurement that allows the researcher to collect data relevant to each construct. To perform the measurement, we will have to define the variable and develop an appropriate measurement technique. Operational definitions define variables by stating the objective and empirical procedures for our measurement. Consequently, creating an operational definition is a critical step in any research (as you will recall from Chapter 1, it is the second step in the scientific method *HOMER*). Operational definitions are different from dictionary definitions, which tend to be descriptive and imprecise. For example, a dictionary definition of *perception* may be *the organization of sensory information*. Can you see the problem with this definition? It really does not address the question the research wishes to answer, that is, whether or not the patient believes the physician to be an expert.

An operational definition, by contrast, refers to observable and/or measurable events that are also open to public verification. We can collect information or data from these observations or measurements and use this information to make inferences about the construct and its relationship to other constructs in our hypotheses or research model.

Take the question about the distance between Cleveland and New York. We need a clear operational definition of distance. What, specifically, do we mean by distance? How should we measure the distance? Should we draw an imaginary straight line between the city halls of the two cities and count the number of miles? Would the mode of transportation (plane, train, boat, or automobile) affect our measurement? One could measure the length of the interstate highways connecting the two cities as a measure of distance. Once we decide on the route we want to measure, what **metric** will we use? Should we record the distance in miles, kilometers, or time spent traveling? The important point to recognize is that an operational definition results in a clear statement of the type of information you will need as a researcher to test the hypotheses in your study.

Now, measuring distance is relatively straightforward, especially if you contrast this with measuring constructs that cannot be directly observed. This is the case with many constructs that you and other researchers are interested in studying. Many constructs, like perceived value, cannot be directly observed. Nevertheless, we want to use these variables to explain and predict behavior. How can we use variables that are not directly observable to explain behavior? One way is to use observable behaviors to infer the presence of events that we cannot directly see. We frequently use **intervening variables** such as behaviors to describe other constructs we cannot directly measure. We also use intervening variables to explain various complicated and unobservable phenomena.

Consider the construct of blood pressure as an example. Figure 3.4 presents an illustration of how we might use perceived value in a hypothesis and how we might operationally define value. The illustration in Figure 3.4 consists of two parts. On the top of Figure 3.4 is the hypothesized construct. On the bottom are observable events that are open to public verification. We use the hypothesized construct, or intervening variable, to help explain the link between one set of variables and another. For example, we may note a positive correlation between the diet and heart rate. From this correlation, we would infer in this case that a person's blood pressure may be influenced by diet and may ultimately impact the functioning of that person's cardiovascular system.

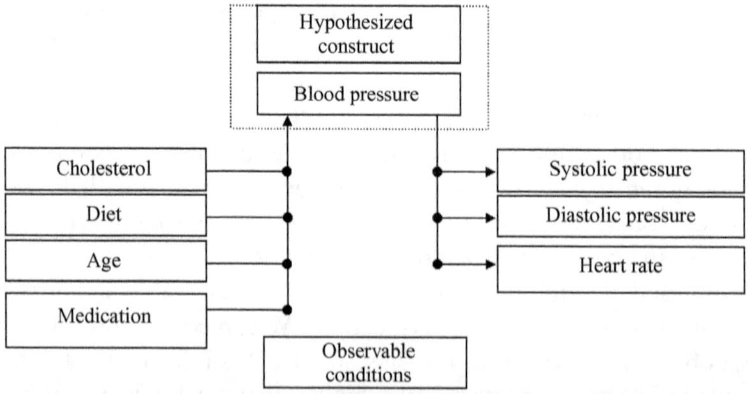

Figure 3.4 Depiction of the relationship between a hypothetical construct, or intervening variable (perceived value), and various observable conditions and reactions.

Measurement Scale

Most people do not give much thought to the meaning of numbers. The general attitude seems to be that numbers are numbers regardless of their use. However, there are different classes, or types, of numbers. More accurately, when we measure things, we can use numbers in different ways. It is, therefore, important to recognize the differences among the types of measurement scales. Researchers and statisticians often speak of four types of measurement scale. Each scale has unique features and presents information in a special manner. In addition, the type of scale we use will determine the type of inferences we can make from our measurement. The four types of measurement scales, in order of increasing complexity, are (1) nominal, (2) ordinal, (3) interval, and (4) ratio.

Nominal Scale

Nominal scales are the most basic form of measurement. We can use nominal measurement to classify things into mutually exclusive categories based on some qualitative property. The purpose of a nominal scale is only to indicate some sort of qualitative difference among things or people. Examples of nominal scales include biological sex, religious preference, marital status, or occupation. Taking religious preference a bit further, in a survey you might ask people to indicate their religious preference by selecting one of three options, such as atheist = 1, Baptist = 2, Catholic = 3. This type of scale allows us to differentiate among people regarding their religious beliefs. The scale does not imply that one perspective is better than the others or that membership in one group represents more or less of something than membership in another group. The number of potential categories depends on the variable of interest and the characteristics of your sample. For

Table 3.2 Example of Operational Definitions for Nominal Scale

Type of care	Home care	Relative or friend	Unlicensed day care
Definition of category	Child stays with a parent	Child stays with an unpaid relative or friend	Parent purchases child care from an unlicensed provider

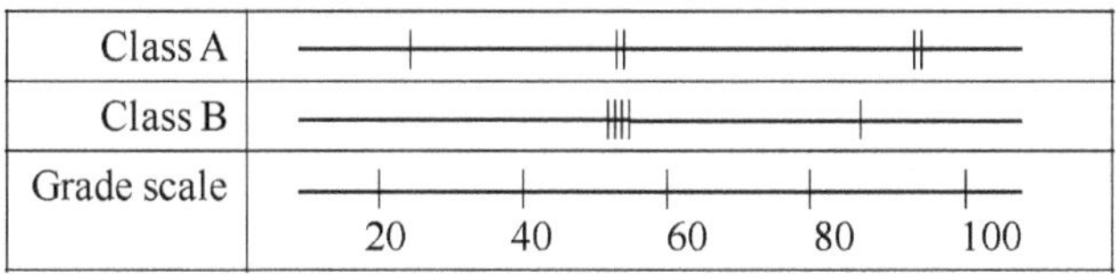

Figure 3.5 An ordinal scale. *Note:* The vertical lines (|) represent the students' scores.

example, depending on your sample you might need more than these three religious preferences and include variables such as agnostic or Zoroastrian (to name only two of many other possibilities).

Nominal scales show up in many healthcare research projects. If you were studying the child-care business in your state, you might want to know where parents take their children during work. You could create a nominal scale definition as shown in Table 3.2. In this example, you created three types of child care. They are nominal categories to the extent that you do not wish to imply that one type of day care is better than another, only that the three types are different. Your goal will be to estimate the number of parents who select one of these options for their child or children.

Ordinal Scale

With an **ordinal scale**, the numbers refer to a quantitative scale that is rank ordered. In other words, whenever you rank something from highest to lowest, you are using an ordinal scale. In a horse race, for example, we rank the horse's order of finish as first, second, third, . . . , last. Movie critics use the number of stars to rate the quality of movies. A five-star rating usually means that the critic enjoyed the film and recommends it without reservation. By contrast, a one-star rating means that someone should burn all copies of the movie. We hope you recognize that, although intuitively appealing, ordinal scales are imprecise. In a horse race, a horse can win by a nose or 14 lengths. The critical feature of the ordinal scale is that there is no consistent magnitude of difference between points on the scale. Consequently, we cannot say that the difference between a 1 and a 2 is the same as the difference between a 3 and a 4. What we can assume is that $1 < 2$ and that $2 < 3$. Figure 3.5 provides an example to consider.

Imagine that two classes took the same exam. The vertical lines (|) represent the scores of individual students. In each class, one student can brag that he or she earned the highest score. Does this mean that both students did equally well? As you can see, the best score in Class A is in the 90s, whereas the highest score in Class B is in the 80s. Also, look at the difference between the first and second scores in the two classes. In Class A, the difference between the two highest scores is slight. In Class B, the difference between the two highest scores is large. The ordinal scale lacks the precision and accuracy found in other forms of measurement. There are many times, however, when the ordinal scale is the most appropriate or best-quality scale available to the researcher.

Interval and Ratio Scales

The interval and ratio scales represent the most comprehensive type of measurement. These scales are quantitative because a one-to-one correspondence is assumed between the construct and the numbers on the measurement scale. In addition, the difference among the numbers along the scale represents equal units. These two types of measurement scale do differ with respect to the meaning of zero (0) on the scale. On an **interval scale**, the 0 is an arbitrary point selected for the sake of convenience. On a **ratio scale**, the 0 represents the absence of the construct of interest. We can use scores on a math test as an example of the difference between an **arbitrary 0** and an **absolute 0**.

A typical classroom math test covers only a small range of math skills, perhaps a student's skills with basic algebra. A student who earns a 0 on this specific test may not know much about algebra but may know something in other aspects of math. By contrast, a comprehensive assessment of math-related knowledge or ability would have a "true 0" score that reflects a complete absence of math-related knowledge by the student. Thus, we cannot assume that a student who scores a 0 on an algebra test knows absolutely nothing about math more broadly defined. Because the test measures only a small portion of all math skills, we cannot infer that a 0 means that the student knows nothing about math. Therefore, the 0 on the math test is an arbitrary number relative to the broader construct, knowledge of math. What about two students who take the math test, one who scores an 80 and another who scores a 40? Does the student with the 80 know twice as much math as the student with the 40? Although $80 = 2 \times 40$, we cannot conclude that the student with the higher score knows twice as much math. If you look at the math knowledge scale, the difference between the two grades is minor relative to the amount of math both students know. Therefore, we would conclude that scores on the math test represent an interval scale for measuring math aptitude.

Examples of ratio scales include grams of food consumed, heart rate, and number of white blood cells. As a guideline, ask yourself what the 0 on a scale actually means. If it indicates the absence of the construct being measured, then the 0 represents an absolute 0. For example, if a special meal includes 0 g of fat, we can infer that none of the calories in the meal are associate with fat. In addition, eating 20 g of fat is twice as much as eating 10 g. Table 3.3 summarizes the attributes and differences among the four scales.

Table 3.3 Characteristics and Examples of the Four Measurement Scales

Scale	Characteristics	Examples	Mathematical operations
Nominal	Unordered and mutually exclusive categories, qualitative differences among groups	• Sex: male vs. female • Psychiatric diagnosis • Religious preference • Political party	= vs. ≠ Number of observations in each group $f(x)$
Ordinal	Ranking or ordering of measurements	• Rank in class • Level on a personality dimension (high vs. low self-esteem) • Rank of quality: 1 = poor, 10 = excellent	< vs. >
Interval	Order, equal intervals, and arbitrary zero	• Temperature in Celsius • Scores on most psychological tests: intelligence, personality, depression	$A - B$ vs. $B - C$
Ratio	Order, equal intervals, and absolute zero	• Height, weight, age, speed • Time to complete a task • Number of problems solved	A/B vs. B/C

RELIABILITY OF MEASUREMENT

In terms of measurement, **reliability** refers to the consistency with which observations or measurements are made. Stated differently, assuming the construct you are measuring is not expected to change over time (e.g., general intelligence), a reliable test or measure will generally produce similar results each time we use it to assess a participant's standing on that construct. As you might guess, all researchers strive to develop and use the most reliable measurement devices available. Measures that are not reliable have very little value to researchers as they can lead to the downfall of a research project. Unfortunately, no test is perfectly reliable because **measurement error** affects all measurement techniques. For our purposes, measurement errors are random events that can increase or decrease an individual score on a given scale or assessment.

Measurement error is not the same thing as bias. Measurement error is a random variable that changes each time we use the test. In contrast, **bias error** is systematic or constant, and is therefore present each time a test is used. A reliable measure can be biased. A bathroom scale, for example, may reliably report your weight, although it consistently adds five pounds to your true weight. This is an example of a systematic bias that inflates your true weight every time you use this measurement device (the scale). In classical test theory, an observed score is assumed to be the sum of the true score plus or minus the effects of any measurement error (random and systematic forms). In formula form, this theory can be stated as *observed score* = *true score* ± *measurement error* ± *bias*. The ± symbol in the equation indicates that random and systematic error can add to and remove

Table 3.4 Illustration of Reliability

Testing	Test with higher reliability	Test with lower reliability
1	107 = 110 − 3	118 = 110 + 8
2	112 = 110 + 2	105 = 110 − 5
3	111 = 110 + 1	107 = 110 − 3
4	109 = 110 − 1	111 = 110 + 1
5	109 = 110 − 1	109 = 110 − 1
	 110	 110

Table 3.5 Sources of Error That Reduce the Reliability of a Measurement Technique

Source of error	Examples
Instrument error	• Equipment stops working. • Questions on a test are poorly worded. • Measurement scale is not precise.
Participant variability	• Participant becomes fatigued or bored. • Participant does not understand the instructions. • Participant is unwilling to cooperate. • Participant cannot respond to some items.
Researcher variability	• Researcher makes errors recording the data. • Researcher does not follow procedures consistently.
Environmental variability	• Distractions occur in the environment, such as noise, or participant discomfort. • Measurement conditions are inconsistent across different participants.

from the true score of the construct that we want to measure. For example, imagine that you administer an intelligence test to a student who has a true intelligence quotient (IQ) of 110. Due to measurement error, each time you administer the test, you may get different results.

Table 3.4 illustrates the effects of measurement error and the difference between two tests, one with higher reliability and one with lower reliability. A person completes two intelligence tests on five occasions. For both tests, there is random error. The random error is smaller for the test with the higher reliability. As you can see for both tests, measurement error raises and lowers the observed test score. Although measurement error affects both tests, the variability among the test scores is smaller for the test with higher reliability. The relation between reliability and measurement error is simple: More reliable tests have less measurement error. Many factors contribute to measurement error. For the sake of simplicity, Table 3.5 lists the general classes of measurement error and several examples of each. As a researcher, your job is to determine what factors could affect the reliability of your measurement device and then to design research procedures that can reduce the chances that this error will emerge. For example, if you believe that environmental distractions will corrupt the data, you will need to conduct your research in a setting where the testing conditions are consistent from subject to subject.

One of the more common methods to improve the reliability of the data is to increase the number of measurements. Carpenters have an old saying: *Measure twice, cut once*. Nothing is more frustrating than cutting a board and then finding that it is too short. To be careful, a good carpenter will double-check the necessary measurements to have greater confidence before cutting a board or drilling a hole. Patients are often advised to "get a second opinion." The same (or at least similar) advice is important to remember with all empirical research: *Measure more than one way*; *interpret once with more confidence*. As an example, measures of intelligence, clinical depression, and classroom performance all usually depend on multiple measurements in order to improve reliability. As you can see in Table 3.4, an individual measurement may be a bit off the mark, but the average of several measurements will tend to hit the target.

Determining Reliability

Reliability is commonly determined statistically using some form of correlation statistic such as Pearson's product moment coefficient (r). It can also be identified with the use of other statistics, such as Cohen's kappa (κ), that show the agreement or consistency between two or more scores or sets of ratings. More details on common procedures for demonstrating reliability are presented in Chapters 8 and 9.

VALIDITY OF MEASUREMENT

Validity actually has less to do with elements of the measure, test, or assessment itself and more to do with the way in which a researcher interprets the data. To be more specific, validity is a property of the conclusions and interpretations that a researcher makes after studying the data collected with some measure. When people say a test is valid, what they are really saying is that a researcher using that instrument is likely to make accurate inferences based on test scores. In other words, accurate conclusions are drawn about the relationship of the construct the measure represents and other constructs that are the focus of the specific research study. When working with a measure that assesses something other than what you are interested in studying, it is common to refer to that measure as not valid. The implication is that using this measure will hinder your ability to make accurate inferences regarding the construct(s) you are trying to assess.

Figure 3.6 represents the relationship and distinction between reliability and validity in measurement. The picture represents three targets with arrows. The targets represent what we want to measure and the arrows represent the measurements we make. When the arrows are scattered across the target, as is the case in

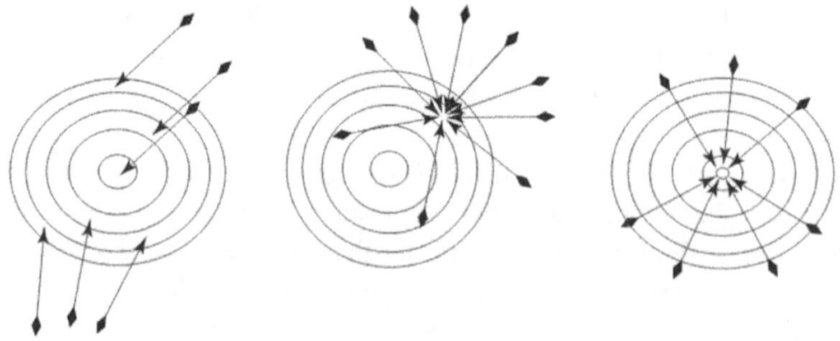

Figure 3.6 The relationship between reliability and validity. *Note:* The arrows represent separate measurements.

the left example, we say that the test is neither reliable nor valid. High reliability is not a guarantee of validity, however, as you can see in the middle picture. In this scenario, all the arrows hit a similar point, but this point is far away from the target construct (the bull's eye). Thus, information from this measure is consistent but biased away from the target construct. Thus, *reliability is a necessary but insufficient precondition for validity*. The scenario on the right illustrates a situation in which the measure is reliable and valid. All the arrows are close to each other and come close to hitting the bull's eye.

Determining Validity

Validity is difficult to quantify because there is no single statistic or index that tells us whether something or some inference is valid. In the broadest sense, validity is a value judgment based on empirical evidence and the interpretations we want to make from the data. As Kaplan (1964) noted, validity refers to what we can accomplish using the data. From a slightly different perspective, all data are potentially valid until we attempt to interpret them. The definition of validity is fuzzy and vague because validity refers to the interpretation of the test score, not the test or measurement instrument itself.

If we gave you a test that measured your knowledge of rules of grammar, that test might contain questions about your knowledge of gerunds, dangling modifiers, relative pronouns, infinitives, subordinate conjunctions, participles, and active voice. Assuming that the test is reliable, what inferences can we draw from your score? Would it be appropriate to assume that you recognize all these terms and the associated rules of grammar? Depending on your score, such an interpretation may be correct. Can we use the test score to determine whether you are a creative writer? Probably not, regardless of your score. Good creative writing may be independent of knowledge of the formal rules of grammar. Thus, the same score from this one assessment of your knowledge of grammar can lead to valid or invalid conclusions, depending on how we choose to interpret the findings.

Researchers generally rely on several criteria to establish or demonstrate the potential validity of a measure or assessment. First, the reliability of that measure is a must. Once that is satisfied, attention then turns to identifying several additional subforms of validity that together help us to understand the full **construct validity** of a measure, or the extent to which that measure allows a researcher to make accurate inferences from observed data about the original construct. The four subforms of validity we will discuss here are (1) face validity, (2) content validity, (3) predictive/concurrent criterion-related validity, and (4) construct validity.

Face validity is determined by the degree to which a participant or test taker looks at an assessment and says to himself/herself, "This seems to clearly be an assessment of X," when X is the construct you are hoping to measure. Face validity is an important consideration during the test development process because when people do not believe that a measurement procedure is face valid, they may not respond honestly or fully.

As the name implies, **content validity** refers to the degree to which the measurement technique adequately samples behaviors relevant to the construct you are trying to measure. If you want to determine a person's proficiency in mathematics, you will ask the individual to solve math problems, not recite Lincoln's Gettysburg Address. In most cases, we determine the content validity of a test by having experts on the subject review the test. Therefore, you might ask several professors of mathematics whether a math test adequately assesses the student's ability to understand and apply specific mathematical concepts. Pay careful attention to the difference between face validity and content validity as they have been presented. It is very easy to confuse the two. The major difference is that face validity deals with the layperson's or test taker's perceptions, while content validity is evaluated based on the opinions of experts.

One of the important components of any good measure is that it allows us to accurately predict other behaviors or outcomes. This type of validity is referred to as **criterion-related validity** because we are trying to predict something (i.e., the criterion). In some cases, we hope that measurements that we take now will allow us to predict events later. If a test allows us to make these predictions, then we say that the test has **predictive validity**. **Concurrent validity** means that a test score correlates with other relevant behaviors or outcomes assessed at the same time as the target variable test. Predictive validity is predicated in a temporal distance between predictor and outcome. This is not the case for concurrent validity.

As an example, consider what is meant when a health science professional refers to a person's overall health. The concept of health is complex and multifaceted. How would we know if we had a good measure of overall health? In general, we would want to show that the components of any measure of general health provide useful information. As a simple example, we might ask a person to rate his/her health overall and relative to others who are of the same age and sex. Using this information, we might be able to construct a scale of overall health. If the measure is valid, we should find that a person's score on this measure of general health is correlated with other indicators of health such as morbidity, cardiovascular health, and life satisfaction.

POPULATIONS AND SAMPLES

Previously, we noted that hypotheses help us to link our expected relationships among measured variables to relationships between constructs that exist at the population level. When we do research, we have to settle for sample-level data to test the proposed relationships that are stated in our hypotheses. Chapter 7 provides a more detailed discussion of populations and samples, but it is important to review now how we can use data collected from samples to make inferences about a population. Therefore, we need to examine the relation between target populations, sampling populations, and samples. The research hypothesis refers to variables in the population.

Because most populations are extremely large, it is impractical or impossible to truly sample from the population. Consequently, most researchers use a sample drawn from a **sampling frame** or a subset of the larger target population. Ideally the sampling frame would be a complete list of every member of the population, but in reality, it is usually a subset of the target population to which the researcher has or can gain access. The final sample itself is often a random selection from this sampling frame or another subsample of the population that is determined based on research access or convenience. Ideally, the sample that a researcher settles on is a representative subset of individuals drawn from the sampling frame, which is also hopefully a representative subset of the more general population.

For example, assume that you want to study smoking. Your population may be all adults. Chances are that you will use students from your college as participants in this research. Therefore, students at your college represent the sampling frame. The students who participate in your study represent the sample. Figure 3.7 represents the relation among populations, sampling frame, and samples.

The logic of sampling is deceptively simple. First, take a representative sample from the sampling frame. The representative sample should be manageable in size and share the same characteristics as the sampling population. Next, conduct the study and collect the data. Once you have analyzed the data, you can generalize from the sample to the sampling frame. If your sampling frame is representative of the population, you can then assume that what is true of the sample will be true of the population. This description of using samples to generalize about the population sounds simple and straightforward. Unfortunately, this advice is similar to

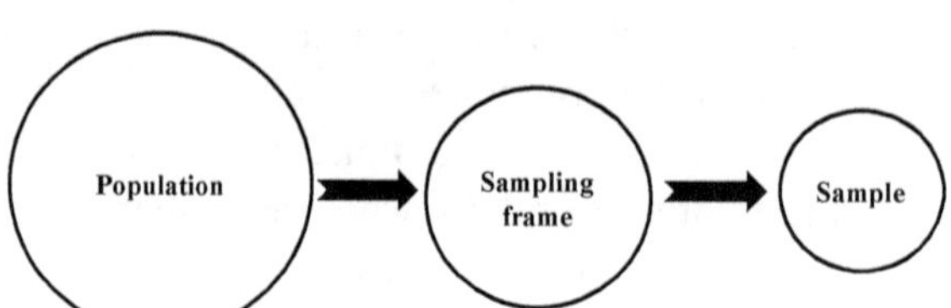

Figure 3.7 Relation between population, sampling population, and the sample.

saying that to build a house all you need to do is buy the right materials and follow the blueprints. Although technically correct, the instructions gloss over many important points.

The fundamental question that we need to ask is, "How do we know that the sample is representative of the population?" We can answer that question by examining two interrelated characteristics of samples. The first characteristic of a sample is its size. The second characteristic is its ability to represent the population. How large should the sample be? We will revisit this question several times throughout this book. Research requires time, money, and energy, and each member of the sample requires these resources. Therefore, to control costs and not spend an entire lifetime on a single project, we want to keep the sample to a reasonable size. At the same time, the sample has to be large enough to adequately represent the population. If the population is very diverse on many characteristics, then the sample may need to be larger to fully reflect this diversity. A large sample, however, is not a guarantee that the sample will accurately reflect the population.

This leads us to the second characteristic of a sample, its ability to represent the population. Many popular magazines publish the results of surveys of their readers. Sometimes the editors will tout that more than 20,000 readers responded to the survey published in a previous issue. Do these results provide a good and useful source of information? Can we consider the data to be representative of the population or even the sampled population? After a moment's consideration, you should say, "No!" There are many potential problems with these data.

First, only people who read the magazine had the opportunity to respond. Second, although 20,000 is a big number, it is not an automatic protection against bias. The people who responded may represent a minority of the readership. Perhaps those who responded are also somehow qualitatively different from those who chose not to respond. As we progress through this textbook, we will examine methods researchers use to determine how large the sample should be. In some cases, we will be able to use one or two people for our research. In other cases, our sample will number in the hundreds or even thousands. Determining the size of the sample depends on what we know about the population and the type of behavior we are studying. As we begin to explore specific types of research, we will show you ways to determine the size of the sample.

Apart from the size of a sample, to be useful, samples must also actually represent the population. Because we want to generalize about the population, we need to ensure that the sample is an accurate model of the population. The **representativeness of a sample** is the consequence of the procedure the researcher uses to collect the data. We can determine whether a sample is representative of a population by examining the methods used by the researcher to collect those data. One of the most common ways to improve the representativeness of a sample during the designing phase of a study is to use **simple random sampling**. Simple random sampling means that each member of the sampling frame has an equal probability of being selected or becoming a participant. Following this procedure, laws of probability ensure that a sample of a sufficient size will accurately reflect the overall population.

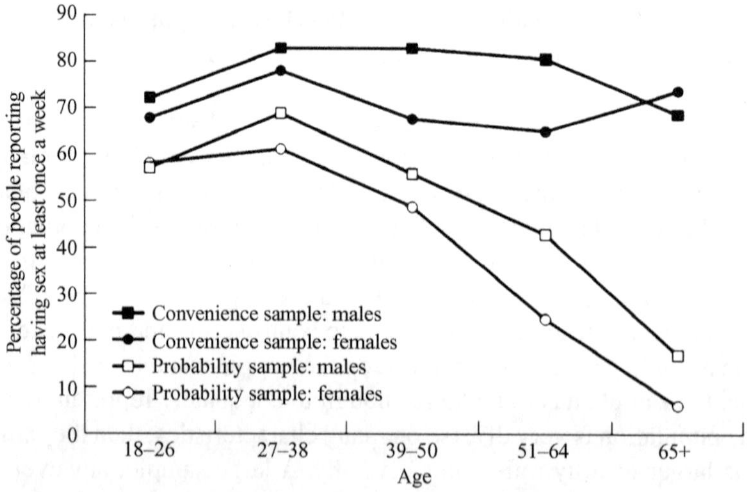

Figure 3.8 The difference in data produced by a convenience sample and a probability sample. *Note:* Data as presented in Hyde and DeLamater (1998).

Conversely, failure to use simple random sampling can produce questionable results. Consider the following example. Hyde and DeLamater (1998) presented the results of two surveys of men's and women's sexual behavior. The first survey was administered to a sample determined by simple random probability sampling. The second survey was administered to a sample gathered via **convenience sampling**. For convenience sampling, the researcher uses members of the population that are easy (convenient) to recruit. A convenience sample is not a random sample because some people in the population have no chance of being selected. Figure 3.8 presents the results of the two studies that examined the frequency of sexual relations among adults.

Hyde and DeLamater (1998) noted that the convenience sampling represents data collected from clients of a clinic for the treatment of sexual dysfunctions and their friends. These people may not represent the typical person in the population. The other set of data represents research conducted by researchers who used random sampling of the population. We cannot trust the results of the convenience sample because the sample (people attending the clinics) is not representative of the population. Instead, we have greater confidence in the results of the probability sample, which is likely to provide responses that are more representative of how the overall population would respond.

Role of Random Sampling in Research

Many students are surprised to learn that few health sciences research studies actually use random sampling. Within the health sciences this is especially true in many cases because the issues and topics studied do not necessarily apply to a large subset of the population, sometimes affecting only very specific types of individuals or groups of individuals in very specific situations. We highlight this because there seems to be a general impression that all research *must* use random sampling to be valid. In some cases, this assumption is correct. In other cases, we may not need to use random sampling to conduct valid research. Our choice of whether to use random sampling depends on the type of research we conduct and on the goals of the research.

When Random Sampling Is Essential

Random sampling from the target population is essential when we want to estimate, with as much precision as possible, specific characteristics of a larger population. Political polling is an example of research for which random sampling is imperative. To predict who will win an election, or which policy issues are most important among registered voters, a pollster will take great pains to ensure that he or she collects a random sample from the target population. The random sample will allow the researcher to predict, for example, that "Strickland will receive 57% of the votes, whereas Hollister will receive 43%, ±3 percentage points." If a true random sample of likely voters was used for this poll, then the pollster/researcher can safely predict that Strickland will receive between 54% and 60% of the vote.

Apart from politics, there are many cases when conducting research where we want to ensure we have a representative sample from the target population. For example, imagine you want to study obesity among adolescents. What is your population and sampling frame? What conditions represent obesity? Using a random sample of adolescents will allow you to develop an unbiased estimate of the prevalence of obesity among teenagers.

When Random Sampling Is Not Essential

Many researchers conduct research that examines the relationships among variables and do not use random sampling. Consider an experiment to study the effectiveness of a new drug to treat depression. Many people suffer from depression, but the researcher may limit the research to only people who live in a particular city. For the study, the researcher will most likely randomly assign some people to a control group that receives no treatment and others to a test group that does receive treatment. The researcher then compares the effect of the drug on the two groups.

For this example, the researcher may hope to find that people in the treatment group improved more than people who receive no treatment. Notice that the focus of this research is on testing for the possible existence of a logical relationship between two variables. For this type of research, the researcher is focused on describing the *qualitative* relationship between two variables and not on estimating the *quantitative* values that would be expected in the broader population of individuals struggling with depression. When we conduct descriptive research, we examine the general characteristics of the population. Using the current example, we assume that all people who are depressed and use the new drug will show improvement, although the size of the effect may vary from person to person. In these types of situations, it is not always necessary to randomly sample from the larger population because it is assumed that the phenomenon or relationship you are studying will exist for all or most people in that population anyway. The goal is to determine if this relationship exists, not to determine the prevalence or universal strength of this relationship. These latter questions would require a more representative sample from the broader population.

Most researchers conduct studies using people they can easily find. They may use children in local day-care facilities or schools. They may advertise in the local newspaper for volunteers to participate in their research. They will use students enrolled in courses at a local college or university. We are taking a risk in making generalizations from a nonrandom sample. The students at your college or university may be radically different from students at other colleges. Children raised in the northeastern region of the country may be much different from children raised in the Deep South. People who respond to newspaper advertisements may be different from the typical person. Thus, significant cultural or geographic factors may bias the data. Although this is a legitimate concern, we must remember that researchers strive to replicate important findings. If a researcher discovers an interesting finding, then other researchers, from different parts of the country, will attempt to replicate those findings. This is why we never rely on only one study. As you will recall from earlier in this chapter, research and theory building represent a cyclical, dynamic process. The more data we have supporting a hypothesis, the more trust we have in the validity of the findings.

RESEARCH IN ACTION: CREDIT OR CASH?

How much does a consumer's method of payment—credit versus cash—influence grocery store purchases? Are people more or less likely to buy "junk food" when paying with a credit card or cash? To answer these questions, Thomas et al. (2011) conducted a series of studies examining the relation between consumer purchases and methods of payment. In one of their studies, they conducted an experiment to determine how payment method influenced purchase behavior. Specifically, they wanted to know if the method of payment would influence the purchase of regular food products and "junk food."

To conduct the study, the researchers used 70 female and 80 male undergraduate students enrolled at Cornell University. The students were brought into the laboratory and were told that they were participating in a "Food Shopping Study" being conducted by a large grocery store corporation. All the students sat at a computer which then displayed a familiar food product and were told that if they wanted to purchase the product, they would hit the "Add to Shopping Cart" button. If they did not want to make a purchase, they would hit the "Next" button. A random half of the participants were told to assume they were making the purchase with a credit card; the other students were told to assume they were using cash to make the purchase.

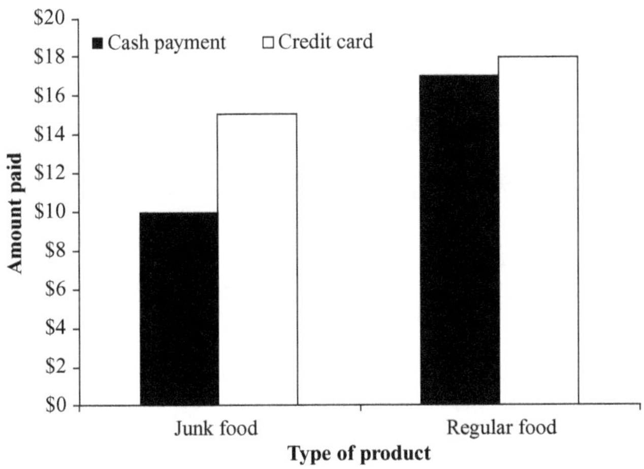

Figure 3.9 Amount spent depending on type of payment and type of food.

In all, each participant viewed 20 products. The researchers classified 10 of the products as regular food products (e.g., pure water—six pack \$6.99, whole bread \$3.85, diced peaches \$3.95 . . . fat-free yogurt \$1.45). The researchers classified other 10 products as junk food (e.g., chocolate chip cookies \$3.91, soda \$4.99, hot cocoa mix \$6.39 . . . cheesecake \$8.99). The average price of the two groups was equal. Based on other studies they had conducted, Thomas et al. (2011) predicted that method of payment would have no effect on purchase of regular food purchases but would influence purchase of junk food. Figure 3.9 presents the results of their study.

The following graph represents the data collected. The horizontal axis of the graph represents the two categories of food product. The vertical axis represents the average amount of money spent by the students in the two groups. The dark black bar represents those students assigned to the cash payment group, whereas the clear bar represents those students in the credit card group. Look at the different heights on the bars. Is there a difference in the amounts spent on the different types of food? Are there differences due to the method of payment? It appears that overall, people spent more money for regular food rather than junk food, and that the method of payment had no influence on the amount spent on regular food. Indeed the amount spent on regular food seems to be identical for both the credit and cash groups. By contrast, when given the option to buy junk food, the students, on average, purchased more when paying by credit than when paying by cash.

Thomas et al. (2011) repeated this experiment but made a few changes. First they used 125 volunteers who had agreed to participate in an online survey regarding preferences for food. Rather than being college students, these participants were working adults (average age of 42) with a median income of approximately \$30,000. The participants completed the same buying exercise as in the previous experiment. The only difference was at the end of the study when participants were asked to answer the following question:

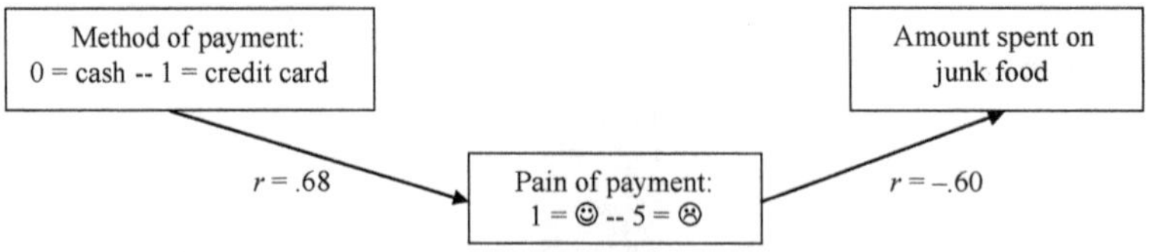

Figure 3.10 Relationship among payment type, pain of purchase, and purchase of junk foods.

How did you feel about spending money on this shopping trip?

1	2	3	4	5
☺				☹

Thomas et al. (2011) believed that the question would measure the participants' perceived "pain of payment." Specifically, they hypothesized that people would experience greater pain of payment when paying for cash than for credit card and that the greater the level of pain of payment would also influence the purchase of junk food. Figure 3.10 presents their results. In essence, they found that the mode of payment did not directly predict impulse purchases. They did, however, find a positive correlation, $r = .68$, between mode of payment and pain of purchase: Those who paid cash experienced greater pain of payment. Thomas et al. also found a negative correlation, $r = -.60$, between that pain of payment and the amount spent on junk food: Those who experienced greater pain of payment spent less on junk food.

KNOWLEDGE CHECK

Use the Thomas et al. (2011) study to answer the following questions:

7. What is the population and sampling population for this research?

8. Is random sampling a critical issue for this research?

9. Can you identify the independent and dependent variables in this study?

10. Are the independent variables manipulated or subject variables?

11. Which scale of measurement best represents
   a. the amount spent on the food?
   b. the food type?
   c. the method of payment?
   d. pain of payment?

12. Thomas et al. used a cover story for their first study. How does the cover story relate to face validity?

13. Why is "pain of payment" an intervening variable?

14. What is the operational definition of pain of payment in this study?

CHAPTER SUMMARY

In this chapter, we reviewed the foundations of all empirical research. In any scientific research project, the researcher will state a hypothesis, determine how he or she will measure the relevant variables, and then collect the necessary data. As we reviewed in this chapter, a hypothesis is more than an educated guess. The hypothesis is a statement that directs the course of the research by describing the variables to be examined, how the data are to be collected, and the relation between the variables. Some researchers develop hypotheses from specific theories in the hope of confirming the utility of the theory. In other cases, the researcher conducts exploratory research to learn more or to explore interesting topics.

Most hypotheses refer to dependent and independent variables. Dependent variables refer to the focus of the study—the variable that the researcher hopes to be able to describe, predict, and explain. The independent variable is the variable that the researcher hopes will allow him or her to describe, predict, and explain the dependent variable. The independent variable may be a characteristic of the participant. Characteristics such as sex, intelligence, and personality are examples of subject variables because they are conditions that the researcher cannot manipulate. A manipulated independent variable is a type of independent variable that the researcher can control or manipulate.

There are four types of hypotheses, including descriptive, correlational, population differences, and cause and effect. For the descriptive hypothesis, the researcher hopes to collect data that will allow him or her to estimate or infer characteristics of a population. A marketing researcher may want, for example, to know how much the average male spends on sports equipment each year. When a researcher poses a correlational hypothesis, he or she predicts that changes in one variable correspond to changes in another variable. An example of a correlational hypothesis is the prediction that there is a positive correlation between a patient's education level and adherence to a physician-prescribed treatment plan. The third type of hypothesis allows the researcher to predict that members of one group are different from another group. This type of hypothesis allows us to compare two or more groups. This type of hypothesis might be used if the researcher was interested in testing whether people who use credit cards are more likely to make impulse junk food purchases than people who use debit cards. The final hypothesis refers to cause and effect. To show cause and effect, one must show (1) that the cause precedes the effect, (2) that there is a correlation between the two variables, and (3) that all other explanations have been accounted for. A researcher, for example, may want to show that using a celebrity in an advertising campaign causes people to be more likely to purchase the product.

All research depends on measurement. In this chapter, we examined the foundations of measurement, which include operational definitions, measurement scales, reliability, and validity. Operational definitions allow us to link observable behaviors to inferred constructs. Depression, for example, is a construct that cannot be directly observed. We can operationally define depression by identifying the behaviors that we believe indicate this condition. When we operationally define a construct, we refer to a measurement scale. The measurement scale can be nominal (putting observations into categories), ordinal (ranking the construct from lower to higher), interval (a scale with a consistent scale, but no absolute 0), and ratio (a scale with a consistent scale and an absolute 0). Reliability refers to the consistency of measurement. A reliable measure is one that provides consistent answers.

Reliability is not the same thing as validity, however. To consider a test valid, we must show that data produced by the test allow us to make useful predictions. In the final section of the chapter, we reviewed populations, sampling populations, and samples. Researchers use samples to evaluate hypotheses. The hypothesis, however, refers to the population. Therefore, researchers hope that what is true of the sample will also be true of the population. Random sampling is not frequently used in most health sciences research because it is difficult to gain access to the entire population and because many researchers working with these issues are interested in describing basic relationships rather than estimating population-level parameters. In addition, researchers will attempt to replicate interesting findings. The more we can replicate a finding, the more confident we are in the finding.

CHAPTER GLOSSARY FOR REVIEW

Absolute 0 A zero point on the scale that represents the absence of the construct.

Arbitrary 0 A zero point on the scale selected for convenience. The 0 does not represent the absence of the construct.

Bias Error A consistent error in measurement that reduces the validity of the test.

Confirmatory Research Conducting research to determine the relative merits of a theory by testing hypotheses derived from the theory.

Constant A measurable characteristic that does not change or take on different values.

Construct Validity The ability of a test to measure the variable or construct it was designed to measure.

Content Validity The degree to which a test appears to be valid to experts familiar with the variables being measured.

Convenience Sampling A sampling technique where the researcher uses only those members of the population most easily obtained or incorporated into the research.

Criterion-Related Validity Empirical data demonstrating the relationship between the predictor and criteria.

Data The information we collect in research that is a product of measurement.

Dependent Variable The variable in the research that the researcher wants to explain, describe, and predict.

Descriptive Statistic or **Statistic** A number that allows us to organize, summarize, and describe the data. Statistics refer only to samples.

Exploratory Research Conducting research to examine the relation among variables and to understand a particular phenomenon.

Face Validity The degree to which a test appears to be valid to the people taking the test.

Hypothesis Any statement or prediction that describes the relation among two or more variables based on theory or previous research.

Independent Variable The variable that the researcher uses to explain, describe, and predict the dependent variable.

Interval Scale A quantitative measurement scale that uses an arbitrary 0 and consistent differences between values.

Intervening Variable A hypothetical variable created within a theory and used to explain the link between two or more observable behaviors.

Manipulated Independent Variable An independent variable that the researcher can directly control or manipulate in an experiment. The researcher can assign individuals to different levels of the independent variable. Other variables in the study do not affect the independent variable.

Measurement Error Errors in measurement due to random and unpredictable events that reduce the accuracy and reliability of the test.

Metric Refers to a standard of measurement, or the specific scale, used to make measurements.

Negative Correlation The description of a systematic relation between two variables—as one increases the other decreases.

No Correlation There is no linear relation between the two variables.

Nominal Scale A qualitative measurement scale that places individuals or observations into mutually exclusive categories.

Operational Definition A rule used to define a specific construct empirically and objectively and to determine how the researcher will measure the construct.

Ordinal Scale A quantitative measurement scale that ranks the observations for a particular characteristic. The differences among the numbers need not be equal.

Parameter A number that summarizes or describes a characteristic of a population. Parameters refer only to populations and are estimated by statistics.

Population A group of people or things that share one or more publicly verifiable characteristics.

Positive Correlation The description of a systematic relation between two variables—as one increases the other increases.

Predictive/Concurrent Validity The ability of a test to accurately estimate current behavior or to predict future behaviors.

Ratio Scale A quantitative measurement scale that uses an absolute 0.

Reliability The degree to which a test produces consistent measurements of the same thing.

Representativeness of a Sample A condition that occurs when the sample data mirror the corresponding parameters of a population. The representativeness of a sample is the product of the sampling procedures used to create the sample.

Sample A group of people or things, drawn from the population, that represents the population from which it was drawn.

Sampling Frame/Population A subset of the population from which the researcher draws the sample. The researcher assumes that the sampled population represents the target population.

Simple Random Sampling A method for creating samples. There is an equal chance that the researcher will select any member of the population.

Subject Variable A variable that represents some preexisting characteristic of the individual in the research. The researcher cannot control the subject variable but may use it to describe or predict the dependent variable.

Theory A broad statement about the general causes of a phenomenon. Theories are more expansive than hypotheses and are often the source of specific hypotheses.

Validity The degree to which a test measures what it is supposed to measure.

Variable Any measurable characteristic that can have different values.

REFERENCES

Bacon, F. (1994). *Novum organum* (P. Urbach & J. Gibson, Trans.) Chicago: Open Court (original work published in 1620).

Hyde, J. & DeLamater, S. (1998). *Human sexuality*. Boston: McGraw-Hill.

Kaplan, A. (1964). *The conduct of inquiry: Methodology for behavioral science*. San Francisco: Chandler.

Mill, J.S. (1986). *System of logic*. New York: Classworks (original work published in 1843).

Rodriguez, H.P., Rodday, A.C., Marshall, R.E., Nelson, K., Rogers, W.H., & Safran, D.G. (2008). Relation of patients' experiences with individual physicians to malpractice risk. *International Journal of Quality Health Care*, *20*, 5–12.

Thomas, M., Desai, K., & Seenivasan, S. (2011). How credit card payments increase unhealthy food purchases: Visceral regulation of vices. *Journal of Consumer Research*, *38*, 126–139.

Chapter 4

An Overview of Empirical Methods

There are only a handful of ways to do a study properly but a thousand ways to do it wrong.

—D.L. Sackett

INTRODUCTION

In many ways, conducting research is a creative process much like painting a picture. Using many brushes, colors, and painting techniques, the artist creates a unique piece of art. Each piece presents its own technical challenges that the artist must solve. The same is true of a researcher who must decide which research technique will yield the best data to allow the researcher to address the target research question(s).

Many students seem to think that learning research methods should be like reading a cookbook. If you want to make chicken soup, find a recipe and follow the directions. If you want to conduct an experiment, find the appropriate chapter and follow the steps. Good research is not this simple. There is no such thing as the "single factor experiment recipe" nor are there five easy steps to conducting a research project. Each research project presents unique technical challenges that the researcher must address. Selecting from a range of techniques, researchers create a **research design** that allows them to find the answer to empirical questions. Therefore, a research design is a plan for collecting data using empirical

Understanding and Conducting Research in the Health Sciences, First Edition. Christopher J. L. Cunningham, Bart L. Weathington, and David J. Pittenger.

techniques. As you will learn in this chapter, there are many research techniques, each with notable merits and limitations.

As the quotation at the start of this chapter suggests, the research design process requires careful planning. As Sackett (1986) recognized, there are a lot of ways to do poor research, but only a few ways to do it well. This chapter will help you learn how researchers take an interesting hypothesis and develop it into a research design that produces clear and useful data.

In Chapter 3, we reviewed the concept of validity as it relates to measurement. As you learned, a measure, assessment, or test is commonly called *valid* if it measures what we want it to measure. To be more accurate, a measure is valid if it allows the researcher to make accurate and specific interpretations or conclusions based on its scores. We can extend this definition of validity to empirical research. Specifically, we can ask whether inferences made from the data support specific conclusions. Just as an assessment measures some constructs, but not others, some research designs allow us to answer some questions better than others. We can evaluate the validity of a research project in terms of its internal, statistical, and external validity.

INTERNAL, STATISTICAL, AND EXTERNAL VALIDITY

Researchers collect data to make interpretations, or inferences, about the relationships among variables. As you will learn, interpretation is one of the more critical steps in the research process and it is one of the primary reasons why the human mind is still needed for research involving social and behavioral issues. Unfortunately, when describing research results, many people use phrases such as "The research proves . . ." or "The data clearly show that . . ." The problem with these phrases is that research and data cannot speak for themselves. *Research* is a verb that describes the work that researchers perform, specifically the act of seeking information. Data are the pieces of information that researchers collect to help them answer a research question. Once the researchers have the data, they must interpret them. Because research requires interpretation, no data are valid or invalid until someone interprets the data. The *researchers* tell us how they *interpreted* the data. Thus, it is better to use the phrase "Researchers tell us . . ." or "The researchers concluded that. . . ."

Here is an overly simplified example. What is the average Scholastic Assessment Test (SAT) score at Mythical College? Two possible answers are "The average SAT score of all Mythical College students is 1067" and "This average score indicates that students at Mythical College are better prepared for college than the typical student." Are these statements valid? The first statement is a straightforward answer to the question. Within the scope of the question, the answer is valid because the mean represents the typical SAT score of students at the college (assuming the SAT scores are normally distributed—for more on this issue, check out Appendix A regarding measures of central tendency).

Is the second response valid? Although the data are the same, it is doubtful that you would so easily agree with this second interpretation. To verify the validity of that statement, we must first obtain the answers to several additional questions. Is an average score of 1067 better than averages for other colleges? Is the SAT test really a measure of students' preparation for college? Finally, what do we mean by *the typical student*? Is the typical student (1) anyone who graduates from high school, (2) only students who attend college, or (3) only students who complete college?

Some statements about the data in this example are valid, whereas others are not. Remember, when we say that an assessment is valid, we really mean that the *interpretation* of the scores from that assessment is valid. The same is true for interpreting the data collected in an experiment. We use validity to describe our interpretation of the data, not the data specifically. Now it is time for more information about internal and external forms of validity.

Internal Validity

A research study is said to have **internal validity** if it is designed in such a way as to facilitate accurate inferences regarding cause-and-effect relationships between the independent variable (IV) and the dependent variable (DV). You should recall from Chapter 3 that we can examine the hypothesis IV→DV only under certain circumstances. Internal validity means that changes in the independent variable, which the experimenter controlled, actually caused changes in the dependent variable. If factors other than the independent variable caused changes in the dependent variable, then we cannot conclude that the study is internally valid because the researcher has not ruled out **alternative explanations**. An alternative explanation is a reasonable explanation of the results that does not depend on the specific hypothesis, IV→DV. When we state that a study is internally valid, we have concluded that the research design has allowed us to control for and rule out alternative explanations for the relationship between the IV and DV that we are interested in studying.

Threats to Internal Validity: Unintended Sequence of Events

There are four broad categories of factors that can threaten or limit a study's internal validity. In many cases, we can avoid these threats through good research design and attention to detail while collecting the data. Threats to the internal validity of a study can be referred to as **confounding variables**. In general, a confounding variable is any extraneous or uncontrolled condition related to the independent variable. Specifically, a confounding variable is any variable not controlled by the researcher that is also correlated with the independent variable. Because the researcher cannot control the confounding variable, he or she cannot be sure whether the results reflect the effect of the independent variable or the effect of the confounding variable. The four major types of confounding variables that we will examine include (1) unintentional sequence of events, (2) nonequivalent groups, (3) measurement errors, and (4) ambiguity of cause and effect.

The first threat to internal validity is the **unintended sequence of events**. As the name implies, the data may reflect a sequence of events that the researcher did not or could not control. There are some research situations in which a researcher will want to study how a sequence of events changes a participant's behavior. For example, a physical therapist may want to examine how a new intervention helps people overcome a specific type of injury. Although this may sound like a simple exercise, there are many opportunities for a sequence of events to confound the results of the study. The following are examples of such unintentional sequences.

Carryover effects (also known as *testing effects*) refer to the influence of participants' earlier experiences/exposures on their present or future performance. One common example of when carryover effects are likely to emerge is when the same participants are asked to respond to the same assessment twice. Knowledge gained by responding to the assessment the first time may influence the participants' scores the next time they complete the same assessment. Consequently, if we used this procedure in a research project, we would not know whether the changes in participants' scores reflect the effect of the independent variable or that participants had more time to think about the answers to the questions.

Maturation is another example of an unintended sequence of events. If the research project is *longitudinal*, spanning a long period of time, it is possible that the observed changes represent the passage of time or the maturation of the person and not a real effect of the independent variable on the dependent variable(s). Consequently, the researcher must ensure that the observed improvements are greater than those that would be due to general maturation alone. For example, in any healthcare study, we can expect the participants' condition to change over time. Depending on the focus of the study, the health of some participants may generally improve or naturally deteriorate. We might also find that as people are exposed to a particular drug, their tolerance for the treatment changes, causing the drug, at the same dosage, to decrease. Finally, other factors associated with the passage of time may affect the dependent variables unrelated to the purpose of the study. Therefore, the researcher needs to show that the changes observed in the experimental conditions are different than those in the control condition. One tactic to address this matter is to test all participants at the start of the study and then at several points during the research project.

Intervening events are related to maturation and are sometimes also called *history effects*. We all live in a complex world. As time passes, you as a researcher and your participants will all be exposed to a variety of environmental factors and general life events. These can all influence your behavior in ways that are completely separate from the influence of the target independent variable in your research. For example, participants in a study could talk to each other or other people about the research, influencing future participants' behaviors. Similarly, a host of random events (e.g., accidents, natural disasters, war) could occur with dramatic influence on the behavior and cognitions of all participants. These outside factors, over which you have little control, may greatly influence your ability to draw valid inferences regarding the link between your stated independent variable and your target dependent variable.

Threats to Internal Validity: Nonequivalent Groups

The second broad category of threats to internal validity is the problem created by **nonequivalent groups**. A common goal for behavioral and social research in any discipline is to identify differences among groups of participants. From these differences, it can then be inferred that an independent variable has some sort of effect or relationship on some dependent variable. Identifying these types of differences is much easier if the groups that are being compared are essentially identical in terms of the dependent variable before the study begins (and before the independent variable is introduced). A problem of inference arises when the multiple groups in a study differ from each other due to factors other than the differing levels of the independent variable that the researcher controls. Two specific conditions that may cause treatment groups to be nonequivalent are mortality/attrition and the uncontrolled effects of subject variables.

Mortality is a rather morbid word that refers to any loss of participants from your sample of participants. Another word that is also commonly used to describe this loss is attrition. Participants may drop out of a study for many reasons—they may become bored or embarrassed, or they may even move to another city (hopefully not because of your research). Participants may even become sick or die over the course of a study that spans a long period of time. The consequence of participant mortality is that the differences among groups when the study is finished may be, in part or entirely, due to the effect of removing a subset of participants from further consideration and not because of any real influence of the independent variable. The reason for this is that the people who drop out of a study may represent an important subgroup of people in the larger population. Because of this, there may be important differences between those individuals that drop out of a study and those that remain.

Subject variables are another important source of potentially nonequivalent groups. For many research projects involving people, the independent variable is a subject variable, otherwise known as an individual difference characteristic. Because the researcher does not have direct control over this variable, he or she can examine the relationship between the independent and dependent variables in this type of study but cannot automatically assume a cause-and-effect relationship exists. Imagine, for example, that we compared two teaching methods, one applied to a course taught at 8:00 AM and the other to a section of the same course taught at 5:30 PM. Students' choice of lecture section may reflect an underlying subject variable among the student participants in that those students who elected to take the early-morning section of the class may differ in some important way from those students who opted for the evening section of the course. Consequently, we will have a hard time answering our original research question, "Are the differences between the two course groups due to the impact of the teaching method or to characteristics of the students themselves?"

Threats to Internal Validity: Measurement Error

A third general threat to internal validity includes problems created by **measurement error**. Using an assessment or measurement instrument that is not a reliable and potentially invalid measure of the independent or dependent variable raises questions about the internal validity of any cause-and-effect conclusion. Measurement errors may also influence the internal validity of a study to the extent that the assessment has low reliability. As discussed in more detail in Chapter 3, reliability is often conceptualized as consistency in measurement. If an assessment produces different results across multiple iterations or test administrators, then the validity of inferences based on this assessment is in question. Although we may find that a measurement technique has suitable reliability and is a valid measure of the variable, there are several special cases of measurement error that require attention when designing a study.

Ceiling and floor effects are important measurement errors to recognize and avoid. Sometimes a researcher may select an assessment that is not sensitive to changes in the dependent variable. Before progressing with a health sciences research project, the researcher should be sure that the test that will be used to measure the dependent variable has a broad range of scores and is sensitive to small differences in the underlying construct. Consider a test of depression that contains five questions (e.g., "Sometimes I feel blue."). This test's questions are so general that anyone will probably respond with an affirmative answer. This, in turn, might create a *ceiling effect* in that all scores may be toward the high end of the possible range of scores. In contrast, using an assessment that produces a large number of low scores could create a *floor effect*.

Instrumentation becomes a source of measurement error if changes are made to the actual measurement instrument over the course of an experiment or study. Most often this threat to internal validity is associated with malfunctioning equipment, which, although rare, can be extremely disruptive to a data collection process. Instrumentation issues can also arise with survey research, especially when Internet surveys are being used to gather the data.

Another source of measurement error is the statistical phenomenon of **regression to the mean**. In some circumstances, the change in a participant's scores from before to after exposure to an independent variable represents a normal return to the mean level of the dependent variable under study. This effect is especially likely to occur when people initially receive exceptionally high or low scores on an assessment. When tested a second time, the scores for the same people will tend to be closer to the average. For example, some people may initially receive a poor score on an assessment because they do not feel well or for other reasons unrelated to their knowledge of the material. When we measure or observe them again a few weeks later, their scores will probably be closer to normal. Consequently, these types of changes in assessment scores may reflect regression to the mean and are not true effects of the independent variable.

It is important to note that extremeness in scores is often based at least in part on measurement error and its random impacts on individual participants. This being the case, if you are able to randomly select participants into your study and randomly assign them to treatment conditions or groups, you can avoid issues of regression to the mean. In most behavioral and social sciences research, however, this level of control is not practical. In these quasi-experimental situations, the next best alternative strategy for minimizing the risk of measurement error as a threat to internal validity is to ensure all measures used have demonstrated reliabilities and good variability in scores (i.e., not demonstrating ceiling or floor effects).

Threats to Internal Validity: Ambiguity of Cause and Effect

The fourth threat to internal validity involves the difficulties inherent in establishing evidence for a clear cause-and-effect relationship. According to British philosopher John Stuart Mill (1986), before we assume cause and effect, we must address three issues: (1) that the cause occurs before the effect, (2) that there is evidence of a correlation between the independent and dependent variables, and (3) that alternative explanations for changes in the effect can be ruled out. There are many cases when the method for collecting the data by itself does not allow us to address one or more of these criteria. When designing a study that is testing a causal hypothesis, therefore, we must pay attention to issues of **temporal order problem** and the possible **third variable problem**.

Table 4.1 summarizes the preceding general threats to internal validity, their characteristics, and an example of each threat.

Statistical Conclusion Validity

Statistical conclusion validity is associated with the use of the proper statistical or analytical methods in a given study (Shadish, Cook, & Campbell, 2002). Proper in this case refers to the use of methods that best allow a researcher to (1) demonstrate that the independent and dependent variables covary and (2) identify the strength of this covariation or relationship. All statistical analyses are based on specific assumptions. This means there is no one-size-fits-all analysis that can be used with every set of data that a researcher may collect. This being the case, to achieve acceptable statistical conclusion validity, it is the researcher's responsibility to use the best possible statistical techniques for a given analysis. It is also important for the researcher to avoid some of the most common threats to this form of validity.

Table 4.1 Threats to Internal Validity

Threat and its description	Examples
Unintended sequence of events: Changes in the dependent variable that are the results of experiences with the testing procedure (carryover effect), changes in the participant (maturation), and intervening events that may occur between phases of the experiment.	• Carryover effects: Taking the same or similar tests allows the participant to practice or become more skillful at taking the test. • Maturation: The participant grows older or changes in some way unrelated to the independent variable. • Intervening events: The participant learns something about the experiment from another participant.
Nonequivalent groups: The difference among the groups is due to one or more variables related to characteristics of the participants that are unrelated to the independent variable.	• Subject variables: The researcher assigns the participants to different groups based on a subject variable (e.g., sex) that also influences the dependent variable. • Participant mortality: Participants in one group are more likely to withdraw from the study.
Measurement problems: The measurement used to assess the independent or dependent variables does not adequately measure the variable, or it introduces additional variance to the data, thus hiding the effect of the independent variable.	• Ceiling/floor effects: Test may not be sensitive to differences among participants. • Measurement error and reliability: Test may create error because of low reliability or validity. • Instrumentation: People responsible for collecting the data change the measurement criteria for different groups. • Regression to the mean: Participants scoring very high or low may score closer to the mean if the test is administered more than once.
Ambiguity of cause and effect: Inability to determine that the IV preceded the DV, or an inability to rule out a third or alternate variable that could have caused the same changes in the DV.	• Temporal order: When the IV and DV are measured at the same point in time, it is difficult to identify whether the IV is truly causing changes in the DV. The IV needs to precede the DV in time. • Third variable: A relationship between an IV and a DV does not necessarily mean that another variable is the real cause of changes in the DV.

Threats to Statistical Conclusion Validity

Multiple factors can prevent a researcher from making valid inferences about the relationship between the independent and dependent variables in any given study. In many ways, these threats should look rather familiar to you as they closely parallel the threats to internal validity presented earlier in this chapter. Much more detail on these and other threats can be found in Shadish et al. (2002), among other sources.

Low statistical power can prevent a researcher from detecting a relationship between an independent and dependent variables. This means that a researcher may miss a relationship that is really present. There are several strategies that can be taken to boost the statistical power in a study. One common technique is to ensure that the sample of participants is large enough based on rules of sample size estimation (see Cohen, 1992, for rough examples or see your statistics book for the full skinny on this). In general, these sample size estimates are based on the size of effect that a researcher expects to observe—large expected effects require smaller sample sizes to detect them than small expected effects (i.e., it is easier to find a pitchfork than a needle in a haystack). Another power-boosting strategy is to make the differences between levels of an independent variable as large as possible. This way, if there is an effect of that variable on the dependent variable, it should be more noticeable. A third general strategy is to improve the quality of measurement by reducing error and restriction of range. Successfully doing this will improve your chances of detecting an effect if it is really there.

Violating assumptions of statistical tests is another major threat to statistical conclusion validity. The fix for this threat is very simple—follow the rules for your statistical tests and use the ones that are most appropriate for your particular situation. The difficulty is that this also requires researchers to have a solid working understanding of the basic underlying assumptions for basic statistical analyses (e.g., analysis of variance [ANOVA], correlation, regression). With more advanced designs and research questions, it can also be valuable to seek the advice of an expert in statistics during the design and analysis phases of a project to ensure that the assumptions for the desired statistics are being met in the actual data.

Unreliability in measurement and setting are two additional threats to a researcher's ability to make valid inferences from statistical analyses. Recall that the validity of an assessment tool depends, in part, on its reliability. Therefore, researchers strive to use highly reliable assessment tools. Along similar lines, if the setting in which data are collected and the administration of the study are unreliable or inconsistent, this can increase the overall amount of error present in any eventual statistical analyses. This inflated error due to unreliability can cloud a researcher's ability to make clear and accurate (valid) inferences from the results. This is one major reason why careful and systematic design and administration are so critical. The direct impact of unreliability in measurement is on the error present in the resulting statistical analyses of data that you collect. An increase in error can make it more difficult for a researcher to detect an effect of the independent variable on the dependent variable (an issue of statistical power), which can then lead to an incorrect inference being made about the data (an issue of validity).

External Validity

External validity refers to the degree to which researchers can generalize the results of an existing study to another setting, sample, or the broader population. When researchers are more confident that their samples and research settings represent the population and broader environment, then it is easier to conclude

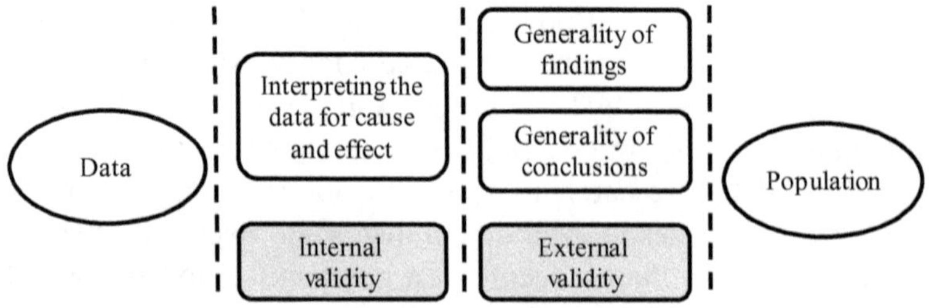

Figure 4.1 Difference between internal and external validity.

that the findings of research based on these representative samples in representative settings will generalize; this would mean that the study has external validity. There are two types of external validity.

The **generality of findings** refers to the link between our sample and the target population. If we assume the sample adequately represents the population, then we can generalize conclusions from the sample data to the population. The **generality of conclusions** refers to our ability to generalize the findings from one population to other populations. In many cases, we conduct a study using a specific group of people, but we may want to extend the results to other groups. For example, a surgeon may find that a vascular stent used in cardiac surgery is very effective in treating a vessel blockage and now wishes to generalize the results to all vascular surgeries that require the repair of a vascular blockage. In such a situation, we need to question whether the results found in a sample of one population—heart surgeries—will generalize to a second population, such as patients with different vascular issues.

Figure 4.1 illustrates the difference between internal and external validity. Both forms of validity come between the data and the population, indicating that validity is a process that allows us to link the sample data to the population. As you can see, internal validity refers to the interpretation of the data as they allow the researcher to draw a cause-and-effect relationship. External validity refers to generalizing the results to the target population. External validity consists of two parts, generality of findings and generality of conclusions.

Threats to External Validity

What are the factors that limit or reduce the external validity of a research project? Cook and Campbell (1979) identified three broad threats to external validity: (1) recruitment of participants, (2) effects of the research situation, and (3) effects of history.

Recruitment of participants is the first step researchers take when initiating a research project. In most cases, we gather these participants by simply asking people to volunteer. At many colleges, students in introductory courses receive extra credit for participating in one or more research projects. Accordingly, many researchers use college students for the sake of convenience. In some cases, this type of sampling method may limit the number and range of target populations to which we can generalize our eventual results. Consider a few examples. Imagine that a researcher wanted to study a new physical therapy treatment. Because the researcher works in the PT clinic of a large state university, the researcher decides to conduct the student with the patients in the clinic who are mostly athletes. Can the researcher generalize the results of this study to all patients from whom this treatment might be appropriate?

We hope you see the potential threat to external validity here: How well does a college-operated PT clinic represent a conventional clinic? Do college athletes represent the typical patient requiring physical therapy? These questions raise a potential threat to the external validity of the results.

How do we overcome this threat to external validity? In the present example, one way would be to repeat the study in a different location or with a different sampling population. By conducting a clinical trial at multiple clinics, the researcher is better equipped to determine the extent to which patient type—athlete or nonathlete—influences the effectiveness of the treatment.

A second threat to external validity comes from the **effects of the research situation**. For this threat, we must ask if the location in which the research was conducted may influence the results in one way or another. As an example, is it possible that research conducted at a highly prestigious teaching hospital creates a type of "institutional placebo" for people participating in the research? What about research conducted in a prison—is it possible that the stresses of living in such a setting will bias the results? Consider a research project examining public health campaigns that use radio advertising. Might it be possible that research conducted in an urban area differ from research in a rural area where access to radio stations is much different? In essence, the careful researcher will consider with care the potential that the location of the research may in some way affect the degree to which the results will generalize to other locations.

Another threat to external validity can come from the **effect of history**. Can the results of an experiment conducted 50 years ago generalize to current situations? What about an experiment conducted during a series of unique political and social conditions? Does the time during which we conduct a study affect the results and thereby compromise the external validity of the results? There are cases where the results of a study are relevant to a particular era and do not necessarily generalize to contemporary situations. For example, many strains of bacteria are now resistant to antibiotics developed decades ago. In other cases, the results of older studies may still be valid today.

Table 4.2 summarizes these basic threats to external validity.

SURVEY OF EMPIRICAL METHODS

The following sections review at a high level a variety of different empirical research methods. In each section, we will give you a brief description of the method, its role in contemporary research, and its potential advantages and

Table 4.2 Threats to External Validity

Threat and its description	Examples
Recruitment of participants: Participants in your sample do not represent the population.	• Generalizing results from one age group to another. • People who volunteer in response to a newspaper recruitment ad may not represent the typical member of the population you wish to study.
Effects of situation: The environmental context for your research is unique and does not generalize or match well with other settings.	• The location of the research may affect the results of the research.
Effects of history: Unique events that only affected members of your sample limit the generalizability or applicability of your findings to other samples in other times and places.	• Results from a study you conducted 10 years ago may not generalize to the present population due to changes in political, social, and economic conditions.

disadvantages. We will also add a few points about issues related to internal and external validity associated with each method.

True Experiments

We begin by reviewing the **true experiment** because it is the strongest possible design technique we have when attempting to determine cause-and-effect relationships between two or more variables. The value of the true experiment is that it allows researchers to control for and remove as many potential threats to internal validity as possible. Specifically, the true experiment allows us to ensure that the independent variable comes before the dependent variable, that the independent and dependent variables are related, and that we can account for alternative explanations for relationships between the independent and dependent variables. Three critical features set a true experiment apart from other research methods: (1) The independent variable is a manipulated variable under the researcher's control; (2) the researcher uses random assignment to place individuals into the different research conditions; and (3) the researcher can create one or more control conditions.

The hallmark of a true experiment is that the researcher directly controls or manipulates the level of the independent variable. In other research designs, the researcher might use subject variables that he or she cannot control as the independent variable under study. By contrast, in the true experiment, the researcher can select a truly controllable or **manipulated independent variable** that can be applied over different experimental conditions to examine its effects on an outcome of interest.

A **categorical independent variable** represents a variable that is measured with a nominal scale. For example, in drug research, the type of drug represents a categorical independent variable in that each drug has a different chemical structure. Although different drugs may have a common affect, the chemical reactions may be much different. Therefore, the type of drug represents mutually exclusive groups for study and comparison. A **quantitative independent variable** is a variable represented by an ordinal or more sophisticated type of measurement scale. Using our drug example, the dosage the patient received is a quantitative variable as the dosage can be prescribed in terms of weight or volume. Whether the independent variable is categorical or quantitative, the researcher has control over the **treatment condition** that participants experience. In the example of advertising media, each form of media represents a level of the independent variable.

Random assignment is another necessary component of the true experiment. Random assignment occurs when each participant in an experiment has an equal chance of experiencing any one of the possible research conditions. The value of random assignment is that it reduces alternative explanations of the results. Figure 4.2 illustrates the logic and steps of a true experiment.

The researcher randomly assigns participants from the sampling frame to the various treatment conditions. The researcher then exposes the groups of participants to specific levels of a manipulated independent variable. In the last stage, the researcher examines the differences among the groups.

In this example, the researcher begins by (we hope) randomly selecting 30 participants from the broader sampling population and then randomly assigning them to one of the three treatment conditions. At this point, the three groups should

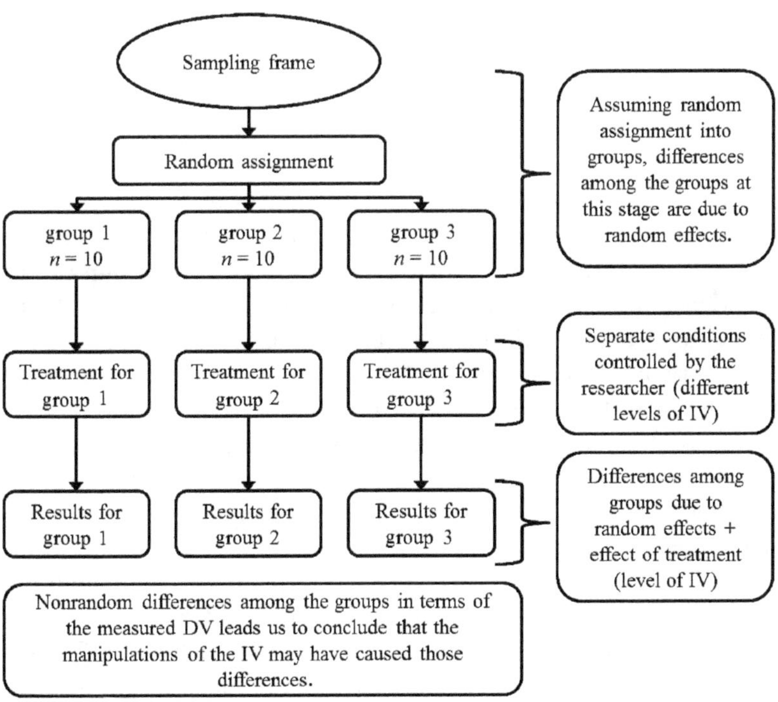

Figure 4.2 Design of a true experiment.

be similar to each other (i.e., homogeneous). Will they be perfectly identical? No; if we were to compare the three groups, we would find minor differences among them. If the researcher uses random assignment to form these groups, however, these differences will be trivial and should not influence the relationship between the independent and dependent variables. This is because any remaining source of difference between groups after random assignment is due to chance or random effects. Once the researcher forms the three groups, the participants in each group will experience a different level of the independent variable. The final step in the illustration is examining the differences among the groups. These differences will reflect the effects of the independent variable and random factors on the dependent variable. In most cases, researchers use statistical techniques to determine the proportion of the difference due to the independent variable and the proportion of the difference due to random factors.

Using random assignment is an important design technique to protect the internal validity of the experiment. By randomly assigning participants to the treatment conditions, we prevent **confounds** from developing due to the placement of participants within particular experimental conditions. Researchers rely on random sampling to reduce the potential confounds created when different treatment groups in a study are not equivalent at the start of the research. If the sample sizes are sufficiently large and if there is no systematic bias in how the participants are placed into the various groups, we can generally assume the groups are equivalent at the start of the study. As you read research reports, you may find that some researchers will describe in detail the procedures they use to ensure random assignment. These researchers may also report a statistical analysis comparing attributes of the groups (e.g., sex, age, and other relevant characteristics) demonstrating that there were no systematic differences among the groups before the intervention began. Such analysis increases the confidence that the subsequent differences among the groups were due to the effects of the experiment.

The final requirement of a true experiment is the use of a control or comparison group. **Control groups** (sometimes referred to as comparison groups) are essential to the true experiment because they help us rule out a host of alternative explanations. In its basic form, a control group is the group of participants who experience the same treatment process as the other participants but are not actually exposed to the independent variable that is expected to influence the dependent variable. Although conceptually simple, the use of control groups can become quite sophisticated. First, there are many types of control groups. In addition, it is possible to use several control groups in a single experiment. Selecting the appropriate control condition or conditions allows the researcher to make a more convincing case that the independent variable, and not some other variable or variables, caused the differences among the groups.

A **placebo treatment** is likely to be one type of control condition you may have heard about somewhere in the past. Researchers frequently use placebos to determine the effectiveness of a drug. A placebo is a treatment condition, such as a sugar pill or a saline solution injection, which does not include the active ingredient or drug under study. When using this type of control condition, the researcher will randomly assign some of the participants to the placebo treatment group. If the researcher finds a meaningful difference between scores of participants in the treatment and placebo groups, then he or she can make a convincing argument that the drug caused the difference. Placebos can also be used in non-drug-related studies, but those examples are a bit less simple to work with.

Utility

The true experiment is one of the most important tools that researchers have to study behavior because it allows us to most clearly test cause-and-effect relationships. Although other research techniques may indicate the meaningful relation between the independent and dependent variables, only the true experiment offers an unambiguous opportunity to determine cause and effect. True experiments provide researchers with this utility because of the high level of control researchers have in the design and execution of experiments versus other widely used research methods.

Potential Limitations

The true experiment is without equal for determining cause and effect. The true experiment does, however, have limitations. First, researchers cannot manipulate some potentially important independent variables because of ethical or practical concerns.

Second, it is impossible to manipulate many existing subject variables. For example, genetic heritage, medical history, and sex are unique characteristics of participants that may help us understand why they do the things that they do. These characteristics are variables that researchers cannot control. Therefore, we must use other research techniques to collect and examine the data.

Intact Group Designs and Quasi-Experimental Studies

There are many times when it is impossible, impractical, or unethical for a researcher to manipulate the independent variable(s) in a research study. In these cases, although we may not be able to control the independent variable, we can still study its relationship with the dependent variable. Two commonly used research methods for this purpose are the **intact group design** and the **quasi-experimental design**. The common characteristic of both designs is that the researcher does not control the independent variable. Instead, the researcher uses the fact that there may already exist different groups of people who have experienced different conditions associated with the independent variable that is the focus of the research. For the intact group design, a critical variable defines categories or groups people who can then be prepared. For the quasi experiment, two or more similar groups of people experience different conditions over time. The difference in condition can then be considered a type of independent variable.

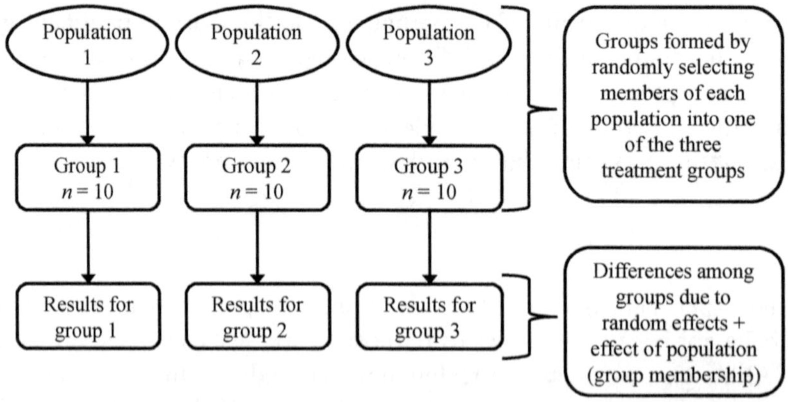

Figure 4.3 Intact group design. *Note:* In this design, the researcher randomly selects participants from different sampling populations.

Figure 4.3 presents an example of a simple **intact group design**. There are some important differences between this design and the true experiment.

First, the researcher does not select participants from a single population. Instead, there are several populations each defined by a different set of subject variables. For example, a researcher may want to compare the effectiveness of a drug on patients diagnosed with different types of cancer. In this case, the type of cancer will define membership in one of the groups.

In this type of situation, we cannot use random assignment to create these groups because the cancer type already exists and cannot be manipulated by the researcher. All we can do is assume that each cancer type sample or group adequately represents the broader sampling population of other people who have that type of cancer. From here the researcher can proceed to collect data and compare scores across the different groups. As with the true experiment, we assume that differences among the groups represent random variation as well as factors associated with the subject or grouping variable. Unlike the true experiment, we cannot use these data to assume that the subject variable caused the differences. We can compare, however, the group averages, testing the hypothesis that the means of the three groups are not equivalent (i.e., all μs will not be equal).

Quasi-experimental designs are different from the intact group design as it compares two similar groups when one of these has been affected by some external circumstance. A *quasi experiment* is similar to a true experiment except that we cannot randomly assign participants to the treatment and control conditions. Figure 4.4 presents an example of a hypothetical quasi experiment.

The researcher uses two preexisting groups of participants and measures each for one or more relevant dependent variables. The researcher then exposes one group to a treatment using the other group as a control condition and then reevaluates both groups.

Because this is a quasi experiment, we have not randomly assigned participants to the groups; the grouping existed before the research began. We measure both groups for the dependent variable and then expose one group to a treatment

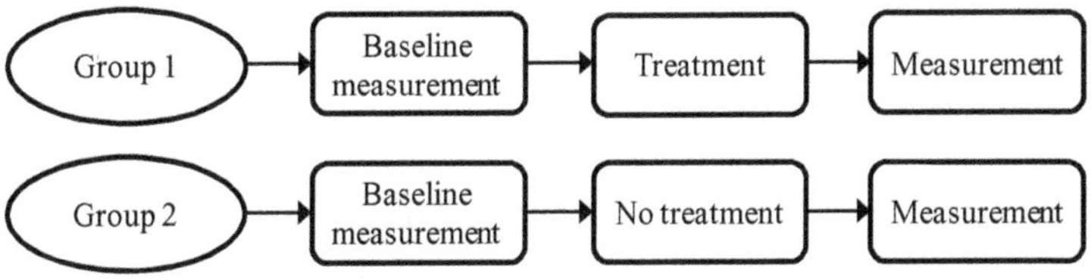

Figure 4.4 Quasi experiment.

condition. After the treatment, we measure both groups again. Therefore, this study consists of a grouping variable as well as a manipulated variable. In this type of study, we call the independent variable a **quasi-independent variable** to indicate that the researcher can manipulate the variable but cannot randomly assign participants to the treatment conditions.

Consider the following example of a quasi experiment in which a researcher wants to study the effect of a college's alcohol policy on student binge drinking. The researcher finds two equivalent colleges. After assessing the amount of drinking on both campuses, one college establishes a "dry-campus" policy. The introduction of the policy represents the treatment. The researcher can then return to both campuses and measure students' actual alcohol consumption. For this example, the researcher wants to test the research hypothesis that the average level of drinking at one campus will be less than the other campus, or $\mu_1 < \mu_2$.

What conclusions could the researcher make if she found that after adoption of a dry-campus policy there was a radical reduction in drinking among the students? Can the researcher conclude that the introduction of the treatment caused the change? The answer to this question leads us to examine the utility and limitations of the intact group design and quasi experiments.

Utility

There is no doubt that the intact group design and the quasi experiments are useful to researchers. Many of the variables that affect human behaviors in groups and organizations are preexisting subject variables. In many cases, business researchers are interested in how people who belong to different populations (e.g., generations, sex, income level, area of country, level of education) spend money, make decisions, and respond to economic conditions. We can address these interesting topics with an intact group design. Similarly, the quasi experiment allows us to conduct research when random assignment of participants to treatment conditions is impossible, impractical, or unethical.

Potential Limitations

The most important limitation of the intact group design and the quasi experiment is the inability to form definitive conclusions regarding cause-and-effect relationships between the independent and dependent variables in a study. We cannot assume cause and effect because we do not have direct control over the

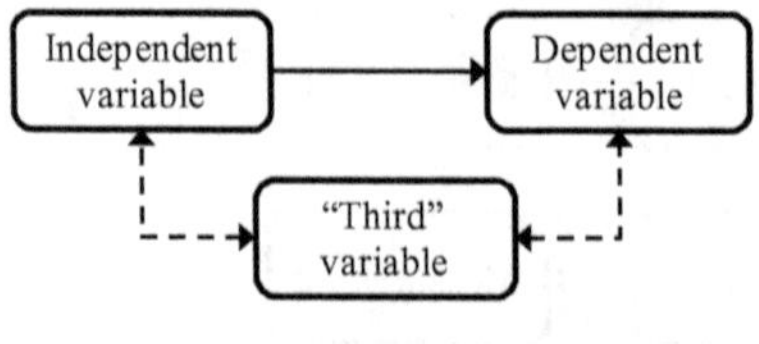

Figure 4.5 Third variable problem.

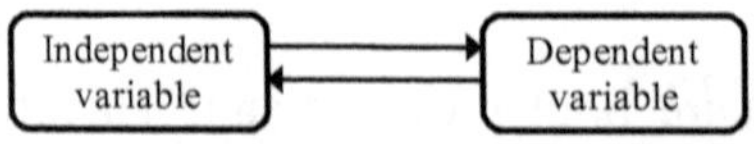

Figure 4.6 Temporal order problem.

independent variable. In other words, these designs cannot ensure the internal validity of the link between the independent and dependent variables. There are two major threats to internal validity that are especially relevant to this type of design: (1) the third variable problem and (2) the temporal order problem.

The third variable problem is that some factor or condition, other than the independent variable, may affect the dependent variable. Figure 4.5 illustrates this problem.

In this example, we assume that there is a direct link between the independent and dependent variables. But there may be some other, third variable that is correlated with both. The third variable is a confounding condition because it correlates with both the independent and dependent variables. The third variable prohibits us from assuming a cause-and-effect relation between the independent and dependent variables.

As you can see in Figure 4.5, the third variable is associated with both the independent and the dependent variable. Because of these relationships, we cannot be sure whether the independent variable or the third variable causes changes in the dependent variable. Assume that a researcher took a random sample of men and women and compared their cholesterol levels. If the researcher finds that women have lower cholesterol levels than men, can he or she assume that sex causes the difference? No, other variables may account for the difference. For example, the women in this study could have, by chance, different diet habits than the men. Therefore, the diet, not the sex, accounts for the difference.

The temporal order problem is a fancy description of the classic question, "Which came first, the chicken or the egg?" Figure 4.6 illustrates this problem. Because we do not control the independent variable, we cannot be sure whether the independent variable caused the dependent variable or whether the opposite is true. This problem occurs when one uses a subject variable or measures both variables at the same time. We cannot be sure which variable produces or causes the other variable.

Imagine that you compared the alcohol use and depression and found a high positive correlation between the two variables (people who are more depressed consume more alcohol than people who are not depressed). Can you use these data to assume that being depressed causes alcoholism? The problem is that in this type of research, the researcher measured the two variables at the same time. Therefore, we do not know which came first, the depression leading to increased alcohol use or the increased use of alcohol bringing about depression.

Surveys

Researchers use **surveys** to estimate population parameters from sample data. As you may recall from Chapter 2, we often denote this type of hypothesis as $M \approx \mu$. Researchers use survey data for many purposes. Surveys are typically constructed to measure multiple different variables at one time. Many healthcare providers use surveys to examine the health-related behaviors of people. The results of these surveys often inform public policy, patient care, and other components of the healthcare system.

Utility

Surveys are extremely useful ways of obtaining information, and they give us accurate estimates of important population parameters. Surveys also allow us to collect information quickly and with minimal expense.

Potential Limitations

We must always be mindful about the reliability and validity of questions asked in a survey. Because surveys and questionnaires are critical to business research, we have written Chapter 8 to review the best practices for developing high-quality survey research instruments and procedures.

Correlational Studies

Correlational studies allow researchers to test the hypothesis that two or more variables are related to one another. Here are two simple examples: (1) Is there a correlation between a particular blood enzyme and the risk for developing a disease? And (2) does the length of treatment affect a patient's chances of making a full recovery? These types of questions can be addressed with a correlational study. Figure 4.7 presents the steps involved in conducting a correlational study. As you can see, the researcher selects participants from the sampling population and measures two or more variables. The researcher then examines the correlation among the different variables.

In the first step, we identify a sample from our target population and then collect data on two or more variables from our participants. The essential characteristic of a correlational study is that we collect two or more bits of information from each participant. For our first example, we might find that there is a negative correlation between the level of a particular enzyme and the development of sepsis in trauma patients. At the same time, we may also find that the length of time the

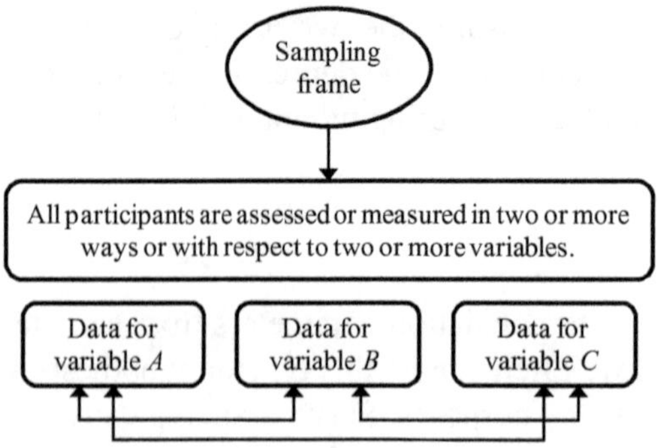

Figure 4.7 Correlational design.

patient has been in the trauma condition correlates with both the level of the enzyme and the probability of developing sepsis.

In Figure 4.7 we used lines with arrows on both ends to connect the variables. This reflects the fact that in a correlational study we cannot usually test for a cause-and-effect relationship. Can we, using the current data, assume that lowering of the level of the enzyme causes the person to develop sepsis? Because there are many things that can cause sepsis, we cannot assume one causes another. The fact that there is a correlation, however, indicates that there may be an important link that requires separate study. To prove the causal link, we would need to show that increasing levels of this enzyme corresponds with decreasing the risk of sepsis.

Utility

The correlational design is very popular in behavioral and social science research and is increasingly used in the health sciences. By examining the correlation among variables, we may be better able to understand complex medical conditions.

Potential Limitations

As with the intact group design and the quasi-experimental design, the correlational design cannot support conclusions of cause and effect. Specifically, the correlational design cannot resolve the third variable problem or the sequence-of-events problem.

Interviews and Case Studies

When most people think about research, they have a vision of vast groups of people assigned to different treatment conditions. The researcher then combines and examines the data as a group. Although this is a popular image of the type of research conducted in the health sciences, it is not completely correct. There are many instances in which a researcher will use focus on only a small group of people or a single person (open any medical journal and you will most likely find a "case study" section that reports about the treatment of a single patient). We will review **case studies** in more detail in Chapter 15.

Utility

Case studies also serve as important teaching tools and exemplars of basic principles. Within healthcare settings, case studies can provide useful information that will then become the basis of more comprehensive research.

Potential Limitations

The problem with this research method is that we are looking at the behavior of a single person or a small group of people. The case study, for example, does not allow us to make sweeping generalizations about the population. Although we may learn much about the treatment of one person, we cannot conclude that what was true for the person in the case study is true of all people.

Thus, although case studies offer interesting information, we must interpret the information with some care. What are some of the potential limitations of case studies? The following are two of the more severe problems:

1. **Verification of Facts:** Many times, the person collecting the information for the case study has no way of confirming the facts that are a part of a person's history. Elizabeth Loftus (Loftus & Ketcham, 1994) provided compelling evidence that the questions people ask us can distort our memories. Therefore, a major weakness of any case study occurs when we must rely exclusively on the individual's memory for critical facts.
2. **Potential for Bias in Examples:** In many cases, an author will present a case study because the case fits his or her theory. Howard Gardner's (1995) book, *Leading Minds*, is an example. In the book, Gardner presented an interesting theory of leadership. He then presented case studies of 10 great leaders from history (e.g., Margaret Mead, George Marshall, and Eleanor Roosevelt). Although the theory and case studies are interesting, we must ask whether they are proof of the accuracy of the theory. One problem is that Gardner may have selected people who demonstrate his theory while ignoring great leaders who do not fit his theory of leadership. Similarly, because Gardner is recounting the history of another person, he may have selected only those episodes of the person's life that fit his theory.

Meta-Analysis

Listening to news reports of research findings can sometimes be frustrating. You may hear that a group of researchers claims that they found the cure for a terrible disease. Later, another team of researchers reports that the treatment is not effective. These experiences with conflicting information lead many people to think that science must be only one step removed from witchcraft. The problem is not necessarily with the scientists but with how we look at individual research reports in isolation from the broader literature base.

No single experiment can offer definitive proof or support for a hypothesis or theorized relationship by itself. Each experiment is subject to a host of random factors that affect the results. Specifically, an experiment is a sample, and only a sample, of what you can expect in the broader population. Hence, there is always a chance that a single experiment or research study, no matter how well conducted, may produce misleading results. This is the reason replication and extension are such an important part of good scientific research. If a researcher reports an interesting finding, the chances are that he or she will conduct more studies to examine the effect in detail. Other researchers will join in and publish their results. How will we evaluate the accumulated collection of research findings?

Many researchers write **literature reviews** that summarize the findings of many research projects. The author reads all the relevant research and then describes what he or she believes to be the general findings. Literature reviews are important because they allow us to take a step back from the individual studies and look for the broader significance of the research. There are some problems with these types of reviews, however. Literature reviews tend to be subjective, and they are not quantitative. When the author reads the results of a study, he or she has to judge whether the results are meaningful. There are better ways to examine the outcomes of empirical research.

In 1978, Glass developed a method to more objectively analyze the outcomes of many studies. This technique, called **meta-analysis**, allows the reviewer to quantify the results of the individual studies. In other words, meta-analysis allows us to statistically analyze multiple results from multiple different studies. This type of analysis results in an overarching summary statistic that is more reliable and capable of leading to potentially more valid inferences than what you would have found in any one of the studies that were included in the meta-analytic sample. The logic of meta-analysis is rather straightforward. The more often one samples from a population of participants, settings, and measures of constructs, the greater the accuracy of the overall estimate of the observed effect or relationship. The goal of meta-analysis is, therefore, to offer a quantitatively based and objective review of the research literature.

Utility

Since its introduction, meta-analysis has become a popular review method across many research disciplines. You will find that researchers in psychology, medicine, political science, sociology, education, and other empirical sciences use meta-analysis extensively. It is now an almost essential component of any literature review when the topics of interest have received significant attention by other researchers.

Potential Limitations

Although meta-analysis uses objective mathematical techniques to examine the data, there remains a subjective component—selection of which studies will be included in the sample of studies used for the analysis itself. To conduct a meta-analysis, one must first determine whether the researcher used proper research methods to collect the data. The reviewer will examine the method and determine whether the data afford valid interpretations. If the reviewer finds an error in the study, he or she will exclude the results from the meta-analysis. Many people complain that this step biases the results of the meta-analysis. This is one reason for the now widely used axiom that with meta-analysis, garbage in = garbage out. Thus, meta-analyzing a bunch of questionable studies will not result in some sort of magical truth—indeed, meta-analyses depend on multiple, high-quality initial studies if they are to be of any use at all. How would you go about identifying good studies to include in a meta-analysis on a topic of your choosing?

COMPUTERS AND STATISTICS

Charles Babbage (1792–1871), a British mathematician who attempted to invent the first mechanical calculator, once exclaimed, "I wish to God these calculations had been [conducted] by steam." Many researchers and students who have calculated statistical tests by hand have often expressed the same sentiment.

Computers have had a profound effect on statistics and research methods by making it easier for researchers to use sophisticated statistical procedures to analyze their data. Before the advent of the computer, most researchers did not use many of the statistical procedures developed by statisticians because these tests required an overwhelming amount of calculation (Pedhazur, 1997). Even simple descriptive statistics, such as the standard deviation or correlation coefficient, are labor intensive because of the many repetitive calculations required to complete the work.

The first practical computers became available in the late 1950s and early 1960s. Although these machines were huge and expensive, they gave researchers the opportunity to conduct sophisticated statistical tests on large data sets. The current generation of personal computers is far more sophisticated than many of the multimillion-dollar mainframe computers built in the 1970s. At the same time, statistics software has also evolved. Like the computer, these programs have become easier to use and less expensive, and provide access to many sophisticated statistical tests. As useful as computers and computational software are, however, there are several potential problems that you must consider when using them.

Garbage In–Garbage Out

No computer program or statistical procedure can spin gold from straw; the quality of any statistical analysis depends solely on the quality of the original data. Unfortunately, many people treat the computer and statistics as magical machines that somehow tell the truth. Computers do exactly what you tell them to do, even if it is the wrong thing. The phrase "garbage in–garbage out" means that the output of any computer program is no better than the data you ask it to analyze. Statistics packages cannot recognize when you have entered the wrong numbers or request a statistical test that does not address the hypothesis you want to test. These observations mean that we need to proceed with caution when using software packages.

Data Issues

The first problem to address is the accuracy of the data to analyze. Software packages treat the numbers .987 and 987 as legitimate values, even if one of them is entered incorrectly or represents an impossible value of measurement for the scale you are utilizing. Therefore, you will need to spend a little extra time proofreading the data you have entered in the data file. At the very least, you should run a frequency analysis of all variables to ensure that all values you have entered conform to the expected minimum and maximum values allowable for each item and each scale. In addition, it is a good idea to check the distribution of scale scores based on your data to ensure you do not have a ceiling or floor effect present on any of your research variables.

Selecting the Right Statistical Test

A separate and far more important problem that we need to address is your knowledge of the statistical tests that the software package allows you to conduct. Just because the computer will let you use a statistical procedure to analyze the data does not mean that it is the right statistical test to use. Consequently, we cannot emphasize strongly enough the importance that you understand the conceptual foundation of any statistical test that you calculate. Whenever you use a statistic, you should be able to justify why you selected it, show how it will help you interpret the data, and know what conclusions you can and cannot draw from its results.

Statistical Fishing Expeditions

Because computer software makes complex statistical tests easy to conduct, many researchers are willing to go on a "fishing expedition" with their data. These researchers conduct statistical test after statistical test in the hope of finding one that will indicate a statistically significant effect. You should select your statistical procedures as a part of your research design and stick to that plan while analyzing the data. Exploratory analysis, also called **data mining**, of the data is acceptable only as long as it is within the framework of the original research design or if it helps generate new hypotheses for new research projects.

RESEARCH IN ACTION

As you have learned in this chapter, internal and external validity are central concepts in all research. In this section, we will review a research project that examined the effectiveness of psychotherapy.

Washing machines, tires, compact cars, and psychotherapy: Do they work—are they worth the money? *Consumer Reports* is a magazine devoted to rating the quality, safety, and economic value of different products that consumers buy. The editors take great pride in their well-deserved reputation for scrutiny, objectivity, and fairness. In 1995, the magazine published an article that examined the effectiveness of mental health treatment. We can use this research to further examine internal and external validity. To do so, we must first examine the methods the researchers used. We can begin our analysis of the study by exploring how the authors collected the data.

Each year, *Consumer Reports* sends questionnaires to its over 4 million subscribers (Kotkin, Daviet, & Gurin, 1996). In 1994, *Consumer Reports* sent 184,000 subscribers a comprehensive survey that examined their experience with different forms of psychotherapy. Of the 184,000 subscribers contacted, 22,000 people completed the questionnaire. Of the returned questionnaires, 6900 people indicated that they had sought professional help for emotional problems within the past 3 years.

Figure 4.8 presents a summary of the data. The graph presents several pieces of important information. First, the figure displays the data by the type of professional: psychiatrists, psychologists, social workers, marriage counselors, or physicians. The figure also distinguishes between people who saw the professional for less than 6 months and those who saw the professional for more than 6 months.

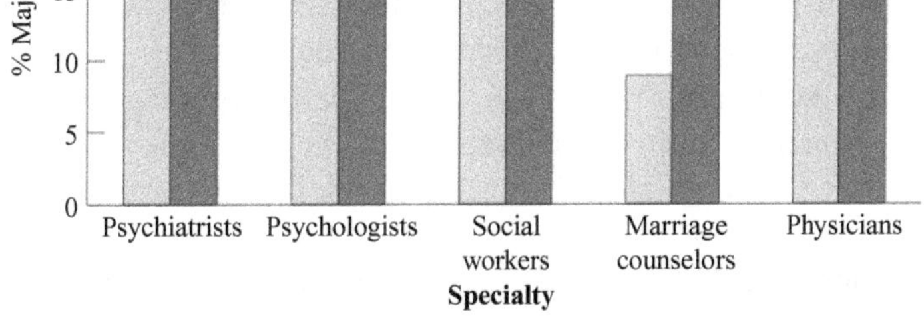

Figure 4.8 Percentage of individuals reporting a major improvement in their mental health depending on the type of psychotherapist they saw and the length of treatment. *Note:* The data are based on a figure presented by Seligman (1995).

We can treat these variables (type of provider and time in therapy) as independent variables. The dependent variable is the percentage of people who reported a "major improvement" in their condition.

The results of the survey, based on this sample, led some people to make several conclusions:

1. People benefited from psychotherapy.
2. Long-term treatment was more effective than short-term treatment.
3. Psychotherapy alone was as effective as medication and psychotherapy combined. This conclusion presumes that psychiatrists prescribed medication as a part of their treatment.
4. Mental health professionals were more effective than marriage counselors and physicians (Seligman, 1995).

Now comes the tricky part. What confidence can we have in conclusions drawn from these data? For the sake of argument, we can focus our attention on the first conclusion: *People benefited from psychotherapy*. We can rewrite the statement in several different ways:

a. People responding to the questionnaire reported being satisfied with the psychotherapy they received.
b. People receiving psychotherapy are generally satisfied with the treatment they received.
c. Psychotherapy caused people's mental health to improve.

Can you see the differences among these statements? What are the implications for these differences? Statement a is a report of fact. The focus of the statement is the sample of respondents. In addition, there is no particular conclusion other than a summary of the results. Statement b takes a step away from the sample and makes an inference about the population. The focus of the generalization is descriptive in that it implies that because the sample of respondents expressed satisfaction, we can assume that the population will express the same level of satisfaction. Statement c presents a dramatic shift from the first two statements. Not only does the conclusion generalize from the sample to the population but it also assumes that there is a cause-and-effect relation between psychotherapy and improved mental health.

Earlier in this chapter, we said that the data themselves are not valid or invalid—only the researcher's interpretations. Can you use what you have learned so far to determine the types of inferences we can make from these data? Specifically, evaluate the internal and external validity of the conclusions. What type of research design does this research represent? At one level, the research is a survey. The staff of *Consumer Reports* mailed a detailed questionnaire to some of its subscribers and then reported the results of the survey. From a different perspective, you can see that we have an intact group design because we have different groups of people receiving treatment from different providers. Can we use these data to assume a cause-and-effect relation as implied in statement c?

What are the conditions necessary to assume cause and effect? Are these conditions present in the *Consumer Reports* study? The three characteristics of a true experiment are (1) manipulated independent variable, (2) random assignment of participants to treatment conditions, and (3) control conditions. Are these conditions present in the study? First, the researchers could not manipulate the independent variables. The people responding to the survey selected the type of therapist they saw. In addition, the researchers could not control how long the people spent in therapy. Therefore, the type of psychotherapist and the time in therapy are subject variables. In addition, the researchers could not randomly assign the people to the different categories of treatment provider or the length of treatment.

Second, are there control conditions? In a true experiment, the most extreme control condition would be a group of people who receive no psychotherapy during the course of the study. We need control groups because some mental health problems improve spontaneously. Like the old saying, "time heals all wounds," some people may feel better as they put some distance between themselves and the situation that caused their distress. They may also find ways of coping with the problem. Thus, the purpose of a control group is to determine if psychotherapy has a greater effect than the passage of time. Clearly, there is no control group in this study.

Because the design of the study does not meet the criteria of a true experiment, we cannot conclude with any degree of certainty that psychotherapy caused people's mental health to improve. In making this observation, we have criticized the conclusion, not the data. The specific objections we raised concern the internal validity of the interpretation based on these data. We determined that we cannot assume that the independent variable (psychotherapy) caused the dependent variable (improvement in mental health).

What about the external validity of conclusions drawn from this study? Statement b implies that the data are representative of the population of people who seek psychotherapy. The statement also implies that clients are generally satisfied with the treatment they receive. Strupp (1996) argued that evaluating a patient's sense of well-being is an important part of any research on the effectiveness of psychotherapy. He, like many psychotherapists, believes that research of this type should examine the client's adaptive behavior, personality structure, and sense of well-being. Therefore, the information concerning the client's evaluation of the treatment they received is an important part of the research examining psychotherapy. Consequently, the *Consumer Reports* study has the potential of telling us how people reacted to the treatment they received.

Do the data allow us to generalize from the sample to the population? Several researchers criticized the external validity of this conclusion (Brock, Green, & Reich, 1998; Brock, Green, Reich, & Evans, 1996; Jacobson & Christensen, 1996). These researchers raised several questions about who responded to the questionnaire. In an ideal world, we would prefer a random sample of all people who received psychotherapy. The current data reflect only people who received psychotherapy, who subscribe to *Consumer Reports*, and were willing to participate in this study by responding to the questionnaire. Although we may agree that *Consumer Reports* is a good magazine, we must ask whether people who subscribe to the magazine represent the typical person seeking psychotherapy or the typical person who may require psychotherapy.

Another issue to examine is the return rate of the questionnaires. As Brock et al. (1998) noted, only a small proportion of the original sample indicated seeking help from a therapist. There may be some self-selection bias in who returned the surveys. We cannot be sure whether the data represent the population or only those satisfied with their experiences. Brock et al. argued that the sample did not represent the population and therefore do not support inferences about the population of people seeking psychotherapy.

Where does all this criticism lead us? Can we conclude that psychotherapy does not work? No, there is no information that would warrant such a reckless conclusion. Many research reports confirm the effectiveness of psychotherapy. Jacobson and Christensen (1996), for example, offered a readable account of this research. Our review of this case merely illustrates that you must interpret the *Consumer Reports* data with caution.

KNOWLEDGE CHECK

1. Explain why internal and external validity refers to a researcher's interpretation of the data and not the data.
2. A researcher claims that there is a cause-and-effect relation between two variables. You believe that there is a significant threat to the internal validity. Can the researcher continue to claim that the interpretation has external validity?
3. A college instructor wants to examine the effectiveness of a new teaching method. On the first day of classes, she administers to all the students a 100-item multiple-choice test that reviews the major content areas of the course. She uses a similar test as the final exam. List the potential threats to internal validity that you believe could apply to this study.
4. Miranda found an article published in the early 1960s that examined the treatment of ulcers but decided not to read it because it is over 40 years old. She told her friend, "Why read the paper? The results are no longer relevant today." Comment on her conclusion.
5. Describe how one's hypothesis will determine the research design that one will use.
6. How does the inclusion of one or more control groups help us determine cause and effect in a true experiment?
7. How does random assignment help ensure the internal validity of our interpretation of the data?
8. There is a strong correlation between whether a mother smokes and the health of her children. The children of mothers who smoke have more

health problems than the children of mothers who do not smoke. Do these data allow us to conclude that smoking causes poor health in children?

CHAPTER SUMMARY

This chapter provided a broad overview of the research methods that behavioral researchers use to conduct their research. A research design is a plan to collect data. More specifically, the purpose of a research design is to allow the researcher to collect data that answer a research hypothesis.

Whenever we prepare to design a study, we need to consider two important forms of validity: internal validity and external validity. As you learned, internal and external validity refer to the interpretations one can make from the results of the study.

Internal validity refers to the researcher's ability to show a cause-and-effect relation between two variables, the independent and dependent variables. Cause-and-effect relations are difficult to demonstrate because of the number of threats to internal validity. Careful attention to research design can help the researcher avoid the threats to internal validity, including (1) unintentional sequence of events, (2) nonequivalent groups, (3) problems with measurement, and (4) ambiguity of cause and effect.

External validity refers to our ability to use the results to make inferences beyond the study. Specifically, we can ask how the findings generalize to the target population. We can also ask whether the results reached with one study will generalize to different populations and different applications of the variables.

There are many threats to external validity, including (1) bias in the methods used to recruit the participants for the study, (2) situational factors unique to the study, and (3) historical events that may be unique to a group of participants.

The remainder of the chapter reviewed the major research designs, including the true experiment, intact group design and quasi experiments, surveys and questionnaires, correlational studies, single-subject methods, and meta-analysis. The defining characteristic of a true experiment is the ability to manipulate systematically the independent variable and to randomly assign participants to the levels of the independent variable. In addition, true experiments include a control group. Participants in the control group experience the same things as the other subjects except for exposure to the independent variable.

Although true experiments are useful for inferring causality, they are sometimes impractical or impossible to perform. The intact group design and the quasi experiment are alternatives to the true experiment. The distinguishing feature of these designs is that the researcher cannot randomly assign the participants to the different levels of the independent variable.

In our review of the quasi experiments and intact group design, we examined two important threats to internal validity, the third variable problem and the temporal order problem. The third variable problem is that some condition, other than the independent variable, affects both the independent and dependent variables. The temporal order problem is that the research design does not allow us to determine whether the independent variable preceded the dependent variable.

Surveys and questionnaires are useful because we can measure an interesting construct by asking participants objective questions. Correlational studies allow us to examine the relation among two or more variables. These research designs are useful for studies of personality or business administration, where we are interested in the relation between one set of behaviors and other behaviors.

As the name implies, single-subject designs allow us to study one or two people. In some single-subject designs, we can use experimental techniques to study the behavior of one person. For other situations, we examine the individual's history to create a case study.

Meta-analysis is the study of research studies. Researchers who use meta-analysis use statistical techniques to combine the results of many studies. The goal of meta-analysis is to come to some uniform conclusion about a phenomenon.

We concluded by examining the role of computers in statistical analysis. Computers are essential when conducting even moderately complex statistical tests. However, computers and statistical software are not foolproof. Misuse of the software, failure to understand the statistical tests being run, and data entry errors can lead to meaningless or even misleading results.

The "Research in Action" section used a report on the effectiveness of psychotherapy to examine how researchers can apply different research techniques to address specific empirical questions. The review of the research illustrated that the research design affects the internal and external validity of the data.

CHAPTER GLOSSARY FOR REVIEW

Alternative Explanation Another method of explaining the results that does not depend on the independent variable.

Carryover Effects A form of unintentional-sequence-of-events threat to internal validity wherein the participants' experiences during one part of the research affect their performance in subsequent parts of the research.

Case Study An empirical research method for which the researcher examines the history and behavior of a single person. In some situations, the researcher will use a case study to describe a patient's reaction to treatment.

Categorical Independent Variable An independent variable best described by a nominal scale.

Ceiling and Floor Effects A form of measurement-error threat to the internal validity wherein the measurement device cannot adequately measure high-level (ceiling) or low-level (floor) performance.

Confounding Variable A variable related to the independent variable that the researcher cannot control and that threatens the internal validity of the cause-and-effect hypothesis, IV → DV.

Confounds External and uncontrolled variables that can reduce the representativeness of treatment groups, even when such groups are created through random assignment.

Control Group A treatment condition in which the participants experience all aspects of the experiment except for the independent variable.

Correlational Studies An empirical research method for which the researcher takes two or more measurements of characteristics for each participant and then examines the correlation among the variables.

Data Mining The process of using a large array of statistical tools to discover interesting relationships among important variables that are hidden in typically large sets of data.

Effect of History A threat to external validity that arises when a time-specific cultural, social, or environmental event, unique to a group, influences the results.

Effects of the Research Situation A threat to external validity that occurs when characteristics of the research environment influence the results.

External Validity The degree to which we can generalize the results and conclusions reached with a sample to the population.

Generality of Conclusions The degree to which the conclusion reached in one population will generalize to different populations.

Generality of Findings The degree to which we can use the sample data to generalize about the population from which the data were drawn.

Instrumentation Changes in the measurement instrument over the course of an experiment.

Intact Group Design A form of research for which the researcher divides participants into separate groups on the basis of one or more subject variables. The goal of the research is to determine whether there are differences among the populations represented by the groups.

Internal Validity The degree to which we can assume that changes in the independent variable caused changes in the dependent variable.

Intervening Events A form of unintentional-sequence-of-events threat to internal validity wherein the participants experience extraneous events, outside the researcher's control, that affect their behavior.

Literature Review A general summary of existing research- and theory-related literature, written by the author of a research manuscript to provide context for the present study.

Low Statistical Power A threat to statistical conclusion validity that occurs when characteristics of the data or research design do not allow the researcher to detect the relation among variables.

Manipulated Independent Variable An independent variable whose primary characteristic can be altered to create various treatment conditions.

Maturation A form of unintentional-sequence-of-events threat to internal validity wherein the participants grow older and consequently change their behavior.

Measurement Error Errors of measurement that occur when the testing procedure is not reliable and produces inconsistent results.

Meta-Analysis A collection of statistical techniques used to combine the results of separate research projects to help one determine the relation between the independent and dependent variables.

Mortality A form of nonequivalent-groups threat to internal validity wherein the participants withdraw from the research.

Nonequivalent Groups A threat to internal validity created when the formation of the groups creates a difference unrelated to the independent variable.

Placebo Treatment Used in a controlled experiment to mimic the treatment condition, but without exposing participants to the actual independent variable being studied.

Quantitative Independent Variable An independent variable described by an ordinal, interval, or ratio scale.

Quasi-Experimental Design A form of research for which the researcher can identify an experimental group that is exposed to the variable of interest and a control that is not exposed to the variable of interest but cannot randomly assign the participants to the two conditions.

Quasi-Independent Variable A variable that the researcher may be able to manipulate but cannot randomly assign participants to different levels of the variable.

Random Assignment Each participant has an equal probability of assignment to one of the treatment conditions.

Recruitment of Participants Methods used to enlist participants in a research project. The methods used will influence the external validity of a research project.

Regression to the Mean A form of measurement-error threat to the internal validity wherein participants who obtain exceptionally high or low scores when first tested will tend to have average scores when tested a second time.

Research Design A procedure or plan for collecting data that will answer one or more empirical questions.

Statistical Conclusion Validity The inferences or conclusions a researcher draws from a statistical test that are appropriate given the data collected and the test used.

Subject Variables A form of nonequivalent-groups threat to internal validity wherein the researcher uses a subject variable to group participants. The researcher will not be able to determine whether differences among groups reflect the independent variable or selection variable.

Survey or **Questionnaire** An objective method of obtaining information from members of a population. In most cases, the researcher will ask participants to answer a series of questions.

Temporal Order Problem A threat to internal validity. Because the researcher measures two variables at the same time, or is unable to control one of the variables, there is no way to determine which is the cause and which is the effect.

Third Variable Problem A threat to internal validity. The problem arises when a third variable correlates with both the independent and dependent variables.

Treatment Condition A level or setting of the independent variable. The differences among the treatment conditions may represent categorical or quantitative differences.

True Experiment A form of empirical research in which the researcher randomly assigns participants to different independent groups, uses one or more control conditions, and uses a manipulated independent variable.

Unintended Sequence of Events A category of confounding variables that threaten the internal validity of the research. The threat arises when the researcher cannot control extraneous events that occur during the research.

REFERENCES

Brock, T.C., Green, M.C., Reich, D.A., & Evans, L.M. (1996). The Consumer Reports study of psychotherapy: Invalid is invalid. *American Psychologist, 51*, 1083.

Brock, T.C., Green, M.C., & Reich, D.A. (1998). New evidence of flaws in the *Consumer Reports* study of psychotherapy. *American Psychologist, 53*, 62–63.

Cohen, J. (1992). A power primer. *Psychological Bulletin, 112*, 155–159.

Cook, T.D. & Campbell, D.T. (1979). *Quasi-experimentation: Design and analysis issues for field settings*. Chicago: Rand-McNally.

Gardner, H. (1995). *Leading minds: An anatomy of leadership*. New York: Basic Books.

Glass, G.V. (1978). Integrating findings: The meta-analysis of research. *Review of Research in Education, 5*, 351–379.

Jacobson, N.S. & Christensen, A. (1996). Studying the effectiveness of psychotherapy: How well can clinical trials do the job? *American Psychologist, 51*, 1031–1039.

Kotkin, M., Daviet, C., & Gurin, J. (1996). The Consumer Reports mental health survey. *American Psychologist, 51*, 1080–1082.

Loftus, E. & Ketcham, K. (1994). *The myth of repressed memories: False memories and allegations of sexual abuse*. New York: St Martin's.

Mill, J.S. (1986). *System of logic*. New York: Classworks (original work published in 1843).

Pedhazur, E.J. (1997). *Multiple regression in behavioral research: Explanation and prediction* (3rd ed.). Fort Worth, TX: Harcourt Brace.

Sackett, D.L. (1986). Rational therapy in the neurosciences: The role of the randomized trial. *Stroke, 17*, 1323–1329.

Seligman, M.E.P. (1995). The effectiveness of psychotherapy: The Consumer Reports study. *American Psychologist, 50*, 965–974.

Shadish, W.R., Cook, T.D., & Campbell, D.T. (2002). *Experimental and quasi-experimental designs for generalized causal inference*. Boston: Houghton Mifflin Company.

Strupp, H.H. (1996). The tripartite model and the Consumer Reports study. *American Psychologist, 51*, 1017–1024.

Part Two

Nuts and Bolts of Research

Chapter 5

Writing the Research Report

Writing is not like painting where you add. It is not what you put on the canvas that the reader sees. Writing is more like a sculpture where you remove, you eliminate in order to make the work visible. Even those pages you remove somehow remain.

—Elie Wiesel

INTRODUCTION

The ultimate goal of any scientist or researcher is to share one's findings with others. Within the health sciences arena, this is efficiently accomplished by summarizing the research in conference presentations/papers and written reports that inform others about the details of you research and its implications. However, just because the report is a summary of the finished research does not mean you should wait to write any details down about your study until it is finished! On the contrary, writing is first and foremost a process that takes time. This is why we are taking this entire chapter, early in this textbook, to introduce you to the process of preparing a high-quality research report.

One caveat before we dig in: Because health sciences research is presented and published in a wide variety of conference and journal-type outlets, we have attempted to tailor this chapter so that instead of emphasizing a singular editorial

Understanding and Conducting Research in the Health Sciences, First Edition. Christopher J. L. Cunningham, Bart L. Weathington, and David J. Pittenger.

style (e.g., American Psychological Association [APA] or American Medical Association), we emphasize key features of the writing process and of manuscripts that present social science research in general. We also attempt to highlight where specific editorial guidelines are most likely to apply.

In many ways, writing a research report requires you to tell a story with a clear beginning, middle, and end. You begin the report with a description of an important and at least partially unresolved question. Then you turn to details that you uncovered regarding your hypotheses and use these clues to help steer the review toward a set of logical deductions (i.e., the hypotheses). In the next part of the report, you tell the reader how you went about collecting the data for the study. Remember, good science can be repeated by others and you want others to know exactly how you tested your hypotheses. After discussing the results of the project, the last portion of this type of report is a space for you to pull information together and explain how this study has answered or addressed the initial question or issue of interest.

Writing is a unique intellectual challenge. Your goal is to influence the behavior or thinking of anyone who might choose to read what you have written. In most cases, you write to share information and persuade the reader that your interpretation of the facts is correct or, at the very least, logical and defensible. The same is true in a good research report. At the start of your report, you introduce readers to the purpose of your research and then convince them that you are asking an important and interesting question. By the end of the report, your goal is to convince the reader that your analysis and interpretation of the data were appropriate and have addressed the focal research questions.

There is no mystery behind becoming a good writer. Similarly, there are no excuses for why any intelligent person cannot write well. Writing is a skill that is learned and refined through practice guided by self-discipline. The best writers write often. These individuals also tend to read a great deal of what other good writers write. Reading exposes you to many styles of writing and can help you pick up good writing habits. You can learn a great deal about how to present your research findings and ideas by learning from the example set forth by other good writers in your field. In other words, it is a good idea to read journals and edited books of research not just for content but also for style and form.

Good writers also do not work in isolation; they have others read drafts of their work, and they seek constructive or helpful criticism whenever it is available. Many aspiring writers seem to believe that no one should read their writing until they submit it for publication or grading. This is a terrible strategy. As professors, all three of us can quite honestly agree that we hate it when we receive "first-draft papers" from our students that are not as polished as they should or could be. Asking someone you trust to read a draft of your paper and provide feedback *will* help you become a better writer (as long as you take to consider that feedback as you continue to revise the paper before finalizing). As we wrote this book, each of us reviewed the others' work and offered extensive comments, corrections, and constructive criticism on each and every chapter (and there's still room for improvement!). Other researchers also provided feedback on our final drafts before everything was submitted to the publisher. Certainly there were times when the feedback from these reviews was frustrating and less than positive, but overall, these comments helped us to fix areas that were confusing and to strengthen areas that were weak.

All this being the case, we hope you can see that simply reading this textbook cannot teach you how to be a good writer any more than watching a video can teach you how to paint like a professional. What we *can* do is highlight several of the core elements and characteristics of good scientific writing about behavioral and social research. Studying these characteristics will help you to critically evaluate your own writing and the writing of others. For this reason, we hope to achieve two things in this chapter. First, we will review general stylistic conventions that are central to good writing. Second, we will introduce you to common editorial guidelines that you will need to follow when preparing a typical research report.

WHAT DO READERS APPRECIATE IN GOOD WRITING?

Good writing, especially scientific writing, involves three essential elements: focus, organization, and integration. It may be difficult to believe, but it is *not* a requirement of scientific writing that research reports be dull and boring. In truth, writing about complicated technical and theoretical issues is difficult, and this is what prevents many researchers from allowing their storytelling abilities to shine through in their writing. With time, practice, attention, and multiple drafts, we are confident that anyone can prepare a research-based report that captivates the interest and attention of serious readers.

Focus means that there is a clear purpose for the writing and definite boundaries established for what you will and will not discuss. Some writers fall into the trap of covering too many topics. Consequently, their papers are rambling expositions of many unrelated topics. For a research report, you should focus on your research questions.

In addition to being focused, good writing is *organized*. Readers working through a well-written paper will experience a logical presentation of ideas that lead to a series of reasonable conclusions. Along the way, each paragraph leads smoothly into each subsequent paragraph—there is a sort of flow from section to section that does not disrupt the reader's attention.

Finally, with focus and organization in place, a good writer is able to *integrate* the information into a coherent message. All writing, even within research reports, requires storytelling. Your job as a writer is to bring together a series of facts and observations and to show the reader with words how this puzzle fits together. Simply listing all the means, standard deviations, and results of the inferential statistics from your statistical analyses is not integration. Far more interesting is learning how the results of your research address your specific and well-supported hypotheses.

ELEMENTS OF STYLE

As mentioned earlier, there are many different editorial style guidelines that authors must adhere to when preparing articles and books for publication. All of these styles are designed to facilitate clear and consistent communication of information to interested readers in a particular discipline. In this section, we review several core elements of writing style that transcend most specific editorial styles and apply to most scientific writing.

Conveying What Is Necessary

Writing a research report requires that you strive for accuracy, brevity, and clarity. You will find that writing a research report is much different than other forms of writing. We do not wish to imply that one form of writing is better than another, only that different styles serve different purposes. Consider an example from the opening pages of Mark Twain's (1876/1986) *The Adventures of Tom Sawyer*, wherein Twain describes Aunt Polly:

> *The old lady pulled her spectacles down and looked over them about the room; then she put them up and looked under them. She seldom or never looked through them for so small a thing as a boy; they were her state pair, the pride of her heart, and were built for "style" not service—she could have seen through a pair of stove-lids just as well. (p. 9)*

In this vivid passage, Twain used metaphor and other literary devices to introduce the reader to Aunt Polly. Twain used the glasses to describe Aunt Polly's personality and her relation to Tom.

Although these and many other techniques make for good literature, they are not typically appropriate for technical, research-based writing. In research reports, literary techniques such as ironic use of words and intentional ambiguity are distractions. These literary devices require too much inference and subjective interpretation on the part of the reader and should be avoided in scientific writing unless clearly explained. In a similar fashion, when writing about research, you should avoid using vague and colloquial expressions (e.g., "After *quite a long time*, the effects of the independent variable *kicked in*") and opaque jargon and euphemistic phrases (e.g., "The individual's behavior created much *collateral disruption* among *interrelated members of the vertically organized unit*"). Such expressions and phrases do nothing to improve the clarity of the message you are trying to share with the reader.

Now, compare the earlier passage from *Tom Sawyer* with the opening paragraph of a research article by Wilker et al. (2012):

> *Extreme temperatures have been associated with hospitalization and death among susceptible populations (Basu et al. 2008; Bell et al. 2008; Koken et al. 2003; Zanobetti and Schwartz 2008) and individuals with congestive heart failure may be particularly vulnerable (Cui et al. 2005). Cold temperatures have been linked to hospital admissions for heart failure and incidence of cardiovascular events (Martinez-Selles et al. 2004; Stewart et al. 2002), but increases in morbidity and mortality have been observed during periods of hot weather as well (Aronow and Ahn 2004). (p. 1083)*

This paragraph is clear and direct. It is an example of scientific writing rather than literary prose. The researcher authors presented a series of facts that make clear that they have selected an important topic. Furthermore, it does not take a vivid imagination to interpret the meaning of this paragraph.

Writing for an Audience

You may have learned in a composition course that you should always consider your audience when writing. This advice is especially important when writing about your research. So, who is your audience? When you are a student, your "audience" usually consists of your course instructor who told you to write a research report. This is not a very detailed or clear understanding of your real audience for this type of work. As a general rule, when writing a research paper, you are typically writing for an audience of other researchers who understand general principles in the discipline, but who may not be an expert on the specific topic you chose to study.

Knowing your audience ahead of time should help you determine what elements of your report will require the most detailed explanation. Although fellow researchers are often the most immediate audience for research-based writing, it is increasingly the case that researchers must be able to write about ways in which their research findings can be translated into practical applications. In either case, readers are typically expected to understand basic issues in the discipline, common research design, and standard statistical tests. For example, if you use a factorial analysis of variance (ANOVA) to analyze the data, you will not have to explain the difference between a main effect and an interaction. You can also assume that the reader understands the logic of conventional statistical tests. Always keep in mind, however, that your research may be interesting to a secondary audience as well, especially if your findings have potential practical implications within society, families, or workplaces. For this reason, it is also important for you to provide a clear and accurate discussion of the logic leading to your hypotheses and of the findings and their practical implications. If you can successfully do this, you will greatly increase your chances of making your research useful to the largest number of people. All of these reasons are also important for you to remember as you work through this text—even if you do not intend to do research yourself in the future, you will likely need to interpret and learn from other researchers' work quite frequently.

Although most immediate readers of research reports may understand basic behavioral and social science research principles, you cannot assume that the reader knows the history and the important questions that are a part of your research as well as you do. Look again at the Wilker et al. (2012) paragraph from before. They have few preconceived notions concerning what the reader already knows about the effects of extreme temperatures on individuals who are susceptible to health problems. Although other researchers who are somewhat familiar with environmental health may find Wilker et al.'s opening comments somewhat obvious, we appreciate that they clearly emphasized the important connection between extreme temperatures and cardiovascular risks. Reading their introduction quickly teaches you as a reader a great deal about that particular area of research. This type of set up makes it easy for the reader to stay on track with the rest of what is presented in the report.

Value-Free Descriptions of Others

Most editorial guidelines strongly discourage biased descriptions of people or groups of people when writing about research. Writing is considered to be "biased" when it includes any description of an irrelevant characteristic of a group of people or of a person. A bias such as sexism can emerge when one draws attention to a particular sex for no particular reason. Doing so implies that there is a meaningful difference between men and women when one may not exist. Consider the word *mothering*. One connotation of the word refers to caring for others; another connotation is being overly protective. Mothering is a sexist term because both men and women have the capacity to care for others and to be overly protective. Unless the author's intent is to focus on women, better words to use here might be *parenting*, *nurturing*, or *protectiveness*. All three words are value-free because they describe the target behavior and do not link the behavior to one sex.

There is no need to refer to a characteristic of a person or group of people unless that characteristics or trait is the focus of the research (i.e., is one of the variables that was studied). In these instances, the descriptive terms used should be as objective as possible. The best strategy is to write with words and phrases that describe the relevant characteristic of the individual's behavior rather than the person. Therefore, it is preferable to replace the terms *fireman*, *mailman*, and *waitress* with *firefighter*, *mail carrier*, and *server*. The second set of words describes the work of any individual in those roles, without implying some sort of connection with a person's being male or female.

Writing sex-neutral sentences is not as easy as it may seem at first. Here are some common traps you should consciously try to avoid when writing. For example, some writers use complicated combination terms such as *s/he*, *he/she*, or *him/her* to avoid sexist language. Although well intentioned, such terms are awkward, making writing very difficult to read smoothly. Fortunately, there are two relatively simple ways you can avoid this type of lexical faux pas. First, you can convert a single noun to a plural noun (e.g., change "Each *participant* received *his* . . ." to "The *participants* received *their* . . ."). This option has the advantage of being simple to implement and does not change the meaning of the sentence. Second, you can use the singular noun and use "his or her," "he or she," or an equivalent. The only time you should use a gender-specific noun, pronoun, or adjective is when the behavior is clearly sex specific (e.g., "Each woman completed the survey soon after learning that she was pregnant").

Editorial policy regarding sexist language is also extended to the people who participate in our research. For instance, the APA editorial guidelines recommend that we use descriptive nouns when detailing the people involved with our research. This is one reason why all the people who are involved with a study are referred to as *participants*. You can also use nouns such as *children*, *college students*, *parents*, *clients*, *individuals*, and *respondents* to refer to the participants of a study. Just as we avoid sexist language by describing a person's behavior, we are challenged to avoid biased language when describing a person's condition as well. Consequently, health sciences researchers typically refer to research participants as participants, patients, and/or persons. While we are discussing the need to avoid bias when describing participants, keep in mind also that it is also important to avoid age-related biases when describing participants in your research. As such, it is typically not a good idea to identify people within your study as "senior citizens" or "elderly"; they are people "over the age of 65" or "over the age of 85."

SPECIAL GRAMMATICAL ISSUES

For the most part, editorial guidelines follow conventional rules of grammar. Thus, what you have learned in other writing composition courses, concerning subject–verb agreement, dangling modifiers, and other elements of grammar, applies to writing a research report. Certain editorial guidelines, including APA's, do emphasize several special grammatical conventions that require some explanation.

Active versus Passive Voice

Active and passive voice refers to the order of the subject, verb, and object of a sentence. Active voice sentences place greater emphasis on the subject, or the actor, of the sentence than on the object of the sentence. In an active voice sentence, the subject precedes the verb. In an active voice sentence, you are also able to identify yourself using personal pronouns (e.g., "I" or "we"). Consequently, active voice sentences make clear who or what is responsible for the outcome. In contrast, passive voice sentences place the object before the verb. Passive voice sentences are, therefore, more difficult to interpret. Because of this, most editorial styles prefer writing that is active rather than passive in its construction.

The goal of active voice is to make clear who did what to whom. For example, we can revise the passive voice in the following sentence: "The experiment was designed using the Solomon four-group control procedure" to be active voice: "We used the Solomon four-group control procedure for the experiment."

Examples of active voice sentences are (subject, verb, object):

Mary caught the football.
Participants completed the baseline survey before beginning treatment.

Examples of passive voice sentences are (object, verb, subject):

The football was caught by Mary.
Surveys were administered to participants before they begin their treatment.

Active voice is commonly recommended for two main reasons. First, active voice sentences clearly identify who or what is responsible for the outcome. In passive voice sentences, the cause is either hidden at the end of the sentence or implied and sometimes missing. Second, active voice sentences tend to use fewer words and are therefore easier to read and understand.

Past and Present Tense

It is good practice, and required within most editorial styles, that a writer use past tense when referring to the work of another author and/or data from a study that is already completed. As an example, you might write, "Freud (1917/1957) believed . . ." or "Freud (1917/1957) argued . . ." because Freud expressed his beliefs long before you began to write your paper. Similarly, you would use the past tense to describe the data you collected as a part of the completed research you are now summarizing. Accordingly, you might write,

Our results supported the hypothesis that . . .

or

As expected, there was a significant interaction among the treatment conditions.

When you describe an event or action that did not occur at a specific time, or if the condition still exists, you may use the present perfect tense. Examples of the present perfect tense are

Since the late 1940s, business researchers have used the ANOVA and other inferential statistical techniques for their management studies.

Marketing researchers have long had an interest in the effects of media on consumer behavior.

Proper Use of Pronouns

Most editorial guidelines emphasize specific rules concerning the use of pronouns. Therefore, you should pay special attention to these types of guidelines when you are finalizing your writing for publication in a particular journal or book outlet. Here are a few common pronoun-related traps to watch out for when writing about research.

Who versus That

When speaking of humans, and especially individual people, use pronouns such as *who*, *him*, or *her*. When speaking of nonhumans, use the neuter pronoun *that*. To illustrate the difference, you would write, "The nurses *who* participated in the study . . . ," or "The medicine *that* the research team tested. . . ." Similarly, you should write, "The doctor *who* treated the patients . . ." because a doctor is a person, not a thing.

Vague Pronouns

All editorial style guides warn against the use of vague pronouns. Unfortunately, many people regularly use the pronouns *this*, *that*, *these*, and *those* without an obvious referent. Consider the following passage:

> *Many surveys indicate that most people endorse the stereotype that women make better nurses than men. This may prevent men from being hired into nursing roles. That is an example of sex-based discrimination.*

What do *this* and *that* refer to in the second and third sentences? Does the second sentence mean that the survey results prevent the hiring of male nurses, or is it the stereotype that creates the discrimination? For both sentences, the pronoun has no clear referent. Consequently, the object, or referent, in the sentences is vague. A few simple modifications can clarify these sentences:

> *Several surveys of people's attitudes have consistently demonstrated that most people endorse the stereotype that women make better nurses than men. Belief in these stereotypes may cause employers to hire more female nurses than men. This type of discrimination is an example of how stereotypes can be detrimental to men as well as women seeking employment.*

Commonly Misused Words

Some authors use specific words indiscriminately and incorrectly. In Box 5.1, we list many of these commonly misused words and attempt to illustrate their correct use. As you proofread your writings, look out for these problem terms; many spelling and grammar checkers built into word processors will not catch these slipups.

ACADEMIC INTEGRITY

Writing with academic integrity means ensuring that you give complete and fair credit to the sources of ideas that you incorporate into your writing. As a general rule, if you state a fact, share an observation, or report the conclusion from another writer, you need to acknowledge the origin of that fact, observation, or conclusion. Following this rule is what prevents you from plagiarism. *Plagiarism* comes from a Latin word meaning to kidnap. Thus, someone who plagiarizes is essentially stealing ideas from others. Most colleges and publishers have strict rules against and penalties for plagiarism. Depending on your college's academic policies, you may receive a failing grade for the assignment or the course if the instructor discovers that you have plagiarized. This issue is becoming increasingly apparent as colleges and universities increasingly make use of new software programs that

BOX 5.1 *Commonly Misused Words*

Affect vs. Effect	*Affect* means to influence or to cause a change. "The independent variable affected the participants' behavior." *Effect* represents the result or the consequence of something. "There were large treatment effects in the experimental groups."
Among vs. Between	Use *among* when discussing more than two people or objects. "There were minor differences among the 30 participants." Use *between* when discussing only two people or objects. "There are minor differences between Hull's and Spence's theories."
Amount vs. Number	Use *amount* to refer to a general quantity. "The amount of reinforcement each participant received depended on the schedule of reinforcement." Use *number* to refer to countable items. "The number of patients seeking treatment decreased after 6 months."
Data	*Data* is the plural form of the word *datum*. As a rule, nouns ending with "a" are plural (e.g., phenomena is the plural form of phenomenon; criteria is the plural form of criterion). "The data are consistent with the theory."
Ensure vs. Insure vs. Assure	*Ensure* refers to procedures that minimize the occurrence of some event. "We called all participants to ensure that they would return to the lab for the second part of the experiment." *Insure* refers specifically to the protection against financial loss. "We instructed the medical students to save the money they would have spent for coffee 2 days out of each week, as a way to begin building a savings account and insure against future unknown financial responsibilities." *Assure* means to convince, persuade, or to affirm a pledge. "The researcher assured the participants that their responses to the research questions would be confidential."
Few vs. Little	*Few* refers to a countable quantity of objects. "Few people continue to question the importance of this theory." *Little* refers to a general quantity. "The research attracted little attention when first published."
Its vs. It's	*Its* is a possessive pronoun. "The value of this theory is its ability to make novel predictions." *It's* is a contraction of "it is." Never use contractions in technical writing.
That vs. Which	*That* is a relative pronoun used in clauses essential to the meaning of the sentence. "The author reviewed the evidence that supports Miller's theory." Use *which* in clauses that are not essential to the meaning of the sentence but provide additional information. "The real purpose of the experiment, which was not shared with participants, was to determine the degree to which people will conform to behavior of a role model."
Utilize vs. Use	*Utilize* is a transitive verb that adds little to a sentence; *use* is sufficient in most cases.

BOX 5.1 (Continued)

While vs. Since	*While* refers to the simultaneous passage of time for two or more linked events. "While the participants were involved in filling out the questionnaire, the confederate began to perform various distracting behaviors." *Since* refers to the passage of time between the past and present. "Since Snyder published his paper in 1974, many researchers have become interested in the self-monitoring construct."
Although vs. Because	*Although* refers to a contradiction with a statement or fact. "Many people believe in extrasensory perception, although psychologists have long questioned the existence of the phenomenon." Use *because* to indicate the reason for some condition or event. "The external validity of the experiment is questionable because the researcher did not use a random selection procedure."

automatically cross-check student work against massive electronic libraries of journal, book, and other manuscript publications from all over the world. Now is definitely the time to learn how *not* to plagiarize.

There are many forms of plagiarism. The most obvious is copying word for word another author's words and presenting them as your own. Other forms of plagiarism may not be as obvious but are just as wrong. For example, presenting another person's ideas or logical arguments as yours without citing the source is also considered plagiarism. We can use the following passage to look at different forms of plagiarism.

Wilker et al. (2012) wrote:

> *In our analyses of associations between temperature and biomarkers of heart failure, we observed positive associations with higher air temperature moving averages but not with relative humidity or ozone . . . Our findings suggest an association with BNP after only two days, but it is possible that longer integrated averages of temperature over several days, which reflect a period of prolonged elevated temperature, are necessary to observe upregulation of CRP. (p. 1086)*

What if a researcher or student wrote the following in his/her own research paper?

> *Analyses of associations between temperature and biomarkers of heart failure have been positively associated with higher air temperature moving averages but not with relative humidity or ozone.*

This passage is an example of plagiarism for several reasons. Although the writer changed a few words and phrases, many of the original authors' words remain in the sentence. All the researcher or student has done is a form of editorial recycling. Perhaps more importantly, the writer has not properly credited the original authors for their contribution. To avoid plagiarism in this scenario, a better approach could have been the following:

> *Higher air temperatures have been associated with an increase in biomarkers associated with heart failure (e.g.,Wilker et al., 2012, p. 1086).*

This sentence gives clear credit to the origin of the idea and is more of a paraphrasing and isolating of a core idea than the originally copied snippet that constituted plagiarism.

Citations of Sources in the Text

In the preceding example, you saw how you can credit others for their work directly within the body of your written report. Depending on the particular editorial style that governs the publisher you are working with to publish a written research report, this citing can be accomplished in a number of ways. Within psychology and sociology, it is most common to list the author, year, and sometimes page number(s) associated with the referenced material in parentheses at the end of a statement that you are crediting to an original source (as is illustrated in the preceding section). Within most biomedical journals, however, in-text citations are handled with superscript numbers that correspond to detailed reference information at the end of the document. Other editorial styles have slightly different conventions. The shared point of consistency across the different styles is the fact that a complete reference citation includes information about the author(s) of the work being cited, the year in which that work was published, the publication or source in which that work was published, the volume (issue, when available), and page numbers associated with the work being cited.

Additional Comments on Quotations

Extensive quoting of other authors' work is a symptom of laziness. Researchers rarely use quotations because the focus of the research paper is on their analysis of ideas, not just on how previous researchers might have expressed those ideas. Your responsibility, as the author of a paper, is to read the work of others and then to synthesize that work into a concise statement that supports your argument.

That said, there are at least two instances when you can and should use quotations. First, use a quotation when it is impossible to paraphrase a passage without losing or significantly altering its meaning. These instances are rare, and you should avoid the temptation of assuming that you cannot accurately describe complex ideas in your words. The second, more appropriate occasion for using a quotation occurs when the author's ideas and expression of those ideas is the focus of your argument. Imagine that you are writing a paper wherein you examined the effects of positive reinforcement on intrinsic motivation. As a part of your paper, you want to illustrate how another author expressed his or her opinion. You could write something like the following:

> *In contrast, some authors believe that positive reinforcement reduces intrinsic motivation. These authors typically make the unqualified claim that reinforcement has detrimental rather than beneficial effects. For example, Kohn (1993) asserted, "But the use of rewards for reading, writing, drawing, acting responsibly and generously, and so on is cause for concern. Extrinsic motivators are most dangerous when offered for something we want children to want to do" (p. 87). This sentiment is common . . .*

In this case, the quotation is evidence that supports your argument. Using the quote allows you to share with the reader Kohn's tone and sentiment regarding the effects of reinforcement.

If you must use a quotation, try to keep it short, typically no more than a few sentences. In addition, you should reproduce the quotation exactly as it appeared in its original form. In this example, several words are italicized because they were italicized in the original text. If we had added the italics, we would have included the following note after the italics: "[italics added]." If you drop words or sentences from the original text, replace the missing words with ellipsis points (. . .). To indicate a misspelling in the original text, use "[sic]."

Although there are slight differences across the different editorial styles, in most cases, a brief quotation (e.g., fewer than 40 words) can be included as part of the sentence or paragraph, while longer quotations are often set apart in some way, often as an indented paragraph. As an example,

> *In contrast, some authors believe that positive reinforcement reduces intrinsic motivation. These authors typically make the unqualified claim that reinforcement has detrimental rather than beneficial effects. For example, Kohn (1990) asserted,*
>
> *But not only are rewards less effective than intrinsic motivation—they actually undermine it. You started out doing something just because you found it fun. But once you were rewarded for doing it, you came to see yourself as working mostly to get the reward. Your fascination with the task* mysteriously vanished *[italics added] along the way and now you can't be bothered to do it unless there's some reward at stake. (p. 32)*

Note that we indented the longer quote and separated it from the rest of the paragraph. The example also shows how you can add emphasis to the quotation to draw the reader's attention to a specific phrase in the passage. Whenever you add one of these editorial notes, place them within brackets. Bottom line: Keep your use of quotations to a minimum. Research papers should not read like grocery lists of quotations embroidered together with a few transitional sentences. Tell the reader what *you* know, how *you* would synthesize the information you have read. Your work is far more interesting to read than a string of other researchers' thoughts.

PARTS OF THE RESEARCH REPORT

All research reports contain essentially the same elements, often including a title page and an abstract that prefaces the essentially mandatory sections for introduction, methods, results, and discussion. In the following pages, we will review the material that typically is included in each of these core sections. Once you understand these basic elements, it is relatively easy to set up this type of structure within a word processor software program. The following guidelines are good places to start, but always remember to check the particular requirements of the publisher that you might be working with in the future (there are subtle differences across publishers that are best addressed up front rather than after reviewers get a hold of your manuscript).

Title Page

When included, a standard title page includes four core elements: (1) the running head, (2) the title of the paper, (3) the names of the researchers, and (4) their professional affiliations. As for the title itself, think short and descriptive—most titles are between 10 and 12 words. Writing a good title requires careful planning. You can get a good impression of how to write your title by looking at the titles of the articles you use in your reference section. Your title should be a short and self-explanatory description of the research. For example, the title "A Study about Weight Loss" tells the reader nothing about the purpose of the research as it describes almost any research project dealing with weight loss. In contrast, the title "A Study of the Impact of Daily Low-Intensity and Moderate-Intensity Activity as a Means of Reducing Excess Body Weight in Otherwise Obese Individuals" is very descriptive but far too long. An example of a better title in this scenario might be "Weight Loss Associated with Low- and Moderate-Intensity Exercise." The actual formatting of title pages differs dramatically from one editorial style to another. Make sure to follow each publisher's guidelines carefully when preparing a manuscript.

Abstract

When included, an abstract is a relatively short description and summary of the entire research study. Readers use the abstract of a research article to determine whether the article interests them and deserves their attention. Newer computer programs for managing references also search within abstracts often to categorize published articles. Therefore, useful abstracts briefly describe the purpose of the research, the population under study, the method used to collect the data, the primary findings, and the researcher's primary conclusions. The best model for you to follow when writing your own abstract is a published abstract from the research articles you read in peer-reviewed journals. Many journals are moving toward structured abstracts, which require authors to say a little about the background, methods, results, and discussion as structured subsections within the brief abstract. Other journals and publishers are much less prescriptive about what goes into an abstract. Regardless of the form, remember that first impressions matter—the abstract is like a first impression for the rest of the manuscript, so make sure it is clear, detailed, and concise as possible.

Introduction

The introduction to a standard manuscript begins after the title page and abstract. The objectives for writing the introduction are to capture the reader's interest and to offer a compelling overview and a rationale for the research. As we noted

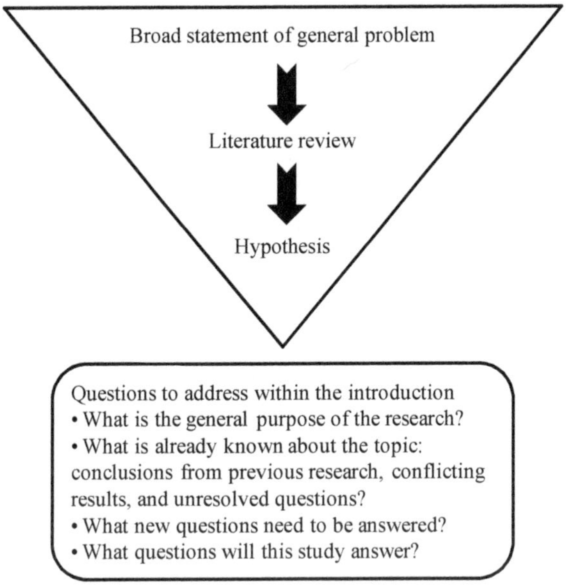

Figure 5.1 Inverted triangle model of the introduction.

previously, you should consider the reader to be a well-informed researcher or practitioner who may or may not know much about the phenomenon that you chose to study. Consequently, your introduction must help the reader understand the phenomenon and why there is a need for more research in this area or application of your findings.

Although there is no universal formula for writing the perfect introduction, you can follow some useful guidelines. Figure 5.1 presents such a guide.

Think of your introduction as an inverted triangle. The paper begins with a broad statement of the general problem. As you progress through the introduction, you should focus the text on issues directly related to your research project. In the literature review section, you can describe the findings and conclusions drawn from previous research. This portion of the introduction allows you to help the reader learn about the focus of your study and the necessity for the research. Finally, you should end the introduction with a review of the hypothesis or hypotheses that you will examine.

We can use the Wilker et al. (2012) published article to illustrate the application of this inverted triangle. These researchers were examining whether increases in ambient temperature might correspond with increases in biomarkers indicative of heart problems. The opening sentence of Wilker et al.'s introduction was "Extreme temperatures have been associated with hospitalization and death among susceptible populations (Basu et al. 2008; Bell et al. 2008; Koken et al. 2003; Zanobetti and Schwartz 2008) and individuals with congestive heart failure may be particularly vulnerable (Cui et al. 2005)" (p. 1083). This sentence broadly introduces the topical area within which Wilker et al.'s research fits.

Following their opening statement, Wilker et al. (2012) narrow the focus of their paper to a testable hypothesis, as illustrated with this snippet from their introduction:

> *It is possible that the biological mechanisms underlying cardiovascular health effects may vary across the range of ambient outdoor temperatures observed throughout the year. However, only a few studies have examined these associations in heart failure patients (Barclay et al. 2009; Goldberg et al. 2009). Moreover, it is not clear if short-term changes in weather patterns can be linked to preclinical changes in heart failure status . . . The evaluation of a panel of biomarkers may provide insight into underlying disease pathology and targeted therapeutic responses (Braunwald 2008). For this analysis, we investigated the association between ambient temperature and four biomarkers of heart failure in a population of individuals with stable systolic heart failure using a repeated-measures approach. We hypothesized that higher ambient and apparent temperature would be associated with elevated levels of these measures, which are associated with heart failure severity and prognosis. (p. 1083)*

Method Details

The method section provides the reader with a detailed description of how you collected the data for your research. The goal of writing the method section is to allow the reader to evaluate the appropriateness of your data collection techniques. This information will help the reader understand how the data relate to the research hypothesis and evaluate the internal and external validity of your conclusions. Most method sections have at least three subsections in which you describe (1) how you identified and obtained the sample you studied; (2) the materials, equipment, and measures you used to collect the data; and (3) the specific procedures you followed during the research.

Participants

The first method subsection pertains to your participants (the term typically used with the population you are studying is human in nature). In this section, you tell the reader the relevant details of your sample and how it was created. As you will learn in Chapter 7, sampling is a critical component of any research project. Consequently, you need to define the sampling population and the procedures for creating the sample. Similarly, you need to define the relevant characteristics of the sample. The purpose of your research usually dictates the characteristics of the sample that you should describe. At the very least, you need to indicate the number of men and women in the study as well as the average age of the participants. If your research depends on specific subject variables (i.e., characteristics within the person, such as ethnicity, level of education, annual household income), then you should summarize those characteristics as well.

In addition to indicating how you recruited and selected the participants, you should describe whether and how you compensated or rewarded them. Similarly, you should summarize any known reasons for missing or incomplete data for any member of the original sample—mistakes happen, sometimes the equipment does not work, or the participant does not understand the instructions or refuses to complete the study. Report errors like these if they occur. For example, you may have to write, "Of the 120 surveys distributed, 5 could not be scored due to incomplete responses" or "The data for three participants could not be used because of a malfunctioning computer."

Materials, Measures, and Manipulations

This method subsection includes the description of the devices you used to collect the data. As with all parts of the method section, your description of the materials must be sufficient for another researcher to repeat the research. Because a research report is typically a rather short document, you cannot describe every nuance of the materials that you used. Rather, this section should offer the reader a short, yet detailed account of the materials and where you acquired them (so that the reader could also acquire them if so desired).

There is no need to describe standard research or laboratory equipment (e.g., videotape recorders, stopwatches, or projectors) readily available to any researcher. However, if the equipment is highly specialized, then you should indicate the name of the manufacturer and the model number of the device. Similarly, if you built a special apparatus for the research, you should offer a brief description of the equipment. Depending on the sophistication of the device, you may want to include a scale drawing or a picture of the apparatus.

You can treat measures, scales, and tests the same way you treat a piece of equipment. If you used a published test, then you should include a brief overview of the test (e.g., number of questions and measurement scale used) as well as a reference to the original source of the test. Most researchers also include a reference to research that has evaluated the reliability and validity of the test. Although it is preferable to utilize existing and validated measures, sometimes researchers need to create their own scales to gather the necessary data for testing specific hypotheses. If you develop your own test or questionnaire for a particular study, you should make sure to describe the general nature of the questions included in the survey in this section of your manuscript. When describing the measures used in your study, it is helpful to the reader if you clearly state what a high score on each measure represents. In addition, it is common to report reliability and validity information for measurement scales and methods whenever this type of information is available (e.g., Cronbach's α).

Design and Procedure

This method subsection describes the design of your study and details the sequence of significant events that all participants experienced during the study. Consequently, you should describe how you assigned participants to various conditions of the research as well as your control procedures. You should also describe the specific instructions you gave to the participants. In some cases, the procedure section may indicate that you distributed a questionnaire to randomly selected classes of students after you had described the purpose of the research. In other cases, the procedure section will be long if there were a series of important stages in the research that require careful description.

Remember, a complete method section provides sufficient detail that interested readers can evaluate the quality of your data collection procedure and repeat the study if desired.

Results Section

The results section of a research report presents your summary of the data and a description of the statistical tests you used to examine the data. This section, like the method section, should be based on data gathered and analyzed in the study you conducted (not previous studies by other researchers). Specifically, you will use this section to lay before the reader the major findings of the research. You can save a more complete interpretation and evaluation of the implications of these findings for your discussion section. The outline of the results section should follow the outline you used to describe the research questions in your introduction. Begin your results section with the general predictions (hypotheses) you made for your study. For each hypothesis, you can then provide a general description of the results followed by the specific statistical tests used to test your hypothesis.

As a side note, there is no reason why you must absolutely stick to the path implied by your introduction when writing your results. If, after carefully examining your data and testing your hypotheses you find interesting patterns in the data that you did not initially expect, you may need to share these insights with readers. These serendipitous results can often reveal interesting perspectives on the phenomenon that you are studying and raise important questions for additional research. These departures from your core proposed study, however, should not cloud your clear summary of hypothesis tests. For this reason, it is common to reserve such exploratory analyses and thoughts for the discussion section that follows your presentation of primary results.

As you prepare your results section, there are a number of editorial issues you must address, including the best way to summarize the data, the level of detail to include, and the best strategy for presenting results from specific statistical tests. By now you should have read a sufficient number of research articles to know that there are many ways to summarize data. Most researchers use a combination of narrative, simple and descriptive statistics, and graphs/figures and tables. Combining these techniques will help the reader quickly understand your primary findings and how the data relate to your research questions. The narrative is relatively straightforward. You describe in words the primary findings of the study with respect to each hypothesis. In many cases, you will find that a table or a graph will do much to augment your description of the data and help the reader visualize the results.

Reporting Statistics

Once you have described the general results, you then need to present the statistical tests that justify your observations. Because your audience consists of fellow researchers, you need only indicate which statistical tests you conducted and the results of the analysis. As with all other parts of the paper, you will present the statistical results in narrative form. The general format is to indicate the statistic that you used and then to describe the outcome.

Consider the following example from an experiment conducted by Eisenberger and Armeli (1997):

> *Planned comparisons revealed the large monetary reward for high divergent thought produced more creative drawings than either no reward or a small reward for high divergent thought, $t(281) = 2.57$, $p = .005$, and $t(281) = 2.25$, $p = .025$, respectively. Further, the large reward for high divergent thought produced subsequent drawings of greater originality than did the same reward for low divergent thought, $t(281) = 3.32$, $p = .001$. (p. 657)*

Notice how the authors stated their conclusion and then indicated the statistical test that supported their claim. Eisenberger and Armeli used a *t*-ratio for the inferential statistic. The report of the statistic included the degrees of freedom (281 in this example), the observed *t*-ratio, and the probability of the *t*-ratio being due to chance. They did not explain the meaning of the *t*-ratio as they assumed that the reader understands how to interpret this statistical test.

Another important characteristic of the Eisenberger and Armeli (1997) passage is that they chose to report the exact probability of the statistic (e.g., $p = .025$). This is a common practice among many researchers. The alternative is to establish a universal criterion for defining statistical significance and to report this if the statistical test meets the criterion.

For example, an author may include a sentence similar to the following early in their results section: "I used the criterion $\alpha = .05$ to establish statistical significance for all inferential statistics." Later, in the results section, the writer will use $p < .05$ if the test is statistically significant. Specifically, the author would write, "$t(281) = 2.25$, $p < .05$" rather than "$t(281) = 2.25$, $p = .025$," even if the statistics software output indicated a *p*-value of .025.

It is increasingly the case that editorial guidelines and most publishers are encouraging researchers to include measures of effect size, such as η^2 or ω^2, along with the results of their statistical analyses. Reporting such effect size information helps the reader interpret the relation between the independent and dependent variables. Table 5.1 presents a list of commonly used descriptive and inferential statistics and the format for reporting them in the text of the results section. For the inferential statistics, the general format is to use a letter associated with the statistic (e.g., *t* for a *t*-test ratio and *F* for an ANOVA), the degrees of freedom, the observed statistic, and the probability of the result being due to chance. Note that you should italicize the letter representing the statistic.

The narrative summary of findings is often supplemented with a few tables and/or figures. As a general guideline, the use of graphs/figures and tables is reserved

Table 5.1 Commonly Used Statistics and the Format for Presenting Them in the Text of the Results Section

Statistics	Presentation in text
Descriptive	
Mean	$M = 25.67$
Median	$Mdn = 34.56$
Mode	Mode = 45.00
Standard deviation	$SD = 1.23$
Standard error	$SE = 12.01$
Standard error of measurement	$SEM = 0.98$
Inferential	
ANOVA	$F(2, 25) = 9.32, p < .05$
Chi-square	$\chi^2(4, N = 55) = 34.10, p < .05$
Mann–Whitney U	$U(12, 15) = 5.67, p < .05$
Pearson's r	$r(98) = .87, p < .05$
Spearman's r	$r_s(74) = -.95, p < .05$
Student's t-ratio	$t(123) = 2.31, p < .05$

Note: Use italics to present the abbreviation for the statistic (as illustrated here).

for times when you cannot present the same information as efficiently in the narrative text. Another general guideline to keep in mind is that any graph/figure or table should be able to stand on its own. This means that the title, labels, and descriptive notes provided with each graph/figure or table should be sufficiently detailed that a reader can examine this component and understand it without having to refer back to other portions of the manuscript.

Figures and Graphs

Most statistics textbooks and editorial style guides provide extensive details regarding how to prepare appropriate figures for publication. Therefore, we will only highlight the essentials of a good graph or figure. Graphs for research reports have several distinguishing features. The first significant feature is simplicity. Good graphs contain only essential visual information and are free of distracting and unessential information. Unfortunately, many popular computer programs allow you to use 3-D effects, fancy shading, and other features that add little information to the graph and can make the graph difficult to interpret. Figure 5.2 is an illustration of this type of useless fluff, which can actually distract from the core meaning of what the author is trying to present. Tufte (1983) called much of the unnecessary graphic elements in this type of figure *chartjunk*. Chartjunk refers to anything in the graph that adds nothing to the interpretation of the data.

For example, 3-D effects add nothing to the display of information. In general, figures should be drawn in two dimensions and kept as simple as possible. Other conventions include that the vertical or Y-axis presents the dependent variable,

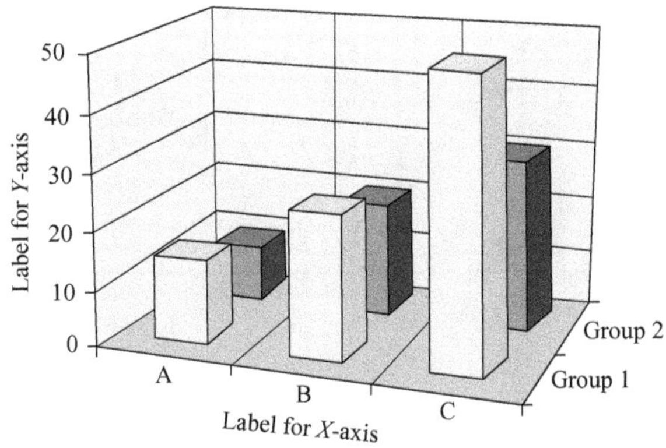

Figure 5.2 A chartjunk-laden figure. *Note:* The 3-D effect, shading, and background grid are unnecessary elements of the graph that are distracting and should not be used.

whereas the horizontal or X-axis presents an independent variable. The lines or bars in the graph should clearly represent the pattern of the data and should not be confused with distracting information. There is considerable art and science in constructing good graphs, far more than we can present here. If you want to learn more about preparing good scientific graphs, we strongly recommend that you check out Kosslyn (1994), Tufte (1983, 1990), and Wainer (1997). These authors make clear that graphs have a grammar, syntax, and style of their own.

The most commonly used graphs are bar graphs, scatter plots, and line graphs. Figure 5.3 presents an illustration of each. Use bar graphs when the independent variable represents a nominal or categorical scale. Scatter plots present the correlation between two variables. Finally, line graphs present the relation among two or more quantitative variables that are typically continuous in nature. For scatter plots, each dot represents the two scores measured for each participant. In Figure 5.3*b*, we have added a regression line to help the reader visualize the directionality of the relationship between the two variables. This addition is optional. Use the option only when you believe that it helps the reader understand the data. For the bar graph and the line graph, the data represent the descriptive statistic used to summarize the dependent variable. If the reported statistic is the mean, you can add error lines that represent the standard error of the mean. This additional information can help the reader determine which group means are statistically different from each other.

Tables

Graphs and figures present pictures that allow us to quickly interpret the results of the data. Although graphs are useful as a research tool, there are many times when we need to present numbers. Constructing useful and easy-to-read tables is like constructing good graphs. There are some simple rules to follow that will help you present your data well; for starters, we will highlight Ehrenberg's (1977) six basic features for constructing good tables.

First, round the numbers in your table to a meaningful value. Every editorial style has its own requirements, based on the level of precision needed for that

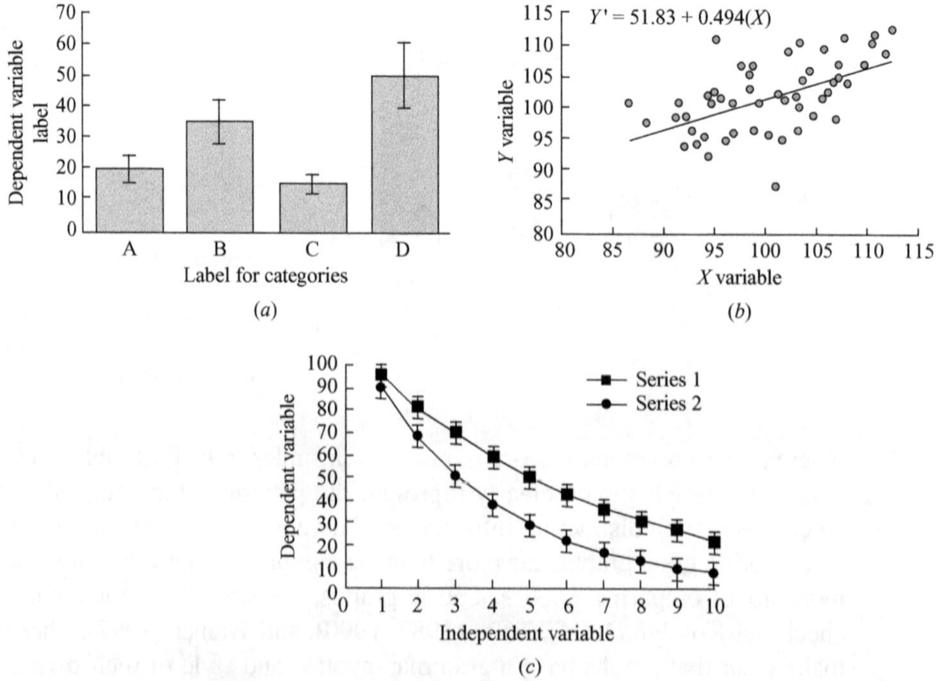

Figure 5.3 (*a*) Bar graph, (*b*) scatter plot, and (*c*) line graph. *Note:* For the bar graph, the heights of the bars represent the statistic used to describe the dependent variable. The lines extending from the bars are optional and represent the standard error of the mean or some other measure of variability that helps the reader interpret the difference among the means. For the scatter plot, each dot represents the two observations recorded for each participant. The regression line drawn through the data is optional. If you include the regression line, you should also include the equation that defines its intercept and slope. For the line graph, each point on each line represents the statistic used to describe the dependent variable. The lines extending from each point are optional and represent the standard error of the mean or some other measure of variability that will help the reader interpret the difference among the groups.

particular field of science and practice. Second, if possible, include row and column averages or totals. These summary statistics help the reader discern general trends in the data and differences among groups. Third, orient the most important data within the columns—readers find it easier to scan a column of numbers than to scan a row of numbers. Another helpful tip is to rank the data from largest to smallest or smallest to largest, if appropriate. Ranking the data helps the reader find the extremes in the data. The fifth recommendation is to keep row-and-column spacing relatively constant. This is easy to do within most popular computer word processing or spreadsheet programs. Finally, use tables only when it is essential to present the quantitative data or when there is no alternative for presenting the data.

Table 5.2 presents an example of a typical table that summarizes the proportion of men and women who reported different forms of cheating within a study. Each row represents the target behavior. The table includes the proportions and the result of the test for the difference between the proportions. This table allows the researcher to present much information in a concise format.

Table 5.2 Example of an APA-Style Table

Table 1. Proportion of different forms of cheating reported by men and women

Form of cheating	Men (n = 48)	Women (n = 124)	Difference between proportions
Cheat on a test	.28	.16	$z = 1.75, p < .05$
Plagiarize in a paper	.21	.24	$z = -0.32, p > .05$
Copy homework	.38	.33	$z = 0.58, p > .05$

Discussion Section

During the introduction of the paper, you explain why your research questions were important. In the discussion section, you tell the reader whether your research answers the research questions, and you discuss the implications of your research for future researchers. The following passages come from the discussion section of the Wilker et al. (2012) article. These authors began the discussion section with

> *In this analysis of the association between temperature measures and biomarkers related to inflammation and cardiovascular function in a population of heart failure patients, we observed higher levels of BNP beginning with 2-day moving averages. (p. 1085)*

This passage complements statements made by these researchers in their introduction, when they present the purpose of their research.

Moving further into the discussion section, the scope of the discussion broadens as Wilker et al. (2012) review the implications and limitations of their research. By the end of their discussion, however, these authors have brought everything to a close, as they summarize the overall study in an efficient manner:

> *In this study of 100 patients with heart failure and systolic dysfunction, we observed significantly higher BNP and CRP, which are biomarkers related to heart failure symptoms and prognosis, with 3- to 4-day moving averages of apparent temperature. These findings suggest that changes in temperature and meteorology may alter underlying physiologic responses in this vulnerable population. (p. 1086)*

In these excerpts fromWilker et al. (2012), we have presented only a glimpse of a longer and tightly reasoned set of arguments. Nevertheless, these passages illustrate the editorial triangle illustrated in Figure 5.4. The first portion of the discussion provides a narrative review of the results as they related to the original research question. Specifically, Wilker et al. concluded that their research largely supported their hypotheses. Working through the discussion section, these researchers noted how their results were important and had potential practical applications. In presenting this information, these researchers presented arguments to convince the reader that their research and conclusions were strong and useful even beyond the bounds of the research context.

As we noted at the start of the chapter, reading many research articles will help you see how other authors construct their discussion sections as well as other

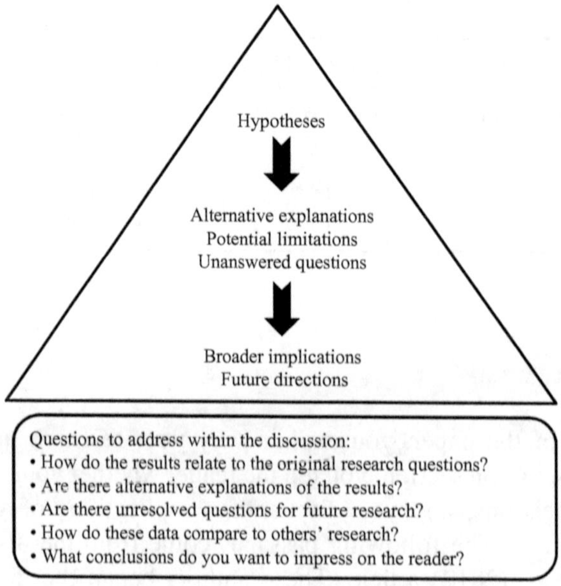

Figure 5.4 Triangle model of the discussion. *Note:* The discussion begins by focusing on how the analysis of the data relates to the original research questions. Depending on the outcome of the study, the scope of the discussion broadens to consider alternative explanations of the data, potential limitations of the current results, and lingering questions. The discussion ends by considering the reasonable conclusions that one may draw from the research.

parts of the research reports. When you read research reports, pay attention to the information the authors present in the paper. At the same time, examine how the authors present their ideas. You can learn a lot by studying the writing style of others.

Reference Section

The references section provides detailed reference citations for all work by others that you have cited within the body of your research manuscript. As mentioned already in this chapter, the specific formatting details for this section vary widely across different editorial styles, but typically, all include information regarding the author(s), publication date, source, and page numbers. You are strongly advised to follow the detailed reference reporting requirements of whatever publisher or journal you are working with to publish your findings. These guidelines are easily accessible online, usually directly through the journal or publisher website. Many researchers also find it helpful to take advantage of computer software programs that facilitate the organization and management of references. There are many such programs available and several are free of charge for basic users. These programs often automatically format references for a variety of common editorial styles, so it may be worth your while to check these out.

PROOFREADING

Few writers in any genre (especially scientific writing) can sit down and write a perfect article in one attempt. Most authors write and revise their work multiple times before submitting it for consideration by a publisher. This process of revision is often referred to as proofreading and it is most certainly an essential component of good scientific writing. Now, for the bad news and the good news: The bad news is that proofreading is difficult. It is easy to overlook mistakes in your own writing. The good news is that there are many resources you can use to proofread and correct your work.

Computer Programs

Most word processors have built-in spelling and grammar checkers. These are useful features, but they have many limitations. For example, spell-checking programs will identify any word as misspelled if it is not in the software's dictionary. This is a problem with many of the field-specific terms associated with any social science discipline. In addition, although this feature can help you to ensure all basic words are spelled correctly, it cannot determine whether you used the correct word in the correct way. For example, you could correctly spell *chose* when you should have written *choose*.

The same advantages and disadvantages apply to software-based grammar checkers. Grammar checkers catch obvious errors and often skip others. For example, my word processor's grammar checker quickly catches passive voice, subject–verb agreement problems, and other common errors. The program also ignores some real whoppers. Consider the following sentences:

The green bachelor's wife snored furiously against the winter of our discontent.
May grammars checker does no catch errors a frequently an I wants it too.

Our grammar checkers found no problem with either sentence, and indicated that they were easy to read. You, by contrast, should see immediately that both sentences are gibberish. The implications of this example are clear. Spelling and grammar checkers can help you as long as you recognize the limits of these tools and that there are superior methods for proofreading your work.

One very helpful, but underused, tactic is to read your paper aloud to yourself or someone else who is willing to listen. Read it as if you were giving a speech. In most cases, you will quickly hear the faults. Missing words will become apparent, awkward phrases will stand out, and vague language will sound obscure. As you read your paper, mark and revise any awkward phrases or words. Another tactic is to ask someone else you trust to read your paper. Most colleges and universities have a writing resource center. Typically, the staff will gladly proofread your paper for clarity and style. Although they may know nothing about your topic, they can adopt the role of your intended audience. Therefore, they will point out parts of your paper that are not clear and that require refinement.

CHAPTER SUMMARY

This chapter provided an overview of the writing process for technical/scientific research reports. We worked through basic elements of APA style, including the various components of a complete research report. Also discussed were issues of integrity in reporting of research. Finally, tips were provided to improve basic proofreading abilities prior to the finalization of any research report.

REFERENCES

EHRENBERG, A.S.C. (1977). Rudiments of numeracy. *Journal of the Royal Statistical Society A, 140*(Pt. 3), 277–297.

EISENBERGER, R. & ARMELI, S. (1997). Can salient reward increase creative performance without reducing intrinsic creative interest? *Journal of Personality and Social Psychology, 72*, 652–663.

KOHN, A. (1990). *You know that they say . . .* New York: Harper Collins.

KOHN, A. (1993). *Punished by rewards: The trouble with gold stars, incentive plans, A's, praise, and other bribes*. Boston: Houghton Mifflin.

KOSSLYN, S.M. (1994). *Elements of graph design.* New York: Freeman.

TUFTE, E.R. (1983). *The visual display of quantitative information*. Cheshire, CT: Graphics Press.

TUFTE, E.R. (1990). *Envisioning information*. Cheshire, CT: Graphics Press.

TWAIN, M. (1986/1876). *The adventures of Huckleberry Finn.* New York: Penguin.

WAINER, H. (1997). *Visual revelations: Graphical tales of fate and deception from Napoleon Bonaparte to Ross Perot.* New York: Springer-Verlag.

WILKER, E.H., YEH, G., WELLENIUS, G.A., DAVIS, R.B., PHILLIPS, R.S., & MITTLEMAN, M.A. (2012). Ambient temperature and biomarkers of heart failure: A repeated measures analysis. *Environmental Health Perspectives, 120*, 1083–1088.

Chapter 6

Reviewing the Literature

INTRODUCTION

In this chapter, we will review how to use the library and other resources to find information that can help you with your research. In addition, we will examine how researchers share information with others and how you can find information to help you design your research project. Learning these skills will help you develop ideas for your research hypotheses. The second half of the chapter will introduce you to the concept of statistical inference as well as forming and testing hypotheses.

BIBLIOGRAPHIC RESEARCH

To paraphrase an age-old riddle: *If a scientist conducts an experiment, but does not share the results, did the research really happen?* Progress in the sciences occurs because researchers share their work with others. As you learned in Chapter 1, almost all research evolves from previous research. You also learned that no research study by itself offers complete and final answers. At best, a good study will answer one or two questions. However, a good study will raise as many questions as it solves. Reading what others have discovered allows us to know how we should proceed with our research.

Understanding and Conducting Research in the Health Sciences, First Edition. Christopher J. L. Cunningham, Bart L. Weathington, and David J. Pittenger.

You can develop a research hypothesis by reading existing research papers. As you learn more about a particular topic, you may find unanswered interesting questions that require additional research. At the same time, you may find that you have a different way of looking at a problem. This chapter will help you learn where to get the best and most current information related to contemporary research. In the first section, we will consider three levels of bibliographic sources called *tertiary*, *secondary*, and *primary* sources. Each has a different role in your bibliographic research.

Tertiary Bibliographic Sources

Tertiary bibliographic sources (also called *third-level* sources) provide the most general and nontechnical review of a topic. Almost all textbooks and articles in popular newspapers or magazines are examples of tertiary sources. The essential feature of tertiary sources is their generality and casual review of a topic. Unfortunately, the authors of these sources do not have the luxury of time and space to review any one topic in detail. For example, consider an introductory biology textbook. In a book of several hundred pages, the author(s) will probably dedicate only a few pages to population ecology. However, if you take an advanced course in population ecology, you may find that the author(s) of the textbook will devote entire chapters to topics that were covered in a few sentences in the introductory book.

Authors of general textbooks must review a nearly inexhaustible amount of information quickly and for readers not particularly familiar with the topic. Therefore, the advantage of a textbook is that it gives you a quick and easy-to-understand introduction to a topic. This advantage is also its limitation. Textbook authors review the topics that are already well understood and typically do not discuss current trends in research. Although authors work hard to keep their books current and accurate, several years can pass between a new discovery and its description in a textbook.

Dictionaries and encyclopedias are also tertiary sources. As with textbooks, these sources have a specific but limited function. These sources are useful to someone who wants a quick answer for a general question and does not have the time and inclination to investigate a topic at length. You should not use textbooks, dictionaries, or encyclopedias (either in print or online) as authorities on a topic. Although they may help you understand the meaning of specific terms or phenomena, they are too general to serve as the foundation of your research. When you write a research paper, you will most likely use primary and secondary sources.

Secondary Bibliographic Sources

As you can infer from the name, **secondary bibliographic sources** stand between tertiary and primary sources. Secondary sources are more in-depth than textbooks or other tertiary sources. In general, secondary sources are comprehensive reviews written by an expert on the topic. There are many outlets for secondary sources. The two categories that we will review are books and literature reviews.

Each year, many researchers publish insightful and useful commentaries on different phenomena, often in the form of specialized books in a given field of interest. General textbooks are necessarily broad overviews. In contrast, a secondary source book may exclusively review a single topic or review the state of the art regarding current theories and trends in research. For example, consider *Medicine, Mobility, and Power in Global Africa: Transnational Health and Healing* by Dilger, Kane, and Langwick (2012). This 358-page book is dedicated to understanding the influence globalization is having on medicine, health, and the delivery of health care.

There are limitations to secondary sources. Although secondary sources offer more focused reviews of a particular topic than textbooks, they are still less focused than primary sources. In addition, there remains the time lag between the publication of primary sources and their discussion in secondary sources. Although secondary sources tend to be more up to date than textbooks, nothing is as current as a primary resource.

A literature review shares some features with focused books and other secondary sources. There are some important differences, however. First, you will find most literature reviews published in specialized professional journals. Second, literature reviews tend to be more topic specific. Like secondary sources, literature reviews provide a comprehensive summary of the current research. The author or authors summarize the relevant research reports and then describe the important factors or variables that explain the phenomenon.

There are several good sources of literature reviews. In most fields of study, you will find journals that publish mostly literature reviews. You will also find that other journals will publish a combination of original research articles and literature reviews.

An additional valuable resource can be found in the "annual review" books published by Annual Reviews (http://www.annualreviews.org). These books are focused on specific fields like microbiology, pathology, psychology, and public health. Although these reviews tend to be broader than literature reviews published in scientific journals, they are a good resource to help you learn more about a specific topic.

Primary Bibliographic Sources

Primary bibliographic sources are the original research reports that you will find published in research journals. A **research journal** is a special type of magazine that publishes scientific research articles. Many professional societies publish journals. For example, the American College of Physicians, the American Medical Association, the American Psychological Association (APA), the British Medical Association, and the Massachusetts Medical Society (to name only a few professional societies) each publish one or more scholarly journals. In other cases, corporations (e.g., Elsevier and Springer) publish scholarly journals. In all cases, the editorial board of these journals consists of prominent researchers in their respective fields. You will find that these journals list the names and academic affiliations of all editorial board members, and that the editors are well-regarded professionals working in academic or other professional institutions.

Research articles represent the greatest level of focus and detail of all the different types of bibliographic sources. The advantage of the primary source is that there is no filtering of information. For other sources, the author must condense and remove much critical information to keep the review short and focused. Consequently, you may not learn everything about a particular experiment even when you read a secondary source. Thus, only the primary source will provide a detailed account of the research methods, the data analysis, and the complete research results.

Although primary sources contain a considerable amount of information, the reading is sometimes difficult. Professional researchers use the primary resource to communicate with other researchers. Therefore, you will find the primary source filled with pages of technical language. Do not despair! Any person who can gain entry into college can read and understand original research articles. This is an important consideration because secondary and tertiary sources are someone else's interpretation of existing research. Everyone knows how the game telephone works and can screw up the final message—a similar consequence is likely if we rely only on other people's interpretation of existing theory and findings.

Peer Review

Before moving on, we want to comment on an important characteristic of research journals and professional books, **peer review**. The goal of the peer review process is to filter out bad research and to make the good research as clear and compelling as possible. Peer review is a part of the normal editing process during which experts read a manuscript to ensure that the researchers used appropriate methods to collect and analyze the data, and then made reasonable inferences from the data.

When a researcher submits a manuscript to a scientific journal, the editor will ask other researchers, who are experts in the area, to read and comment on the merits of the report. Using the reviewers' comments, the editor can reject the manuscript, require that the author revise it, or accept it for publication. If the editor accepts the manuscript for publication, it means that the reviewers and editor agreed that the study met the basic requirements for sound scientific research and that other researchers will be interested in the paper.

Although the peer review system works well, errors can and do occur. There are cases of poorly designed research or faulty conclusions finding their way into the published literature. Similarly, one editor will sometimes reject important and insightful research that another editor will publish. In this way, the peer review process involves a large subjective component that no amount of tinkering can remove. Despite its flaws, the peer review system does have a very important essential feature; it is self-correcting and it forces researchers to focus on quality in the work they present through publication outlets.

Sometimes, published articles become the center of considerable controversy. As a part of this controversy, many question whether the peer review system works. In 1998, the journal *Psychological Bulletin* published the article "A Meta-analytic Examination of Assumed Properties of Child Sexual Abuse Using College Samples" (Rind, Tromovitch, & Bauserman, 1998). Some readers, especially several politically motivated individuals, objected to the article, arguing that its authors advocated pedophilia. (The authors made no such claim. Rather, they argued that several commonly held beliefs regarding the effects of pedophilia are not accurate and required clarification.)

In response to intense political pressure, the APA requested that the American Association for the Advancement of Science (AAAS) conduct an independent review of the peer review process that the journal had used in evaluating the original article. The AAAS declined the request. In their letter to the APA, the AAAS noted that

> *We see no reason to second-guess the process of peer review used by the APA journal in its decision to publish the article in question. While not without its imperfections, peer review is well established as a standard mechanism for maintaining the flow of scientific information that scientists can refer to, critique, or build on [italics added]. . . . Uncovering [a problem with a manuscript is] the task of those reviewing it prior to publication [and] to the reader of the published article [italics added] . . . (Lerch, 1999)*

The AAAS offered an important message in their response. Science depends on an open and free exchange of ideas. The peer review process should be a filter for faulty and flawed work but should not become a barrier that censors challenging or unpopular ideas. Furthermore, the reader has the ultimate responsibility to critically evaluate any and all information they gather and read, regardless of the source of that information. You, as the reader, must assume the responsibility to analyze and critique the ideas presented in a published article.

Because of this built-in critical evaluation process, errors in the published literature tend not to last for long. If an author publishes a paper with an error, other researchers will be quick to note the mistake. In most cases, when researchers disagree with the findings printed in an article, they will conduct new research to support an alternative perspective or interpretation. The researcher will then put the new results into the publication cycle. Therefore, bad ideas, poor research, and faulty inferences do not last long in the scientific community. Researchers quickly ignore and forget findings and conclusions that cannot withstand the scrutiny of systematic and direct replication. In the same vein, novel and unique results that researchers can replicate receive considerable attention in the research community.

THE INTERNET AS A SOURCE

> *Wikipedia is not a primary source!*
>
> *—Bart L. Weathington, Christopher J.L. Cunningham, and David J. Pittenger (and many other professors and editors)*

Although many people find the Internet a fascinating and interesting resource, it has several serious limitations. First, there is no peer review for the vast majority of Internet web pages. There are currently few, if any, restrictions on what one may post on a web page. In many ways, the Internet is the modern soapbox. Any person with Internet access can publish his or her ideas on the Internet.

A second problem with the Internet is that it is an ever-changing medium with no permanent record. Books, journals, and other print media are tangible, permanent products. There is no exact equivalent for web pages. An attempt has been made to address this issue with the use of **digital object identifiers** (DOIs). DOIs are character strings used to uniquely identify an object such as an electronic document. Metadata about the object is stored in association with the DOI name and this metadata may include a location, such as a URL, where the object can be found. The DOI for a document is permanent, even though the location of the document and other details about the source may change over time. Most guidelines for citing material from the Internet require both the address of the website and the date the material was downloaded. Many now also indicate that, when available, a DOI should be cited.

Although the Internet is a wonderful resource, you should use it with care and a strong dose of skepticism. You cannot trust everything you read on the Internet. In fact, you should be careful with everything you read, regardless of its source. There are some important differences, however, between what you will find on the Internet and in professional journals. This raises the question, "How should we use the Internet?"

Click, Check, and Double-Check

If you use the Internet, you should be aware of some core principles of smart Internet research. First, search engines and subject directories are not complete and exhaustive reviews of all existing research in a particular area, nor even all current web pages (Brandt, 1996, 1997). Therefore, when you are searching the Internet, you should recognize that any set of search results will be incomplete and that different search engines will produce different results. Second, you should be cautious about the material you find on the Web. As with any research project, you should not rely on one source of information. When you read something, from the Internet or some other source, you need to determine whether the information is credible. Here are a few tips you should follow as you examine a web page:

- **Check the Credentials of the Author:** Be skeptical of web pages that have no clearly identified author or if the author does nothing to indicate his or her credentials or professional affiliation.
- **Check How Well the Author Annotates the Web Page:** If the web page contains lots of claims and conclusions but no supporting information, be suspicious. Serious authors include a bibliography with their text. The

bibliography allows the reader to double-check the original sources to ensure that the author has presented the information accurately. You should use this practice for any information you use, Internet or print.

- **When in Doubt, Doubt:** There is an old saying about free advice, you get what you pay for. You should not rely on a web page as your primary source of information. Find corroborating evidence from other sources before you make an assertion and run the risk of embarrassing yourself with incorrect information. Related to this principle, consider the content author's reason for publishing the information you are reading. Many Internet authors are writing with a very strong bias, and as a researcher, you need to make sure that what you are about to cite as substance really represents the whole picture (and not just one person's strong opinion, no matter how well written or presented).
- **Print a Copy of the Web Page:** If you plan to use information from a web page, it is a good practice to print a copy of the web page. This tactic will allow you to have a permanent record of the information in case the author revises or removes the material from the Internet or if your reader wants to read the reference.

DEVELOPING A SEARCH STRATEGY

A common observation is that there are more scientists living now than at any time in history. Along with the Internet and increased globalization, this means there is more research being published now than at any point in history. This growth trend is expected to continue. The future of science is also cross disciplinary. People are complicated and it is becoming increasingly recognized that cross-disciplinary collaboration is needed to address many research questions within all domains of human involvement. This means there is an incredible amount of information available you will need to develop a strategy for finding articles that interest you. The following subsections provide our basic suggestions to anyone who is just starting out on a research endeavor.

Find a Topic That Interests You

What topics do you find fascinating? Can you remember reading about an experiment in one of your courses and thinking that it was interesting? If you are looking for a research topic, it is a good idea to begin with your interests. Sometimes, when beginning a research project, it is useful to sit down as a group and to talk about the topic areas group members find interesting. As you and your colleagues talk, you may find that you have common questions about why people behave as they do. Another way to find interesting topics is to read textbooks. Chances are that you will find one or more interesting topics for your research. Keep in mind that a decent research project will take a large amount of time and effort—it is much easier to stay invested in research when you are seriously interested in the topic that you are studying.

Read Tertiary and Secondary Sources

Once you find a topic that you find interesting, keep reading. Begin with tertiary sources; they are the easiest to read and offer the broadest coverage. Assume, for example, that you want to learn more about altruistic behavior. As you read the general sources, you may find that altruism is a broad topic and that it includes many subcomponents. While you are reading, take note of several things. Specifically, look for specialized terms, researchers' names, notable experiments, important independent and dependent variables, and recommendations for new research. This information will help you later when you begin to search the primary literature. Also keep track of any and all terms, keywords, or strings that you use when searching for articles in any database—this will allow you to build on and/or recreate a search in the future if necessary.

Use Specialized Terms

Specialized words or phrases are important because they often describe specific aspects of research. If you read a chapter on altruistic behavior, you are likely to read terms such as *diffusion of responsibility*, *empathic concern*, *negative-state relief*, and *norm reciprocity*. Any one of these concepts can serve as a focal point for research on altruism. Consider *empathic concern* as an example. The phrase refers to feelings of pity for another person in distress. We can ask many empirical questions about this concept:

- Are there ways to increase and decrease the level of empathic concern that people experience, and how do these changes affect helping behavior?
- Are people more likely to feel empathic concern for people like themselves?
- Are people more likely to feel empathic concern if they feel good about themselves or are happy?

Use Researchers' Names

When searching for information on a topic, why not look for people who seem to be doing the interesting research? For instance, if reading about altruism, you might learn that Batson is a leading researcher on helping behavior. If interested in the related topic of helping behaviors, you might come across researchers like Cialdini and Dovidio, among other researchers who have conducted innovative studies on helping behavior. Therefore, we will want to see whether these people have done additional research on the topic.

Review Notable Experiments

You will often find reference to one or more classic experiments in tertiary reviews. It is a good idea to read these articles as they provide a historical context and starting point for any new research. If you wanted to study helping behavior, for example, you might want to read Darley and Latané's (1968) research on the diffusion of responsibility phenomenon. You may also want to find a literature review on the topic. For example, Latané and Nida (1981) reviewed the literature on the diffusion of responsibility research and found that many researchers have verified Darley and Latané's conclusion. This background reading may help you understand the purpose of more recent research.

Talk to Your Instructors

Another good source of information is your professors; talk to them about your research interests. Faculty typically enjoy talking with others about research, especially when there is excitement about a particular topic and a passion for learning more about a particular issue. As you talk about your interests, your professor may be able to give you invaluable guidance by recommending additional readings or by helping you refine your research questions.

SEARCHING THE LITERATURE: THE LIBRARY

Many people have the misconception that a library is simply a place that stores books and that may be a convenient exam-preparation spot. This is a gross oversimplification. A library is a place to find information and to learn. Furthermore, the size of a library's collection does not matter; access to information is the essential service of any contemporary library. Even if the physical space of a library is small, via the Internet, interlibrary loan, and other library resources, it is typically possible in time to obtain any piece of published information that you desire. All you need to do is learn how to ask the right questions. Thus, the focus of this section is to help you understand how to get the most out of your college or university's library.

Ask the Library Staff

It has been our experience that students typically overlook one of the greatest resources in the library, the professional staff. All libraries use complex storage and retrieval systems that allow librarians to store and find information efficiently. Most librarians have a graduate degree in library science, which means that they know a lot about how to find information. Many librarians also have an additional graduate degree in a specialized academic area. At a large university, there may be a reference librarian for each general subject matter.

Utilize Search Tools

Most libraries now offer many computer tools to conduct professional searches. The chances are that your college subscribes to several specialized search tools such as PubMed™, Medscape™, ProQuest©, PsycINFO®, MDLinx®, and others. These specialized search tools allow you to find specific research articles quickly.

Search the Online Catalog

All libraries keep their bibliographic material in a well-organized system that you can use to find information quickly. Specifically, most colleges use the Library of Congress classification scheme that uses a combination of letters and numbers to classify each book, document, and other materials such as films and records. This complete set of numbers and letters is the **call number**. Each book, journal, recording, and other reference in the library has a unique call number.

Because the health sciences are broad and interdisciplinary subjects, you will find many related books spread throughout the library. For instance, librarians place books on abnormal psychology, or psychopathology, in the internal medicine section (RC) and books on animal behavior in the zoology section (QL). Therefore, you can expect to roam throughout the entire library as you search for bits of information.

Most colleges allow you to search for books using an online catalog. There are many different catalog systems, too many to describe in detail. The common feature of these systems is that you can search by the author's name, the title of the book, the topic, or a keyword. Most students find that these systems are easy to use and provide much useful information. For example, most systems will indicate whether the book you want is on the shelf or has been checked out. In some cases, you can use the program to browse through the titles on the same shelf. This is a useful feature. Because the librarians store the books in logical categories, you might find surrounding books that are interesting and related to your research topic.

Leverage Key Databases

There are many other databases to which your library may have access. Additionally, any computer with a connection to the Internet will have access to the broad database, Google Scholar™. Each of these databases differs slightly in terms of its content, format, and depth. However, they all have similarities, and learning to use one database will help you with using all of them. For example, two of the most common and popular of these databases for behavioral and social science research that is relevant to the health sciences are PubMed and PsycINFO.

Research databases are comprehensive and contain references to both books and topic relevant journals. Using these systems, you can find research articles and literature reviews on specific topics. Each college, university, or hospital has a slightly different way of gaining access to research databases. The following information will apply regardless of the system you use. Ask the reference librarian to show you how to use the resource if you have questions.

Most systems allow you to search the database using commonly used words or keywords. A **keyword** can be any term or phrase you want to use for the search. It can be a commonly used word or a technical term used by psychologists to describe a specific phenomenon. You can also use **Boolean operators** to condition your search. The two primary Boolean operators that you will use are **AND** and **OR**. For the sake of an example, assume that you are interested in learning more about integrity testing and employee selection. You can use the terms *employee selection*, *integrity*, and *honesty* as keywords.

The AND operator causes the search process to collect references that contain two or more keywords. Therefore, the AND operator is a fast way to scale down the number of references in a search. In some cases, it is useful to increase the breadth of the search and to examine more rather than fewer topics. In this case, the OR operator is useful. The OR operator causes the search process to collect references that contain any combination of the keywords. To summarize, AND *decreases* the number of articles included in the search, whereas OR *increases* the number of articles.

Try a Thesaurus

You may be wondering what keywords you should use. There are two answers to this question. First, if you followed our previous recommendations, you will read tertiary and secondary texts and note terms that the authors used to describe the phenomenon you want to study. The second answer is to use a thesaurus. For example, all of the keywords used to classify articles in PsycINFO are listed in the *Thesaurus of Psychological Index Terms* (American Psychological Association, 2007). If you are having trouble conducting your search, such as getting too many unrelated references or not enough relevant references, you should use the thesaurus relevant to the search engine you are using to find and select alternative keywords. Using the thesaurus will help you to select the most representative articles. Each entry of the thesaurus has a brief definition of the term and a series of related terms.

Pay Attention to All Fields

Each entry in the PsycINFO database consists of a number of search fields. A field represents a specific type of information. There are fields for the title of the article, the author's name, and the journal that published the article. You can use several of these fields to your advantage to help make your search more productive. Consider, for example, the language field. At first glance, this field seems rather trivial until you realize that there are many foreign-language journals. When you conduct a search on a popular topic, it is common to find articles written in Russian, French, German, or Japanese. Therefore, unless you can read these languages, you should filter them out of your search. The easiest way to do this is to use the Advanced Search option and select English in the language field. You can use the same logic and other search strategies to prune your search.

Consider Practical Matters

Conducting an effective search of the literature is as much an exercise in art as it is a scientific process. Much of the success in using this valuable tool comes from planning, practice, and patience. Searching through a database is like panning for gold. In some cases, you might have to use many filtering techniques to find the right set of articles. You should not expect to find five great articles in five minutes. Rather, plan to spend quality time at the computer searching for the articles you need. It is important to note that the first articles you find may not be the ones that ones that would be most central to your developing research idea; keep looking until you stop seeing new material and new perspectives on the topic at hand (i.e., until you reach the point of saturation).

Thanks to the online availability of most journal articles and even some books, it is possible for researchers to quickly amass a large electronic library of materials. These collections of relevant information can be easily transported or managed using a thumb/flash drive or a free Internet-based e-mail account. We urge you to make use of the technology that is available to you and to become comfortable with your various options for having the necessary research literature at your fingertips whenever you need it. Just be sure you abide by copyright laws when printing or sharing files.

Selecting Articles

Once you have a list of potential articles, you need to select the articles that you will want to read in detail. The best way to make this judgment is to carefully read the abstract for the article. All articles published in scholarly journals have an abstract that databases allow you to read. In most cases, the abstract is a short paragraph that describes the purpose or thesis of the article, a brief review of the author's method, and a synopsis of the major findings. Reading the abstract should allow you to determine whether you want to invest the time and energy in reading the entire article.

Reading the abstract provides a fair review of the content of the article. Depending on your interests, you may be willing to read the entire article or move on to the next abstract. If you want to find similar articles, you can use several tricks. Look at the key concepts and keywords fields. These contain words and phrases that characterize this article and that may help you search for similar articles. For example, the phrases "emotional intelligence" and "employee selection" combined with "validity" may focus your search on those articles that examine the effectiveness of tests of emotional intelligence for selecting employees. As useful as an article's abstract is, however, please realize that an abstract does not convey all the details within an article, nor does it necessarily provide you with sufficient information to judge the quality of the actual study or the generalizability of the study's findings—this information can only be found within the article itself.

RESEARCH IN ACTION: DOES LISTENING TO MOZART MAKE YOU SMARTER?

In this section, we will review an experiment conducted by Steele, Ball, and Runk (1997) as an example of how to read a journal article. Steele and his colleagues examined what has now become known as the *Mozart effect*. In essence, some people believe that listening to Mozart's music can help improve one's IQ. Steele et al. conducted an experiment to determine whether they could replicate the effect. We will begin with the introduction of the article.

Begin by reading the text in Box 6.1. The text is from the introduction of the Steele et al. (1997) article.

KNOWLEDGE CHECK

1. What is the problem being studied?
2. What is the authors' hypothesis?
3. How was the hypothesis developed?
4. How does this study relate to the problem?
5. What type of research design did Steele et al. use?

Steele et al. (1997) provided a short and clear rationale for their study. They began by reviewing the previous research that has examined the Mozart effect and found that the results are inconsistent. In their review, they noted that several research teams did not replicate the Mozart effect when using different dependent variables. As a part of their review, Steele et al. focused on Rauscher et al.'s (1995) hypothesis that the Mozart effect works when the task incorporates "spatial and temporal transformations" (p. 1180). The focus of this hypothesis provides the foundation for the study. In the last paragraph of the introduction, Steele et al. made clear that they conducted the experiment to determine whether they could replicate the Mozart effect following Rauscher et al.'s (1993) procedures.

The next section (Box 6.2) is the method section from Steele et al. (1997) article, in which the authors described the people who participated in the study, the materials they used, and the systematic procedures they followed.

BOX 6.1 *Introduction from the Steele et al. (1997) Experiment*

Rauscher, Shaw, and Ky (1993) reported that 36 undergraduates increased their mean spatial reasoning scores the equivalent of 8 or 9 IQ points on portions of the Stanford-Binet Intelligence Scale: Fourth Edition (Thorndike, Hagen, & Sattler, 1986) after listening to 10 min. of Mozart's Sonata for Two Pianos in D Major, K448 (hereafter labeled the "Mozart effect"). The Mozart effect was temporary, having disappeared within 10 to 15 minutes. Rauscher, Shaw, Levine, Ky, and Wright (1994) reported further that short periods of music education in school produced both a temporary effect, immediately after music training, and a permanent increase over a school year in performance by preschool children on the Object Assembly portion of the Wechsler Preschool and Primary Scale of Intelligence–Revised (Wechsler, 1989).

The hypothesis that musical experiences of short duration can have powerful effects on IQ scores on both a short-term and long-term basis is important for both practical and theoretical reasons. However, attempts to replicate the original report by Rauscher et al. (1993) have been unsuccessful. Kenealy and Monsef (1994) were unable to produce a Mozart effect on performance using portions of the Stanford-Binet test, the Paper Folding and Cutting task, and the Matrices task. Studies by Newman, Rosenbach, Burns, Latimer, Matocha, and Vogt (1995) and by Stough, Kerkin, Bates, and Mangan (1994) did not yield a Mozart effect when items from the Raven's Progressive Matrices (Raven, 1986) were the dependent measure. Carstens, Huskins, and Hounshell (1995) reported no Mozart effect when the dependent measure was the Revised Minnesota Paper Form Board Test, Form AA (Likert & Quasha, 1948).

Rauscher, Shaw, and Ky (1995) have reported a replication of the Mozart effect, using elaborations of the Stanford-Binet Paper Folding and Cutting task as the dependent measure. Further they specified that an appropriate task was one that involved not just spatial recognition but that it should incorporate spatial and temporal transformations. This observation was the basis for the dependent measure used here, a backwards digit span task. A backwards digit span task requires that a person listen to a string of digits and then reproduce them in reverse sequence. Theoretically, the backwards digit task is of interest as a spatial reasoning task because it requires rotation or transformation of the sequence (Carroll, 1993; Das, Kirby, & Jarman, 1979). Empirically, performance scores on this task correlate strongly with scores on memory for designs (Schofield & Ashman, 1986), performance with Raven's Progressive Matrices (Banken, 1985), and is a good predictor of performance with the Rod and Frame task (Haller, 1981). Right-hemisphere dysfunction reduces backwards digit span performance while left-hemisphere dysfunction reduces forward digit span performance (Rapport, Webster, & Dutra, 1994; Rudel & Denckla, 1974), although this difference does not occur in all types of disorders (Gupta, Mahto, Tandon, & Singh, 1986).

KNOWLEDGE CHECK

6. What are the independent and dependent variables?

7. Why do you think that Steele et al. used three conditions?

8. What controls did Steele et al. use to control for confounding variables?

9. Why did Steele et al. use the backwards digit span task?

The next section of the paper is the results section, wherein Steele et al. (1997) offer a narrative account of their analysis of the data. Note how they combine the use of descriptive text, tables of descriptive statistics, and inferential statistics to describe their data (see Box 6.3).

BOX 6.2 *Method Section for the Steele et al. (1997) Experiment*

Participants

Thirty-six Euro-American upper-division university students (28 women and 8 men) from two sections of a psychology course volunteered and received course credit for their participation.

Apparatus

Two stimulus tapes of approximately 10 min. duration were created. One contained the Mozart Sonata for Two Pianos in D Major (K448) and the other contained the sound of a gentle rainstorm ("Spring Showers") from an environmental sounds recording. Sequences of digits were recorded on separate tapes for the digit span task. Tapes were played on a good quality portable system.

Procedure

The experiment took place in a room reserved for that purpose. The participant was told that the experiment concerned the effect of relaxation on recall and was instructed to sit in a large, comfortable, recliner chair. The chair faced away from the experimenter who operated the tape player that had been placed on a table by the left arm of the recliner chair.

Each participant listened in turn to the Mozart tape, the rainstorm tape, or sat quietly following the verbal instruction "to relax." The order of stimulus conditions was counterbalanced across participants using a Latin square design. Following exposure to a stimulus condition, each participant listened to three nine-digit sequences. Digits were presented on the tape at the rate of one every 2 sec. After each nine-digit sequence, the participant attempted to repeat that sequence in reverse order. The score recorded was the sum of number of correct scores across the three sequences, the maximum score being 27. Each participant heard nine sequences of digits across the experimental session, three per stimulus condition. Digit sequences were created by a random-number generator and no sequence was repeated in a session to a participant. Three different units of digit sequences were created and assigned in a balanced fashion across participants.

The number of digits correctly recalled in reverse order was recorded for each subject for each condition. A correct recall was defined as the correct digit in the correct serial location. For example, if the original sequence was 7–5–3–1–9 and the recalled sequence was 9–1–3–4–7, then the score would be 4 correct. The Rauscher et al. prediction is that the number of digits correctly reversed in recall should be enhanced in the Mozart condition relative to both the silence and the rainstorm condition.

KNOWLEDGE CHECK

10. Did Steele et al. find evidence of the Mozart effect?

11. Why did Steele et al. also examine the "order of task" and "first stimulus" effect?

The final section of Steele et al.'s (1997) paper is the discussion section, presented in Box 6.4 for you to review how these authors explained the implications of their research.

KNOWLEDGE CHECK

12. How do the results of this experiment relate to previous research?

13. Steele et al. noted that the Rideout and Laubach (1996) study did not incorporate a silence control group. What is the relevance of that observation?

BOX 6.3 *Results Section for Steele et al. (1997) Experiment*

Table 6.1 shows three descriptive measures of mean recall on the backwards digit span task. The headings under "Stimulus Condition" show mean performance as a function of the type of stimulus that immediately preceded the recall task. There was no difference overall in mean recall as a function of the preceding stimulus condition, $F(2, 70) = 0.03$, $p = .97$. The outcomes of specific inferential contrasts were consistent with this observation, Music versus Rain, $t(35) = 0.03$, $p = .98$ and Music versus Silence $t(35) = 0.21$, $p = .83$.

The lack of differences in performance among stimulus conditions was not due to unsystematic variability. For example, a clear practice effect overall was observed $F(2, 70) = 21.92$, $p < .001$. Although serial position was completely counterbalanced in stimulus presentation, we calculated performance as a function of serial position. The headings under "Order of Task" in Table 6.1 give mean recall as a function of the serial position of the stimulus condition. The data indicate that mean recall was improved by additional experience in the task. This observation is confirmed by inferential tests, First versus Second $t(35) = 4.24$, $p < .001$ and Second versus Third $t(35) = 2.41$, $p = .02$.

All three stimulus conditions were administered in a single session as was done by Rauscher et al. (1993). Although the effect of music is supposed to be short-lived, it is possible that there was some carryover effect of music onto the other stimulus conditions or the reverse. Therefore we compared performances after the first stimulus condition only, when there would be no such effects. The headings under "First Stimulus" in Table 6.1 indicate recall following a stimulus condition presented first in the session. Overall there was no significant difference among treatments $F(2, 22) = 1.26$, $p = .30$. The mean recall after music is not different from that after silence, $t(11) = 0.38$, $p = .71$. Although mean recall after the rainstorm condition was lower than after music, the difference was not statistically significant, $t(11) = 1.26$, $p = .23$.

Table 6.1 Mean Scores on Backwards Digit Span

Condition	*M*	*SD*	*n*
Stimulus condition			
Music	18.53	4.14	36
Rain	18.5	6.07	36
Silence	18.72	5.09	36
Order of task			
First	15.64	4.7	36
Second	19.14	4.87	36
Third	20.97	4.29	36
First stimulus			
Music	16.67	2.77	12
Rain	14.17	5.7	12
Silence	16.08	5.13	12

Note: Maximum score = 27. *n* is the number of scores in comparison.

BOX 6.4 ***Discussion Section for Steele et al. (1997) Experiment***

Exposure for 10 minutes to a recording of the Mozart Sonata for Two Pianos in D Major (K448) was not followed by an enhancement in performance on a backwards digit span task, a task chosen because it required a temporally extended quasispatial solution as did the paper folding and cutting task. The lack of effect here is inconsistent with the findings of Rauscher et al. (1993, 1995) but is consistent with reports from other laboratories (Carstens et al., 1995; Kenealy & Monsef, 1994; Newman et al., 1995; Stough et al., 1994). This difference is made more puzzling by the observation that Rauscher et al. have reported large effects in their studies while both Newman et al. (1995) and Stough et al. (1994) conclude confidently that there was no Mozart effect in their experiments. One explanation for the failure of this and other experiments to obtain a Mozart effect could be related to the use of different dependent measures. But different measures cannot be the entire explanation because Kenealy and Monsef (1994) did not obtain a Mozart effect even though they used a paper folding and cutting task as did Rauscher et al. Kenealy and Monsef (1994) used silence as their control condition. Rideout and Laubach (1996) reported recently a positive effect with a paper folding and cutting task but they compared exposure to a Mozart sequence against exposure to a progressive relaxation tape only. The lack of a silence-only control condition means that one cannot state whether listening to Mozart improved performance or listening to the progressive relaxation tape reduced performance. Rauscher et al. (1993) reported a Mozart effect relative to both silence and a relaxation-tape control condition.

There seems to be some important methodological difference between Rauscher et al.'s work and that of other experimenters that has not yet been elucidated. The nature of this difference constitutes a puzzle since the experimental design seems straightforward. Rauscher et al. (1994) emphasized the potential beneficial effects of increases in time and money allocated to music education in the grade-school curriculum. These practical considerations add to the importance of the solution of this scientific puzzle.

14. Have Steele et al. resolved the inconsistencies in the Mozart effect?

15. Why do you think Steele et al. commented on the requests to allocate more money to grade-school music education?

STATISTICAL INFERENCE AND HYPOTHESIS TESTING

Once you have reviewed the existing literature and narrowed your focus to a specific research question, it is important to consider how this information will be used. Simply gathering articles is not enough. An integrative review of the literature is needed to form the foundation of a quality research project. In Chapter 3, we talked about the basic types of research hypotheses. Now that you have done a literature review, you can decide on the specific hypotheses you wish to evaluate and begin to consider the appropriate statistical techniques that will be necessary in order to test your hypotheses.

The importance of carefully establishing your hypotheses cannot be overstated. Once you have read a sufficient number of articles and other sources related to your topic of interest, you will begin to feel comfortable with the general subject. At this point, you will need to begin to formulate your general research question, shifting from its most general statement to a more specific question. This may require you to identify additional sources to help you specify the variables that you will study and the nature of the relationship that you expect to observe between those variables. Eventually, you will need to further boil down this specific research question into the form of a testable hypothesis, which is a mathematical representation of an expected relationship between two or more variables.

Statisticians and researchers invented inferential statistics during the early part of the twentieth century (Oakes, 1986). Over time, general rules and principles for using inferential statistics became common practice. For example, most researchers now routinely include a null and alternative hypothesis as a part of their statistical test. In this section, we will examine the specific steps involved in hypothesis testing.

One important feature of the standard approach to research (null hypothesis significance testing) is that we form the hypotheses *before* we collect or examine the data. All of us can describe the past with accuracy; predicting the future, however, requires good information and some level of talent. This being the case, we are more impressed by and have much greater confidence in a prediction of the future that comes true, rather than an observation of something that has already occurred. As you will see in the following steps, much of the work associated with hypothesis testing comes before we collect or analyze the data.

State the Null and Alternative Hypotheses

The null hypothesis, H_0, is the mathematical statement that we use as a comparative standard when we conduct a research study. More specifically, the typical null hypothesis states that the independent variable (what we manipulate or are studying as a predictor) is not related to, or has no effect on, the dependent variable (the outcome we are observing). If we find evidence through our research that there is in fact some relationship between these variables (or a difference between two or more groups in terms of the dependent variable), then we reject H_0 and accept instead the alternative hypothesis, H_1, which is typically the focus of the study. If our evidence is not strong enough to reject H_0, then we retain this statement of no relationship between variables. Typically, the alternative hypothesis is the motivating rationale behind the study—why would you conduct a research study if you expected to find no effect or difference?

It is important to note that although it is true that the no-relation/difference hypothesis is integrally a part of the way social sciences test for relationships among variables, it is sometimes not accurate, especially outside of the experimental setting. This error in understanding has led to a great deal of controversy in the research methods literature over the past few decades regarding the future of null hypothesis testing. For a good discussion of these issues, the interested reader is referred to Kline (2004).

Despite this ongoing debate, this approach to research is still the norm in most research involving social and behavioral issues. Within this paradigm, the null hypothesis is the hypothesis to be nullified or disproved. As conceived by Fisher, it is the hypothesis that represents our current understanding of the world. If we are genuinely ignorant of any relation between two variables, then the no-relation hypothesis (hypothesis of ignorance) is correct. However, in most cases, the no-relation hypothesis is demonstrably false based on existing research, and rejecting it does nothing to advance knowledge.

The implication here is that the null hypothesis can be correct only if there is no evidence to suggest that there is or should be a relationship between two or more variables. In most cases, researchers base their research on the expectation of a relationship between variables, and as such, they are asserting that they expect there to be a relationship. This being the case, most researchers in the social and behavioral sciences engage in research to collect data that they expect will support their alternative hypotheses. If these data do not reflect the expected relationship, then these researchers will revert to the null hypothesis as the next best explanation for their lack of identifying a relationship between the study variables. This is why research is theory driven. Simply disproving the null hypothesis does not mean the alternative hypothesis is true. It simply indicates there is a relation or difference among variables or groups, and if a thorough literature review has been conducted, then the alternative hypothesis is likely to be our next best explanation for what we have observed.

The null hypothesis can be either nondirectional or directional. The nondirectional hypothesis for a two-group comparison study is that H_0: $\mu_1 = \mu_2$. This version of the hypothesis states that the two population means are identical and that any observed difference between the group means is due to random events unrelated to the independent variable. By contrast, the directional null hypothesis can be either H_0: $\mu_1 > \mu_2$ or H_0: $\mu_1 < \mu_2$. The directional hypothesis predicts that one population mean is less than the other mean with the exception of the effects of random sampling error. The alternative hypothesis is always the mathematical complement of the null hypothesis. The complementary relational signs are $=$ and $\neq$, $\geq$ and $<$, and $\leq$ and $>$. As a rule, each null hypothesis has only one alternative hypothesis.

Identify the Appropriate Statistical Test

This step may seem obvious but is often the downfall of many students and researchers. Part of the problem is that technically, it is possible to apply almost any statistical test to any set of data. One question you could ask is "Can I use this statistic to analyze these data?" The answer will nearly always be "Yes," because you can use many statistical tests to analyze the same set of data. A better question is "Will this statistical test allow me to answer the question I posed in my hypothesis?"

Different statistical tests provide different perspectives on the data you have collected. Some allow you to compare groups, whereas others examine the correlation between groups. Therefore, your null and alternative hypotheses will determine the type of test you use. Other factors also influence the selection of the appropriate statistical test. Many inferential statistics require that the data meet specific criteria to produce valid results. For example, the *t*-ratio for independent groups (more on this in Chapter 11) assumes that the observations for the two groups are independent, that the data are normally distributed, and that the variances of the two groups are equivalent. Failure to meet these criteria can produce spurious results.

As you plan your research, you should identify the appropriate statistical tests for analyzing the data and identify the requirements for using that test. In addition, you should identify alternative strategies for analyzing the data in the case that your data do not conform to the requirements of a specific test. There are many ways to analyze the data from a research project.

CHAPTER SUMMARY

The advance of any science depends on an exchange of information. Medical researchers, like other scientists, share the results of their research in professional research journals. Because human behavior and social interaction are popular and complex topics, there is a monumental amount of information about various behavioral and health-related topics. Libraries are, therefore, important resources to help you find what researchers have discovered about phenomena.

There are three classes of bibliographic source: tertiary, secondary, and primary. Tertiary sources are the most general resource but are a useful introduction to different topics. Secondary resources are more specific and offer a more detailed review of the research that has been conducted on a specific topic. Primary resources are the original articles written by researchers.

An important component of professional research journals is peer review. Peer review means that fellow professionals review a researcher's manuscript to determine whether it is worthy of publication. In other words, peer review is a form of quality control to ensure that a journal publishes only good-quality research.

The Internet is a fascinating resource but is still in its infancy. There are few checks and balances for reviewing information posted on the Internet. Consequently, you must be cautious of information that you gather from the Internet.

We reviewed many ways that you can develop your interest and develop a research hypothesis. One of the more important resources is reading the research that has already been conducted. This background work will help you understand what is known about a specific topic and what issues remain a mystery.

There are many resources you can use to locate useful information. You can search the database using a search engine. These search engines use Boolean operations that allow you to find specific topics. The AND operator causes the search engine to select articles where two or more specific keywords describe the research. The OR operator allows you to conduct a broader search of related topics.

When you find an interesting research article, you will find that it is divided into several sections. The introduction offers a brief history of the research on the topic and a review of the researcher's hypotheses. In the method section, the researcher describes the techniques used to collect the data. The results section provides a detailed account of the results the researcher obtained. Finally, the discussion section includes the researcher's commentary of the meaning of the data with regard to his or her hypotheses.

In the "Research in Action" section, we reviewed a published article to illustrate how to read a research article.

The chapter ended with a discussion on forming hypotheses and the importance of selecting the appropriate statistical technique(s).

CHAPTER GLOSSARY FOR REVIEW

AND A Boolean operator that requires that two or more keywords be present in the citation.

Boolean Operators Logical terms that control the search process for keywords.

Call Number A combination of letters and numbers used in a library classification system to store and locate recorded information.

Digital Object Identifier Unique character string associated with electronically retrieved documents designed to provide a form of persistent identification.

Keyword A word or phrase used to search PsycINFO or other databases.

OR A Boolean operator that requires that either keyword be present in the citation.

Peer Review Part of the editorial process for professional books and journals in which professionals in the area review a manuscript for its scientific merits.

Primary Bibliographic Sources A class of text resource that includes original research reports.

Research Journal A periodic publication that contains original research articles, summaries, and commentaries. The editorial board consists of fellow researchers who have expertise on the primary topic of the journal.

Secondary Bibliographic Sources A class of text resources that includes specialized books and articles and offers a comprehensive review of the literature for research on a specific topic.

Tertiary Bibliographic Sources A class of text resource that includes textbooks and general reviews of a topic.

REFERENCES

American Psychological Association (2007). *Thesaurus of psychological index terms* (11th ed.). Washington, DC: Author.

Brandt, D.S. (1996). Evaluating information on the Internet. *Computers in Libraries*, *16*, 44–47.

Brandt, D.S. (1997). What flavor is your Internet search engine? *Computers in Libraries*, *17*, 47–50.

Darley, J.M. & Latané, B. (1968). Bystander intervention in emergencies: Diffusion of responsibility. *Journal of Personality and Social Psychology*, *10*, 202–214.

Dilger, H., Kane, A., & Langwick, S.A. (2012). *Medicine, mobility, and power in Global Africa: Transnational health and healing*. Bloomington, IN: Indiana University Press.

Kline, R.B. (2004). *Beyond significance testing: Reforming data analysis methods in behavioral research.* Washington, DC: American Psychological Association.

Latané, B. & Nida, S.A. (1981). Ten years of research on group size and helping. *Psychological Bulletin*, *89*, 308–324.

Lerch, I. (1999, November). Reprint of letter to Dr. Richard McCarty. *Psychological Science Agenda*, *12*, 2–3.

Oakes, M. (1986). *Statistical inference: A commentary for the social and behavioral sciences.* New York: John Wiley & Sons.

Rauscher, F.H., Shaw, G.L., & Ky, K.N. (1993). Music and spatial task performance. *Nature*, *365*, 611.

Rauscher, F.H., Shaw, G.L., & Ky, K.N. (1995). Listening to Mozart enhances spatial-temporal reasoning: Towards a neurophysiological basis. *Neuroscience Letters*, *185*, 44–47.

RIDEOUT, B.E. & LAUBACH, C.M. (1996). EEE correlates of enhanced spatial performance following exposure to music. *Perceptual & Motor Skills, 82*, 427–432.

RIND, B., TROMOVITCH, P., & BAUSERMAN, R. (1998). A meta-analytic examination of assumed properties of child sexual abuse using college samples. *Psychological Bulletin, 124*, 22–53.

STEELE, K.M., BALL, T.N., & RUNK, R. (1997). Listening to Mozart does not enhance backwards digit span performance. *Perceptual & Motor Skills, 84*, 1179–1184.

Chapter 7

Sampling

It is a capital mistake to theorize before one has data.
—Sir Arthur Conan Doyle

INTRODUCTION

Much of what we know we learn from studying samples of larger "populations." Consider some common examples. When someone asks you whether the lunches at a restaurant are good, you use your experiences to answer the question. The meals you have eaten at that restaurant represent a sample of all the possible meals that you and others could have eaten at that restaurant. The establishment may produce thousands of meals and has served hundreds of customers since its opening. You have no way of being sure that each meal was prepared well and enjoyed. Yet, based on your experience with the restaurant, you make a single and confident conclusion, "The food here is good."

Understanding and Conducting Research in the Health Sciences, First Edition. Christopher J. L. Cunningham, Bart L. Weathington, and David J. Pittenger.

The same is true when you meet a person for the first time. Social psychologists tell us that we make long-lasting conclusions based on our first impressions of people. Think for a moment of what happens in such situations. After a mere five minutes, many people are willing to state confidently that they have formed a good impression of another person; that is, an extremely small sample of behavior creates a lasting impression of that person. Given the complexity of human behavior, this is an amazing conclusion based on such a limited sample.

As you will recall from Chapter 1, Bacon warned about the idols of the cave, relying too much on our personal experience for seeking truth. Objective or scientific sampling is a way of overcoming the idols of the cave because it ensures that our experiences are as broad and representative as possible. Moreover, as you learned in Chapter 3, having a representative sample of the population in a study helps to ensure the external validity of the conclusions we draw from the data to other members of the population.

What are the common features of samples and sampling? First, we use samples to describe and compare things. Consider, as an example, the challenge of describing someone's personality. Each of us is affected by many things, ranging from situational factors to interactions with others, and myriad other variables too numerous to list. Even for an individual, it would be impossible to observe all the things he or she would do under different situations. However, based on a representative sample of that person's behavior, we may be able to predict how he or she will react to specific situations. This observation leads us to the second feature of sampling.

We use samples to help us make predictions and conclusions about other things or conditions. As the quotation at the start of this chapter suggests, our theories are the product of the data we collect. Perhaps, by extension, we can also suggest that no theory is any better than the data that either inspired its creation or are used to test it.

How can we be sure that what we observe in a sample accurately represents the broader population? This is a critical question. As you learned in Chapter 3, such a question is at the heart of external validity. If the sample is not representative of the population, then the data will be of little value to us because they do not support inferences about the population.

Therefore, the purpose of this chapter is to examine the foundations of samples and sampling. In the following pages, we will examine the methods researchers use to collect accurate samples of the population. In addition, we will review how statistics, based on sample data, allow us to make inferences about population parameters. Hence, we will begin with a more detailed analysis of the characteristics of samples. We assume that you have had a course in statistics and understand the foundations of descriptive statistics such as measures of central tendency (e.g., mean and median), measures of dispersion (e.g., variance and standard deviation [*SD*]), and basic standard or *z*-scores. If you are not familiar with these statistics, or wish to brush up on your skills, you might benefit from reviewing Appendix A.

THE NATURE OF SAMPLES

There are several ways that we can define samples. The primary feature of a "good" sample for research is that the sample actually represents that population from which it was selected. One way to define a sample is to contrast samples with things that are not samples. We can also define samples by the methods used to create them. As you may recall from our previous discussions of validity, we cannot directly assess the external validity or the representativeness of a sample. Rather, we examine the methods used to create the sample to infer its validity. Consequently, researchers work hard to ensure that their sampling techniques produce useful data.

Scientific Samples versus Anecdotal Evidence

What is the difference between sample-based data and other types of information? We can begin by considering personal experiences and anecdotal evidence and then compare them to a scientific sample. Imagine a person who traveled to Paris for a short vacation and then pronounces the French to be rude and inconsiderate people. Such a conclusion is indefensible. A moment's thought will reveal that many factors bias this unflattering conclusion. Specifically, a stranger traveling abroad, spending a short time in a small portion of the country, and interacting with a minuscule proportion of the population, cannot come to a meaningful conclusion about the manners of a nation of people.

Anecdotal evidence and personal experience are not sampling in the technical sense of the word. Our hypothetical traveler probably did meet several rude and inconsiderate people, an experience likely to occur when traveling to any large city. Therefore, we cannot deny the person's experiences. We do deny the validity of the conclusion, however, because we do not consider that one person's encounters to be representative of the population.

We can use the criteria of representativeness to distinguish anecdotal evidence and personal experience from scientific samples and empirical conclusions. The goal of scientific research is to identify and collect data from samples of participants that represent the population that we study. In an experimental scenario, researchers are also interested in creating treatment groups that are equivalent on all variables other than the independent variable(s) of interest. From these samples and treatment groups, researchers collect data and then have to generate reasonable and valid conclusions regarding these data. When we collect the data, we use procedures that ensure that the samples will be free of bias and thereby represent the population. By contrast, anecdotal evidence is typically a haphazard collection of personal experiences that may be biased.

Unfortunately, anecdotal evidence too easily sways many people's opinions, an observation that Bacon noted when he described the idols of the tribe and idols of the cave. Social psychologists have long studied the poor decision-making processes that many people follow (Plous, 1993). A well-known phenomenon is the "person-who" effect that occurs when someone uses anecdotal evidence to discount a statistical generalization. For example, a smoker may dismiss the risk of smoking by noting that his or her father smoked two packs of cigarettes a day and lived to be 75. In this case, the person seems to ignore the larger body of evidence that people who smoke have, on average, a shorter life expectancy and an increased risk of health problems.

The problem we all must confront in research and in life is the pervasive nature of anecdotal evidence and its influence on decision making. As researchers, we must remain vigilant to ensure that our data are representative of the population. Finally, you should recognize that the plural of *anecdote* is not *data*. Anecdotes, even a whole collection of them, like rumors, cannot be trusted to be accurate, reliable, or valid.

OVERVIEW OF PROBABILITY SAMPLING

Before we delve into the topic of sampling any further, we must define an important term. To say that an outcome is **random** means that each possible outcome has an equal probability of occurring and that the outcome of one event has no influence on the probability of subsequent outcomes. Imagine a balanced six-sided die. Each time you throw the die, the probability of any one side landing faceup is always one-sixth. In addition, each time you throw the die, the results of the previous tosses have no effect on the current toss. That one of these events has no effect on the other possible events is an illustration of the principle of **independence**. Random sampling is also an example of independence because when used in sample selection, selecting one object from the population will not influence the selection of other objects.

Understanding randomness is important for sampling because it allows us to link individual samples to the population. According to the **law of large numbers**, any large number of items chosen at random from a population will have, on average, the same characteristics as the population. This law is the cornerstone of probability sampling and allows us to infer that what is true of a randomly selected sample is also true of the population. In this section, we will examine how we can use random events and other procedures to create samples that represent the population.

Populations and Samples

In Chapter 3, we introduced you to the concepts of the population and the sample. As you should recall, the population consists of all the individuals or things that the researcher wants to describe. Researchers define the population by developing the criteria that determine membership in the population. Defining the population is essential as it determines the conclusions that the researcher may draw from the data (Wilkinson, 1999).

When referring to a population, many researchers refer to the **target population** or the **universe**. For example, a medical researcher may want to study the rates of obesity among teenagers between the ages of 13 and 18. In this example, the target population is children between the ages of 13 and 18, inclusive.

In some cases, it is impractical or impossible to draw a representative sample from the target population. Consequently, researchers will draw their sample from a smaller subset of the population called a **sampling population**. For our purposes, a sampling population consists of an accessible group of people who share the same characteristics as the target population. In most health sciences research, the sampling population consists of people who live or work near the researcher. When they report their results, researchers define the characteristics of the sampling population. Doing so allows the reader to determine whether the sampling population is representative of the target population.

Some researchers pull their study samples from an existing **subject pool**, a group of people to whom the researcher has easy access. At many colleges and universities, the academic departments maintain a subject pool of students enrolled in introductory-level courses who may receive extra credit for participating in a study. In other cases, people who have a known medical condition, for example, may agree to have their name added to a list of potential participants for controlled studies.

A related concept is the **sampling frame**, which is the set of individuals within a population who can actually be reached for a specific research purpose, for example, a list of names that identifies the members of the population is a sampling frame (Babbie, 1998). Consider a study conducted by Washington, Sun, and Canning (2010) in which they created a sampling frame to better examine medical care received by veterans. As they noted in their study, most veterans do not receive health care from the Department of Veterans Affairs (VA). Therefore, using data drawn exclusively from VA data sets may be incomplete and biased. To address this potential problem, they created a sampling frame using data from the Veterans Health Administration, Veterans Benefits Administration, and Department of Defense. Washington et al. found that the sampling frame provided a wider range of cases than would have been examined had they used only the VA data set. Therefore, samples generated using this sampling frame provides researcher access to data that are more likely to generalize to the health care and status of male and female veterans.

Detailed descriptions of the characteristics of the population, sampling population, and sampling frame help us evaluate the external validity of inferences made from a sample. As you have seen in the previous example, Washington et al. (2010) were concerned that research based exclusively on VA data would be incomplete and not representative of the population of veterans. By expanding the sampling frame to include other data sets, they demonstrated that samples from this sampling frame would be far more inclusive and representative of the population of American veterans.

The connection among external validity and the sampling population, subject pool, and sampling frame is important. If any of these subgroups is not representative of the population, then this threatens the external validity of any conclusions that are based on the data collected from the sample. For these reasons, researchers spend considerable time designing careful sampling plans and analyzing and describing the sources of their data. Doing so allows the researcher to establish a connection between the sample statistics and the population parameters. With these concerns in mind, we can turn our attention to the business of creating samples.

Characteristics of Probability Sampling

There are many ways to secure a representative sample of a population. We can use techniques known as simple random sampling, stratified sampling, systematic sampling, or cluster sampling, to name just a few. Although there are many types of sampling procedures, all of those mentioned earlier have one thing in common: They are statistical or **probability sampling** techniques.

All probability sampling techniques share common features. First, they treat each member of the population as a potential member of a sample. This is a critical feature as it ensures that the sampling technique does not systematically exclude portions of the population from the eventual study sample, thereby biasing the results. By implication, the procedures employed in true probabilistic sampling procedures are objective and systematic. All sampling procedures involve following a set of rules, known as a *protocol*, for selecting members of the population. In some cases, the protocol for sampling may be minimal; in other cases, the sampling protocol may be extremely elaborate. Researchers follow the sampling rules specified in these types of protocols to ensure that the data they ultimately collect are not biased.

The second feature of probability sampling techniques is that when used appropriately, they make it possible for researchers to determine the probability that any one sample of a population would be selected. In addition, these techniques make it easier for the researcher to test whether the sample that was studied really is representative of the broader population of interest (Cochran, 1977). This may sound a bit strange, but it is an important assumption that allows us to connect the results of the sample to the parameters of the population. Consider the following example of probability sampling.

Assume that you have a population containing the whole numbers 1–6. Using sampling without replacement, and with a sample size of 2, you could create the 15 potential samples listed in Table 7.1. If we use probability sampling, we can

Table 7.1 All Potential Samples of Size 2 Drawn from a Population Consisting of Values 1–6

S_1 (1,2)	S_2 (1,3)	S_3 (1,4)	S_4 (1,5)	S_5 (1,6)
S_6 (2,3)	S_7 (2,4)	S_8 (2,5)	S_9 (2,6)	S_{10} (3,4)
S_{11} (3,5)	S_{12} (3,6)	S_{13} (4,5)	S_{14} (4,6)	S_{15} (5,6)

determine the probability that we will select any one of these samples. In this example, there is an equal likelihood that we will select any one of the samples. Specifically, for each sample, $p = 1/15$ or $p = .0667$. The probability (p) that we will select any one of these samples is approximately 6.67%.

You may object and say that we rarely know the true nature of the population, and if we did, why would we need sampling? You are correct; we rarely know the parameters of the population. This illustration shows the logic of probability sampling. For any given population, we can draw an infinite number of samples. Using mathematical techniques that we will soon review, we can estimate the probability of selecting any one of those samples. Armed with this knowledge, we can then make several interesting inferences about the population. You read about these inferences all the time. For example, various research agencies report "consumer confidence" based on samples of consumers that they can judge to be representative of the broader population of consumers.

PROBABILITY SAMPLING METHODS

There are many methods of drawing samples from the population. Each method shares a common goal: to ensure that the sample is an unbiased depiction of the population. In this section, we will review several of the more frequently used sampling methods that are based on the probability theory just discussed, including simple random sampling, sequential sampling, stratified sampling, and cluster sampling. This section can provide only a general introduction to sampling methods. Sampling itself is a science, and there are many sophisticated techniques researchers use for specific purposes (see also Babbie, 1998; Cochran, 1977; Salant & Dillman, 1994).

Simple Random Sampling

Simple random sampling is the most basic of the sampling procedures. Simple random sampling occurs whenever each member of the population has an equal probability of selection into the sample. The steps involved in conducting a simple random sample are clear-cut. First, estimate the size of the population. Second, generate random numbers to associate with each person in the population, and third, use those numbers to determine which members of the population are included in the sample. In the final stage, collect and analyze the data.

Figure 7.1 illustrates random sampling conceptually. The block of circles on the left represents members of a population. The block of circles on the right represents a random sample of the population. In this example, there are five types of circle, each represented by a different shade. Because there is no bias in the sampling procedure, the sample should be similar to the population, and any difference between the sample and the population represents chance or random factors. In simple random sampling, each circle in the block on the left has an equal chance of selection into the sample (the block on the right).

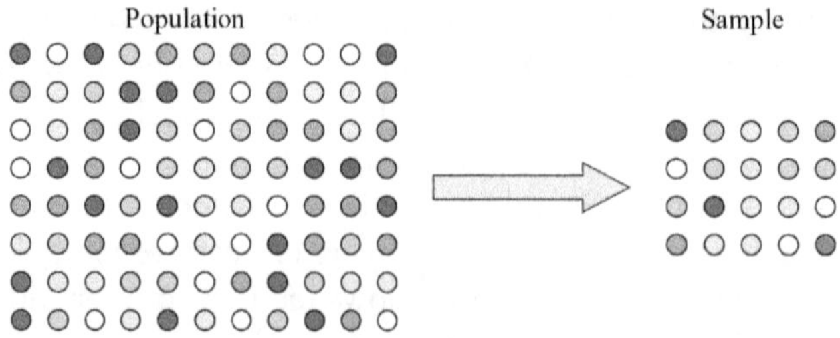

Figure 7.1 Random sampling.

Imagine that you want to conduct a survey of student eating habits at your college. The college's institutional review board approved your research, and you are ready to collect the data. The first step is to determine the size of the college's student population. This should be relatively easy. Go to the college's records office and ask the registrar for the number of students currently enrolled at the college. For the sake of the illustration, assume that there are 7584 students enrolled at your college and that you have decided that you will pull a random sample of 250 students for your research.

For the next step, you will need 250 **random numbers**. Random numbers are a series of numbers with no order or pattern. Technically, random numbers are independent of each other. Independence means that each number has an equal chance of selection and the selection of one number has no effect on the selection of another number. Before the advent of personal computers, many researchers used a random number table created by the Rand Corporation (1955) that contains 1 million random digits. The creators of that table went to great pains to build and test a machine that produced random digits.

Many researchers now use their personal computers to generate random numbers. Different computer programs can produce a string of random numbers between specific values. In the current example, we need 250 random numbers between 1 and 7584, inclusive. Once you have generated the 250 random numbers, you could sort the entire data file in order of these numbers and select the first 250 individuals in the sorted list of all the students in the population at this college. If everyone in this randomly selected sample cooperates with your request to complete the questionnaire, your sample will be an unbiased representation of the population.

In case you are interested, many computer programs generate random numbers, but most do this based on a preset random number "seed" or starting point. Researchers call these numbers *pseudorandom* numbers because of this, but in most research cases, using pseudorandom numbers is sufficient to ensure representativeness within a sample (Pashley, 1993).

Sequential Sampling

Although simple random sampling is the gold standard of probability sampling, many researchers find other techniques, such as **sequential sampling**, more practical (Babbie, 1998). For a sequential sample, we list the members of the sampling population and then select each *k*th member from the list.

The steps for sequential sampling are straightforward. First, we need to determine the **sampling interval** that identifies which members of the list to select. The sampling interval is merely

$$\text{Sample interval} = \frac{\text{Size of population}}{\text{Desired size of sample}} \tag{7.1}$$

In our example, we wanted to create a sample of 250 from a population of 7584 students. Therefore, the sampling interval is 30 (7584/250 = 30.336). Now all you need to do is obtain a current list of all students and then select every 30th student on the list. This procedure assumes that the list you select students from represents the entire sampling frame.

Stratified Sampling

Stratified sampling is another variation on simple random sampling. This technique is especially useful in situations where the population contains different subgroups that the researcher wants to compare or study. A fine example of stratified sampling is in a report by Robb et al. (2010), which examines differences among ethnic groups regarding breast and cervical cancers in England. To conduct the study, the researchers used stratified sampling of the six largest ethnic groups—Indian, Pakistani, Bangladeshi, Caribbean, African, and Chinese—to identify 1500 participants who then completed a number of surveys regarding cancer awareness. Robb et al. found that there were notable differences among the groups regarding knowledge of the forms of cancer and previous screenings for the cancers.

Figure 7.2 illustrates how stratified sampling works. In this example, the population contains five distinct subgroups of different sizes. The researcher wants to include a sample of each subgroup in the sample. In order not to bias the sample, the size of the subgroups in the sample should equal the relative size of the subgroups in the population.

The box on the left represents a population consisting of smaller subgroups of strata. The researcher randomly selects from each stratum to create the sample.

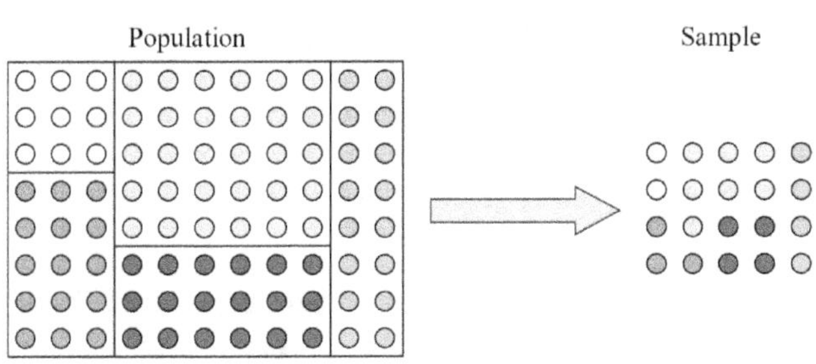

Figure 7.2 Stratified sampling.

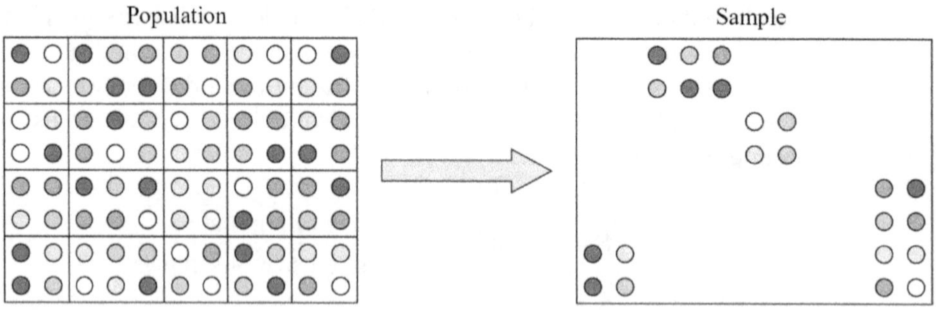

Figure 7.3 Cluster sampling.

The sample will have approximately the same proportions of the subgroups as the population.

Figure 7.2 offers a hint of how a researcher would conduct stratified random sampling. First, the researcher would identify the specific subgroups in the population and attempt to estimate their relative size. Second, the researcher can use simple random or sequential sampling within each subgroup. The result is a representative sample. If all goes well, the size of the subgroups in the sample will be the same relative size as the population.

Cluster Sampling

Researchers use **cluster sampling** when it is not convenient to pull one or two people out of their environment for the research or when other methods of sampling are impractical. Researchers often use this technique when working with intact groups. An example would be a research project examining the effectiveness of different teaching techniques.

Imagine that Figure 7.3 represents students enrolled in 20 different sections of the same course. The researcher may want to know whether differences in the teaching method affect students' performance in the course. It may be impractical to randomly select students and create special sections of the course. As an alternative, the researcher can select different sections at random for the research. In Figure 7.3, the researcher randomly selected five different clusters.

The small blocks of individuals in the left square identify separate clusters of individuals. The researcher selects several clusters for the sample.

In other cases, the clusters may represent cities, neighborhoods, schools, or other naturally occurring clusters. The researcher can then select clusters randomly or systematically. If the researcher knows that particular clusters contain important portions of the population, he or she may then purposely select those clusters.

NONPROBABILITY SAMPLING

Although probability sampling is the ideal, it is often beyond the budget, time, and personal resources of the researcher. Therefore, in many situations, researchers will resort to **nonprobability sampling**. You will also see this type of sampling referred to as *judgmental* sampling since it is not statistical or based on statistical rules. These methods are often more convenient for the researcher, but the results of studies using samples of this sort need to be interpreted with caution. As you should recall from your reading in previous chapters, using a nonprobability sampling method may limit the inferences that you can make from the data. However, these methods can produce useful data when collected and interpreted under the right conditions (Cochran, 1977).

Convenience Sampling

The most common type of nonprobability sampling is **convenience sampling**. In contrast to probability sampling, convenience sampling means that the researcher uses members of a population who are easy to find. Interviewing people coming to a clinic or visiting a patient at a hospital is an example of convenience sampling. In these cases, the researcher allows the individual's behavior to determine who will, and more importantly who will not, be a part of the study.

Do you see the difference between probability sampling and convenience sampling? For all forms of probability sampling, the *researcher's method* of sampling determines who will be a potential member of a sample. For convenience sampling, the *individuals' behaviors* determine whether they could become part of a study's sample. Consequently, convenience sampling can bias the results and interpretation of the data. Imagine that you wanted to conduct a survey of students at your college. Do you think you would get different results if you surveyed students lounging around the college's student union versus students studying in the library? In many ways, convenience sampling suffers the same problems as anecdotal evidence because the sample does not represent the population. Indeed, when researchers use random sampling techniques, they will review characteristics of the sample in the hope to confirm that those who volunteered appear to match the general characteristics of the sampling population.

Despite these drawbacks, however, there are times when a convenience sample may be the only realistic sampling strategy. Many good research studies have been conducted using a convenience sample. However, the specific results of any one particular study may be an artifact of the sample. The growth of science is dependent on a body of knowledge that builds on preexisting research. This is especially true for convenience samples, and the results of individual studies should be examined in light of existing and future research.

Snowball Sampling

Sadler, Lee, Lim, and Fullerton (2010) outline the benefits of snowball sampling for a population of participants that may be otherwise difficult to contact using traditional sampling techniques. Sometimes the members of a cohort are difficult to find and recruit for research. The members of the cohort may wish to remain anonymous, or there is no list identifying the members of the cohort. Imagine you wanted to study active users of a controlled substance such as methamphetamine or examine the health of homeless teens. These people may not be difficult to find using standard techniques. Consider another example, people who seek medical treatment from alternative sources (e.g., folk remedies). Indeed, as Sadler et al. note, probability sampling will most likely not produce a representative sample of these populations.

The concept of **snowball sampling** is relatively simple but may require considerable skill to accomplish. In its simple format, the researcher gains the trust of one or more members of the population. A healthcare provider may know that women in a specific ethnic group use indigenous or folk medicine during their pregnancy. In this example, the researcher would first need to find a suitable method to contact women in this group and to develop a level of trust such that the women will introduce the researcher to other women in the group. Building upon this trust, the additional participants can then be encouraged to recruit additional participants from the population. Snowball sampling has been used successfully to study human behavior in populations as diverse as gangs, illicit drug users, and other hard-to-reach populations. For those interested in other techniques researchers use to study difficult-to-find populations, we recommend Kendall et al. (2008).

KNOWLEDGE CHECK

1. What are the differences between anecdotal evidence and scientific sampling?
2. What are the potential problems with relying on personal experience to make decisions?
3. Describe the similarities and differences among simple random sampling, sequential sampling, stratified sampling, and cluster sampling.
4. Anne is selecting courses for next semester and must take an English literature course but does not know which professor she should select. To solve her dilemma, she decides to ask three of her friends, who are English majors, for their recommendations.
   a. What type of sampling is Anne using?
   b. Describe the factors that may bias the information Anne receives.
5. The dean of academic affairs at a small college wants to examine the study habits of students attending the college. He randomly selects 50 students from the college and asks them to record the number of hours they work on homework each day for 1 week (assume that all the students participate and answer honestly). According to the results, the typical student studies an average of 3.2 hours a day.
   a. Can we conclude that the dean has a valid estimate of the students' study habits?
   b. Are there factors that can bias the results of the data?

6. Professor Smith distributes a survey to 45 students in his English Composition 101 course. Professor Jones distributes the same survey to 45 students in her Organic Chemistry 250 course.
   a. What are the factors that may bias the results of these samples?
   b. Is one sample more likely to be representative of the students at the college?
7. A national magazine publishes the results of a survey of its readers. The magazine argues that the results are valid because of the large number (e.g., 12, 592) of readers who responded to the questionnaire. Do you agree that the results are valid?
8. A researcher in a large city wishes to conduct a survey and ensure that the sample includes a representative sample of each of the five major ethnic groups that live in the city. Describe how the researcher could use each probability sampling technique to create such a sample.
9. A researcher wants to examine the birth control practices of members of different Christian denominations (e.g., Baptist, Catholic, Episcopal, Lutheran, and Presbyterian). What might be the most cost-effective method of generating a representative sample?
10. Do you agree with the statement "A sample of convenience is little better than anecdotal evidence"? Defend your answer.

CENTRAL LIMIT THEOREM

You may be wondering how we can continue to assert that a sample statistic is an unbiased estimate of the population parameter. How do we know that the mean and standard deviation of a sample accurately estimates the mean and standard deviation of a population (e.g., $M \approx \mu$ and $SD \approx \sigma$)? To answer that question, we need to examine an important principle in statistics known as the **central limit theorem**. The central limit theorem is important because it supports our use of a sample as a representation of a population.

The central limit theorem describes the distribution of certain sample statistics from probability samples. In other words, the central limit theorem describes the sampling distribution. As we noted previously, one of the characteristics of probability sampling is the ability to predict or estimate the probability of selecting a specific sample. We can use the central limit theorem and a sampling distribution to estimate the probability of obtaining, at chance, any potential sample from the population. Knowing this can allow us to infer whether the data we have is extraordinary or "significant" in some way. Before we proceed, however, we need to define some terms.

A **sampling distribution** represents the result of taking many samples of a specific size from a population and plotting the frequency of a sample statistic, such as the mean, that is computed within these samples. A sampling distribution is the theoretical distribution of the potential values of a sample statistic that would occur when drawing an infinite number of equal-sized random samples from the population. We can use a thought experiment as an illustration.

Imagine that you have access to a population of college students you want to survey. You draw a random sample of 25 students, test each person, and then calculate the mean score. You can now repeat these steps to produce the mean for a second sample. Chances are that the two sample means will be slightly different. If you continue to collect samples from the population, you will eventually have a large collection of sample means. When you create a graph of the frequency of the values of the means, you will form the sampling distribution for the means.

Not all the sample means will be equal to each other; there will be variability among the sample means. Specifically, many of the sample means will be slightly greater or less than the population mean, and some will be much different from the sample mean. We can describe this variability among potential sample means using the statistical concept of **standard error** (σ_M), or the average standard deviation of the sample distribution of the same means. Because we used random sampling to create the samples, the difference among the sample means represents random or chance factors. The central limit theorem allows us to describe the shape of the sampling distribution of sample means and the amount of standard error.

The central limit theorem makes several specific propositions or statements about the shape of the sampling distribution of the sample means. The propositions are

1. As the size of the individual samples increases, the shape of the sampling distribution of sample means will become progressively normal regardless of the shape of the population.
2. The mean of the sampling distribution of sample means (μ_M) will equal the mean of the population (μ), $\mu_M = \mu$.
3. The standard deviation of the sampling distribution of sample means (μ_M) will equal the standard deviation of the population (σ) divided by the square root of the sample size ($\sqrt{n}$), $\sigma_M = \sigma / \sqrt{n}$.

What are the implications of these propositions? Why are they important for statistics? We will consider the answers to these questions in the following sections.

Shape of the Sampling Distribution

One of the most interesting predictions supported by the central limit theorem is that the shape of the sampling distribution of sample means will be normal, especially as the size of each sample increases. This is true even if the actual population-level distribution is not itself normal in shape. Figure 7.4 illustrates this prediction. The top panels of the figure represent the shape of three

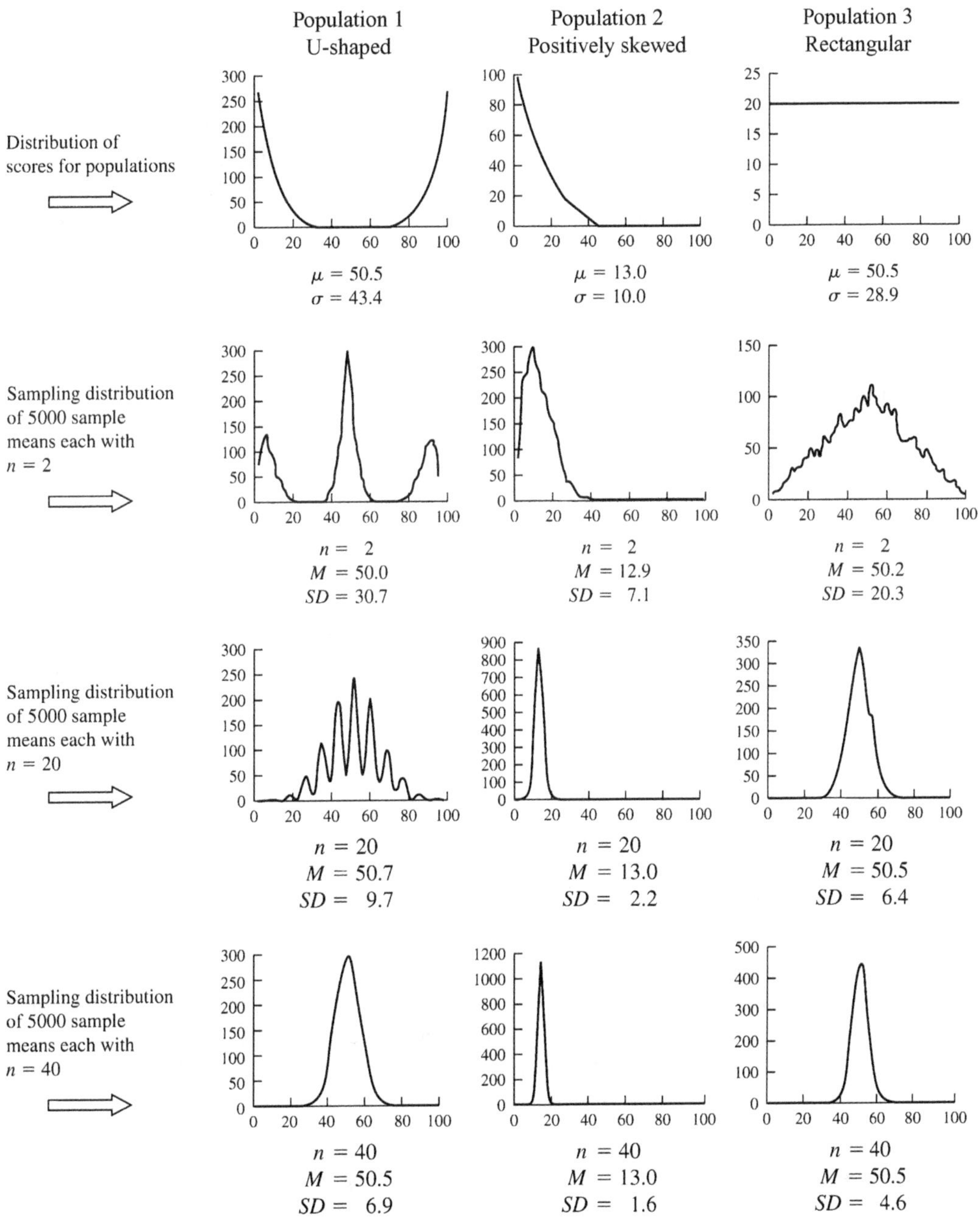

Figure 7.4 Central limit theorem illustration.

populations: U-shaped, positively skewed, and rectangular. The lower panels represent the sampling distributions of means for sample sizes of 2, 20, and 40. Each sampling distribution represents 5000 random samples taken from the population. As you can see, as the size of the sample increases, the distribution of means becomes progressively normal in shape.

The top panels in Figure 7.4 represent the shape of the population from which the samples are drawn. The lower three rows of graphs represent the sampling distribution of the mean of sample sizes 2, 20, and 40. Each sampling distribution represents 5000 random samples generated by a computer program. According to the central limit theorem, the shape of the sampling distribution will become normal as sample size increases. In addition, $\mu_M = \mu$ and $\sigma_M = \sigma/\sqrt{n}$. The mean (*M*) of each sampling distribution is close to the mean of the population, and the standard deviation of each distribution (*SD*) is approximately $\sigma/\sqrt{n}$.

Mean of the Sampling Distribution: $\mu_M = \mu$

The symbol μ_M represents the theoretical mean of the sampling distribution of means. In other words, the quantity represents the mean of all possible sample means. As an equation,

$$\mu_M = \frac{\sum M}{N} = \frac{\text{Sum of all sample means}}{\text{Number of samples}} \tag{7.2}$$

This equation indicates that we take the sum of all possible sample means drawn from the population and divide by the number of samples. You should be clear about the difference between *sample size* and *number of samples*. Sample size, represented as *n*, refers to the number of observations in each sample. The total number of samples, represented as *N*, refers to the number of samples drawn from the population. For Equation 7.2, we assume that the sample size is constant and that *N* is an extremely large number. As an example, look at the bottom row of graphs in Figure 7.4. Each graph represents 5000 random samples. Therefore, $N = 5000$. Each of the samples consists of 40 scores. Consequently, $n = 40$.

You should now be able to see that the sample mean is an unbiased estimate of the population mean. Although there is standard error in any sampling distribution, the error is random and nonsystematic. Therefore, any individual sample mean is equally likely to be above or below the population mean. Using these facts allows us to conclude that the sample mean is an unbiased estimate of the population mean, or that $M = \mu$.

Standard Error of the Mean (*SEM*): $\sigma_M = \sigma/\sqrt{n}$

The **standard error of the mean** (*SEM*) is the standard deviation of the sampling distribution of sample means. The equation for *SEM* tells us that as sample size increases, the distribution of sample means will cluster closer around the population mean. As sample size increases, the sample mean will become a more accurate estimate of the population mean. In other words, the lower the *SEM*, the more accurate the estimate of *M* is of the population mean.

To summarize, the central limit theorem allows us to make several inferences concerning the relation between sample statistics and population parameters. First, the sample mean is an unbiased estimate of the population mean, if we use a probability sampling procedure to generate the sample. Furthermore, the sample mean becomes a better estimate of the population mean as the sample size increases. Because of these properties, researchers prefer to use the sample mean as the descriptive statistic to infer the population parameter. Second, the central limit theorem allows us to determine the probability of obtaining various sample means.

APPLICATIONS OF THE CENTRAL LIMIT THEOREM

In this section, we will examine how we can use the central limit theorem to our advantage as researchers. As you will see, we can use the central limit theorem to answer two important questions, "How reliable is our sample mean?" and "How large should our sample be?" We can answer the first question by calculating the standard error of the mean and the confidence interval (*CI*). Once we review these basic statistical tools, we can begin to address the second question.

We use *SEM* to describe the standard deviation of the sampling distribution of means. This value allows us to estimate the range of potential sample means drawn from a population. Figure 7.5 represents the hypothetical results of taking 20 samples at random from a population mean. Assume that the mean of the population equals 100 ($\mu = 100$), that the standard deviation of the population equals 30 ($\sigma = 30$), and that we draw 20 random samples of 9 observations each ($n = 9$). We can calculate the standard error of the mean as $\mu_M = 10.0 = 30/\sqrt{9}$.

The modified box-and-whisker plots in Figure 7.5 represent 20 random samples drawn from a population where $\mu = 100$ and $\sigma = 30$. For each plot, the long horizontal line represents the range between the lowest and highest scores. The rectangle represents the middle 50% of the sample. The vertical line within the box is the sample mean. The sample size of each sample is $n = 9$. The four long vertical lines represent the location of $\pm 1\sigma_M$ and $\pm 2\sigma_M$. For the sampling distribution of means, $\mu_M = 100$ and $\sigma_M = 10$, $10 = 30/\sqrt{9}$. Note: Typically when using a box-and-whisker plot, the median and not the mean is used as the plotted measure of central tendency. However, for this example, the mean was used since *SEM* deals with a sampling distribution of sample means.

As you can see in Figure 7.5, there is variability among the individual sample means. If you look closely, almost all sample means fall between 80 and 120, within two standard deviations of the mean, $\pm 2\ \sigma$. Most fall between 90 and 110, within one standard deviation of the mean, $\pm 1\ \sigma$. One sample is greater than 120, but it seems to be the exception rather than the rule. This figure illustrates how we can use the sample statistics and sample size to determine the accuracy of the sample mean as an estimate of the population mean.

Confidence Interval

A confidence interval (*CI*) represents a sample and potential variability of the sample mean. In other words, and more importantly for our present purposes, a

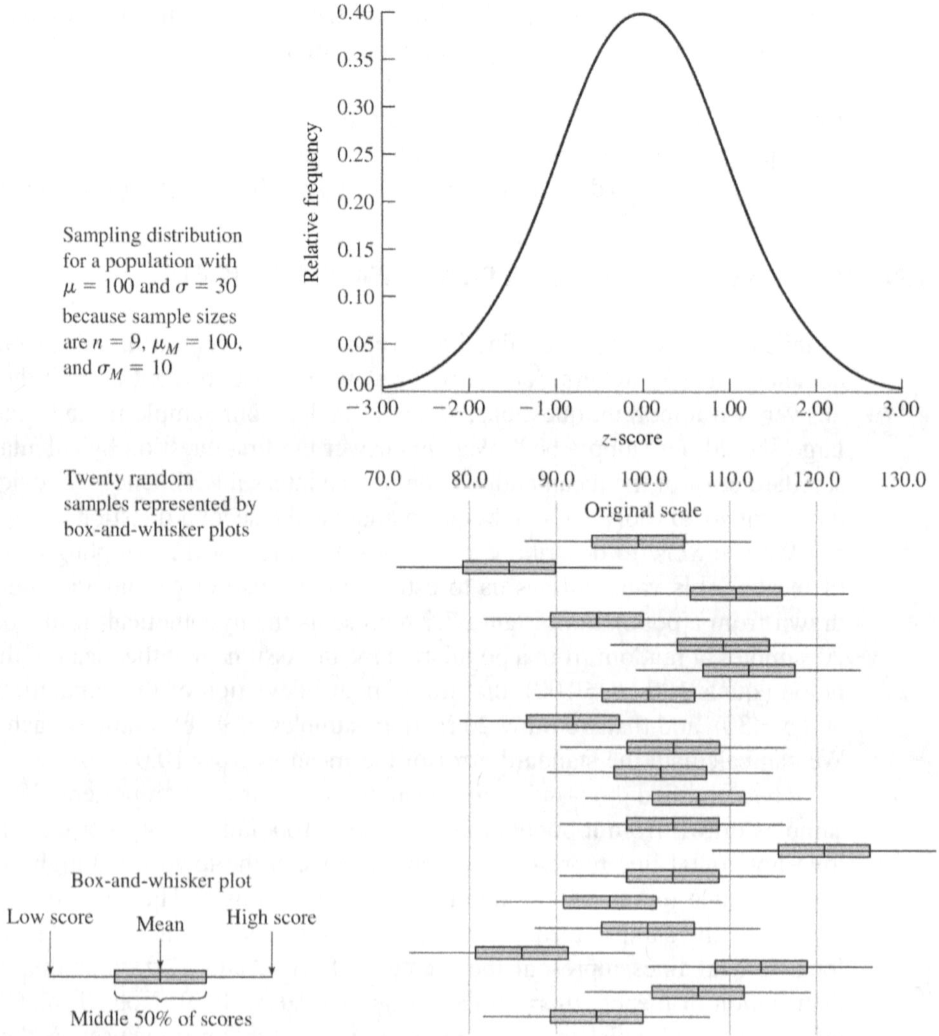

Figure 7.5 Illustration of the central limit theorem.

confidence interval is another way of evaluating the accuracy or precision of the sample mean as an estimate of what really could be expected in the larger population. If the *CI* is relatively small, then the mean is a relatively accurate estimate of the population parameter. If the *CI* is relatively large, then the mean will be a relatively less accurate estimate of the population mean.

Whenever you read or hear about a political poll, you may hear that the results are accurate to ±5 points or ±3 points. This statement reflects a form of *CI*. For example, if Smith and Jones are running for public office and a poll of registered voters shows Smith with 60% of the popular support ±5 points, we can conclude that between 55% and 65% of the registered voters favor Smith. Statisticians call these central values **point estimates** and the low- and upper-bound values **interval estimates**. The point estimate refers to the use of a sample statistic to estimate the corresponding population parameter. In this example, we estimated that 60% of the voters would vote for Smith. The interval estimate allows us to describe the accuracy of the statistic as an estimate of the population parameter.

We can also use point and interval estimates to make inferences about the differences between population parameters. Consider two examples. First, imagine that the estimate for Smith is 60% ±5 points and the estimate for Jones is 40% ±5 points. There is no overlap of the interval estimates for the two candidates. Given these data, we may infer that Smith is the clear favorite.

Now imagine a different situation. In this example, the estimate for Smith is 52% ±5 points and the estimate for Jones is 48% ±5 points. Because the two sample point estimates are within the interval estimate of the other sample, we would infer that the difference between 52% and 48% may represent the standard error. In other words, registered voters do not clearly favor one candidate over the other.

There are several ways to calculate the confidence interval depending on the type of data one collects. Among the many *CI* forms, the most widely reported are often associated with a proportion and a mean. As you will see, they share common elements.

Confidence Interval for Proportions

A proportion (p) is a special type of mean. In the political example regarding the Smith–Jones race, we might find that for Smith, $p = 0.60$. To obtain this number, a pollster may have randomly contacted 400 registered voters who indicated that they are very likely to vote. Of these, 240 indicated that they favor Smith. Therefore, $p = 0.60 = 240/400$. What would happen if we conducted another random sample? Assuming that people have not changed their opinion of Smith, what range of proportions would we expect to find? To answer this question, we first need to calculate the standard error of the proportion using the following equation:

$$SE_p = \frac{\sqrt{p(1-p)}}{\sqrt{N}} \tag{7.3}$$

For this equation, p represents the proportion that we want to examine and N represents the total number of people we sampled. Applying the equation to this example, we proceed with the following steps:

$$SE_p = \frac{\sqrt{(.60)(1-.60)}}{\sqrt{400}} = \frac{\sqrt{(.60)(.40)}}{20} = \frac{\sqrt{.24}}{20} = \frac{.4899}{20} = 0.0245$$

Once we have the standard error of the proportion, we can then determine the confidence interval using the equation

$$CI = p \pm z_{(1-\alpha)/2} \tag{7.4}$$

Table 7.2 Calculating the Confidence Interval for a Proportion

Confidence interval			
68% $(1-\alpha)/2 = .34$	90% $(1-\alpha)/2 = .45$	95% $(1-\alpha)/2 = .475$	99% $(1-\alpha)/2 = .495$
$z_{(1-\alpha)/2}$ scores			
$z_{(1-\alpha)/2} = 1.00$	$z_{(1-\alpha)/2} = 1.65$	$z_{(1-\alpha)/2} = 1.96$	$z_{(1-\alpha)/2} = 2.58$
$CI = .60 \pm 1.0\ (0.0245)$	$CI = .60 \pm 1.65\ (0.0245)$	$CI = .60 \pm 1.96\ (0.0245)$	$CI = .60 \pm 2.58\ (0.0245)$
$CI = .60 \pm 0.0244$	$CI = .60 \pm 0.0404$	$CI = .60 \pm 0.0480$	$CI = .60 \pm 0.0632$
.5755–.6244	.5595–.6404	.5552–.6480	.5367–.6632

For this equation, $z_{(1-\alpha)/2}$ represents the z-score that indicates the lower and upper levels of the confidence interval. To determine the appropriate z-score, you will first need to determine the confidence interval that you want to use. For example, if you set $1 - \alpha = .90$, then you want to have a range that represents 90% of potential means. If you set $1 - \alpha = .95$, then you have established a 95% confidence interval. To determine the corresponding z-score, divide $1 - \alpha$ by 2 (e.g., $(1 - \alpha)/2$), then turn to Table B.1 of Appendix B. Looking down column B, find the proportion that matches $(1 - \alpha)/2$ and use the corresponding z-score for your calculations. Table 7.2 represents the z-scores that correspond with different confidence intervals and the corresponding z-scores.

As you can see in Table 7.2, if we had selected a 95% confidence interval, the confidence interval is $.60 \pm 0.0480$. By multiplying the values by 100, we can convert the proportions to percentages—$60 \pm 4.8\%$. Therefore, if we collected additional random samples of 400 registered voters, we would expect the percentage of people supporting Smith to range between 55.2% and 64.8%.

Confidence Interval for Means

We can apply the same logic used in the previous section to determine the confidence interval for the arithmetic mean. You should recognize some familiar terms in Equation 7.5; M represents the mean of the sample, and $SD/\sqrt{n}$ is the standard error of the mean. In this form of the equation, we use SD to estimate the population standard deviation. The other term in the equation, $t_{(1-\alpha)/2}$, may be new to you. This variable is similar to $z_{(1-\alpha)/2}$ in that it determines the specific values for the upper and lower limits of the confidence interval. The difference between $z_{(1-\alpha)/2}$ and $t_{(1-\alpha)/2}$ is that $z_{(1-\alpha)/2}$ represents the normal distribution, whereas $t_{(1-\alpha)/2}$ represents the sampling distribution of the mean for specific sample sizes. We can work through a problem to illustrate how to use Equation 7.5:

$$CI = M \pm t_{(1-\alpha)/2}\left(SD/\sqrt{n}\right) \tag{7.5}$$

Table 7.3 Calculating the Confidence Interval for a Sample Mean for $df = 15$

Confidence interval		
90%	95%	99%
$(1 - \alpha) = .90$	$(1 - \alpha) = .95$	$(1 - \alpha) = .99$
$t_{(1-\alpha)/2}$ scores		
$t_{(1-\alpha)/2} = 1.753$	$t_{(1-\alpha)/2} = 2.131$	$t_{(1-\alpha)/2} = 2.947$
$CI = 75 \pm 1.753\ (1.25)$	$CI = 75 \pm 2.131\ (1.25)$	$CI = 75 \pm 2.947\ (1.25)$
$CI = 75 \pm 2.191$	$CI = 75 \pm 2.638$	$CI = 75 \pm 3.684$
72.8–77.2	72.4–77.6	71.3–78.7

Assume that you create a random sample of 16 people and give each person a test. According to your calculations, $M = 75.0$ and $SD = 5.0$. What is the confidence interval for your mean? The first thing you will need to do is determine the appropriate value for $t_{(1-\alpha)/2}$. To obtain this value, turn to the appropriate table in Appendix B. In the instructions, you will see that you need to convert the sample size to the degrees of freedom (df) by calculating $n - 1$. For this example, the $df = 15$ (i.e., $16 - 1$). Using the value of df, select the appropriate value of t for the equation. To find this value, use the column of numbers labeled "Level of Significance of a Two-Tailed or Nondirectional Test" and the appropriate level of $1 - \alpha$. This column of numbers represents that we want to estimate a symmetrical area about the mean of the distribution with α split equally between the two extreme ends of the table. For example, if we set $1 - \alpha = .95$, we find that $t_{(1-\alpha)/2} = 2.131$. Table 7.3 presents an example of these calculations.

If we had set $1 - \alpha = .95$, then the confidence interval is 72.4–77.6. Therefore, if we continue to create random samples of 16 from this population, we can expect that 95% of them will be between 72.4 and 77.6. We can also conclude that the probability of obtaining at random a sample with a mean less than 72.4 or greater than 77.6 is 5%.

Interpreting the Confidence Interval

The confidence interval allows us to estimate the potential range of sample means that will include the population parameter k% of the time. Consider the 95% confidence interval, one that many researchers typically use. For the example in Figure 7.5, the 95% confidence interval extends between 80.0 and 120.0, inclusive. The confidence interval indicates that if we were to take an infinite number of samples from this same population, each with a sample size of 9 (as in this example), and calculate their mean and associated confidence intervals, we would expect 95% of those confidence intervals to contain the true population value.

Factors That Affect *SEM*

What are the factors that influence the size of the standard error in a given data set? Are there ways that we can reduce the standard error and thereby increase the accuracy of the sample mean as an estimate of the population mean? According to the central limit theorem, two primary factors influence the standard error of the mean, the standard deviation of the population (σ) and the sample size (n). As the standard deviation of the population decreases, the size of the *SEM* will also decrease, all else being equal.

How can you change the value of σ if it is a constant value and parameter of the population? Technically, you cannot change σ, but you can change the way you define your population. Broad and sweeping definitions of the population tend to correspond with large variability in individual scores. As you begin to refine your definition of the population, the members of the population may tend to become more homogeneous (similar) to each other. This, in turn, will reduce the σ within the population you are really interested in studying. Having a more specific definition of your population may also help you better understand the phenomenon you are studying.

Adult consumers have different spending habits. If you defined your population as any adult consumer, your population would include anyone over the age of 18. By contrast, if you defined your population as young adult (18–29 years of age), women in your population would probably be much more homogeneous; that is, σ will be smaller. You may find that this more narrow definition of the population will allow you to better understand the spending habits of a specific group. In addition, you might be able to make more accurate statements about spending habits in different populations (e.g., men vs. women, older vs. younger people, college graduate vs. high school graduate).

As the researcher, you can directly control the size of your sample. According to the central limit theorem, the sample size affects the *SEM* by $1/\sqrt{n}$. Specifically, as the sample size increases, the *SEM* will decrease. Figure 7.6 presents the relation between the *SEM* and the sample size. The message presented in the graph is clear; as the sample size increases, the magnitude of the *SEM* decreases. Therefore, if you want the sample mean to accurately represent the population mean, you should maximize the sample size.

Another important message in Figure 7.6 is that increasing sample size has diminishing returns. Specifically, greater increases in sample size produce minor reductions in the *SEM*. As you can see on the left side of the graph, small increases in sample size (e.g., $n = 5$ to $n = 25$) produce large drops in the *SEM*. As the sample size increases, however, the changes in *SEM* decrease. Why is this fact important?

Sampling is expensive. Selecting individuals from the population and collecting the necessary information is time consuming. In addition, recruiting and collecting data from each participant adds to the cost of the research. Therefore, it is not always practical or feasible to collect exceptionally large samples. Because of the economic impact that sample size has on the cost of the research,

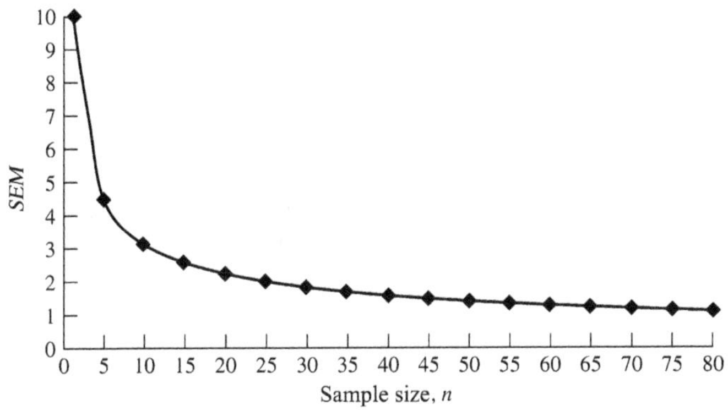

Figure 7.6 Size of the standard error of the mean (*SEM*) as the sample size increases. *Note:* As sample size increases, the size of the *SEM* decreases by $1/\sqrt{n}$.

determining the optimal sample size of the research is an important part of any research plan. We will examine methods to determine the optimal sample size in subsequent chapters when we review specific research methods. Although these techniques vary depending on the type of research, they all depend on the propositions contained in the central limit theorem.

SOURCES OF BIAS AND ERROR: A REPRISE

The focus of this chapter has been sampling. We have examined different methods of obtaining samples that we hope are representative of the population. At different points in the chapter, we have reviewed bias and error. *Bias* refers to nonrandom and systematic factors that cause the sample mean to be different from the population mean. *Error* refers to random events that a researcher cannot control. The goal of research design is to design a study so as to eliminate as many sources of bias and to minimize the risks of random error to the extent possible.

Figure 7.7 presents an outline of the relation between the target population, sampling population, sampling distribution, and individual sample. Each level in the figure indicates a potential source of bias or error, as discussed in the following subsections.

Target Population

The target population consists of the people we are attempting to represent with the sample that we have selected and will ultimately include in our research. The mean of the population, μ, represents the typical score of individuals in the population, and differences among the individuals within the population represent the natural variation among the members of this population. We use σ to represent the naturally occurring differences among individuals. If we draw an unbiased sample from this population, then the mean and standard deviation of the sample should estimate μ and σ, respectively.

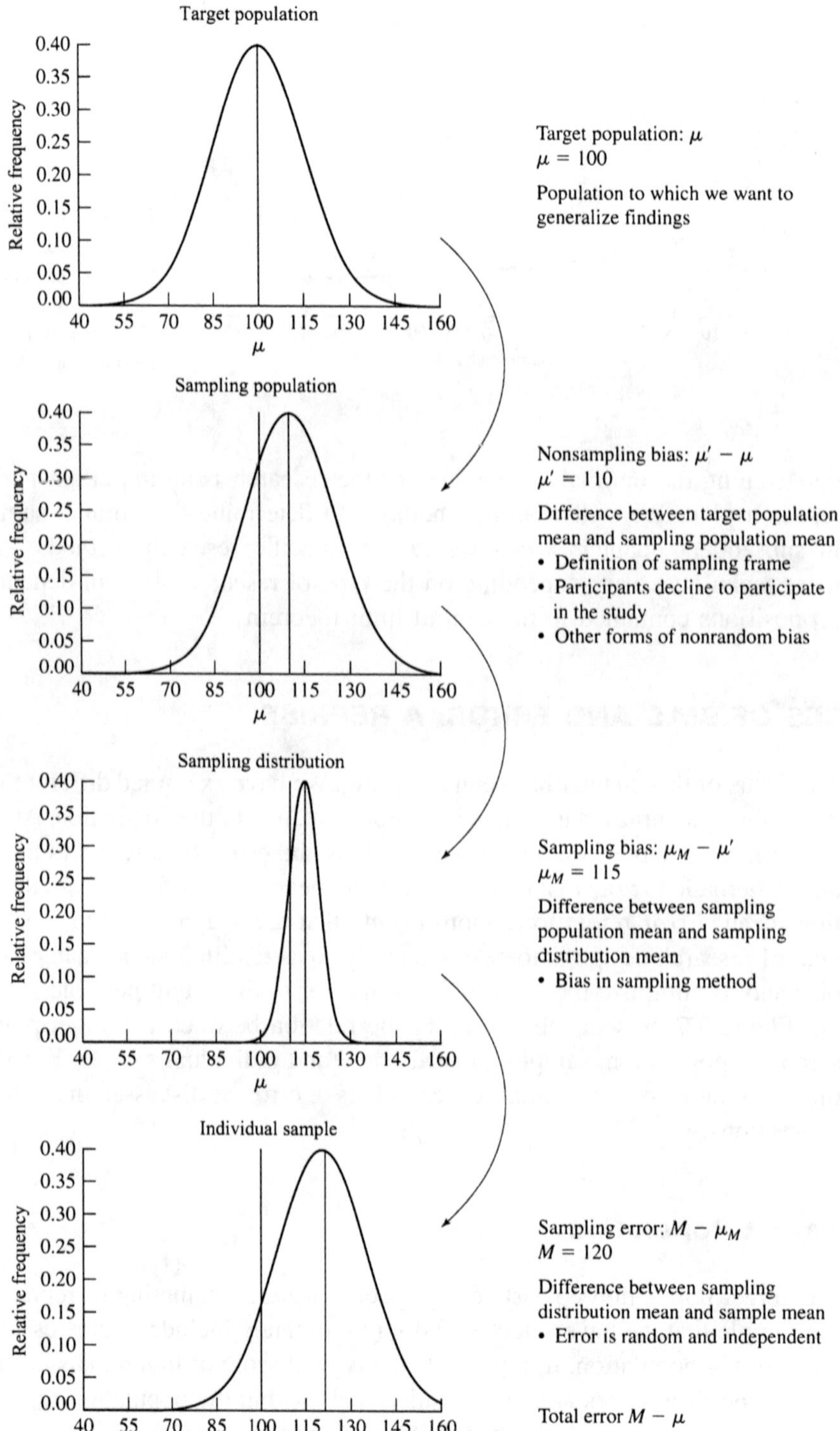

Figure 7.7 Potential sources of bias and error when forming samples.

Sampling Population

The sampling population represents the population of individuals from which we eventually draw the sample. In Figure 7.7, the mean of the sampling population is μ'. In an ideal research situation, μ and μ' will be identical because the sampling population will represent the target population. Consequently, any difference between μ and μ' represents a **nonsampling bias**. Figure 7.7 depicts a nonsampling bias where the mean of the sampling population is greater than the target population. There are several potential sources of nonsampling bias. Our definition of the sampling population and use of a sampling frame could bias our results. As Washington et al. (2010) noted, sampling only from VA records is most likely to produce a biased estimate regarding the health status of American veterans, hence the need to develop a broader and more inclusive sampling frame.

Sampling Distribution

The sampling distribution represents the theoretical distribution of sample means for samples drawn from the sampling population. If the sampling procedure is truly random, then the mean of the sampling distribution will equal the mean of the sampling population, $\mu_M = \mu'$. A systematic selection bias in the sampling procedure will produce a **sampling bias**.

Individual Sample

The last step illustrated in Figure 7.7 is the creation of the individual sample. Any difference between the mean of the sample and the mean of the sampling distribution represents the standard error. The standard error, unlike the different forms of bias, is a random event. Therefore, we consider the sample mean to be equivalent to the mean of the sampling distribution.

Total Error

The total error is the difference between the individual sample mean and the population mean, $M - \mu$. From Figure 7.7, you can see that the total error reflects the total effects of nonsampling bias, sampling bias, and the sampling error.

Inferences from the Sample to the Population

At the start of the chapter, we noted that we cannot directly assess the external validity or the representativeness of a sample merely by inspecting the sample data. Because we never know the true value of μ, we cannot determine what proportion of the total error reflects bias and what proportion reflects the standard

Table 7.4 Proportions of Childhood Abuse among LGBT Adults by Ethnic Group

	African American (n = 49)		Latina/o (n = 52)		Asian American (n = 45)		White (n = 523)	
Abuse type	Sample (%)	95% *CI*	Sample (%)	95% *CI*	Sample (%)	95% *CI*	Sample (%)	95% *CI*
Emotional	59.2	45.2–72.8	52.9	38.4–65.6	57.8	43.6–72.4	56.8	52.8–61.2
Physical	51.1	37.0–65.0	56.9	43.5–70.5	46.7	32.4–61.6	32.5	28.0–36.0
Sexual	52.2	39.0–66.9	62.7	49.9–76.1	29.5	16.6–43.4	35.5	30.9–39.0

error. Therefore, we need to ensure that our sampling and research methods are free of bias.

RESEARCH IN ACTION: SAMPLING MATTERS

To what extent were lesbian, gay, bisexual, and transgendered (LGBT) adults subject to emotional, physical, and sexual abuse, and to what extent does this abuse lead to adult mental health problems? There were the basic questions Balsam, Lehavot, Beadnell, and Circo (2010) addressed in their 2010 study in which they used snowball sampling to recruit participants. Balsam and her colleagues decided to use the Internet and snowball sampling to form their database. More specifically, the researchers recruited individuals to participate via listservs, Internet groups, organizations, and clubs that support people who have identified themselves as LGBT. A part of that invitation was the request that the recipient ask friends and associates to participate as well. In total, 1217 people responded to their request. Of these, 207 participants identified their ethnic group (African American, Latina/o, Asian American, or white). Table 7.4 presents their results.

As you can see in the table, more than half of all participants reported experiencing some form of emotional abuse while they were children. Also notable is the fact that African American, Latina/o, and Asian Americans reported much higher rates of physical abuse than white respondents. Finally, African American and Latina/o respondents reported higher rates of sexual abuse than Asian and white respondents.

These data present a first glimpse at forms of abuse LBGT adults experienced as children. The study is interesting in that it used a combination of technology—the Internet—and snowball sampling in an attempt to acquire data that might otherwise be difficult to collect through other forms of sampling as there are insufficient sampling frames to generate random samples.

Balsam and her colleagues are mindful of the limitations of their data and the inferences that can be drawn. They note, for example that even though the Internet has become commonplace, there may still be large segments of the population who do not have private access to the Internet and that all members of the LGBT community may not choose to participate in Internet mediated groups. Nevertheless, these data do provide a first step for subsequent research.

KNOWLEDGE CHECK

11. What is the difference between a sampling distribution and the distribution of sample scores?
12. Assume that you were to create sampling distributions by creating random samples. Describe the factors that will affect the shape of the distribution.
13. What is the relation among M, μ, and μ_M?
14. What is the relation among SD, σ, and σ_M?
15. For an individual random sample:
    What accounts for $X - M$?
    What accounts for $M - \mu$?
16. Some people are surprised to learn that some samples used to represent national trends consist of "only 500" participants. Based on what you know about sampling, why might a sample size of 500 be sufficient?
17. A researcher wants to conduct a study of people who embezzle. Why would probability sampling be difficult to use in this situation? How could the researcher use nonprobability techniques to create a sample of this population?
18. The president of a prestigious private college reports that the average graduate of the college makes 2.5 times more money that the average graduate of other colleges. When asked about the source of the numbers, the president claims that the staff of the alumni office called a random number of the college's alumni and requested current information, including annual income. The president then says that the average income for other college graduates comes from the most recent census data that indicate the median income of college graduates. Do you accept the president's claim that graduates of the college are more affluent than the typical college graduate?
19. A researcher believes that his or her sampling procedure may be biased. Will increasing the sample size help the researcher collect data to better represent the population?

CHAPTER SUMMARY

The primary focus of this chapter was how researchers use samples to make inferences about populations. Specifically, we examined the purpose of samples, how researchers construct useful samples, and ways to ensure that the sample represents the population.

Sampling is not a haphazard and casual method of collecting information. We examined how scientific evidence is different from personal experience and anecdotal evidence. When we depend on anecdotal evidence and personal experience, we are likely to come to the wrong conclusion. Bacon recognized this problem when he described the idols of the cave. Personal experience is not sufficient for understanding complex behavioral phenomenon. Rather, we need to depend on objectively constructed samples to represent the population.

Probability sampling ensures that the sample will represent the population. According to the law of large numbers, if we randomly select members of the population, the sample will share the same features as the population. Random selection means that every element of the population has an equal and independent probability of being selected. Using probability sampling, we can predict the probability of selecting different samples from the population.

When we describe sampling procedures, we make distinctions between different types of populations. The target population represents the group that we hope to describe with our research. The sampling population is the group from which we create our sample. If the sampling population is representative of the target population, then we can use the sample to make inferences about the target population. The sampling frame represents the list or resource that identifies the members of the sampling population. For all cases, the researcher creates operational definitions of the target and sampling populations.

There are several probability sampling techniques. For random sampling, each member of the population has an equal and independent probability of selection. Sequential sampling allows us to select each *k*th member from the sampling frame. Another sampling tactic is stratified sampling, in which we identify important subgroups in the population and then take random or sequential samples from each subgroup. Finally, in cluster sampling, we identify clusters of individuals and then randomly select several clusters for the research.

In some situations, nonprobability sampling procedures are the only ways of creating a sample. For convenience sampling, we allow the participants' behavior to determine whether they are included in the sample. Convenience sampling means that the participants are easy to locate. Another technique is snowball sampling, where we ask the participants to help the researcher locate or recruit additional participants for the research.

The central limit theorem is a statement about the shape of a sampling distribution. The central limit theorem states that for any population, the sampling distribution of means will tend to be normally distributed, especially when the sample size is large; that the mean of the sampling distribution will equal the mean of the population; and that the standard deviation of the sampling distribution will equal the standard deviation of the population divided by the square root of the sample size. The central limit theorem, therefore, allows us to predict the probability of various samples.

Given the predictions of the central limit theorem, we examined how the researcher can change the standard error by increasing sample size or changing σ by redefining the population.

Creating a sample is not easy to do because there are several sources of bias and error. We examined how different factors, such as a poor-quality sampling frame, can bias our results. Similarly, we distinguished among nonsampling bias, sampling bias, and sampling error. Nonsampling bias occurs when the sampling population differs from the target population. Sampling bias occurs when we use nonprobability sampling procedures. The standard error, by contrast, represents random differences among sample means.

We ended the chapter by reviewing a comprehensive and national research study of risky sexual behavior. The example illustrated the use of various sampling procedures and demonstrated how sample data allow us to make inferences about population parameters.

CHAPTER GLOSSARY FOR REVIEW

Anecdotal Evidence A brief and typically personal account that may not represent true events. Anecdotal evidence is often a biased and unrepresentative sample of a population.

Central Limit Theorem A mathematical postulate that states that the sampling distribution of randomly generated sample means will (1) tend to be normally distributed especially as sample size increases, (2) have a mean equal to the population mean, and (3) have a standard deviation equal to the standard deviation of the population divided by the square root of the sample size.

Cluster Sampling A form of probability sampling in which the researcher identifies natural groupings or clusters within the population that are used to create the sample that is ultimately involved in the research.

Convenience Sampling Selecting members of the population who are easy to find and study.

Independence A condition that exists when each event has no effect on subsequent events.

Interval Estimate An estimate of the accuracy of the sample statistic as an estimate of the population parameter, specifically the potential range of sample statistics that would occur when creating additional samples.

Law of Large Numbers Prediction that any large number of items chosen at random from the population will, on average, represent the population.

Nonprobability Sampling Any method of sampling that does not use random events or probability methods to create the sample.

Nonsampling Bias A nonrandom and systematic set of conditions that causes the mean of the sampling population to be different from the mean of the target population.

Point Estimate Using a sample statistic to estimate the value of the corresponding population parameter.

Probability Sampling A method of creating a representative sample that uses random and independent procedures for selecting individuals from the population.

Random Each potential event has an equal probability of occurring, and the occurrence of one event has no influence on the probability of subsequent events.

Random Numbers A sequence of numbers having no pattern or sequence because there is an equal probability of selecting each number, and the selection of one number has no effect on selecting another number.

Sampling Bias A nonrandom and systematic set of conditions that causes the mean of the sampling distribution to be different from the mean of the sampling population.

Sampling Distribution A theoretical probability distribution of the potential values of a sample statistic that would occur when drawing an infinite number of equal-sized random samples from the population.

Sampling Frame A list or some other resource that identifies the members of the sampling population.

Sampling Interval The number used in sequential sampling to select members of the population for the sample.

Sampling Population An accessible population that shares the same characteristics as the target population and from which the researcher draws the sample.

Sequential Sampling A method of sampling where one selects every *k*th member from a list.

Simple Random Sampling A method of creating a representative sample of a population. Each member of the population has an equal probability of selection, and the selection of one member has no effect on the selection of other members of the population.

Snowball Sampling Asking members of the sample to identify additional members of a population who could participate in the research.

Standard Error The random difference between a sample statistic and the corresponding population parameter.

Standard Error of the Mean The standard deviation of the sampling distribution of means.

Subject Pool A group of individuals available and willing to participate in research projects.

Target Population or **Universe** All members of a group who share one or more common characteristics defined by the researcher.

REFERENCES

BABBIE, E. (1998). *Survey research methods* (2nd ed.). Belmont, CA: Wadsworth.

BALSAM, K., LEHAVOT, K., BEADNELL, B., & CIRCO, E. (2010). Childhood abuse and mental health indicators among ethnically diverse lesbian, gay, and bisexual adults. *Journal of Consulting and Clinical Psychology* [serial online], *78*(4), 459–468.

COCHRAN, W.G. (1977). *Sampling techniques* (3rd ed.). New York: John Wiley & Sons.

KENDALL, C., KERR, L.R., GONDIM, R.C., WERNECK, G.L., MACENA, R.H.M., PONTES, M.K., et al. (2008). An empirical comparison of respondent-driven sampling, time location sampling, and snowball sampling for behavioral surveillance in men who have sex with men, Fortaleza, Brazil. *AIDS and Behavior*, *12*, 97–104.

PASHLEY, P.J. (1993). On generating random sequences. In G. Keren & C. Lewis (Eds.), *A handbook for data analysis in the behavioral sciences: Methodological issues* (pp. 395–418). Hillsdale, NJ: LEA.

PLOUS, S. (1993). *The psychology of judgment and decision making*. Boston: McGraw-Hill.

ROBB, K., WARDLE, J., STUBBINGS, S., RAMIREZ, A., AUSTOKER, J., MACLEOD, U., et al. (2010). Ethnic disparities in knowledge of cancer screening programmes in the UK. *Journal of Medical Screening*, *17*(3), 125–131.

SADLER, G.R., LEE, H.C., LIM, R.S.H., & FULLERTON, J. (2010). Recruitment of hard-to-reach population subgroups via adaptations of the snowball sampling strategy. *Nursing & Health Sciences*, *12*(3), 369–374.

SALANT, P. & DILLMAN, D.A. (1994). *How to conduct your own survey*. New York: John Wiley & Sons.

WASHINGTON, D., SUN, S., & CANNING, M. (2010). Creating a sampling frame for population-based veteran research: Representativeness and overlap of VA and Department of Defense databases. *Journal of Rehabilitation Research and Development*, *47*(8), 763–771.

WILKINSON, L. (1999). Statistical methods in psychological journals: Guidelines and explanations. *The American Psychologist*, *54*, 594–604.

Chapter 8

Assessments, Surveys, and Objective Measurement

The doctor may also learn more about the illness from the way the patient tells the story than from the story itself.

—James B. Herrick

INTRODUCTION

Measurement is at the heart of all the empirical sciences. Without objective measurement, there can be no science. Therefore, the task of this chapter is clear; it will show you how researchers create questionnaires and surveys.

PURPOSE OF MEASUREMENT

There are two goals of measurement. The first is to replace the ambiguity of words and general concepts with operationally defined constructs. In our day-to-day

Understanding and Conducting Research in the Health Sciences, First Edition. Christopher J. L. Cunningham, Bart L. Weathington, and David J. Pittenger.

language, we often say that someone has an "extraverted personality," is a "senior citizen," or is "adept at mathematics." We may also want to know whether rewards and incentives will "increase the prevalence of healthy eating and exercise." Although these phrases convey some information, they also present considerable ambiguity. What, for example, is a *senior citizen*—someone over 65, 75, or 85? Defining an age in this case clarifies our meaning of *senior* by offering an operational definition. As Wilkinson (1999) noted, the value of an operational definition is that it provides a specific method for converting observations to a specified range of potential values. Using operational definitions helps researchers achieve the goal of public verification of observation because other researchers can then use and, if necessary, critique and revise the operational definition for future work.

The second goal of measurement is standardization or consistency in measurement. Consistency of measurement allows us to compare people using a common set of procedures and scales. Standardization also implies that the numbers used in the measure have a constant meaning. For example, a score of 110 on a standardized test of a construct such as intelligence has a meaning that does not change over time, across situations, or across people.

CAVEAT ASSESSOR

There are two common misperceptions of scientific measurement. Interestingly, these two misperceptions operate in opposition to one another. On the one hand, we tend to mistrust measurement, while on the other hand we tend to put too much trust in measurement. With respect to mistrust, how many times have you read or heard something such as "There are some things that you cannot measure" or "Some human behaviors are too complex to measure objectively?" These are common statements made by people who want to either discredit a research finding or argue that science cannot be applied to people. The problem with this type of belief is that it is inconsistent to say that you cannot objectively measure a construct that is describable with words. That we can use words to describe a specific phenomenon means that we must have the ability to recognize that phenomenon in some way. Therefore, translating those points of recognition into objective measurement cannot be too far away.

The second error associated with any measurement is due to our tendency as humans to place uncritical trust in numbers. Kaplan (1964) called this problem the **mystique of quantity**, which he defined as "an exaggerated regard for the significance of measurement, just because it is quantitative, without regard to what has been measured . . . [the] number is treated as having intrinsic scientific value" (p. 172). One consequence of the mystique of quantity is what philosophers of science call **reification**. Reification occurs when we incorrectly treat something, in this case a quantified measurement, as if it represents a real thing rather than an estimate of some more abstract construct. As Dewdney (1997) noted, just because we have a word for something does not mean that this thing exists independent of the word. For perspective, the reification problem is an example of

Bacon's idol of the marketplace, which we reviewed in Chapter 1. As you may recall, we examined how researchers' use of phrases such as "maternal absence" influenced how they conducted and interpreted the results of their research on the effects of child care.

Consider the words *beauty*, *anxiety*, and *intelligence*, words we use routinely. For example, one might say that a painting by the famous artist Mary Cassett is "beautiful." There is nothing in the painting that is, by itself, beautiful. What does exist is Cassett's unique choice of pigmented paints and their arrangement on the canvas; nothing in the picture itself is objectively beautiful. Instead, *beauty* describes a viewer's perceptions of, and reactions to, the elements of the painting. We commonly invent words or constructs to help us make predictions and offer explanations. In most cases, we can operationally define these words for the purpose of our research. In Chapter 3, we used the example of perceived value. *Value* is an abstract concept that we use to pull together different conditions and to explain various behaviors. If we have a clear operational definition of the construct, we might find that the concept helps us explain the relation among several variables.

There is nothing inherently wrong with using hypothetical terms like *intelligence*. We use these words to help us describe, predict, and explain the world. The problem arises when we forget that some of our variables are inventions, and we begin to treat them as if they are real things. Gould (1981) complained that the reification of intelligence has caused many biologists, psychologists, and sociologists to search for its genetic component much as one would search for the genetic foundation of eye color, sickle-cell anemia, Hodgkin's disease, or other heritable conditions. Many believe that such a search is a fool's errand because intelligence is not solely a biologically determined process like eye color but rather something that develops from the interaction of a person and his or her environment and experiences.

Another serious problem that arises from the mystique of quantification and reification is that some people forget that the numbers by themselves cannot capture or express all the important characteristics associated with a particular phenomenon. No measurement, quantitative or qualitative, can represent all the important components of human behavior. While "placing a value on an attribute of a person is essential in representing the complex ambiguity of life" (Weathington, 2011, p. 137), unfortunately, many people believe that a test measures the "real thing," forgetting that any test may ignore critical characteristics of an individual construct. Miles-Tapping (1996) provided a good example of this problem when she noted an important difference in the way people who must use wheelchairs used the word *independence*. Many researchers used this word to mean that a person walks without the assistance of a cane, walker, or wheelchair. Therefore, a person who is wheelchair bound is, by that definition, not independent. By contrast, Miles-Tapping found that persons who must use a wheelchair perceive the chair as a symbol and instrument of their independence.

In summary, we need to view testing and measurement with guarded enthusiasm. With sufficient planning, analysis, and revision, social science researchers in any discipline can develop useful measures of the constructs that they are trying to study. We must remember, however, that although we will be able to create a measure that suits most of our needs, all measures are tentative, subject to error and bias, and forever in need of refinement.

DEVELOPING A MEASUREMENT SCALE AND DATA COLLECTION STRATEGY

In the following subsections, we will show you how to create measurement procedures for various applications including observational research, interviews, and surveys. Although each research study is unique, there is a relatively uniform process to follow when creating a sound measurement approach. We introduce the steps in this process with three questions that researchers must answer when designing a study and measurement strategy. Then we turn to various techniques for the gathering of data. Finally, we examine more specific information about generating the best questions to ask when gathering data.

What Questions Are You Trying to Answer?

The best way to answer this question is to carefully review your hypothesis. If you have a well-crafted hypothesis, you should be able to use it to clearly define your independent and dependent variables. Identifying the variables is half the battle when it comes to measurement as knowing this helps you determine the types of phenomenon (e.g., attitudes, behaviors) you might want to assess. The more specifically you can describe the variables in your study, the better. Consider the hypothesis "Patients who actively participate in their treatments will demonstrate fewer complications and posttreatment relapses than patients who do not participate in their treatments." This is a fairly well-stated hypothesis and a good starting point for identifying an appropriate measurement strategy. From this hypothesis, we know that the independent variable is a patient's "active" participation in treatment. While we would need to operationalize what is meant by active participation, this phrasing gives us an understanding of the gist of the research. The dependent variable has something to do with complications during treatment and issues experienced following treatment. All we need now is an operational definition of these latter two concepts. How will we measure complications? What time frame are we looking at posttreatment and how do we measure relapses?

What Is the Most Convenient Method for Producing the Data?

Identifying observable behaviors related to the dependent variable will help determine how you will collect the data. Using the previous example, the definition of interpersonal conflict may focus on how team members interact with each other. Specifically, do the children from different ethnic groups play together or do they fight? In this case, you may find that an observational technique will supply the data that you need.

However, you may also want to know how the children perceive members of different ethnic groups. Because you are interested in the child's perceptions, you may want to use a questionnaire or an interview. As you will see in the following sections, each method of data collection has its relative advantages and disadvantages. Therefore, you will need to weigh these as you consider your options.

What Is the Most Accurate Measurement Technique?

There is no such thing as a perfect measurement technique nor is there one that is best in all research situations. Because all measurements have some form of error, you will need to consider methods that will help reduce bias and random errors. For observational research, we can use several trained observers to help ensure that we capture all the relevant information. For questionnaire-style research, we can focus on the wording and design of the questions to ensure the best possible data.

INTERVIEWS, QUESTIONNAIRES, AND SURVEYS

There are many ways that we can ask people questions, ranging from face-to-face interviews to questionnaires sent through the mail or administered over the Internet. Although the general hypotheses should be generated before choosing the specific research technique or data-gathering procedure, it is often the case that the data collection method we choose will influence the way we ask the questions during the research process. The questions we ask can be open-ended, allowing respondents to tell us as much or as little as they wish. Our questions can also be closed-ended, requiring a choice from among several options or a simple yes or no response. As with all elements of research design, the selection of specific data collection procedures represents a balancing act between the need for information and the feasibility/appropriateness of the procedure for the given research scenario. What follows is a review of several common and different self-/other-report data collection techniques and their relative advantages and disadvantages.

Personal Interviews

The personal interview, either face-to-face or through the telephone, is a popular and useful way to understand human attitudes and beliefs (Fontana & Frey, 1994). Interviews are one of the most commonly used tools in employee selection, but the usefulness of this technique goes beyond hiring. Personal interviews, especially face-to-face interviews, tend to encourage a high degree of cooperation by participants. In addition, people are likely to answer an interviewer's questions rather than check the "Don't know" box on a questionnaire. This should not be terribly surprising—after all, it is difficult to skip or ignore a question when another person asks you a direct question. Other advantages of this method of

Table 8.1 Review of Various Interview Methods and Their Characteristics

Type	Characteristics	Format	Use
Brainstorming	Unstructured interaction among group members	No set format; interviewer asks questions to facilitate discussion	Generation of ideas for a project that will be examined in detail later
Case study	Intensive study of one person's history and current situation	Unstructured and guided by interaction between interviewer and participant	Demonstration of a general principle or application of a clinical intervention; often the source of ideas for later research
Focus or Delphi group	Structured interaction between interviewer and a small group of people	Interviewer uses a script of questions but can ask follow-up probes for additional information	Gathering of information regarding people's reactions to specific topics, often used for opinion gathering
Formal interview	Highly structured interaction between interviewer and participant	Interviewer follows a script with little or no room for deviation	Collection of specific information typically for a larger program of research

measurement are that the interviewer can ensure that the participant understands the questions and ask follow-up questions to clarify participants' responses.

In sum, personal interviews can yield a great deal of rich information. There is no single format for the personal interview. Interviews can be highly structured or unstructured. Similarly, the interview may be limited to two people or may involve a small group. Fontana and Frey (1994) described many types of interviewing formats, each of which has a specific role for contemporary research. These formats take place in different settings, require different roles of the interviewer, involve different numbers of people in the discussion, and use different formats for the questions. Table 8.1 presents a review of the different types of interview methods, ranging from not at all structured to highly structured.

Potential Limitations of Interviews

Although interviews can produce rich data, this information comes at a price. Interviews are time intensive and expensive to conduct. First, unless you have significant extra time of your own or can find a team of willing volunteers, you will need to pay staff to assist you. Second, conducting a good interview requires significant practice, a bit of acting, and the ability to respond to unpredictable situations. Interviewers need to be trained on how to use the interview technique, how to answer a participant's questions, and how to react to the participant's many comments that may or may not apply to the questions asked.

Although there is no one-size-fits-all approach to good interviewing, in order to prevent experimenter and participant biases from affecting the data, it is often desirable to use at least a semistructured interview format in which all interviewers ask the same questions, in the same order, and record responses in a structured fashion. Developing structured questions and response formats takes time and requires additional interviewer training. The benefits of doing this far outweigh the consequences, however, and it is becoming standard practice in many applied areas of research and practice linked with the social sciences. The following example illustrates some of the pitfalls common when unstructured interview processes are used.

Conducting interviews is especially difficult when there are significant racial and cultural barriers between the interviewer and the people interviewed. Consider the following example. During the 1930s, the Federal Writers' Project funded in-depth interviews of African Americans living in the southern United States, an area in which racial divisions were and sometimes still are especially prominent (Davidson & Lytle, 1992). In one case, two people independently interviewed Susan Hamlin, a former slave. During part of the first interview, she said,

> *Mr. Fuller [Susan's owner] was a good man and his wife's people been grand people, all good to their slaves. Seem like Mr. Fuller just git his slaves so he could be good to dem. He made all the little colored chillen love him. (p. 161)*

During the second interview, she said,

> *. . . but our master ain't nebber want to sell his slaves. But dat didn't keep Clory [a mulatto slave owned by Mr. Fuller] frum gittin' a brutal whippin'. Dey whip' 'er untul dere wasn't a white spot on her body. Dat was de worst I ebber se a human bein' got such a beatin'. (p. 165)*

Clearly, Susan told different stories about her life as a slave. In the first interview, she described the slave owner as a gentleman who treated his slaves well. The second interview tells a story about ruthless behavior. Davidson and Lytle (1992) discovered that the first interviewer was white, whereas the second interviewer was black. Susan was evidently distrustful of white people and unwilling to tell them of her horrible experiences.

The previous example illustrates the need to exercise caution when planning an interview and interpreting the results of an interview. Good interviewing is not just reading the questions from a script. In many cases, the interview is successful only after the interviewer has gained the participant's trust and willingness to answer honestly. This important skill takes time and practice to cultivate. Personal interviews are useful sources of information but require considerable financial support, professional staff, and time.

Surveys

An alternative to the personal interview is the survey. The survey is the model of simplicity; give many people a few questions and ask them to mark their answers on a sheet of paper. This technique can also be used with open-ended questions that can serve, essentially, as written interviews. More detail on response formats for these types of questions is provided later in this chapter.

The advantages to this method of data collection and measurement are obvious. First, the cost of photocopying and distributing a survey is a fraction of the cost of developing and conducting a personal interview. Second, surveys are relatively easy to distribute. We can mail the survey along with a stamped return envelope, hand them out to a class of students or some other well-defined group, or administer through an Internet link.

The real costs associated with surveys arise with respect to the quality of the data they can provide. Many people are likely to ignore or recycle surveys they receive in the mail. Because of this, survey data may be biased if the responses that are returned only represent the perspectives of those who are conscientious enough to respond to questions when asked (a special subset of the broader population, in most cases). In addition, the researcher has little control over a survey once it is first administered to a participant. There is often no feasible way to guarantee that returned responses come from the person who was actually targeted for the survey.

Although there are many potential liabilities with surveys, this method of data collection is extremely popular among social and behavioral researchers. In those situations where many people complete the survey, this method can be a cost-effective means of obtaining data. Researchers can use a couple of simple techniques to maximize compliance with a survey.

Use a Captive Audience

If one of your research goals is to maximize the number of responses from people within a certain larger group, one strategy is to try to distribute your survey within natural subgroupings of the larger group. For example, in studies that involve college students, many researchers ask students to complete a survey in the last 10 minutes of a class meeting. This is the logic behind the ever-popular use of introductory sections as research participant pools within most universities. Such large-enrollment courses provide an opportunity to gather a large amount of data quickly.

Use Behavioral Tendencies to Your Advantage

According to Salant and Dillman (1994), there are many other strategies you can use to increase the return rate of surveys sent through the mail or hosted on the Internet. Plan to send several letters/e-mails to potential participants. The first should be an advance notice that points out that a survey will be arriving within several days. This letter should convey the importance of the research and the need for the person's response. The second mailing is the survey along with a cover letter that reiterates the importance of the research. If necessary, you can send a third and fourth letter to remind the participants to return the completed questionnaire, although there seems to be a diminishing improvement in survey returns when more than three reminders are sent.

Salant and Dillman (1994) also recommend doing what you can to enhance the distinctiveness of your communications with participants. For example, when sending mailings, use a distinctive envelope, priority mail, or some combination of the two. Although you do not want your package to look like a sweepstakes mailing, you do want it to create the impression that it contains an important document. Finally, make your cover letter as personal as possible, and convey the impression that you eagerly await the person's comments.

Another successful method for increasing return rates is to give potential respondents a small token of thanks. One explanation for how this works to increase response rate comes from the social psychological phenomenon of normative reciprocity. Quite simply, if we do something nice for you, you will feel obligated to return the favor. As an example, Salant and Dillman (1994) noted that attaching a $1 bill to a questionnaire increases response rates by 5–8%. Larger denominations ($5 or $20) increase response rates even higher (probably not a practical option for most research).

Finally, perhaps the easiest thing you can do as a researcher to increase the response rate is to make sure you do not give your sample members an easy excuse for not completing the survey immediately. Include a sharpened pencil or a pen with the questionnaire along with a stamped return envelope and make it clear that the deadline for responses is very short.

QUESTION–RESPONSE FORMATS

In addition to selecting the most appropriate method for gathering data from the potential participants in your sample, you also need to identify the right question to ask. These questions need to address the core constructs of your research and, therefore, need to extend directly from your study hypotheses or primary research questions. In general, two main types of questions are asked. The first is the **closed-response question** that requires the person to select a response choice from a set of options. The second is an **open-ended question** that asks participants to generate their own response to the question.

Closed-Response Questions

The primary feature of the closed-response question is that the researcher supplies the response options for the person. You can use many alternative formats for a closed-response question. The following are common examples of these formats.

Nominal Category Response

The answer options for this category represent a nominal scale. In some cases, the question will treat the options as mutually exclusive and force the person to

Table 8.2 Example of Nominal Category Response Items

Mutually exclusive response options:				
Sex:	___ Female	___ Male		
I am:	___ Single	___ Married	___ Divorced	___ Widowed

Multiple nominal responses:		
I use the following sources to learn about national news (please check all that apply):		
___ Newspapers	___ News magazines	___ Television news
___ Radio	___ Internet	

Table 8.3 Example of Forced Choice Alternatives

For each of the following pairs of words, mark the word that best describes you:

___ Liberal versus ___ Conservative
___ Shy versus ___ Outgoing
___ Leader versus ___ Follower
___ Assertive versus ___ Passive
___ Introverted versus ___ Extraverted

Mark the statement that best describes your belief*

(a) ___ In the long run, people get the respect they deserve in this world.

or

(b) ___ Unfortunately, an individual's worth often passes unrecognized no matter how hard he or she tries.

*Item from Rotter's (1966) Internal-External Locus of Control Scale.

select only one category. In other cases, the question will allow the person to select several separate categories. Table 8.2 is an example of nominal category response items.

Forced Choice Alternatives

For these questions, we ask the participants to select a response that best represents their answer to the question. In this case, we assume that the alternatives represent an underlying scale that ranges between the two extremes. Table 8.3 presents examples of forced choice alternative items.

Numerical Response Format

The numerical response format (often referred to as the Likert format) is one of the most popular options for the closed-response format questionnaire for two primary reasons. First, it offers a clear and unambiguous ordinal scale of

Table 8.4 Examples of the Likert Format

Overall, I believe that Mayor Matthews is doing a good job.

1	2	3	4	5
Strongly disagree	Disagree	Neutral	Agree	Strongly agree

If the election were held tomorrow, I would

1	2	3	4	5	6	7
Definitely vote Democrat						Definitely vote Republican

The meals prepared by the food service are

1	2	3	4	5	6	7
Consistently good						Consistently poor

measurement. Second, you can use the same format for many different questions. Therefore, you can combine the responses to multiple questions to yield an overall or average score. Table 8.4 presents several examples of a Likert-type response format. In its most common form, the Likert format consists of an odd number of response options reflecting a range of negative to positive reactions to a specific item or statement. The middlemost response option in this type of scale is a neutral choice. Research on the reliability and validity of scale scores from Likert-type scales has led to the conclusion that 7 or more response options are preferred to the more commonly used 5-option response, but that at least 4–10 response options be provided (Chang, 1994; Preston & Colman, 2000).

Guttman Format

A Guttman scale format, like the Likert format, represents a response continuum that ranges from one extreme to another. For this format, the researcher arranges the answer options by levels of acceptance. When the person responds, we assume he or she agrees with the statement and all the preceding statements as well. Consider the example in Table 8.5. The response alternatives are ranked in such a way that if you agree with one statement, you are likely to agree with all of the preceding statements (e.g., if you agree with option 4, then you likely agree with options 1–3 as well).

Although useful, Guttman response formats can be difficult to prepare because you cannot rely on your own opinion to determine the best ordering of the response options. To create the scale, you should distribute the individual options to a small and representative sample of the target population and ask the participants to

Table 8.5 Example of a Guttman Format Response Alternative

Which of the following best represents your opinion regarding abortion? Select only one option.
(1) ___ Abortion may be performed when the mother's life is in danger.
(2) ___ Abortion may be performed if the fetus has a severe disability.
(3) ___ Abortion may be performed in cases of rape.
(4) ___ Abortion may be performed as a type of family planning.
(5) ___ Abortion may be performed for any reason.

identify the options that they would endorse. The order of the items for the final Guttman format represents the percent agreement for each response option—those with high levels of agreement come first, followed by options that received lower percentages of agreement and are therefore less likely or more extreme (as in the example presented in Table 8.5).

Open-Ended Questions

A primary advantage of an open-ended question is that it may allow you to obtain very rich and complex data that you could not gather with a closed-response question format. Consider the typical course evaluation form that most colleges use at the end of each semester. Most questions on these evaluations tend to be closed-response, asking students to rate the instructor on a five-point scale of greatness from something like "horrible" to "amazing." Following these ratings, it is also common to see at least one open-ended question such as "What do you like most about the instructor's teaching?" or "In what ways can the instructor improve as a teacher?" Although the answers to the closed-response questions tell us what students think about our teaching in general, it is the written responses to the open-ended questions that are often more informative. Every semester, each of us receives comments that are sometimes flattering, sometimes humbling, but always illuminating. The written comments provide information difficult to obtain with closed-response questions.

This does not mean that open-ended questions are always the preferred question-and-response format. First, many people provide minimal or vague answers to open-ended questions, especially when they have to write out their own response. Second, written responses can be difficult to evaluate objectively. For example, how should a professor respond to the written comment, "This course is difficult and requires a lot of reading"? Is the student stating a fact or complaining about the course? Because of these ambiguities, many researchers prefer to use closed-response questions for surveys. One compromise is to ask closed-response questions and then to provide room for written responses. A second approach is to use qualitative research techniques such as thematic coding to assist with the interpretation of open-ended responses (more details on these techniques are found in Chapter 17).

KNOWLEDGE CHECK

1. Describe the "reification problem" in your words.
2. What is meant by the phrase "mystique of quantity"?
3. How are the concepts of "reification" and "mystique of quantity" related?
4. Consider the terms *intelligence*, *personality*, and *anxiety*. What are the differences between the way a researcher would use these terms and the way another person would use these terms?
5. Can a science exist without measurement? Justify your opinion.
6. Describe in your words the goals and characteristics of measurement.
7. Why are operational definitions essential for creating a measurement instrument?
8. Assume that you wanted to collect data concerning students' attitudes toward minorities. What would be the relative advantages and disadvantages of using a personal interview or a self-administered questionnaire?
9. Describe in your words the similarities and differences between the Likert and Guttman formats.
10. What do you see as the differences between conducting a face-to-face interview versus a telephone interview?
11. A researcher wants to conduct a survey by posting the question on a web page. What do you see as the advantages and disadvantages of this option?

WRITING GOOD SURVEY ITEMS

The real work of preparing a good survey is writing questions that sample the opinions, attitudes, beliefs, or behaviors that you want to measure. Whenever possible, researchers are better off using existing scales or measures that other researchers have shown to be reliable and valid. The process of measure development is not easy and often requires multiple rounds of data collection itself. There are times, however, when no adequate measures exist, and this is when the material in this section of the chapter should be most helpful.

Ask Single Questions or Make Single Statements

Example: Do you agree with the college's plan to build a new student recreation center and increase the student fee to $500 to help pay for the building?

This question has two parts: building the new student recreation center *and* increasing the student fee. Some people may agree with both statements, some may disagree with both, and others will agree with one statement but not the other. Because of the inherent ambiguity created by asking two questions in one, the question should be broken into two parts. The results will provide an unambiguous indication of student sentiment.

Alternatives

1. To what extent do you agree/disagree with the college's plan to build a new student recreation center?

Strongly disagree	Moderately disagree	Somewhat disagree	Neutral	Somewhat agree	Moderately agree	Strongly agree

2. How willing would you be to pay $500 in student fees to help pay for a new student recreation center?

Not at all						Absolutely

Ask Specific Questions and Avoid Vague Terms

Example: Do you actively support the college's decision to build a new student recreation center?

What does *actively support* mean? Although people recognize the difference between *support* and *oppose*, we have no way of determining the extent of their support. Therefore, we need to find an alternative wording that will indicate the extent to which students support this construction project.

Alternatives

1. I believe that the college should build a new student recreation center as soon as possible.

Strongly disagree	Moderately disagree	Somewhat disagree	Neutral	Somewhat agree	Moderately agree	Strongly agree

2. Which of the following building projects should be started as soon as possible? Pick one:

☐ an addition to the library
☐ a new arts and sciences building
☐ a new student recreation center
☐ renovations to the Mary Beach and Russell residence halls.

Write Neutral Statements and Avoid a Biased Tone

Examples

1. Many people believe that it is bad to spank children. Do you?
2. Do you believe that all students must complete two semesters of English composition regardless of their writing skills?

Both questions will create a biased answer as they imply a "correct" answer. The first question makes it clear that you should agree with the majority and condemn spanking children. Similarly, the second question makes it clear that the English composition courses are not necessary for all students. By contrast, notice how the following alternatives merely ask for an opinion.

Alternatives

1. Do you believe that parents should spank their children?
   □ Yes
   □ No

2. To what degree do you favor or oppose the new university policy requiring all students to complete two semesters of English composition?

Strongly oppose	Moderately oppose	Somewhat oppose	Neutral	Somewhat favor	Moderately favor	Strongly favor

Ask Questions That Do Not Embarrass or Anger the Participant

Examples

1. Are you still a virgin?
2. How often do you get drunk?

Highly personal questions such as these can offend some people. In addition, these questions seem to imply a moral overtone about one's sexuality and drinking habits. Research questions should be neutral in tone, and the response options should allow the individual to respond in a neutral or null manner if so desired.

Alternatives

1. With how many people have you had sexual intercourse?
   _ 0 _ 1 _ 2 _ 3 _ 4 or more
2. During the typical month, how many times do you drink until you are intoxicated?
   _ 0 _ 1–2 _ 3–5 _ 6–7 _ 8 or more

Use Simple Words and Concepts

Example: Do you believe that you received sufficient time to consider your alternatives before you were required to make a decision about which treatment option to pursue?

This question uses many formal and long words that may be difficult for some readers to understand. Good questions use simple words and sentence structure. Most word processors have a function that examines the readability of a sentence and estimates its reading level or reading score (seriously, try it). The previous question received a rating at the 12th-grade reading level. The following alternatives are easier to understand (second- to fifth-grade reading level).

Alternatives

1. Did you feel rushed to make a treatment decision?
2. Did you have enough time to think about your treatment options?

Ask Questions That the Respondent Can Easily Answer

Example: What percentage of the typical week do you spend studying?

This question asks the participant to provide an estimate and most people do not do a good job of estimating percentages. Consequently, you may not obtain useful information. As an alternative, consider asking participants to estimate the actual amount of something. You can then determine the percentages later.

Alternative

1. During a typical weekday (Monday–Friday), how many hours do you do each of the following (total should equal 24 hours):
   _ = attend classes
   _ = study or work on homework and class assignments
   _ = work at a job
   _ = socialize or party with friends
   _ = sleep
   _ = other activities

   24 = total hours.

Recognize That One Question May Not Be Enough

Example: All in all, I am inclined to feel that I am a failure.

Many psychological constructs, such as opinions, attitudes, and personality, cannot be accurately measured with a single question. In these cases, it is appropriate to ask several highly related questions. The average answer to a group of questions is a more reliable measure of the person's opinion, attitude, personality, or whatever construct that you are measuring. Asking a series of related questions also creates a **priming cue** for the participant. If you ask a series of related questions, the participant will have more time to think about the issue and thereby to offer a more thoughtful answer to the questions.

Alternatives[1]

1. All in all, I am inclined to feel that I am a failure.
2. I feel that I have a number of good qualities.
3. I am able to do things as well as most other people.
4. I feel that I do not have much to be proud of.
5. On the whole, I am satisfied with myself.
6. I certainly feel useless at times.

Consider the Influence of Question Order

You should be cautious of the order of questions because each question sets the stage for how the participant will answer subsequent questions. For example, Dillman (2000) reported that 21% of college students agreed with a policy that a student should be expelled from college for plagiarism. However, 34% of another sample of students endorsed expulsion of the student if the preceding question asked whether a faculty member should be fired for plagiarism. The difference in the response rates raises the question, *which data set really represents the students' opinion?* Results like these represent what researchers call an **order effect** on participants' responses elicited by the sequencing of questions.

There is no easy solution to an order effect other than to recognize that it exists. Dillman (2000) recommended that if you suspect that the order of the questions will influence the participants' responses to critical questions, you should create several versions of the questionnaire with different sequences of questions. The alternative is to identify the order of questions in the report of the data, thus allowing the reader to understand the context in which the participants answered the questions.

Word Questions to Reduce the Risk of a Response Set

Look at the six alternative item statements in the previous section. These questions all assess a person's self-esteem. Although they measure the same thing, look at their wording. If the participant has a high level of self-esteem, then he or she will have to agree with some questions and disagree with others. If all the questions were worded in the same manner, then the person may become lazy and be tempted to give the same answer to each question without really reading the question. Researchers call this type of response tendency a **response set** (Cronbach, 1950). One way to avoid the response set is to reword the questions and statements so that half are stated in the positive (e.g., "I am a good person") and the remaining stated in the negative (e.g., "There is little to like about me"). The response scales stay the same, but before analyzing the data, responses to the reverse-coded items are reversed so that the eventual scale total score or overall mean will accurately

[1]These items are from Rosenberg's (1965) Self-Esteem Scale.

reflect a higher level of the target variable. Whatever technique you choose, just make sure you keep notes about which items are reversed and which response scales are backwards. Otherwise, you will have some nasty surprises when it is time to code and score the data.

Avoid Questions and Statements That Have Obviously Correct Answers

Examples

1. Women are generally not as smart as men.
2. I would not want a woman as my boss.

In some cases, being straightforward is not the best policy. Most people know that the previous statements about women are not popular. Therefore, respondents may tell you what they think you want to hear, not what they believe. One alternative is to reframe the questions to somewhat obscure the intent. For example, Swim, Aiken, Hall, and Hunter (1995) developed a test that measures sexism using subtly worded statements.

Alternatives

1. Discrimination against women is no longer a problem in the United States.
2. Women often miss out on good jobs due to sexual discrimination.
3. On average, people in our society treat husbands and wives equally.

Resources for Tests and Measurements

Why reinvent the wheel when there are many sources for information on existing, high-quality tests and measures? You can find many of these tests in published literature. Therefore, before you create your own test or scale, see whether someone has already done the hard work for you. You can use research databases to search for and identify measures/tests/assessments that have been published and are already in use by other researchers. Often the scales and details about their development and reliability and validity can be easily found within peer-reviewed journals, highlighting yet another good reason to keep reading. Chapter 6 provides much more information on using databases to find existing research.

DETERMINING THE SAMPLE SIZE FOR A SURVEY

In Chapter 7, we introduced you to the central limit theorem and the standard error of the mean. As you learned, we use these mathematical concepts to determine the accuracy of the sample mean as an estimate of the population mean. We can also use these concepts to determine the optimal sample size for survey-type studies. In this section, we will examine how to determine sample size for surveys in which you require the person to select one response from several options. A common example is a political poll wherein you ask potential voters to choose between candidates in an upcoming election. We can treat the answer as a **binary response** because we want to know whether voters will vote for a candidate (yes or no). Specifically, we want to estimate the probability (p) that voters will vote for a particular candidate.

Cochran (1977), in his influential text on sampling, demonstrated that Equation 8.1 allows us to estimate sample size, n', given p and α. We use α to determine the width of the confidence interval of the mean. Also in the equation is $z_{(1-\alpha)/2}$, which represents the absolute value that defines the boundaries of α. We reviewed how to find $z_{(1-\alpha)/2}$ in Chapter 7. Equation 8.1, therefore, allows us to estimate the sample size when given specific conditions:

$$n' = \frac{\left(z_{\frac{1-\alpha}{2}}\right)^{(P)(1-P)}}{\alpha^2} \tag{8.1}$$

There are several steps for estimating the sample size. First, we need to determine what we would consider an acceptable margin of error for our specific research. Smaller values of α will give us greater precision in our estimate. For example, if we accept $\alpha = .05$, then we are comfortable with our margin of error being ±5 percentage points. To increase the accuracy of our estimate, we could use $\alpha = .01$, which would mean that the margin of error becomes ±1 percentage points. Next, we need to select a value of p that predicts what we expect to observe when we collect the data. If you have no way to make this prediction, you can use the conservative option by setting $p = .5$.

Table 8.6 presents the relationships among α, p, and estimated sample size (n'). As you can see, when $p = .5$ and $\alpha = .05$, you need approximately 384 participants in your survey to obtain a relatively reliable estimate of the population. As we reviewed in Chapter 7, the accuracy of any sample statistic depends on the method we use to generate the sample. Table 8.6 also illustrates that increasing sample size will greatly increase your precision, but at a cost. For example, doubling the precision of measurement ($\alpha = .05$ to $\alpha = .025$) requires a considerable increase in sample size (384–2009). It is important to note that in the present case, the required sample size follows a bell-shaped distribution, peaking in the middle at $p = .50$ and dropping off to either side. This is due to the binary nature of the outcome in the present example. One way to look at this is that respondents will be asked to either endorse a candidate or not endorse a candidate. If the results are clearly one way or the other, then a smaller sample will suffice. If there are lots of potential variability in the responses of people who are polled, you will want a larger sample size to increase your confidence in the quality of the resulting estimate of probability.

Table 8.6 Relations among α, p, and Estimated Sample Size

95% Confidence interval $\alpha = .05$ $z_{(1-\alpha)/2} = 1.96$		97.5% Confidence interval $\alpha = .025$ $z_{(1-\alpha)/2} = 2.24$		99% Confidence interval $\alpha = .01$ $z_{(1-\alpha)/2} = 2.58$	
P	n'	P	n'	P	n'
.90	138	.90	723	.90	5,971
.80	245	.80	1,286	.80	10,615
.70	322	.70	1,688	.70	13,933
.60	368	.60	1,929	.60	15,923
.50	384	.50	2,009	.50	16,587
.40	368	.40	1,929	.40	15,923
.30	322	.30	1,688	.30	13,933
.20	245	.20	1,286	.20	10,615
.10	138	.10	723	.10	5,971

Small Populations

If you are working with a small population, you can adjust your estimated sample size using Equation 8.2. You can use this equation whenever you know the actual size of the population. If the population is small, you may find that you need fewer participants to maintain an appropriately narrow confidence interval for your statistical estimates:

$$n'' = \frac{N(n')}{N + n'} \tag{8.2}$$

For this equation, n' represents the estimated sample size from Equation 8.1 and N represents the size of the population. Here is an example. Assume that you want to conduct a survey of students at your college and ask them whether they support the faculty's decision to use a plus/minus grading system. For the sake of the example, assume that you want to be accurate to within ±3 percentage points ($\alpha = .03$), that you set $p = .5$, and that there are 7500 students at your college. Given these facts,

$$1 - \alpha = .97, (1 - \alpha)/2 = .485, z_{(1-\alpha)/2} = 2.17$$

$$n' = \frac{\left(z^2_{\frac{1-\alpha}{2}}\right)^{(P)(1-P)}}{\alpha^2} = \frac{\left[(2.17)^2\right](.5)(1-.5)}{.03^2} = 1308.03 \cong 1308$$

$$n = \frac{N(n')}{N + n'} = \frac{7500(1308)}{7500 + 1308} = 1113.74 \cong 1113$$

Therefore, based on these formulas, you should survey at least 1113 students to meet the objectives of your study.

The preceding example illustrates how to determine sample size for binary responses. Determining sample sizes for other response scales, such as those with Likert response formats, or those that measure quantities using ordinal, interval, or ratio scales, requires a different set of procedures. We will examine these procedures in a subsequent chapter.

OBSERVATION

For many types of research, there are no better tools than a pair of eyes, a sharp pencil, and a sheet of paper (possibly an audio and video recorder). Although good observational research is difficult to conduct, it can produce a wealth of useful information. With proper planning and work, one can conduct observational research that includes systematic manipulation of independent variables and the recording of dependent variables all in a natural setting. In other words, one can conduct experiments and produce data that have clear external validity. This section provides you with a brief introduction to observational methods of measurement. Additional details on these techniques are provided later in this text, when we discuss qualitative methods in more depth.

Observational techniques go by many names, including "field studies," "naturalistic observation," or "natural experiment." The essential feature of this research technique is that we watch and record the individual's ongoing behaviors. Here are three brief examples of how one might use observational techniques.

An Experiment in Anxiety: Many developmental psychologists study separation anxiety, the emotional reaction that children have when separated from their parents (e.g., Ainsworth & Bell, 1970; Ainsworth, Blehar, Waters, & Wall, 1978). This research has many direct applications to working parents and day-care issues. To study separation anxiety, a researcher may invite parents to bring their toddlers to the department's laboratory. The researcher can then observe the children's reaction to different situations through a two-way mirror or with a hidden video camera. In this type of research, the researcher is able to control many of the different variables such as the child's age and the test situation. One can easily use naturalistic observation in a true experiment.

Watching the Bully in the Workplace: Bullying and abusive behavior is not simply confined to the schoolyard. As an occupational health issue, many researchers have begun to examine the causes and consequences of bullying and violent behavior on the job (e.g., Einarsen, 1999). A popular method of studying bullying is to watch employee interactions during breaks or during "downtime." As in the previous example, the researcher finds some inconspicuous place where he or she can observe interactions. As with all observational research, the person collecting the data will watch for and record the duration, frequency, or intensity of specific behaviors.

Who Said What to Whom? In some cases, the target behavior will be what people say to each other. For this type of research, we make a record of what people say in a conversation. DePaulo and Kashy (1998) conducted an interesting example of this type of research by asking participants to keep a daily journal of the lies they told their friends, family, and associates. Using an objective scoring procedure, research assistants who did not know the hypotheses being tested read and scored the episodes described in the journals. DePaulo and Kashy found that the participants did not lie often to friends but occasionally told "white lies" to protect their friends' feelings. By contrast, the participants were more willing to tell self-serving lies to casual acquaintances and strangers.

What to Observe

The steps we take to create a survey are the same steps we can follow to develop a strategy for observing, recording, and quantifying behaviors. First, we need to decide what behaviors we want to observe. Most creatures (well, maybe not slugs . . .) engage in multiple simultaneous behaviors, making complete observation impossible in any one study with only one observer. Therefore, we need to identify a manageable set of behaviors that we can realistically and consistently observe and document. As with all research, we select observable behaviors that correspond to the constructs around which our study has been designed.

The next step is to ensure that there is a clear operational definition of the behaviors to be monitored. Some variables may be easy to define and measure rather objectively. For example, we can count the number of cigarettes smoked in an hour, the time spent watching television and the name of the shows selected, or the number of four-letter curses offered by a cartoon character within one 30-minute episode.

Other potentially observable behaviors are not so easily defined. For example, what do we mean by an aggressive act? Is a tackle in a football game an aggressive act or part of playing a game? If one person accidentally hurts another, is the behavior aggressive? What about verbal threats? Will you include these in your definition of aggression? Consider health and social behavior. Assume that a researcher wanted to study how patients being treated for addictive behavior react when they are the new member of a therapy group. The focus of a study along these lines might be to determine how an individual behaves when he or she is the new member of a group. Will the person immediately contribute to the group, or will he or she be cautious and watch the other group members without speaking up or interacting with them?

Strategies for Collecting Quantitative Observational Data

Once you have defined the behaviors to observe, you will need to select the type of data to collect. As a generality, there are three types of quantitative observational recording techniques: frequency recording, duration recording, and interval recording. As with all aspects of research, selection of the recording technique depends on the definition of the variable and the research hypothesis.

Frequency Recording

As the name implies, for frequency recording, we count the number of times that a behavior occurs during a specified interval. Most researchers use frequency recording to count discrete behaviors that generally have a fixed duration. Therefore, we can count the number of times one student interrupts another student, the number of times a worker on an assembly line shares a tool, or the number of cigarettes a person smokes. Other behaviors, such as studying, reading a book, napping, and watching television are behaviors better measured using other recording techniques because they are ongoing behaviors that may occur infrequently, but last for an extended period of time.

Duration Recording

Duration recording focuses on the amount of time a person spends engaged in one behavior. This method is useful for measuring ongoing behavior that may last for an unpredictable amount of time. For example, a researcher may use duration recording to monitor employee behavior during a fire drill or other disruptive occurrence. Specifically, the researcher could time how quickly the individual stopped working and followed fire drill protocol and instructions and how long it took for the individual to return to work following the drill.

Interval Recording

For interval recording, the researcher observes one or more individuals for a fixed period of time. The researcher then divides the session into intervals of equal lengths. The interval length is short enough that the individual can engage in only one behavior during the interval (e.g., 15 seconds). During each interval, the researcher indicates the presence or absence of the relevant behavior. In most cases, there may be a pause or delay between intervals. Although this technique is often recommended to researchers, it has also been the target of criticism (Altmann, 1974; Murphy & Goodall, 1980; Tyler, 1979). Altmann concluded that the method has no general value for behavioral research because it does not accurately measure the frequency or the duration of behavior. She also criticized interval recording because the method cannot adequately indicate the proportion of time the individual engages in a behavior.

Special Considerations for Observational Research

As with all research procedures, there are special controls that you should consider to collect accurate, reliable, and valid data. The essential question to ask is "Who is collecting the data and do they know what they are doing?" The reliability of the data depends on the vigilance of the person observing and recording the data. The most common method to ensure that the data are reliable is to have two or more observers record the data and then report the average of the observations.

Before you aggregate the data, however, you need to ensure that the observers agree with each other by determining the **interrater reliability**. The interrater reliability is a statistical index that represents how well the records from multiple observers match. If the interrater reliability is high, we can place greater confidence in the data and proceed with their analysis. If the interrater reliability is low, analysis of the data may not be useful because the measurement error is too great. Low interrater reliability can indicate that the observers need to be better trained, that the definitions need to be clarified, that the recording procedure needs to be revised, or that some combination of these solutions may need to be used. There are many ways to assess interrater reliability. The following sections examine the more popular options.

Coefficient Kappa

One of the more popular methods of determining interrater agreement is Cohen's kappa, κ. Researchers use κ to assess interrater reliability when the data represent a nominal or an ordinal scale. As an example, consider the data presented in Table 8.7. In this example, two raters independently observed the same person and counted the number of times one of three behaviors occurred. The numbers in the table represent their assessments. Look down column A of the table. The numbers are 15, 2, and 1. These numbers represent agreements and disagreements. Raters 1 and 2 agreed 15 times on the occurrence of behavior A but disagreed on the classification of three other behaviors. The numbers along the diagonal of the table represent agreements; the numbers in the other cells represent disagreements. Cohen's κ allows us to determine the level of agreement among the raters. To calculate κ, we use the following equation:

Table 8.7 Example of Data from Two Independent Raters Who Classified the Same People or Behaviors into One of Three Categories

		Rater 1 category			
		A	B	C	Row totals Σ
Rater 2 category	A	15	1	1	$R_1 = 17$
	B	2	21	2	$R_2 = 25$
	C	2	1	16	$R_3 = 18$
Column totals	Σ	$C_1 = 18$	$C_2 = 23$	$C_3 = 19$	$T = 60$

$$\kappa = \frac{P_A - P_C}{1 - P_C}, \quad \text{where} \quad P_A = \frac{\sum O_{ii}}{T} \quad \text{and} \quad P_C = \frac{\sum R_i C_i}{T^2} \qquad (8.3)$$

In this equation, P_A represents the proportion of agreement between the raters. To determine P_A, we add the observed agreements, O_{ii}, and divide by the total number of observations, T. The second component of κ is P_C, which estimates the number of agreements that would have happened by chance.

Procedure for Calculating Coefficient Kappa, κ

$$P_A = \frac{\sum O_{ii}}{T} = \frac{15 + 21 + 16}{60} = \frac{52}{60} = .8667$$

$$P_C = \frac{\sum R_i C_i}{T^2} = \frac{(18 \times 17) + (23 \times 25) + (19 \times 18)}{60^2} = \frac{1223}{3600} = .3397$$

$$\kappa = \frac{P_A - P_C}{1 - P_C} = \frac{.8667 - .3397}{1 - .3397} = \frac{.5270}{.6603} = .7981.80$$

Interpreting κ is straightforward; the larger the better. Values of κ can range between 0, which represents no interrater reliability, and 1.0, which represents perfect interrater reliability. Most researchers strive to have κ be greater than .75. When $\kappa < .50$, there is too much disagreement among the judges to produce useful information. Table 8.8 lists general guidelines for interpreting the size of κ.

In some situations, there may be more than two observers. Although the calculations are more cumbersome, we can still calculate κ to determine the interrater agreement among the observers. Another technique that is perhaps more commonly used with more than two raters is to calculate a Pearson correlation coefficient, r, to establish the relationship between multiple sets of ratings. Researchers use this statistic specifically when observations are collected on an interval or a ratio scale. Therefore, the statistic might be used for determining the reliability of frequency and duration measures. Chapter 9 is devoted to this important statistic, so much more detail is provided there. In brief, though, we interpret the r as we do κ. As with κ, the absolute values of r can range between 0 and 1, with larger values representing greater levels of interrater agreement. If you are interested in

Table 8.8 General Guidelines for Interpreting the Size of κ (from Altman, 1991)

Value of κ	Level of agreement
<.20	Poor
.21–.40	Fair
.41–.60	Moderate
.61–.80	Good
.81–1.00	Excellent

learning more about the use of r for establishing interrater agreement, one place to start is a recent review of this approach by Brown and Hauenstein (2005).

Another technique for indicating agreement between multiple raters is to calculate an intraclass correlation coefficient (ICC) that reflects the overall similarity of a set of multiple ratings of the same target person or object. There are several ICC forms that are more or less appropriate, depending on your specific research situation. Thus, a full examination of this technique is beyond the scope of the present book, but we recommend that you review work including Bliese (2000) and other more recent statistics books for assistance in this area.

Improving Interrater Reliability

There are a number of things that you can do to maximize the reliability and validity of observational data. One of the most important is to keep the rater from knowing your hypothesis.

If the raters know what you expect to find in the study, they may be tempted to score their observations in favor of your expectations. Another is to train the observers. Barkhof et al. (1997) provided an example of the need for training to ensure reliable data. The researchers examined the ability of five novice researchers and five expert researchers to read magnetic resonance images (MRIs) and rate the lesions of patients with multiple sclerosis (MS). Lesions are scar tissue in the central nervous system and appear as white spots on the MRI. As you might guess, the interrater agreement for the novice researchers was low compared to the experts (novice $\kappa = .37$ vs. expert $\kappa = .65$). After intensive training, the interrater reliability of the novices and experts improved (novice $\kappa = .65$ vs. expert $\kappa = .74$).

Observing behavior is not easy work. It can be difficult to do well for long uninterrupted periods. Therefore, researchers who do this type of work limit their observation time to short periods or work in shifts with other observers. Altmann (1974) described her experiences observing primate behavior and the difficulty of remaining alert during the observation session. She wrote that "even with two observers, one 15-minute sample per hour was near the upper limit of our capacity when obtaining an accurate record" (p. 246). The implication of this is that observational sessions should be relatively short (e.g., less than 15 minutes) and punctuated with ample rest. Fortunately, video cameras make the task of observing behavior easier. Cameras do not get tired or bored, and they record all the behaviors that occur. Furthermore, the researcher has access to a permanent and rich source of information of all behaviors. Therefore, the researcher can review and score the behaviors at his or her convenience and at a pace that is not exhausting.

Another important feature of the video camera is that it is easy to hide. Hiding the video camera helps to ensure that the people being observed act naturally. If you wanted to study bullying in the workplace, using a hidden camera will record more natural behavior than having a researcher obviously watching employees.

As you learned in Chapter 2, however, there are significant ethical issues that the researcher must confront before conducting this type of research.

Participant Observation Research

A variant of naturalistic observation is participant observer research. There is an important difference between naturalistic observation and participant observation. For the most part, the observer in naturalistic observation is not an immediate part of the ongoing behavior as he or she attempts to record the data unbeknownst to the individuals being observed. By contrast, the participant observer joins the group to study the group's behavior.

Participant observation does allow one to study a group of people, especially if their behavior is covert or underground. This technique is especially popular among anthropologists and sociologists who study small social groups. The participant observation technique has some serious risks, however. An infamous example is Laud Humphreys's (1970) research for his book *Tearoom Trade: Impersonal Sex in Public* Places. Humphreys wanted to study the men who engaged in homosexual acts in public places. To study these men, Humphreys began to hang out at public restrooms, known as "tearooms" among homosexuals, and acted as a lookout while the men had their sexual encounters. He then secretly followed the men to their homes and later interviewed them. Many commentators have criticized Humphreys's tactics because they violated the men's right to privacy. Therefore, conducting participant research cannot be entered into lightly. Participant research can also be dangerous. Many groups of people wish to remain closed to outsiders and react poorly to a stranger in their midst. For example, Thompson (1985), studying the Hell's Angels, received a severe beating from the members of the gang he had attempted to join.

RESEARCH IN ACTION: ANALYSIS OF ASSAULTS

Acquaintance rape, or date rape, is an unfortunate experience that too many people encounter. Although this behavior has been examined for many years, Larimer, Lydum, Anderson, and Turner (1999) believed that much of the research was biased against men. Larimer and her colleagues noted that for many of the previous studies, the researchers examined women's experiences exclusively or used sex-biased survey techniques. Therefore, Larimer et al. attempted to document the prevalence of unwanted sexual encounters experienced by men and women and the role that alcohol consumption played in these encounters.

Larimer et al. (1999) asked participants to complete a series of questionnaires that examined the prevalence of unwanted sexual encounters and the use of alcohol. All of the questionnaires were scales developed and published by other researchers. Table 8.9 presents a portion of two of the scales examining unwanted sexual experiences and the relation between alcohol consumption and unwanted

Table 8.9 Examples of Questions Larimer et al. (1999) Used to Examine the Prevalence of Unwanted Sexual Encounters and the Use of Alcohol during Unwanted Sexual Encounters

1. In the past year, have you been in a situation where your partner became so sexually aroused that you felt it was useless to stop them even though you did not want to have sexual intercourse?
2. In the past year, have you had sexual intercourse with someone who didn't really want to because you felt pressured by their continual arguments?
3. In the past year, have you been in a situation where someone used some degree of physical force (twisting your arm, holding you down, etc.) to get you to have sexual intercourse with them when you didn't want to, whether or not intercourse actually occurred?
4. In the past year, have you had someone attempt sexual intercourse with you by giving you alcohol or other drugs, but intercourse did not occur?
5. In the past year, have you had sexual intercourse when you didn't want to because a person gave you alcohol or other drugs?

Items based on the Sexual Experiences Survey (Koss & Oros, 1982) as revised by Larimer et al. (1999).

1. Has drinking ever gotten you into sexual situations which you later regretted?
2. Because you had been drinking, have you ever had sex when you really didn't want to?
3. Because you had been drinking, have you ever had sex with someone you wouldn't ordinarily have sex with?
4. Have you ever been pressured or forced to have sex with someone because you were too drunk to prevent it?
5. Have you ever pressured or forced someone to have sex with you after you had been drinking?

Items based on the Young Adult Alcohol Problem Severity Test (Wood, Johnson, & Sher, 1992) as presented by Larimer et al. (1999).

sexual encounters. For their first questionnaire, Larimer et al. revised the Sexual Experiences Survey (Koss & Oros, 1982) to ensure that questions were not sex biased. Larimer et al. used 296 (165 males, 131 females) students who were new members of fraternities and sororities at a large West Coast public university. The majority of the students were freshmen and sophomores.

Figure 8.1 presents the results for the Sexual Experiences Survey. One notable feature of these data is the percentage of men reporting that they had experienced unwanted sexual encounters or had felt pressured into a sexual encounter. Also apparent in the data is that women reported higher percentages of some form of coercion (physical force or alcohol/drug use) to engage in sexual intercourse. These data are important. First, the data support previous research by replicating the finding that women often experience some forms of coercion during an unwanted sexual encounter. Second, the data suggest that both men and women experience unwanted sexual encounters. These data are interesting as they conflict with the stereotype that men are always the perpetrators and women are victims of these situations.

Figure 8.2 presents the percentage of participants who had sexual encounters related to their drinking behavior. The data make clear that nearly half of the men and women in the study later regretted a sexual encounter related to their drinking.

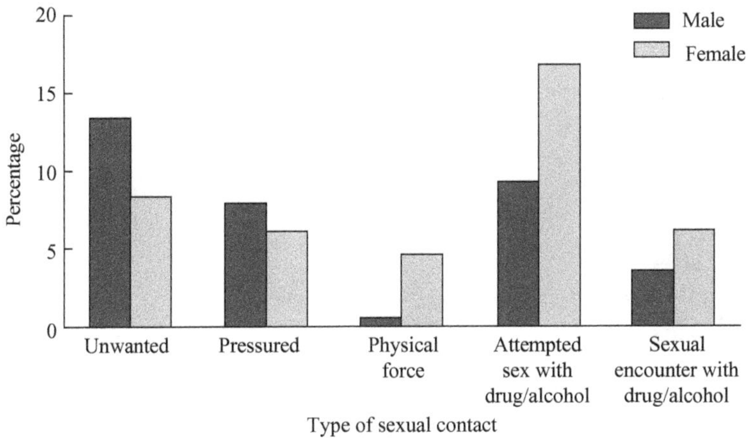

Figure 8.1 Percentage of participants reporting various types of unwanted sexual encounters. *Note:* Graph based on data presented by Larimer et al. (1999).

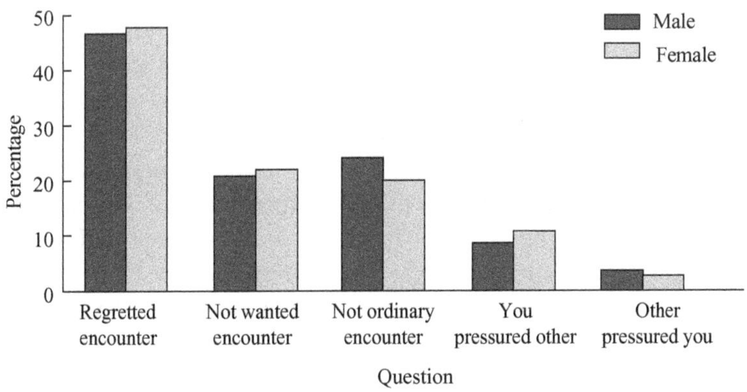

Figure 8.2 Percentage of participants reporting various types of unwanted sexual encounters related to drinking. *Note:* Graph based on data presented by Larimer et al. (1999).

Finally, Figure 8.3 presents the mean number of alcoholic drinks consumed per day during the previous 3 months. Two interesting trends appear in these data. First, men claim to consume more drinks per day than do women. Second, participants who reported an unwanted sexual encounter appear to drink more than those who had not experienced an unwanted sexual encounter. The data led Larimer et al. 1999 to conclude that men and women experience unwanted sexual intercourse. Their data also implicate alcohol as a major contributor to sexual victimization of both men and women. Because sexual coercion and unwanted sexual intercourse can have a dramatic impact on one's emotional well-being, it is essential that researchers better understand this phenomenon. As Larimer and her colleagues concluded, "further empirical study of the emotional consequences of these events and the context surrounding them for both genders is warranted" (p. 307).

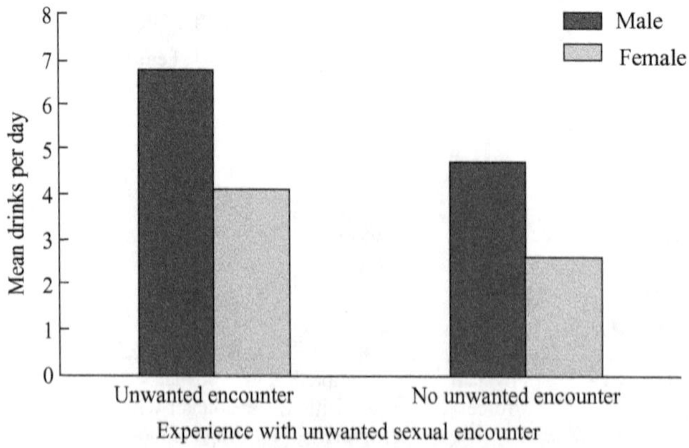

Figure 8.3 Mean number of alcoholic drinks consumed per day during the past 3 months for men and women who had or had not experienced an unwanted sexual encounter. *Note:* Graph based on data presented by Larimer et al. (1999).

Before we embrace the data too quickly, we need to examine several features of the study. To their credit, Larimer et al. (1999) recognized and described many of the problems with their research and called for additional and more refined studies. We can begin by examining the population and the sample.

The sampling population consisted of students who had joined a Greek-letter society. Although fraternities and sororities are popular on many college campuses, their membership may not be representative of the student population. Greek-letter organizations may be attractive to only a segment of the contemporary college student body. Similarly, the life experiences of students living in a fraternity or sorority may be different from life in a residence hall. For example, most Greek-letter organizations sponsor intensive training programs that sensitize their members to the dangers of alcohol abuse and sexual coercion. Thus, we cannot be sure that life in a fraternity or sorority is representative of the typical college student.

A related problem is the age of the participants when they completed the survey. Most of the participants were freshmen or sophomores. Consequently, these students may not have had sufficient time living at college to experience the types of sexual encounters experienced by older students. If the sample had included a broader cross section of all students, we could examine the relation between the age of the student and the risk of assault. We can raise other questions about the sampling population. Are the experiences of students at large state universities similar to those of smaller colleges? Are students who attend West Coast institutions representative of students attending college in other parts of the country? Future research on this topic should attempt to sample from a broader range of students. The sample would include a greater representation of all segments of the typical college population. There is no reason that we must limit our analysis to college students. Although the data may not be as easy to obtain, it would still be informative to examine the experiences of young adults not enrolled in college. We can also question whether the sample is sufficiently large. Although 296 participants is a large sample, it may not be sufficiently large to detect small differences between men's and women's behavior or events that do not occur frequently. Consequently, those who wish to replicate this study should consider a larger sample size.

Finally, we can ask whether the wording of the questions produces the answers we need. For example, do men and women "regret" unwanted sexual relations for the same reasons? Similarly, is the phrase "pressured or forced" as clear as it can be? The phrase seems to require considerable interpretation on the part of the participant. Larimer et al. (1999) studied an interesting and important topic. Moreover, they reported findings that challenge several preconceived beliefs. Their research also demonstrates that one can examine interesting topics using questionnaire methods. At the same time, this research confirms a theme that recurs throughout this book—no individual study is complete by itself. The Larimer et al. study is one of many studies that examine the relation between date rape and alcohol and will surely not be the last such report. In fact, as of the time this chapter was written, Larimer et al. had been cited in more than 40 additional published studies.

KNOWLEDGE CHECK

12. Why is it necessary to pretest a questionnaire before using it in a research project?

13. Can a researcher use closed-response questions in a face-to-face interview?

14. A researcher developed a series of open-ended and closed-response questions. What are three ways the researcher could collect the data? What are the relative advantages and disadvantages of each?

15. Describe in your words the advantages and disadvantages of open-ended and closed-response questions.

16. Imagine that you are planning to conduct a survey and need to estimate the number of participants you need to include in your sample. You decide that for most of the questions, $p = .50$ and that $\alpha = .02$. How many participants will you need in the sample if your population consisted of

a. 1000 people?
b. 5000 people?
c. 25,000 people?
d. 50,000 people?

17. Dr. Blume has trained observers to use frequency recording to record the behavior of 12 schizophrenic clients. According to the analysis of the data $\kappa = .47$, do you think the observers are sufficiently trained to allow Dr. Blume to proceed with the research?

18. Imagine that you have been hired to assess the content of children's television shows. You are to determine how often the different children's shows depict aggressive acts versus altruistic acts.

a. How would go about selecting the children's shows to watch?
b. How would you define aggressive acts?

c. How would you define altruistic acts?
d. Which measurement technique (frequency, duration, or interval recording) would you use and why?

19. Assume that you used two trained observers to watch the episodes from various children's shows. The observers were to watch a segment, typically 20–30 seconds long, and classify the behavior as aggressive, altruistic, or other. The following table presents the data. How well do the raters agree with each other?

		Rater 1		
		Aggressive	Altruistic	Other
Rater 2	Aggressive	121	13	57
	Altruistic	3	67	45
	Other	12	10	93

20. As a researcher, what ethical responsibilities do you have when conducting personal interviews or self-administered questionnaires?
21. If you were going to use a hidden video camera to record the behavior of a group of people, what ethical issues would you need to resolve?

CHAPTER SUMMARY

All the sciences depend on objective measurement, and research on people in the workplace is no exception. Although the primary focus of this chapter was creating and using questionnaires, the concepts reviewed in this chapter apply to any situation wherein the researcher must collect data. Many people do not understand the uses and limitations of measurement. Two common errors include reification of the measurement and denial of the measurement. Reification is an example of mystique of quantity and occurs when people infer too much meaning from a single measurement. In contrast, some people dismiss many psychological tests, believing that the underlying construct cannot be measured. We create tests to help us describe and explain the phenomenon that we are studying.

Creating an appropriate test or measure of the dependent variable depends on the hypothesis we wish to test, the conditions of our research, and the accuracy of various techniques. Several of the more common ways of collecting information include personal interviews and self-administered surveys. Personal interviews, while time consuming and expensive to conduct, can provide valuable forms of information. By contrast, self-administered surveys are inexpensive and can produce much valuable information.

For much of the chapter, we reviewed techniques to create questions and scales to collect data. The goal of these techniques is to prepare simple and objective questions that the participants will answer honestly. When we plan any research project, it is important to determine the optimal sample size for the research. Therefore, we reviewed one method for estimating sample size. This technique allows us to predict, and thereby control, the accuracy of the population estimates. Observational techniques are also common in behavioral research. Two of the more commonly used methods for observational research are frequency recording and duration recording. These research techniques, like the questionnaire, require clear operational definitions of the behavior we wish to observe and attention to the procedures for collecting the data.

Whenever we use a measurement technique, we need to assess its reliability or consistency in measurement. We reviewed how the coefficient kappa allows us to evaluate the interrater reliability among two or more raters. In the "Research in Action: Analysis of Assaults" section, we examined how researchers used a survey method to evaluate the interrelation between alcohol abuse and date rape. We used this research to illustrate how the researchers selected a series of questionnaires to measure the relevant dependent variables. The example also allowed us to practice a critique of the data.

CHAPTER GLOSSARY FOR REVIEW

Binary Response A response that has only two answers, such as yes or no.

Closed-Response Question A question that the participants answer by selecting from among alternative statements provided by the researcher.

Interrater Reliability A descriptive statistic that indicates the degree to which two or more observers agree on the classification of a behavior or an object.

Mystique of Quantity An uncritical trust in the meaning and importance of the numbers produced by measurement.

Open-Ended Question A question that requires the participants to state or write their answer to the question.

Order Effect The effect of asking a sequence of questions on how the participant answers subsequent questions.

Priming Cue Questions that cause the participant to think about a topic and to more thoughtfully respond to subsequent questions.

Reification To treat an abstract construct as if it has material existence independent of the words.

Response Set The tendency to give the same answer to all questions regardless of the true answer.

REFERENCES

Ainsworth, M.D.S. & Bell, S.M. (1970). Attachment, exploration, and separation: Illustrated by the behavior of one-year-olds in a strange situation. *Child Development*, *41*, 49–67.

Ainsworth, M.D.S., Blehar, M.C., Waters, E., & Wall, S. (1978). *Patterns of attachment*. Hillsdale, NJ: Erlbaum.

Altman, D.G. (1991). *Practical statistics for medical research*. Boca Raton, FL: Chapman & Hall.

Altmann, J. (1974). Observational study of behavior: Sampling methods. *Behaviour*, *49*, 227–267.

Barkhof, F., Filippi, M., van Waesberghe, J.H., Molyneux, P., Rovaris, M., Lycklama a' Nijeholt, G., et al. (1997). Improving interobserver variation in reporting gadolinium-enhanced MRI lesion in multiple sclerosis. *Neurology*, *49*, 1682–1688.

Bliese, P.D. (2000). Within-group agreement, non-independence, and reliability: Implications for data aggregation and analysis. In K.J. Klein & S.W.J. Kozlowski (Eds.), *Multilevel theory, research, and methods in organizations: Foundations, extensions, and new directions* (pp. 349–380). San Francisco: Jossey-Bass.

Brown, R.J. & Hauenstein, N.M.A. (2005). Interrater agreement reconsidered: An alternative to the r_{wg} indices. *Organizational Research Methods*, *8*(2), 165–184.

CHANG, L. (1994). A psychometric evaluation of 4-point and 6-point Likert-type scales in relation reliability and validity. *Applied Psychological Measurement, 18*, 205–215.

COCHRAN, W.G. (1977). *Sampling techniques*. New York: John Wiley & Sons.

CRONBACH, L.J. (1950). Further evidence on response sets and test design. *Educational and Psychological Measurement, 10*, 3–31.

DAVIDSON, J.W. & LYTLE, M.H. (1992). *After the fact: The art of historical detection* (3rd ed.). New York: McGraw-Hill.

DEPAULO, B.M. & KASHY, D.A. (1998). Everyday lies in close and casual relationships. *Journal of Personality and Social Psychology, 74*, 63–79.

DEWDNEY, A.K. (1997). *Yes, we have no neutrons: An eye-opening tour through the twists and turns of bad science*. New York: John Wiley & Sons.

DILLMAN, D.A. (2000). *Mail and Internet surveys: The tailored design method* (2nd ed.). New York: John Wiley & Sons.

EINARSEN, S. (1999). The nature and causes of bullying at work. *International Journal of Manpower, 20*, 16–27.

FONTANA, A. & FREY, J.H. (1994). Interviewing: The art of science. In N.S. Denzin & Y.S. Lincoln (Eds.), *Handbook of qualitative research* (pp. 75–103). Thousand Oaks, CA: SAGE.

GOULD, S.J. (1981). *The mismeasure of man*. New York: W. W. Norton.

HUMPHREYS, L. (1970). *Tearoom trade: Impersonal sex in public places*. Chicago: Aldine.

KAPLAN, A. (1964). *The conduct of inquiry: Methodology for behavioral science*. San Francisco: Chandler.

KOSS, M.P. & OROS, C.J. (1982). Sexual experiences survey: A research instrument investigating sexual aggression and victimization. *Journal of Consulting and Clinical Psychology, 50*, 455–457.

LARIMER, M.E., LYDUM, A.R., ANDERSON, B.K., & TURNER, A.P. (1999). Male and female recipients of unwanted sexual contact in a college students sample: Prevalence rates, alcohol use, and depression symptoms. *Sex Roles, 40*, 295–308.

MILES-TAPPING, C. (1996). Power wheelchairs and independent life styles. *Canadian Journal of Rehabilitation, 10*, 137–145.

MURPHY, G. & GOODALL, E. (1980). Measurement error in direct observation: A comparison of common recording methods. *Behaviour Research & Therapy, 18*, 147–150.

PRESTON, C.C. & COLMAN, A.M. (2000). Optimal number of response categories in rating scales: Reliability, validity, discriminating power, and respondent preferences. *Acta Psychologica, 104*, 1–15.

ROSENBERG, M. (1965). *Society and the adolescent self-image*. Princeton, NJ: Princeton University Press.

ROTTER, J.B. (1966). Generalized expectancies for internal versus external control of reinforcement. *Psychological Monographs, 80*(1, Whole No. 609).

SALANT, P. & DILLMAN, D.A. (1994). *How to conduct your own survey*. New York: John Wiley & Sons.

SWIM, J.K., AIKEN, K.J., HALL, W.S., & HUNTER, B.A. (1995). Sexism and racism: Old-fashioned and modern prejudices. *Journal of Personality and Social Psychology, 68*, 199–214.

THOMPSON, H. (1985). *Hell's Angels*. New York: Ballantine.

TYLER, S. (1979). Time-sampling: A matter of convention. *Animal Behaviour, 27*, 801–810.

WEATHINGTON, B.L. (2011). Whence applied science in a person-centric work psychology? *Industrial and Organizational Psychology: Perspectives on Science and Practice, 4*(1), 136–137.

WILKINSON, L. (1999). Statistical methods in psychological journals: Guidelines and explanations. *The American Psychologist, 54*, 594–604.

Chapter 9

A Model for Research Design

To call in the statistician after the experiment is done may be no more than asking him to perform a postmortem examination: He may be able to say what the experiment died of.

—Sir Ronald A. Fisher

INTRODUCTION

A research design refers to the methods used to collect data that will decisively answer an empirical question. Consequently, a good research design addresses three concerns. First, it must be efficient; the design should produce the best-quality data for the least amount of time, effort, and money. Second, the design must eliminate factors that bias the results. Third, the research design should produce useful data that clearly address the research question and account for alternative explanations. In this chapter, we examine the details of good research design for projects that incorporate two levels of an independent variable. Using this format, we can review the basic principles of research design that apply to all research, no matter how complex.

Understanding and Conducting Research in the Health Sciences, First Edition. Christopher J. L. Cunningham, Bart L. Weathington, and David J. Pittenger.

A MODEL FOR RESEARCH DESIGN

A common myth is that a researcher is an introverted, white-lab-coat-wearing, dispassionate, and objective umpire of the data who "calls 'em as I sees 'em." Unfortunately, there are too many examples in the history of science to treat this statement as anything but fantasy. Many researchers across disciplines have lost their objectivity and allowed their preconceived beliefs to cloud their judgment and interpretation of the data. Consequently, the purpose of good research design is to protect researchers from lapses in objectivity and the bias inherent in personal judgment.

In any contemporary research that focuses on people, we have an additional source of bias and error: the people. Participants in this type of research bring their own anxieties, biases, and awareness that they are participating in a research study. Rarely do we have any idea what this means to them or how this affects their behavior in the study. Being a health sciences researcher requires knowledge of human behavior *and* research design in order to conduct useful scientific studies.

We could easily fill this book with examples of the factors that can bias the results of a research project. A more efficient and useful alternative approach, however, is to use a model that highlights the phases of research that are most sensitive to different forms of bias. Figure 9.1 illustrates the potential sources of bias arising from the interaction among the elements of researcher, design of the study, and participants. Many of the ideas presented in this illustration come from the work of Hyde (1991) and Rosnow and Rosenthal (1997). In this figure, the column on the left summarizes behaviors of the participants, while the column on the right summarizes behaviors of the researcher. The central column represents the general stages of research. The horizontal lines represent instances where the participants' and researchers' behavior interact.

Purpose of Research

A primary source of bias that can affect research comes from within the researcher. Most research ideas emanate from a researcher's theoretical orientation and personal experience. The researcher's perspective on the phenomena under analysis can have considerable influence over how he or she will view a research question. This, in turn, will most definitely influence how a research project is designed, potentially in a manner that leads to bias and causes the researcher to neglect important alternative viewpoints or factors. Therefore, as a researcher, we urge you to be willing to examine alternative explanations until the data clearly indicate that you need to move on to a different perspective. One way to practice this openness in your own research is to read broadly, stretching beyond the bounds of your particular subfield of study to search for research in other disciplines that may be related to your own interests. You will be surprised at what you can find.

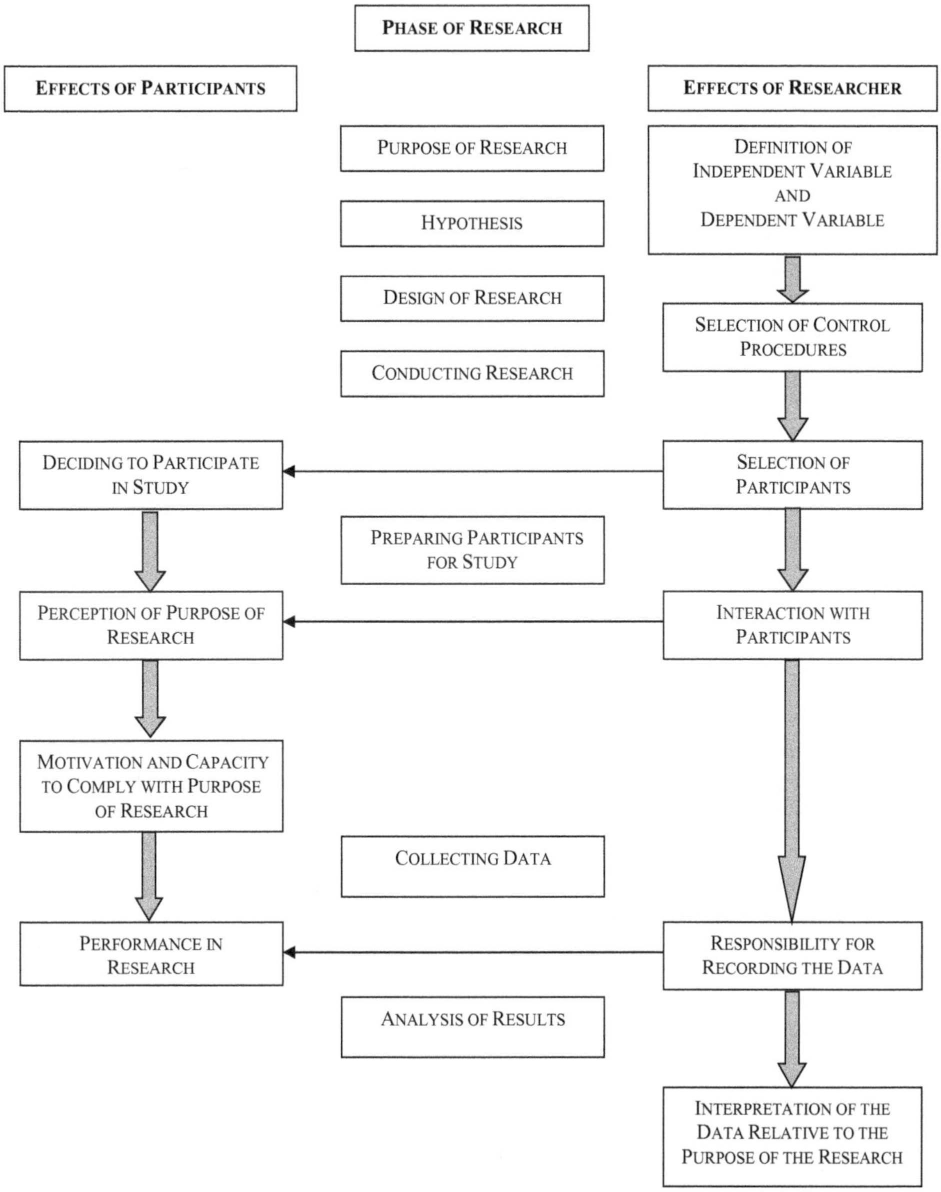

Figure 9.1 Illustration of potential sources of bias arising from the interaction among the researcher, the design of the study, and the people participating in the research.

Design of Research

After identifying the purpose for the research and a hypothesis or at least guiding research question, the next step in the process presented in Figure 9.1 is the design phase, during which the researcher develops controls for potential confounding variables. As we reviewed in Chapter 4, confounding variables are conditions that reduce the internal validity of the study. As we progress through this book, we will examine different research designs that help reduce or eliminate the effects of well-known confounding effects.

Selection of Participants

Once the researcher begins to select participants for a study, the sources for potential bias increase dramatically. In Chapter 7, you learned how to develop a representative sample of the population. As you learned, the need to ensure that your selection procedure will yield a representative sample of the population is a critical factor in determining the eventual generalizability or the external validity of your research. Sampling procedures can often be a source of criticism especially when the sample does not appear to be representative of the population. We cannot brush off these concerns as trivial. Therefore, it is important that you clearly define the target population you wish to study and ensure that your actual sample is a fair representation of that population.

Volunteer Participants

Related to our selection of participants into a sample is the question of which members of the sample ultimately choose to participate in our research. Do people who volunteer to participate in our research represent the typical person? The answer is probably no. Imagine conducting a study examining adults' knowledge of principles related to diet and health. What would happen if you conducted your study only among people who were visiting their physician for a routine checkup? It may well be that people who routinely visit their physician have different knowledge of diet and health than people who seek medical care only in emergency situations.

Preparing the Participant for the Study

Have you ever been driving on the interstate and seen a police car turn onto the road behind you? What do you do? If you are like us, you first make sure you are not speeding. We become self-conscious of driving when we see the patrol car. The same type of self-conscious reaction occurs in people who participate in a research project; they know (or perhaps dread) that a researcher is scrutinizing their health.

In Figure 9.1, you can see that there is a link between the researcher's interaction with the participants and the participants' perception of the purpose of the research. This link indicates that the researcher's behavior influences how each participant behaves during the study. Specifically, the participants' ideas about the purpose of the research may cause them to do what they think the researcher wants them to do. We can review this interaction between the researcher's behavior and participants' perceptions using a classic example.

The **Hawthorne effect** is an example of how the researcher's behavior and the participants' perceptions of the research may affect the reactions of those participants. The name of this phenomenon is linked to the organization in which this research occurred: the Hawthorne plant of the Western Electric Company, beginning in the late 1920s. The researchers in this series of studies wanted to find ways to improve assembly-line productivity by varying working conditions. Much to their surprise, the researchers found that almost any change, even returning to the original working conditions, at least temporarily improved worker productivity (Sundstorm, McIntyre, Halfhill, & Richards, 2000). There are many possible explanations for this outcome, but most come back to the participants' awareness of being the target of researchers' attention during the research process. Performance increased when changes were tried, possibly due to participants' expectations that their performance was being watched closely and might lead to termination of employment or some other negative consequence. Thus, despite what participants may or may not have been told by the researchers, it is possible they believed their performance in the study would have an impact on future job performance evaluations and employment.

The Hawthorne research episode is a frequently told story, with many different versions (Gillespie, 1993). The history of exactly what happened at the Hawthorne plant is murky at best. Nevertheless, many writers use the Hawthorne episode to illustrate how a person's behavior changes when he or she knows that someone else is watching. The lesson from this version of the story is that we need to ensure that our research design and our behavior as the researcher do not create a **demand characteristic** among the participants that could contaminate the results of the study. A demand characteristic is any condition created by the design of the study that leads the participants to infer the purpose of the research and adjust their behavior in line with this inference, thereby biasing the data from what might have otherwise occurred.

Another version of the Hawthorne story is that the researchers inadvertently created a confounding variable in their study. During the study, the workers' pay reflected their productivity; the more components they made, the more money they earned. Therefore, the workers' behavior reflected a well-known behavioral and economic principle: People work harder for valued rewards such as compensation (Parsons, 1974).

The Hawthorne effect and demand characteristics are more than historical episodes and interesting phenomena described in textbooks; they are real threats to the quality or internal validity of any research project. If we study history to avoid making the same mistakes, then we should heed the lesson learned from the Hawthorne studies when we design our research projects. The take-home message from all of this is that researchers need to pay attention to the design and implementation phases of the study to rule out alternative explanations of the results.

Recording and Analyzing the Results

Hamlet: *Do you see yonder cloud that's almost in shape of a camel?*

Polonius: *By the mass, and 'tis like a camel, indeed.*

Hamlet: *Methinks it is like a weasel.*

Polonius: *It is backed like a weasel.*

Hamlet: *Or like a whale?*

Polonius: *Very like a whale.*

—Shakespeare (1604)

In this excerpt, Polonius is quick to see what Hamlet claims to see in the random formation of clouds. A single cloud quickly becomes a camel, a weasel, and then a whale all upon Hamlet's suggestion. Clearly, Polonius is a sycophant who readily agrees with everything Hamlet says. *Are researchers like Polonius? Do they allow their research hypothesis to influence their perception? What about research participants?*

Do researchers see what they want to see when observing a person's behavior? The answer is yes. If we believe something to be true, we tend to interpret things within that context and act in ways that bring about the expected result. Psychologists call this phenomenon the **self-fulfilling prophecy**. One of the classic examples of the self-fulfilling prophecy is the Pygmalion-in-the-classroom effect (Rosenthal, 1994; Rosenthal & Jacobson, 1968). To demonstrate the effects of the self-fulfilling prophecy, researchers told grade-school teachers that several of the students were "late bloomers" and would soon begin to excel in school. However, the late bloomers were actually randomly selected students. Nevertheless, many of these students had moderate increases in their IQ scores by the end of the year. This increase in IQ may have been due to the way the teachers interacted with the students. Because they expected the students to blossom academically, the teachers may have given the late bloomers extra attention, support, and encouragement during the year.

In general, we refer to the self-fulfilling prophecy in research as **experimenter bias** or an **experimenter effect** to denote its influence on how the researcher might treat the participants during the research practice. If the person collecting the data knows the research hypothesis, he or she may treat the participants in the groups differently, thus affecting the participants' behavior and confounding the results.

Even after the data are collected, what will convince us as researchers that the data truly support or refute our hypothesis? Can we look at the data and shout, "Eureka, we have solved the problem!"? Data typically do not really "speak for themselves." Analyzing data is an interpretative process and is therefore prone to a host of personal and subjective biases. Consequently, we tend to rely on statistical analysis of quantitative data and objective reasoning to reduce further opportunities for subjective inference and unsubstantiated conclusions. However, see Chapter 17 for more on alternative approaches to sense-making when data are more qualitative by nature.

Some of the answers to the research challenges identified thus far will come in the following pages and the other chapters of this book. Each research project is unique, however, and no single source can offer a fixed set of rules that covers every potential experiment and the design problems that will arise. The preferable alternative is to work through the framework illustrated in Figure 9.1 and use it to consider potential problems and suitable improvements for each research project.

To help us explain and to help you understand these principles, we want to use two research projects as examples. These hypothetical examples will help us illustrate the many choices and design questions that researchers must resolve. You should note that we present these examples in their most simple form. As you will learn, health sciences research can become extremely complex due to the many variables associated with underlying medical conditions, lifestyle patterns, and general attitudes toward health and self-care. For the purposes of this chapter, we focus on simple examples to help you understand basic concepts of research design; understand that these principles easily generalize to even the most complicated research designs.

EXAMPLE 9.1 ***Sex Differences in Response to Medication***

Do men and women respond the same way to a specific medication? It may well be that men and women respond differently to a drug, with one group being more or less responsive to the effects of the medication. Indeed, it may be that a particular drug could be helpful for men but harmful for women (or vice versa).

EXAMPLE 9.2 ***Effectiveness of Medications to Treat a Specific Condition***

Are all medications equally effective in treating a specific medical condition? With many drugs available to treat a specific condition, which medication has the greatest effect on a patient's underlying symptoms?

WHAT IS THE INDEPENDENT VARIABLE?

It is critical to ensure that we define all of our terms as clearly as possible. Consider the research examining sex differences and drug effectiveness. The independent variable in this instance is a participant's sex: men versus women. Technically, sex is a subject variable because the researcher cannot actually manipulate it or randomly assign participants to be either male or female. Nevertheless, the researcher can use the variable to predict the dependent variable. By the way, it is increasingly common to use the word *sex* when referring to the biological distinction between men and women. The word *gender* is reserved for referencing the underlying identity of individuals or when referring to men and women as members of social groups or traits associated with men and women.

For medical research, some subject variables are relatively easy to quantify. For the current example, the researcher could define the participants as adult men and women. In other cases, the subject variable may be more complex. Consider, for example, a researcher who wants to examine the differences between groups of people diagnosed with various forms of cancer. In this case, the researcher would, perhaps, classify the participants by the type of cancer as well as its current stage of development. In both of these examples, the common element is that the independent variable is a characteristic of the participant (sex or cancer type) that the researcher is able to operationally define and quantify.

What Type of Independent Variable?

The type of independent variable can also have profound effects on the design of the research project as well as the conclusions we can draw from the data. There are several ways we can classify independent variables.

Subject versus Manipulated

For the study about the relationship between sex and reaction to a particular medication, sex is a subject variable because the condition existed before the person's participation in the study. In addition, it is impossible for the researcher to randomly assign people to be men or women. In contrast, the study examining the effectiveness of different medications uses a manipulated independent variable because the researcher can randomly assign participants to receive one of the different drugs to be studied.

This type of independent variable is different than a subject variable because it is manipulated by the researcher who has control over alternative explanations for eventually observed effects involving the dependent variable that is the focus of the research. If we use a manipulated independent variable, we can create one or more control or comparison conditions for our experiment. Consider the comparison among the drugs designed to treat a common ailment. What control or comparison groups could we create and what function would they serve? The simplest control would be to have a group of people who took no drug at all and only be tested as the other participants in the experiment. This control establishes a baseline to determine the effect of no treatment at all. Will this single control group account for other alternative explanations?

The very act of taking a drug may have an effect on the participants' health. You may know this phenomenon as the placebo effect. To avoid this form of bias, many health sciences researchers use some form of "sham" treatment to examine the placebo effect. For example, when testing a medication, participants may be given a colored pill and told that it will treat their symptoms when, in fact, the pill is nothing more than some form of inert compound (e.g., a sugar pill). If participants demonstrate an improvement as a result of taking this "medication," the researcher can assume that the change may be due to the passage of time and/or the placebo effect. For the other drugs to be seen as effective, the amount of

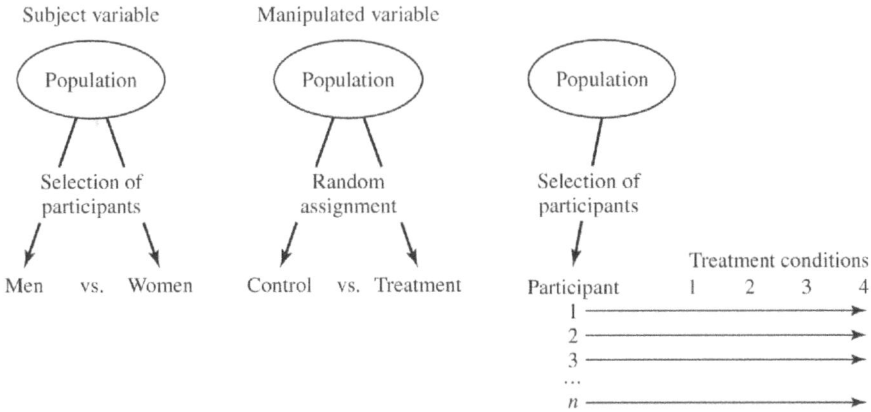

Figure 9.2 Illustration of the difference between a between-subjects variable and a within-subjects variable.

improvement associated with their ingestion will need to be greater than what was observed in the placebo control condition.

Can you think of other types of control conditions? With some time and thought, you might be able to find better examples than ours. The purpose of this discussion was to reiterate the importance of defining the meaning of the independent variable. The exercise will help ensure that you measure the variable correctly and consider alternative definitions of the variable. In addition, the exercise will help you design control conditions that will allow you to account for alternative explanations.

Between-Subjects versus Within-Subjects Variables

We can also classify independent variables as being either a between-subjects or a within-subjects variable. Figure 9.2 illustrates the difference between the two types of variable. The distinguishing feature is the number of times we test the individual.

For a **between-subjects variable**, we test the individual under only one treatment or research condition. By contrast, for a **within-subjects variable**, we test the same person under two or more research conditions. Sometimes, researchers will call a within-subjects variable a **repeated-measures variable**. For example, a person's sex is a between-subjects variable; it does not make sense to test the same person once as a male and then again as a female. An independent variable is also a between-subjects variable if we randomly assign participants to experience only one treatment condition.

If we observe the behavior of the same participant under different research conditions within a single study, that variable is a within-subjects variable. Time can be a within-subjects variable if we test or observe our participants' symptoms on different occasions. In examining the effectiveness of a drug, the researcher will arrange to record the participants' symptoms before the start of treatment and then several times later during the course of the study. In addition, the researcher may plan to increase or decrease the dosage to determine the effect of such changes. We call these variables within-subjects variables because we compare the same person's performance under different conditions or at different times.

Recognize the difference between between-subjects and within-subjects variables because they require special considerations for research design as well as statistical analysis. For now, we focus on between-subjects designs; we review within-subjects research designs in Chapter 14.

WHAT IS THE DEPENDENT VARIABLE?

As with the independent variable, we will need to create an operational definition of the dependent variable. We can use the two example studies in this chapter to examine how we will define the appropriate dependent variable. How will we evaluate the differences between men and women? How will we examine the effectiveness of common drugs? As you should recall from Chapter 3, we are revisiting the concepts of operational definitions, measurement, and validity. Our primary goal is to ensure we have a way or ways of measuring the dependent variable in a meaningful manner.

Reliability and Validity

Reliability refers to the *consistency* of our measurements, whereas validity refers to the accuracy and appropriateness of the conclusions we draw from the data. The downfall of many research projects comes from selecting tests or measurement techniques that were unreliable, invalid, or both. As you might imagine, for example, how will the researcher evaluate the participants? Would you have confidence in a study that used a survey consisting of a series of questions such as "I am feeling better now that I am taking the drug" and "I think the drug is helpful"? What if the researcher collected blood samples, measured blood pressure, and conducted other specific clinical tests? The survey questions are useful, to a point, in that it allows the researcher to understand the participant's perception of the drug. Of course, the researcher would want to verify the effect of the drug by monitoring key diagnostic signs as well.

ARE THERE CONFOUNDING VARIABLES?

As you learned in Chapter 3, a confounding variable is a condition that threatens the ability to assume a cause-and-effect relationship between the independent and dependent variables. In this section, we will examine some or the more commonly used techniques to eliminate confounding variables. Of course, the whole matter of identifying confounding variables is a matter that vexes all researchers. It is impossible to create a remotely comprehensive list of all possible confounds that can occur and associated work-arounds for the problems they create in any given research context. This is why it is so important to read broadly and attempt to understand deeply the phenomenon or phenomena you are studying. It is also critical to carefully read the "Method" and "Discussion" sections of any existing research you might intend to replicate or extend. Researchers will often provide insight into problems with their own research design that you can hopefully overcome when designing your own study. Using this information from previous studies can help you prepare good designs that will lead to success in your research. The following subsection details one strategy for avoiding a common source of confounding variables.

Objective Recording of the Data

It is important that you use consistent and systematic measurement procedures that do not bias the results. Examining the reliability and accuracy of the data recording method is one way to remove or reduce potential bias. One of the most useful methods for reducing confounds due to experimenter bias during data collection is to use what researchers call the single- and double-blind procedures. In this case, *blind* means that specific people involved in the research do not know the purpose of the study or the relevant independent variables that the participant experienced. Almost all researchers use some form of **single-blind** and many use a **double-blind** procedure.

The single-blind procedure indicates that the person participating in the research does not know the specifics of the researcher's hypothesis. The goal of the single-blind procedure is to reduce any demand characteristics that may result from the participants' knowledge of the researcher's hypothesis. In a true experiment, for example, the participants would not know whether they were in one of the treatment conditions or in a control group. If the single-blind approach is successful, the participants' behavior will reflect natural reactions to the research conditions.

How would a researcher use single-blind techniques in the two hypothetical examples we are studying? In the sex-differences study, the participants will know that the researcher is examining the effect of the drug. What the participants will not know is why the researcher is collecting the data (i.e., to compare the effects of the drug on men vs. women). In the second study, the participants will know that they are taking a medication, but will now know if they are receiving an active drug or a placebo.

Some research projects require additional controls to reduce the risk of experimenter bias. The double-blind procedure means that neither the person participating in the study nor the person collecting or scoring the data knows the researcher's hypothesis and the treatment the participant received. In both of our running examples, the persons administering the medication and monitoring symptoms knows only that they are to administer a specific treatment and then collect specific bits of information or samples from the participants.

KNOWLEDGE CHECK

1. Answer these items after reading each of the short descriptions of research projects that follow.
   a. Identify the independent variable(s) and the dependent variable.
   b. What is the hypothesis that the researcher appears to be testing (you will have to infer this from your reading of the scenario)?
   c. Identify the empirical method that the researcher used.
   d. What aspects of the research design would influence our evaluation of the internal validity of the research?
   e. What aspects of the research design would influence our evaluation of the external validity of the research?
   f. What changes would you make to the study to improve the internal and external validity of the study?

I. A researcher wants to test the effects of a new drug to treat depression. For the first part of the study, participants completed a test of depression. Those with the highest level of depression then receive the treatment. Two months later, the researcher retested the participants. All participants showed a marked improvement. The researcher concludes that the treatment is useful.

II. A researcher wants to examine the effectiveness of a new treatment for hypertension. To begin the study, the researcher randomly assigns participants to three treatment conditions. Some received no treatment. Others received a red pill with no drug. Others received a red pill with the new medication. The participants' blood pressure was measured once a week for the next 4 months. At the end of the study, the researcher found that 45% of those in the third group were no longer suffering from hypertension.

WHAT IS THE RESEARCH HYPOTHESIS?

Preparing the research hypothesis is an integral part of preparing a research design. Indeed, the format of the research hypothesis shapes the design of the research and the statistical analysis of the data. When we write a hypothesis, we want to be sure that it clearly states the expected relationship between the independent and dependent variables and the type of results we expect to obtain. For the shopping behaviors and essay-writing examples, we are interested in the differences between two groups. Therefore, we will frame each hypothesis as a statement of the relationship between the different treatment groups.

Directional versus Nondirectional Hypothesis

A research hypothesis describes the relationship among population parameters. There are, however, different ways that we can write the hypothesis. The major difference is whether we use a **directional hypothesis** or a **nondirectional hypothesis**. In a directional hypothesis, we predict a specific relationship between the two groups using either a greater-than or less-than type of prediction. Examples of directional hypotheses include "The drug has greater effectiveness in women than in men" and "Medication A is more effective than medication B for lowering blood pressure." Both predictions use words to indicate the researcher's expectation or belief that the response of participants in one group or condition would be greater than the behavior of the other group.

In contrast, a nondirectional hypothesis predicts that there will be differences between groups but does not specify the type of difference that will occur. We can rewrite the previous hypotheses as "Men and women differ in their reactions to medication A," and "The different medications will have different levels of effectiveness on reducing symptoms." We have retained the emphasis on the relationship and expected differences between the groups, but we have not indicated which group will be greater than the other.

Which type of hypothesis should you use? There is no clear answer to this question, but there are plenty of opinions. Some researchers believe that if you are able to form a clear prediction about the anticipated direction of the results, then you should formulate a directional hypothesis. The justification for this is that if you are confident in your prediction based on existing research and/or theory, then you should expect only one form of outcome. Therefore, if you want to confirm the accuracy of a prediction, then the directional hypothesis may be the better choice. Another matter to consider is that fact that a directional hypothesis will have greater statistical power to detect a significant effect than a nondirectional hypothesis.

The justification for the nondirectional hypothesis approach is that researchers should examine all results, regardless of their direction. From this perspective, we would say that any evidence of a statistically significant difference between the groups is important and worthy of our analysis.

Both directional and nondirectional hypotheses have their place in empirical research. In many ways, selecting between directional and nondirectional hypotheses is much like resolving an ethical dilemma. Specifically, the researcher needs to examine his or her philosophical stance on hypothesis testing and then to develop a rationale for selecting one type of hypothesis. If you believe that it is imperative to test that $\mu_1 > \mu_2$, then you should use a directional hypothesis. By contrast, if you believe that it is essential to explore the differences between μ_1 and μ_2, then use a nondirectional hypothesis, $\mu_1 \neq \mu_2$ (Jones & Tukey, 2000).

Stating Hypotheses Mathematically

It is often easier and more concise to convert a well-written hypothesis into a series of mathematical or logical statements. These mathematical statements then help us interpret the results of the study. At one level, any hypothesis refers to the relationship between the independent and dependent variables in your study.

For example, our hypothesis might imply that one's sex affects reactions to a medication or that different medications have different impacts on patient's conditions. A hypothesis also refers to the parameters that exist within the populations. According to what you learned in Chapter 7, the sample data should be representative of the target population, and we can use the sample data to generalize about the population. Consequently, the hypothesis describes what we believe to be true of the target populations.

When we prepare a **mathematical hypothesis**, we write two complementary hypotheses, the *null* hypothesis and the *alternative* hypothesis. In its basic form, the null hypothesis implies that there is no relationship between the independent and dependent variables by indicating that the separate treatment groups will be equivalent in terms of the data associated with the dependent variable. The alternative hypothesis, in contrast, implies that there is a relationship between the independent and dependent variables and, by implication, that the groups will be different from each other.

We can use the sex-differences study example to illustrate the null and alternative hypotheses. For the null hypothesis, we would say that one's sex is unrelated to symptom change for the medication. As a mathematical statement, we can write the null hypothesis for this study as H_0: $\mu_M = \mu_F$. A translation of the equation is "The population mean of men's symptom equals the population mean of women's symptoms." The alternative hypothesis in this case is H_1: $\mu_M \neq \mu_F$. In writing the alternative hypothesis, we replace the equal sign (=) with the not equal sign (≠). Table 9.1 presents examples of the null and alternative hypotheses for directional and nondirectional hypotheses.

There are several important points to emphasize in Table 9.1. First, as was discussed in Chapter 6, we use H_0 and H_1 to refer to the null and alternative hypotheses, respectively. This practice is a convention that has evolved among statisticians and researchers who adhere to the tenets of null hypothesis significance testing. Second, and more importantly, in the alternative hypothesis, the equality sign (>, <, or ≠) matches the adjective in the research hypothesis (greater than, less than, not equal to). We use the complementary equality sign in the null

Table 9.1 Examples of Null and Alternative Hypotheses for Directional and Nondirectional Hypotheses Using a Two-Group Design

	Directional hypothesis		
Hypothesis	Greater than	Less than	Nondirectional hypothesis
Research hypothesis	"Men will show greater improvement with the drug than will women."	"Men will show less improvement with the drug than will women."	"Men and women will differ in their response to the drug."
Null hypothesis	$H_0 : \mu_M \leq \mu_F$	$H_0 : \mu_M \geq \mu_F$	$H_0 : \mu_M = \mu_F$
Alternative hypothesis	$H_1 : \mu_M > \mu_F$	$H_1 : \mu_M < \mu_F$	$H_1 : \mu_M \neq \mu_F$

hypothesis. For example, = and ≠ are complementary signs as are the signs ≥ and <. Finally, the null hypothesis always implies that there is no relationship or difference between the independent and dependent variables, whereas the alternative hypothesis implies that there is a meaningful relationship or difference between these variables.

EVALUATING HYPOTHESES

Traditionally, hypothesis testing within the health sciences focuses on the null hypothesis and is, as such, referred to as null hypothesis statistical testing. Indeed, the technical question we ask when using inferential statistics is if the data warrant the rejection of the null hypothesis. You may be wondering why we use the null and alternative hypotheses. It seems rather odd that we would use two hypotheses when one would seem to be enough. The practice of using null and alternative hypotheses has evolved during the past century as a way to make inferences about sample data. We can use the example of the differences between men and women's recall of life events to illustrate the reasoning behind null hypothesis testing.

When conducting empirical research, the researcher's primary responsibility is to provide evidence to support his or her claim about the relationship between the variables described in the research hypothesis. The research hypothesis and alternative hypothesis are the same thing, and the researcher must collect data that will convince us of the accuracy of his or her predictions. Therefore, we begin with the null hypothesis as our default expectation regarding the linkage between the independent and dependent variables. The researcher is then required to provide evidence that is strong enough to convince us that an alternative, non-null explanation might be more accurate at describing the relationship between the variables under study. This alternative explanation is only seriously considered if and when the gathered data are strong enough to reject the null hypothesis because it is perceived to be "false."

In the sex-differences study, the null hypothesis states that there is no meaningful difference between men and women. Before this statement of null effect (i.e., null hypothesis) can be rejected, the researcher must present data that clearly suggests something to the contrary, namely, that there is a difference between men and women. If the data that the researcher gathers indicate a difference between the sexes that is large enough to not be explained away as a chance occurrence, then the null hypothesis is rejected and the alternative hypothesis is retained as a more accurate description of the actual relationship between sex and the shopping strategies. We require the researcher to demonstrate that there is a sizable or meaningful difference between men's and women's response to the drug. Note carefully here, however, that the alternative hypothesis in most research of this nature is only one of several possible alternative explanations of the relationship between the variables under study. This is another reason why science is cumulative and requires replication and extension before any findings are received as fully explanatory.

The decision to reject the null hypothesis is an inference; the researcher uses sample data to draw inferences about the relationship as it would be expected to occur at the level of the population. There are two important characteristics of an inference. First, whenever we make an inference, we use the experience created by samples to describe the population. Second, whenever we make an inference, we must assume that there is a probability that we are wrong. As you know, the sample mean is an unbiased estimate of the population mean. The important word is *estimate*. There is always a probability that your inference will be wrong. By implication, then, all hypothesis testing is conditional or based on probabilities.

We use the data to determine whether to reject the null hypothesis. Look at Table 9.1. We use μ to describe the two populations. If we find that the sample means, drawn at random from the populations, are sufficiently different from each other, and we can assume that our sample are representative of the larger population, then we will reject the null hypothesis in favor of its alternative. Will we make mistakes when evaluating the null hypothesis? Yes. Table 9.2 lists these errors.

There are two mistakes we can make when judging the null hypothesis. In the first case, we can be misled by the data and reject a true null hypothesis. Researchers call this error a **type I error**. We prefer to think of this as a "false alarm." Consider the following example to illustrate this type of inferential mistake. Assume that we are a manufacturer that conducted an experiment to the effectiveness of a new drug. For this experiment, our alternative hypothesis could be that $\mu_1 > \mu_2$. What would happen if we made the wrong decision and that drug was not effective? This decision represents a type I error because we concluded that there was a difference when one did not exist. Clearly, there are a number of problems with this conclusion.

A second type of inferential error is a **type II error**. We like to think of this type of error as a "miss." A type II error occurs when a null hypothesis is incorrect, but we fail to recognize it as false. For example, it might be true that men and women react differently to a drug. Indeed the drug might be helpful for women but not for men. In this example, the null hypothesis that $\mu_{\text{Men}} = \mu_{\text{Women}}$ (where μ represents average height) is a false statement. Imagine that you conducted a study using 10 adult women and 10 adult men and found that the effect of the drug was equal. Because the average levels of the symptoms are equal, you decide not to reject the statement $\mu_{\text{Men}} = \mu_{\text{Women}}$. This decision would reflect a type II error because there is a real difference between the heights of men and women that you were simply unable to detect with your data. The type II error occurred due to random error that you selected men and women whose average heights happened to be equal.

A type I error is committed when we rejected a true null hypothesis; a type II error is committed when we fail to reject a false null hypothesis. Table 9.2 presents an illustration of type I and type II errors.

The Greek letters α and β are probability estimates; **alpha (α)** represents the probability of committing a type I error and **beta (β)** represents the probability of committing a type II error. The probability of a type I error is under the direct control of the researcher because he or she can determine the statistical criterion for deciding to reject the null hypothesis. Therefore, if the researcher wants to be

Table 9.2 Decision Matrix Illustrating Type I and Type II Errors Relative to Accurate Research Decisions

	Population parameters ("Reality")	
Researcher's decision	$\mu_1 = \mu_2$ H_0 is correct	$\mu_1 \neq \mu_2$ H_0 is false
Reject H_0 and accept H_1.	Type I error α	Correct decision $1 - \beta$
Do not reject H_0.	Correct decision $1 - \alpha$	Type II error β

conservative and lower the risk of committing a type I error, he or she can use a smaller value of α (i.e., instead of $\alpha = .05$, the researcher could set $\alpha = .01$). Therefore, α is often referred to as the **significance level** (or **criterion**) for a study as it determines the requirement for rejecting the null hypothesis.

The value of β is more difficult to determine and manage. As we will discuss later in this chapter, sample size, the difference between groups, and α all contribute to the size of β. Consequently, researchers can influence the probability of a type II error by carefully planning their research.

When designing studies, researchers attempt to find an optimal balance between α and β. Generally, decreasing α will increase β—lowering the risk of committing a type I error increases the risk of committing a type II error. With careful planning, you can design your research in such a way as to reduce the risk of committing a type I and a type II error, but you will never be able to eliminate both errors.

In many ways, evaluating the null hypothesis is like the decision a jury must make in a criminal case. According to American legal theory, any person accused of a crime is innocent until proven guilty. The presumption of innocence is the legal equivalent of the null hypothesis. The prosecution must provide ample evidence to persuade the jury to reject this assumption in favor of the most supported alternative explanation (i.e., that the accused is guilty). In criminal cases, the jury must find the defendant guilty beyond a reasonable doubt. A guilty verdict means that there is ample and compelling evidence to reject the presumption of innocence.

Does a "not guilty" decision prove the defendant's innocence? No, a verdict of not guilty means that there is insufficient evidence to convict the defendant (i.e., so the "null" is retained); it does not prove innocence. Similarly, we can never really accept the null hypothesis or assuredly prove the alternative hypothesis. To accept means to prove true. Because we use inferential reasoning in hypothesis testing, we can never prove a hypothesis true beyond a doubt. The only decisions we can make regarding the null hypothesis are to reject or not to reject the null hypothesis.

How are type I and type II errors related to a jury's verdict? A type I error in this example is the same as convicting an innocent person. Finding an innocent person guilty is similar to a type I error because we incorrectly reject a true null hypothesis. Similarly, a type II error in this example is the same as failing to convict a real criminal. A type II error occurs because you do not have sufficient evidence to conclude that the null hypothesis is wrong, just as a jury may not be convinced that the defendant is guilty.

EVALUATING HYPOTHESES: PRACTICAL MATTERS

Our decision of whether to reject the null hypothesis depends on the context in which we interpret the data. When we look at the data, we need to answer one question: Is there sufficient evidence to reject the null hypothesis as a false statement? One way to answer this question is to examine the relative difference between the groups. In this section, we will examine the concept of effect size, one of the primary factors in determining whether to reject the null hypothesis (and whether an observed difference is of a sufficient magnitude that it would mean anything in actual practice).

Effect Size

Imagine that we collect data for two groups. The mean of the first group is $M_1 = 100$; the mean of the second group is $M_2 = 105$. What can we say about the difference between the two means other than they are different from each other? What could cause the difference between the two means? Is the difference trivial or meaningful? One way to answer this important question is to examine the **effect size** of the difference between the groups.

Equation 9.1 presents how to determine the effect size for the difference between the means of two groups:

$$d = \frac{M_1 - M_2}{\sqrt{\frac{VAR_1 + VAR_2}{2}}} \tag{9.1}$$

From the equation, d is the effect size for the difference between two sample means, where M_1 and M_2 represent the sample means, and VAR_1 and VAR_2 represent the sample variances of the two groups. We assume that the two variances are equivalent ($VAR_1 \approx VAR_2$) and that the sample sizes are the same ($n_1 = n_2$).

An effect size like that summarized in Equation 9.1 is nothing more than a simple ratio of the differences between group means to the variability of scores on the dependent variable within the different groups. The larger the absolute value of d, the greater the relative difference between the two group means. Figure 9.3 illustrates three examples of effect size for a two-group comparison. Each curve represents the distribution of scores. For all the data sets, the standard deviation of the samples is 10.0.

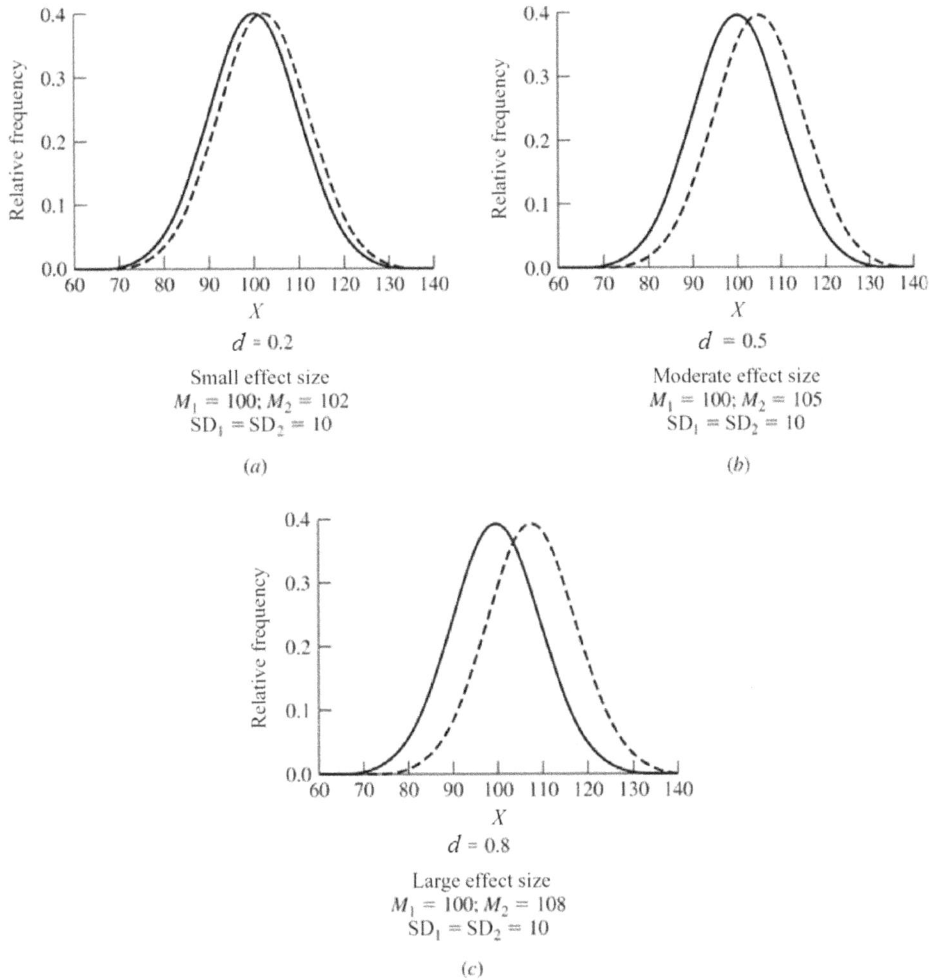

Figure 9.3 Illustration of effect size with $d = 0.20$, 0.50, and 0.80. *Note:* The effect size is the ratio of the difference between the two population means to the standard deviation of the groups.

With all else being constant, as the difference between group means increases, the effect size increases. In his influential book on statistics, Cohen (1988) described the three general guidelines for estimating effect sizes shown in Figure 9.3 as (*a*) small, (*b*) moderate, and (*c*) large.

As you might guess, the size of *d* has a lot to do with our interpretation of the null hypothesis. The larger the size of *d*, the stronger the relationship between the independent and dependent variables (and in this type of example, the more evident the difference between the two groups). Therefore, when we design any research project, we need to find ways to increase the effect size as much as possible.

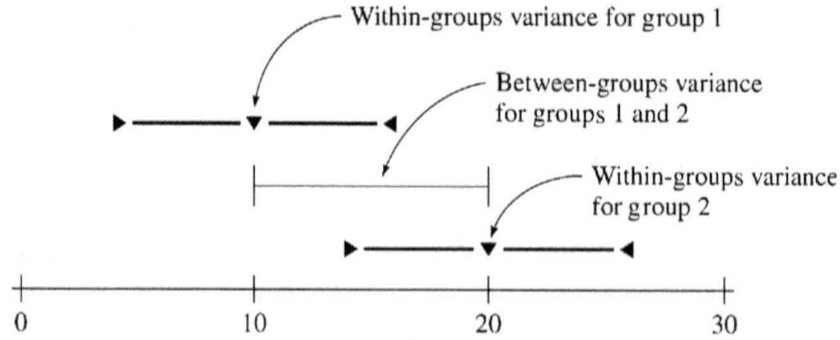

Figure 9.4 Illustration of the within-groups and between-groups variances.

Factors That Influence Effect Size

Two general factors influence effect size: the between-groups variance and the within-groups variance. There are many things that we can do to control these sources of variance and therefore improve the quality of our results. Figure 9.4 depicts the two types of variance. The difference between the two group means in our running example represents the between-groups variance. The variability among scores within each of the two groups represents the within-groups variance, or deviations between each group members' scores and the mean scores in each group.

Between-Groups Variance

As mentioned, the **between-groups variance** represents the difference among the group means. This variability represents the true difference between the population means plus the effect of random sampling or measurement errors. There are several ways that researchers can maximize the chances of identifying a difference between sample means (and therefore identifying a difference between two populations). First, if you are using a manipulated independent variable, you will want to use levels of the independent variable that you believe create the greatest possible effect or difference between groups in terms of the dependent variable under study. For example, to study the effectiveness of a drug, a researcher may want to use several dosages to find the optimal or most effective dosage to treat a medical condition. The goal of these tactics is to increase the chances that the "treatment" (in this case the emphasis on personal reflection in the presence of real facts about the tuition increase) has on the dependent variable of interest (in this case, student's opinion about the tuition increase).

When using an independent variable that is a subject variable, you will want to be sure that the two groups are as clearly distinct as possible. In the sex-differences study, our use of sex as an independent variable does not easily lend itself to further modification because there are only two sexes. There are, however, subject variables that have differing levels. In these cases, we would want to define the sampling populations to maximize the differences between the groups. Imagine that you are interested in studying patients with varying levels of an ailment. When defining the populations, you would want to select participants who clearly present with notable differences in terms of the severity of their ailment or condition. Use these tactics with caution, however, as some procedures along these lines may produce spurious statistical results, particularly if the only way to create separate groups is to split one sample into two based on a median split, discussed next.

Avoid Median Splits

Despite frequent warnings in the published scientific literature, some researchers continue to use a technique called a **median split** to create two (or more) intact groups for further comparison. This technique creates intact groups by using the score of an otherwise-continuous measure of some variable to group people into subgroups. For example, if you wanted to compare the relationship between blood pressure and the risk for a stroke, you could test people's blood pressure and split the students into two groups based on the median of their diastolic blood pressure. You would then label people on one side of the median as low blood pressure and the others as hypertensives. Although this may sound like a good idea, it is typically not the best choice, especially when a researcher is struggling to ensure sufficient statistical power for testing the hypotheses for a particular study.

A moment's thought should reveal to you the problems with the median split procedure. First, this technique results in a loss of potentially valuable information. The median split converts scores measured on an interval or more continuous form of response scale to a less informative ordinal or even nominal scale (depending on how you use it). Losing this continuous information will limit your ability to demonstrate the relationship between your independent and dependent variables.

Second, most measures of subject variables are normally distributed, and the majority of scores are close to the median. Consequently, a small difference between two scores (e.g., 99 and 101) can put people into radically different groups, even if those two groups are not necessarily different in any meaningful way. In sum, median splits can make it difficult to detect differences between groups and thereby lead to bogus conclusions (Bissonnette, Ickes, Bernstein, & Knowles, 1990; Maxwell & Delaney, 1993).

All of this said, median splits are usually not a good idea, so if you use a subject variable to classify people, ensure that it naturally places people into discrete groups. A better technique may be to use correlation analysis techniques. For example, you could examine the correlation between gender-role identity and job satisfaction. Using this tactic, you would create a large representative sample and then examine the correlation between the subject variables and the dependent variable.

Within-Groups Variance

While the between-groups variances is largely due to the influence of the independent variable you are studying, **within-groups variance** represents the differences among individuals' dependent variable scores within each study condition (as illustrated in Fig. 9.4). Two factors contribute to the within-groups variance. The first is the natural variation within the population you are studying. As an example, consider the experiment to test the effectiveness of a drug on hypertension. While the people in the study are all diagnosed as hypertensive, most will have different systolic and diastolic pressures. These differences among people represents with within-groups variation.

The second factor that affects within-groups variance is measurement error. Because no test is perfectly reliable, each test score includes some amount of error. Highly reliable tests contribute little measurement error; tests with lower reliabilities contribute more measurement error. These facts allow us to find ways to reduce the within-groups variance.

Homogeneous Samples

Whenever possible, try to use participants who are similar to one another. Researchers prefer to use **homogeneous samples** rather than **heterogeneous samples** because it allows them to limit irrelevant variability within groups. In other words, while a homogeneous sample refers to a condition wherein all the participants are similar to each other, a heterogeneous sample refers to a condition wherein there is considerable variability among the members of the sample.

There are several ways to increase the homogeneity of a sample. In health sciences research, you may want to sample people who have the same health profile, live in a similar region of the country, and share similarities on other important variables. The goal of the researcher then is to identify the relevant subject variables that will facilitate sampling of people into the desired homogeneous groups.

Reliable Measures

You can also decrease the within-groups variance by increasing the reliability of the measurement procedures. Measurement error is a random variable that increases some scores and decreases other scores. Therefore, anything you can do to increase the reliability of a test will help you detect any meaningful differences between the groups.

RESEARCH IN ACTION

Sex Differences

Continuing with our hypothetical example comparing male and female reactions to a specific medication, assume that the researchers had men and women complete a month-long trial using the same dosage of a drug. At the end of the study, the researchers collected blood samples and conducted three specific tests. The results are presented in Table 9.3.

Table 9.3 Observed Test Results for Male and Female Participants

	Test A		Test B		Test C	
	Men	Women	Men	Women	Men	Women
	32	20	2	3	4	2
	32	20	1	3	4	1
	12	10	0	1	4	2
	14	18	0	3	2	1
	14	12	0	1	2	1
	26	12	1	1	2	1
	24	12	0	1	2	1
	22	16	1	1	2	1
ΣX	176	120	5	14	22	10
M	22	15	0.625	1.75	2.75	1.25
VAR	64	16	0.554	1.071	1.071	0.214
SD	8	4	0.744	1.035	1.035	0.463

Table 9.4 Blood Test Results for Control versus Medication Group Participants

	Control		Medication	
	8	7	13	14
	15	10	8	16
	9	11	11	12
	10	9	12	14
	8	8	17	13
N		10		10
M		9.50		13.00
VAR		5.17		6.44
SD		2.27		2.54

Evaluating a New Drug

For this experiment, the researcher randomly assigned 20 participants who were diagnosed with a disease to one of two groups. The participants in the control group received a placebo. The other participants received the new medication. The dependent variable represents the result of a blood test (Table 9.4).

KNOWLEDGE CHECK

Use the data for each of the above two "Research in Action" studies to complete the following tasks.

2. Select a research hypothesis that best fits the purpose of the research. Describe the hypothesis in words and prepare the corresponding H_0 and H_1.
3. Calculate the effect size for each set of data. What inferences can you draw about the data given this statistic? For each data set, what factors contribute to the between-groups variance and the within-groups variance?
4. For each data set, describe the implications of committing a type I or a type II error.

CHAPTER SUMMARY

The primary theme of this chapter was designing a research project that reduces the effects of unwanted sources of error and bias. Because research is a uniquely human enterprise, there are many opportunities for the researcher's behavior to bias the participants' behavior as well as the researcher's interpretation of the data. Thus, we examined the sequence of events for the research process from defining the purpose of the study to the final interpretation of the data to examine where and how bias can enter the research process. The first source of potential bias is the effect that volunteer participants may introduce to the study. People who volunteer for a research project may not represent the typical person we hope to describe in our population.

Once we begin a study, there is a risk of the participants responding to an unintended demand characteristic. Most people know that they are participating in a research project and may want to modify their behavior to match what they believe to be the purpose of the research. A classic example of a demand characteristic is the Hawthorne effect.

A related source of bias is the researcher's expectations. If the persons interacting with the participants or recording the participants' behavior know the hypothesis under study and the treatment that the participants received, their knowledge may create subtle and not-so-subtle forms of bias. The researcher may bias the results by inadvertently treating the participants differently or may interpret the behavior of the participant to be in keeping with the research hypothesis.

We then examined how the researcher can define the independent and dependent variables. The independent variable may be either a subject variable or a manipulated variable. The critical difference between the two variables is whether the researcher can randomly assign participants to different levels of the variable. If the researcher can randomly assign participants to different levels of the independent variable, then it is a manipulated variable.

The independent variable may also be characterized as either a between-subjects variable or a within-subjects variable. For a between-subjects variable, we compare and contrast different groups of people against each other. For a within-subjects variable, we observe the same person's behavior under several conditions.

There are several ways that the researcher can reduce the effects of the demand characteristic and experimenter bias. These techniques fall under the general heading of a blind. A single-blind study means that the participant does not know the researcher's hypothesis or the specific treatment condition to which he or she has been assigned. In addition, the researcher may use a placebo or a cover story to control what the participants believe about the purpose of the study and the treatment they receive. In a double-blind study, both the participant and the person collecting the data are unaware of the research hypothesis being tested and the participant's treatment condition.

A critical component of any research project is converting the research hypothesis into a mathematical hypothesis. The research hypothesis is a general description or prediction of the relation between the variables in the study. The mathematical hypothesis is a specific statement regarding the pattern of results that the study should produce. A mathematical hypothesis may be either general, as occurs in the nondirectional hypothesis, or specific, as occurs in the directional hypothesis.

Mathematical hypotheses contain two parts: a null hypothesis and an alternative hypothesis. The null hypothesis states that there is no meaningful relation between the variables in the study. The alternative hypothesis is the mathematical equivalent of the research hypothesis. The goal of the researcher is to collect data to allow him or her to reject the null hypothesis and accept the alternative hypothesis.

When evaluating the null and alternative hypotheses, a researcher can make one of two errors. The first error, a type I error, occurs when the researcher rejects the null hypothesis as false when the null hypothesis is correct. The second error, a type II error, occurs when the researcher fails to reject a false null hypothesis. The researcher can directly control the risk of committing a type I error by establishing a criterion called the significance level, or α. The size of α determines the probability that the researcher will commit a type I error if the null hypothesis is a correct statement. The researcher can also influence $1 - \beta$, the probability of correctly rejecting a false null hypothesis.

The effect size is a statistic that reflects the relative difference between two groups and is a statistic that can help us understand how to increase the chances that we will successfully reject a false null hypothesis.

One way to increase the likelihood of rejecting a false null hypothesis is to increase the between-groups variance. The between-groups variance reflects the difference between the group means. If the independent variable does influence the between-groups variance, then selecting levels of the independent variable that create the greatest possible effect can increase the chance for rejecting a false null hypothesis.

The researcher can also attempt to reduce the within-groups variance. The within-groups variance represents the variability among scores within each treatment condition. If there is much heterogeneity of variance, then it will be more difficult to detect a statistically significant difference among the groups. If the researcher can make the data more homogeneous, then the within-groups variance will decrease and the researcher will be better able to detect differences among the groups.

In the final section of the chapter, we reviewed examples of two contemporary research projects. These studies demonstrated the application of good research design and demonstrated how one can collect useful data.

CHAPTER GLOSSARY FOR REVIEW

Alpha (α) The probability of committing a type I error. The level of α is under the direct control of the researcher.

Beta (β) The probability of committing a type II error. The size of β reflects many factors, including the design of the research, sample size, and level of β selected.

Between-Groups Variance The difference or variance among group means. The difference reflects differences among the population means and the effects of sampling error.

Between-Subjects Variable A form of independent variable for which the researcher tests the participant on one occasion or under one level of the independent variable.

Demand Characteristic Any unintended condition created by the design of a study that leads the participants to misinterpret the purpose of the research and to respond accordingly.

Directional Hypothesis A research hypothesis that the researcher uses to predict the relation among groups using either a greater-than or a less-than prediction.

Double Blind A control procedure where the person interacting with the participants and recording the data does not know the hypothesis of the research or the relevant details of the independent variable the participant experienced. In addition, the participant does not know the details of the project or the level of independent variable.

Effect Size A standardized index of the different between-groups means or the relation between two variables.

Experimenter Bias or **Experimenter Effect** When the experimenter's expectancies and knowledge of the anticipated results alter the way he or she treats the participants in the different research conditions or alter his or her objectivity when recording the data.

Hawthorne Effect Used to describe any situation wherein the participants' assumptions about the purpose of the research bias their performance.

Heterogeneous Sample A sample for which the members are dissimilar to each other regarding one or more relevant variables.

Homogeneous Sample A sample for which all the members are similar to one another on one or more relevant variables.

Mathematical Hypothesis A statement that describes the relation between population parameters using relational symbols (e.g., $=$, $\neq$, $<$, $>$, $\leq$, and $\geq$). The null and alternative hypotheses use complementary relations (e.g., $=$ vs. $\neq$, $\geq$ vs. $<$, or $\leq$ vs. $>$).

Median Split A technique for creating two intact groups by first administering a test of a subject variable and then dividing the participants into two groups based on the median score.

Nondirectional Hypothesis A research hypothesis that the researcher uses to predict a general difference among groups.

Self-Fulfilling Prophecy Within the research context, the tendency for a researcher to treat the participants differently based on preconceived ideas about the participants or the treatment condition that the participants receive.

Significance Level or **Criterion** The level of α the researcher selects to determine whether to reject the null hypothesis.

Single Blind Any procedure used to prevent the participants from knowing the researcher's hypothesis or the treatment condition received.

Type I Error When the null hypothesis is correct and the researcher treats it as false.

Type II Error When the null hypothesis is false and the researcher does not reject it.

Within-Groups Variance The variability among scores within a group. This variability represents the natural variability among the members of the population and measurement error.

Within-Subjects Variable or **Repeated-Measures Variable** A form of independent variable for which the researcher tests the same participant on more than one occasion or under more than one level of the independent variable.

REFERENCES

Bissonnette, V., Ickes, W., Bernstein, I.H., & Knowles, E. (1990). Personality moderating variables: A warning about statistical artifact and a comparison of analytic techniques. *Journal of Personality*, *58*, 567–587.

Cohen, J. (1988). *Statistical power analysis for the behavioral sciences* (2nd ed.). Hillsdale, NJ: LEA.

Gillespie, R. (1993). *Manufacturing knowledge: A history of the Hawthorne experiments*. New York: Cambridge University Press.

HYDE, J.S. (1991). *Half the human experience: The psychology of women* (4th ed.). Lexington, MA: Heath.

JONES, L.V. & TUKEY, J.W. (2000). A sensible formulation of the significance test. *Psychological Methods*, *5*, 411–414.

MAXWELL, S.E. & DELANEY, H.D. (1993). Bivariate median splits and spurious statistical significance. *Psychological Bulletin*, *113*, 181–190.

PARSONS, H.M. (1974). What happened at Hawthorne? *Science*, *183*, 922–932.

ROSENTHAL, R. (1994). Interpersonal expectancy effects: A 30-year perspective. *Current Directions in Psychological Science*, *5*, 176–179.

ROSENTHAL, R. & JACOBSON, L. (1968). *Pygmalion in the classroom*. New York: Holt, Reinhart & Winston.

ROSNOW, R.L. & ROSENTHAL, R. (1997). *People studying people: Artifacts in behavioral research*. New York: Freeman.

SHAKESPEARE, W. (1604). *The tragedy of Hamlet, Prince of Denmark, Act 3, Scene 2*. New York: Washington Square.

SUNDSTORM, E., MCINTYRE, M., HALFHILL, T., & RICHARDS, H. (2000). Work groups from the Hawthorne studies to work teams of the 1990s and beyond. *Group Dynamics*, *4*, 44–67.

Part Three

Common Research Designs

Chapter 10

Correlational Research

Give a researcher three weapons—correlation, regression, and a pen—and he or she will use all three.

—Anonymous

INTRODUCTION

Correlational research allows us to study the relationship between two or more variables. We can use the correlation to make predictions about one variable using another variable. This prediction might help us explain the dependent variable using the independent variable. For example, a researcher studying public policy may want to examine the relation between alcohol consumption and controlled access to alcohol (e.g., Brand, Saisana, Rynn, Pennoni, & Lowenfels, 2007). Similarly, a researcher might want to determine if postpartum depression can be

Understanding and Conducting Research in the Health Sciences, First Edition. Christopher J. L. Cunningham, Bart L. Weathington, and David J. Pittenger.

predicted by the mother's perception of her health during the pregnancy (e.g., Zubaran et al., 2010). As you will learn in this chapter, studying the correlations between variables is a powerful tool for understanding various behavioral phenomena. Indeed, we will use several examples from contemporary medical research to illustrate the value of this important statistical technique.

In this chapter, we review several of the more common applications of correlation statistics. We examine the Pearson product-moment correlation (r) and basic regression analysis. Although these are not the only statistical tools for analyzing the relationships among variables, they are among the most frequently used analytical techniques in the social and behavioral sciences. The primary focus of the chapter is on study design and how to use and interpret these statistics.

As a brief preview of what you are about to read, you should understand that there is a distinction between correlation research methods and correlational statistics. When we speak of the *method*, we are describing how we go about gathering the data. The most general account of the correlational method is that the researcher gathers two or more bits of information about participants in the study. In some cases, the researcher may gather the information during one session. For instance, a psychiatrist might administer a standard measure of depression shortly after drawing a blood sample from participants. In this example, the researcher wants to determine if there is a correlation between the test scores and current blood chemistry.

In other situations, the researcher will want to determine whether one variable will predict another variable. For this type of study, the researcher will collect data on two or more occasions. As an example, a researcher may measure the body mass index (BMI) of a large number of children of a specific age and then collect the same data several years later (e.g., Svensson et al., 2011). The goal of the study would be to determine the extent to which childhood BMI score predict subsequent obesity. Knowing that there is a high correlation between childhood BMI and later obesity may allow healthcare providers intervene earlier in a child's health care to reduce the risk of subsequent obesity. In correlation research, the researcher often examines the relation among subject variables and does not necessarily place the participants into different research conditions, as is the case for a true experiment. A researcher uses correlational research to determine whether two or more variables are interrelated.

A researcher can use correlation statistics for any data set regardless of the method of collecting the data. A researcher may conduct a true experiment, a quasi experiment, or a correlational study and then turn to correlation statistics to examine the relationship between the variables. In this chapter, we focus our attention on the correlation coefficient as a tool for statistically identifying the relationship between two or more variables.

CONCEPTUAL REVIEW OF CORRELATION

Although there are many ways to statistically determine the correlation between two variables, all **correlation coefficients** share certain features. First, the correlation is calculated from two or more sets of measurements taken from the same individual. Second, the correlation coefficient is a descriptive statistic that estimates the **linear relation** between two variables. A linear relation refers to a pattern of data best described by a straight line. In addition, correlation coefficients range between −1.0 and +1.0.

Figure 10.1 presents six scatter plots representing correlations that are $r = +1.0$, $r = .80$, $r = .50$, $r = 0$, $r = -.50$, and $r = -1.0$. As the value of the correlation

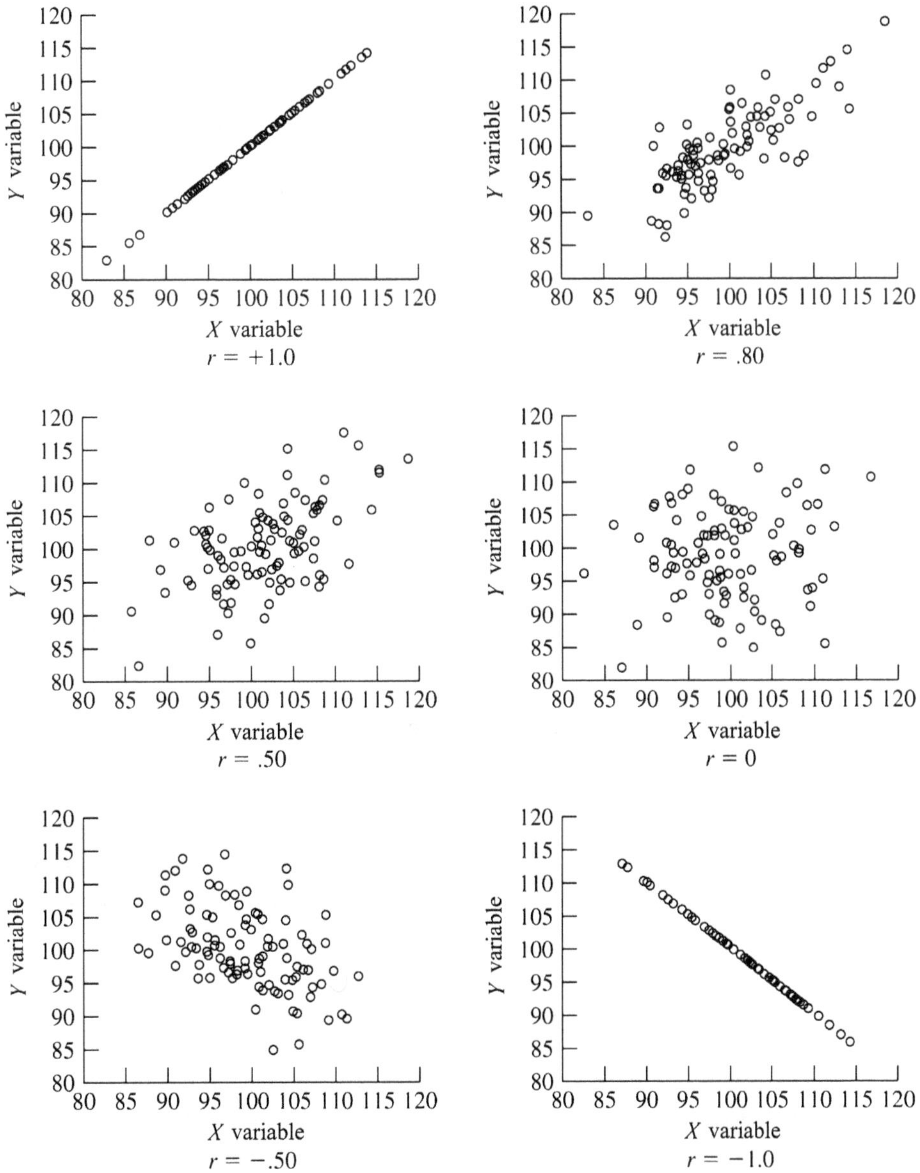

Figure 10.1 Scatter plots for six sets of data representing $r = 1.0$, $r = .80$, $r = .50$, $r = 0$, $r = -.50$, and $r = -1.0$.

becomes closer to +1.0 or –1.0, the data are more likely to fall along a straight line. A **positive correlation,** $r > 0$, indicates that increases in one variable correspond to increases in the other variable (or that a decrease in one corresponds to a decrease in the other). A **negative correlation,** $r < 0$, indicates that increases in one variable correspond to decreases in the other variable (or that the two variables move in opposite directions). Finally, a correlation coefficient, regardless of its size, is not, by itself, evidence of a cause-and-effect relationship. We will revisit this matter later in the chapter.

PEARSON'S *r*

Pearson popularized one of the most commonly used correlation coefficients. The full name for the statistic is the *Pearson product-moment correlation coefficient*, but most researchers refer to it as Pearson's *r*. We can use the definitional equation, presented in Equation 10.1, to examine how this statistic summarizes the relationship between two variables:

$$r = \frac{\sum z_x z_y}{N} \tag{10.1}$$

First, we convert each observed score to a *z*-score and then determine the cross products of these scores by multiplying them both together. The sum of the cross products ($z_x z_y$) divided by the number of pairs (N) is the correlation coefficient. The advantage of using *z*-scores is that they convert the data to a common scale. The mean and standard deviation of *z*-scores are always $M = 0.00$ and $SD = 1.00$ (see Appendix A for a detailed description of this process). Because the *z*-score converts the data to the same scale, we can examine the correlation between any two variables regardless of their respective means and standard deviations.

INTERPRETING THE CORRELATION COEFFICIENT

The correlation coefficient is a descriptive statistic that indicates the degree to which changes in one variable correspond with changes in a second variable. As you have already learned, the correlation coefficient can range between –1.0 and +1.0. In this section, we will examine how to interpret the size of the correlation coefficient.

Magnitude and Sign of *r*

There are two important pieces of information conveyed with the correlation coefficient: its sign and its magnitude. The sign of the coefficient indicates whether the two variables being studied are positively or negatively related. Many people misinterpret the sign of the correlation and believe that positive correlations are somehow better than negative correlations. Nothing could be farther from the truth. A negative correlation indicates only that there is an inverse relationship between the two variables: As one variable increases, the other decreases. An example of a negative correlation is the relationship between one's score for a round of golf and time spent practicing. In golf, the goal is to get as low a score as possible. Hence, we can predict that the more a person practices, the lower his or her score is likely to be.

Perhaps a better way to look at and interpret the correlation coefficient is to examine its absolute value. When the absolute value of r is zero (i.e., $|r| = 0$), there is no linear relationship between the two variables. By contrast, when $|r| = 1.0$, there is a perfect linear relationship between the variables. Cohen (1988) suggested that correlations fall into three general categories: small ($|r| = .20 - .29$), medium ($|r| = .30 - .49$), or large ($|r| = .50 - 1.00$).

You should use Cohen's (1988) size criteria with caution. These guidelines were developed to facilitate description of general behavioral research. You may find that researchers working in different areas have different guidelines for the magnitude of a correlation coefficient.

Coefficients of Determination and Nondetermination

Another way to describe and examine the correlation coefficient is to square its value, an operation that yields the **coefficient of determination** (r^2). The coefficient of determination indicates the proportion of variance shared by the two variables that you have correlated. If, for example, the correlation between two variables is $r = .82$, then $r^2 = .67$. Therefore, we can conclude that if we know X, then the linear relationship between these two variables allows us to account for 67% of the variability in the Y variable.

We can also calculate the **coefficient of nondetermination** ($1 - r^2$). For the current example, $1.00 - r^2 = .33$. The coefficient of nondetermination is an estimate of the proportion of unshared or unexplained variance between two variables. Thus, when $1.00 - r^2 = .33$, this means that 33% of the variability in Y cannot be explained by our knowledge of X.

Figure 10.2 presents a conceptual illustration of the meaning of r^2 and $1.00 - r^2$. Imagine that the circles represent the X and Y variables. When $r = 0$, there is no overlap between the two circles. As the magnitude of the correlation increases, the amount of overlap increases. The overlap of the two circles represents the coefficient of determination between two variables and the unshaded areas represent the coefficient of nondetermination.

Each pair of circles represents two variables, X and Y. The amount of overlap indicates the magnitude of the correlation between the two variables, or the amount of explained variance.

Causality

A large correlation coefficient, positive or negative, is not evidence in itself of a cause-and-effect relationship. The problem is that when we use the correlation

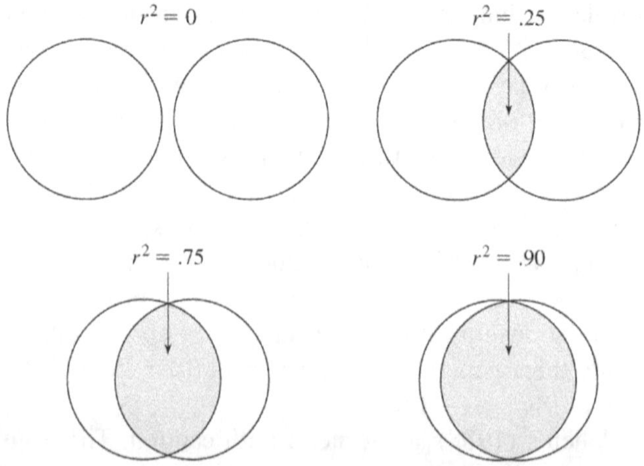

Figure 10.2 Illustration of the coefficient of determination, r^2.

coefficient, there is often no direct way to resolve the temporal order problem or the third variable problem. Recall from Chapter 3 that the temporal order criterion for cause and effect requires the cause to occur before the effect. In much correlational research, we collect the data for both variables without being able to control temporal order. A researcher may examine the correlation between postpartum depression and the woman's perceived overall health by asking women to rate their health during their pregnancy and their emotional state after giving birth; that is, the researcher is assessing the mother's emotional state and perceived prenatal health at the same time. Because the researcher collected the data at one point in time, we cannot be sure whether the mother's prenatal health actually leads to postpartum depression or if the postpartum depression affects the mother's retrospective assessment of her prenatal health.

The third variable problem refers to another (i.e., third) unmeasured variable that may influence the two measured variables. Consider the example of the relation between prenatal health and postpartum depression. It might well be that there is another variable that predicts both variables. For example, could the mother's social economic status influence both prenatal health and risk for postpartum depression? That is, prenatal health and postpartum depression are linked because of the mother's economic situation.

FACTORS THAT CORRUPT A CORRELATION COEFFICIENT

If you are planning to conduct a correlational study, you need to be aware of several factors that can corrupt the magnitude of r, causing it to be artificially high or low. These factors include nonlinearity of the data, truncated range, extreme populations, outliers, and multiple populations. In this section, we will examine how these factors affect the size of the correlation coefficient, how to detect their effects in the data, and how to avoid these problems.

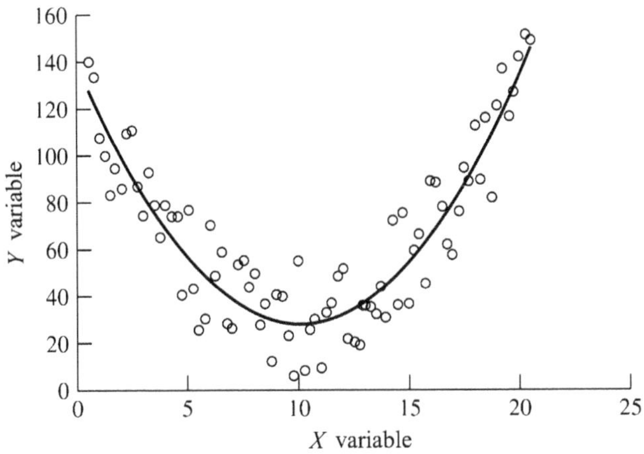

Figure 10.3 Illustration of data where a curved line represents the relation between the *X* and *Y* variables.

Nonlinearity

A primary assumption for correlational analyses is that a straight line best represents the relationship between the two variables. Knowing the correlation coefficient allows us to draw a straight line through the scatter plot that describes the relation between the variables. A **curvilinear relationship** would be characterized by a curved line that better describes the relationship between the two variables. Consider Figure 10.3 as an example. Clearly, there is a systematic connection between *X* and *Y*, but the relationship is not appropriately represented by a straight line. The relationship between *X* and *Y* in this example is U-shaped. Using Pearson's *r*, the correlation between the variables is $r = .14$, which suggests a small linear relation between the two variables. Nonlinear relationships like these often occur in health sciences research, as in Koren et al.'s (2011) study of plasma glucose levels and length of sleep. Koren et al. found that people who slept for 7.5 hours had the lowest glucose levels, whereas people who slept less or more than 7.5 hours had much higher glucose levels.

If we did not consider a scatter plot of the data (as in Fig. 10.3), we would have assumed that there is a small or trivial relation between the two variables. Looking at Figure 10.3, it appears that there is a strong relationship between the *X* and *Y* variables, but that this relationship is nonlinear. Consequently, we need to use different statistical tools (not just *r*) to analyze the data. Fortunately, many computer statistics programs can quickly and efficiently perform the analyses necessary to appropriately analyze these data.

Therefore, a nonlinear relationship between two variables is not a problem unless you fail to recognize its existence. Finding a curved rather than a linear pattern in your scatter plot may lead to useful insights about the phenomenon that you are studying. That curvilinear relationships exist demonstrates that you cannot rush the data through a simple-minded computer analysis. A carefully planned research design implies a careful statistical analysis of the data as well.

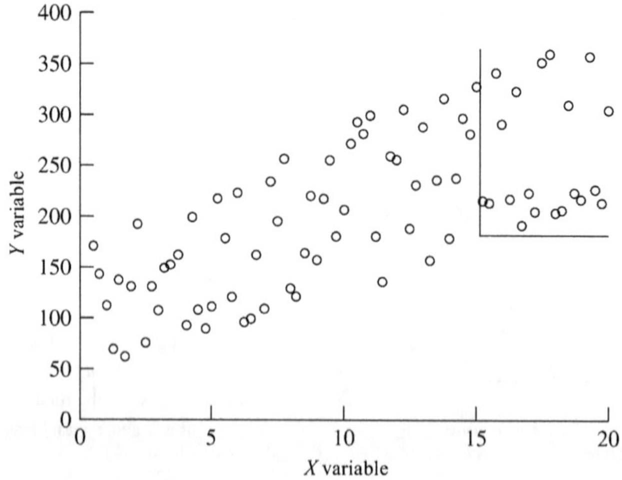

Figure 10.4 Illustration of the effects of truncating the range of scores on the correlation coefficient.

Truncated Range

The **truncated range** problem refers to a form of bias created by poor sampling. A representative sample is one in which the characteristics of the sample match the parameters of the broader population. A truncated range occurs when the variance among scores of members in the sample is much smaller than the variance of the population. Figure 10.4 represents what happens to the correlation coefficient when we have a truncated range of data to consider.

Assume that the entire scatter plot in Figure 10.4 represents what would happen if you created a truly representative sample of the population. For all the data, the correlation is $r = .71$, a strong linear relationship. What would happen if your selection procedure was biased and you ended up selecting only those people who have X scores of 15 or greater? The smaller $x - y$ axis in the upper right corner of Figure 10.4 represents this biased sample. For those data, the correlation is $r = .10$, a small correlation that does not well represent the true correlation between the variables.

The problem here is that the correlation coefficient represents the shared variance of the two variables. If your sampling procedure restricts the variance of the data, then the correlation statistic generated from these data cannot provide an accurate estimate of the true relationship in the population. The best solution to this potential problem is to avoid it by following proper sampling procedures when designing your study, to ensure that your sample will capture the natural range of potential scores for both variables in the population.

Extreme Populations

Another problem due to nonrepresentative sampling that can affect your correlation coefficient is selecting groups of participants that represent extreme ends of

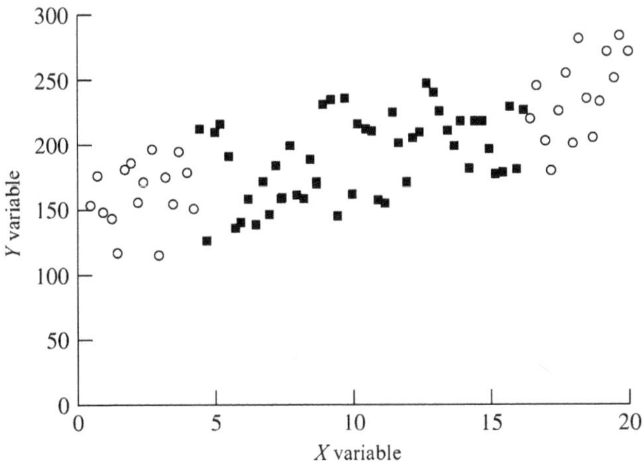

Figure 10.5 Illustration of the consequence of sampling from extreme ends of one variable.

the possible score distribution for either X or Y. Consider Figure 10.5 as an example of this problem. In this figure, there are two subsets of data, one represented by circles and the other by squares. If we examine only the circle data, the correlation is $r = .83$. However, if we examine all the data presented in the scatter plot, the correlation is lower, $r = .69$. In general, analyzing data from the extreme ends of the variables tends to inflate the correlation coefficient. The correlation coefficient for the data represented as circles is $r = .83$. However, the correlation for all the data is $r = .69$.

Outliers or Extreme Scores

As with any statistical test, outliers or extreme individual scores can also affect the value of the correlation coefficient. Hence, it is important that you review your data to ensure that you have accurately recorded all the data. Studying a scatter plot of your correlated variables can help you see a score or scores that stand apart from the rest of the data.

Multiple Populations

Another problem that can create artificially high or low correlation coefficients occurs when the sample contains data from two distinct populations in which the relationship between your two variables of interest may be very different. In Figure 10.6, there are two populations, one represented by squares and the other represented by circles. If we ignore the two populations, the correlation of the entire data set is $r = .58$. The correlations for the separate groups are higher than for the combined data. For the squares alone, the correlation is $r = .79$; for the circles alone, the correlation is $r = .82$. The important lesson here is that careful sampling is critical to the designing and conducting of correlational research. As you analyze the data, it is also important to determine whether there are relevant

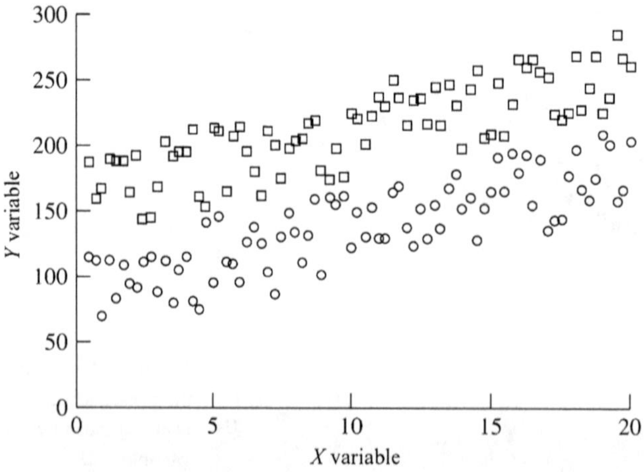

Figure 10.6 Illustration of the effect of having two populations within one data set.

subject variables that need to be incorporated into your analyses to more fully explain the data.

The correlation for the entire data is $r = .58$. The correlation for the data represented by circles is $r = .82$. The correlation for the data represented by the squares is .79. As a practical example, consider the research of Salvadori et al. (2003) who found that there is a significant correlation between plasma potassium levels and resting heart rate. More importantly, they found data similar to those presented in Figure 10.6, namely, that obese patients had higher heart rates than normal patients.

SAMPLE SIZE AND THE CORRELATION COEFFICIENT

The correlation coefficient is a descriptive statistic. It is also considered to be an unbiased estimate of its corresponding population parameter, ρ (rho), the correlation between populations. Thus, as we use the sample mean, M, to estimate the population mean, μ, we can use r to estimate ρ. As you learned in Chapter 7, the confidence interval about any estimate of a population parameter is a function of the sample size within your research. As sample size increases, the width of the confidence interval decreases, and for this reason, narrower confidence intervals imply that the sample statistic more closely estimates the corresponding population parameter.

Determining the confidence interval for the correlation coefficient is complicated because the sampling distribution for the correlation coefficient is not symmetrical. The skew of the sampling distribution for r increases when it moves toward −1.00 or +1.00. A section at the end of this chapter includes the formula for determining the confidence interval for Pearson's r. Although we will not examine those calculations here, we can use them to help guide us in determining the relation between sample size and width of the confidence interval.

Table 10.1 presents the confidence intervals for population correlations ranging between $\rho = .20$ and $\rho = .90$. For each correlation, the table presents the 95%

Table 10.1 Estimated 95% Confidence Intervals for Different Correlation Coefficients and Sample Sizes

	$\rho = .20$	$\rho = .40$	$\rho = .60$	$\rho = .80$	$\rho = .90$
N	r_{low}–r_{high}	r_{low}–r_{high}	r_{low}–r_{high}	r_{low}–r_{high}	r_{low}–r_{high}
50	−.08–.45	.14–.61	.39–.75	.67–.88	.83–.94
100	.00–.38	.22–.55	.46–.71	.72–.86	.85–.93
150	.04–.35	.26–.53	.49–.69	.73–.85	.86–.93
200	.06–.33	.28–.51	.50–.68	.74–.84	.87–.92
250	.08–.32	.29–.50	.51–.67	.75–.84	.87–.92
300	.09–.31	.30–.49	.52–.67	.76–.84	.88–.92
350	.10–.30	.31–.48	.53–.66	.76–.83	.88–.92
400	.10–.29	.31–.48	.53–.66	.76–.83	.88–.92
450	.11–.29	.32–.47	.54–.66	.76–.83	.88–.92
500	.11–.28	.32–.47	.54–.65	.77–.83	.88–.92
600	.12–.28	.33–.47	.55–.65	.77–.83	.88–.91
700	.13–.27	.34–.46	.55–.65	.77–.83	.88–.91
800	.13–.27	.34–.46	.55–.64	.77–.82	.89–.91
900	.14–.26	.34–.45	.56–.64	.78–.82	.89–.91
1000	.14–.26	.35–.45	.56–.64	.78–.82	.89–.91
2000	.16–.24	.36–.44	.57–.63	.78–.82	.89–.91

Note: The sampling distribution of the correlation coefficient is not symmetrical, especially when the sample size is small. Therefore, the confidence intervals will not be symmetrical around ρ.

confidence interval for specific sample sizes. We can use an example to illustrate how the table works.

Imagine that you plan to conduct a study to examine the correlation between two variables and believe that the correlation will be moderate, say, $\rho = .40$. According to Table 10.1, if you collect data from 50 participants, you would predict that 95% of the sample correlations would range between $r = .14$ and $r = .61$, a wide range of values. What would happen if you increased your sample size? As you would expect, the width of the 95% confidence interval will decrease. Using 450 participants, for example, the 95% confidence interval is $r = .32$ to $r = .47$. Samples larger than 450 produce marginal reductions in the width of the confidence interval. If you could reasonably expect to have a higher population correlation, then you may not require as many participants to estimate accurately the correlation coefficient.

KNOWLEDGE CHECK

1. Explain in your words how we can examine the correlation between two variables even when the means and standard deviations of the two variables are radically different.

2. Why does a correlation of –.75 represent a stronger relation between two variables than a correlation of +.60?
3. An instructor distributes a questionnaire to students asking them to estimate the number of hours they study each week and their current GPA. According to the data, the correlation between the two variables is +.93. Can we use these data to show students that studying causes improved grades?
4. Paula conducted a correlational study using 100 pairs of scores and found that $r = .10$. This correlation is far less than she expected. Describe in your words what could have caused this small correlation.
5. Devin is Director of Nursing at a large hospital and wants to study the relation between job stress and "burnout," the frequency of job changes among nurses. He plans to select randomly 35 nurses working in the emergency surgery unit and asks each to complete a standardized test of perceived job stress and desire of the nurse to be reassigned to another unit in the hospital. He then examined the correlation between stress score and desire for reassignment. Do you believe that Devin has created a research design that will allow him to reach reasonable conclusions about the relation between job stress and burnout?

APPLICATIONS OF THE CORRELATION COEFFICIENT

Like most statistical tests, the correlation coefficient has many applications. In the following sections, we will see how we can use the correlation coefficient to examine the reliability and validity of tests and other measures that we create.

Using Correlation to Determine Reliability

In Chapter 8, you learned how to use coefficient kappa (κ) to determine the inter-rater reliability for nominal and ordinal scales. As you might guess, we can use Pearson's r to determine the reliability of a test that uses an interval or ratio scale. The comparison between κ and r is straightforward; the closer either statistic is to 1.00, the greater the reliability of the test or measurement device. In this section, we will examine the many ways that we can use the correlation coefficient to examine the reliability of a test.

Test–Retest Reliability

One of the most common measures of reliability is test–retest reliability. To determine test–retest reliability, the researcher will administer the same test, or similar versions of the test, on two occasions. Consider a researcher who wants to develop a "fitness" index by sampling a range of physiological processes including resting heart rate, heart rate after mild exercise, and time for heart rate to recover to resting levels. In this case, the researcher would assess a wide group of people on one occasion and then again sometime later. Assuming that the overall health of the people did not change over time (say, 6 months), the correlation coefficient between the two scores indicates the stability of the measure over time. We can use test–retest reliability information to determine the degree to which random events that we do not want to measure affect the test scores. If the test–retest reliability is 0, we must conclude that the test scores are not stable over time and are subject to considerable fluctuation. By contrast, if the test–retest reliability is 1.00, we can conclude that the test measures the construct consistently and is unaffected by random events. This type of reliability information is most useful when measuring constructs that are not expected to change over time (e.g., general intelligence).

Inter-rater Reliability

In the previous chapter, we showed you how to use coefficient κ to determine inter-rater reliability when there are two or more observers and they are using either a nominal or an ordinal scale. In this section, we will see how we can use the correlation coefficient to calculate the inter-rater reliability among observers if the measurement scale is interval or ratio. For inter-rater reliability, the pairs of scores represent the scores given by the two observers. Specifically, X_1 would represent the score assigned by the first rater, and X_2 would represent the score assigned by the second rater. As with Cohen's κ, the larger the correlation coefficient, the greater the inter-rater agreement. If the inter-rater reliability is sufficiently large, then you can average the ratings of the two reviewers as the final observation. If the inter-rater reliability is very small, then it may be necessary to bring the raters together to work out a more consistent set of ratings.

Internal Consistency Reliability

Internal consistency reliability is similar to inter-rater reliability. The only difference is that we are looking at the correlation among participant responses to items or measurements within a test rather than the correlation between observers' ratings of participants. **Internal consistency**, like inter-rater reliability, refers to the degree to which responses to items in a test agree with one another. If a test has high internal consistency, then each test item produces the same or similar score for each person taking the test. Remember that internal consistency refers to how responses to items in a test correlate among each other, not to the variability among overall test scores. For our purposes, internal consistency is a way of determining the reliability of a test. If the internal consistency is low, then the test may not be measuring the variable we want to measure. Therefore, researchers strive to ensure high levels of internal consistency.

A common measure of internal consistency is Cronbach's (1951) **coefficient alpha (r_α)** or more often simply designated as α. This statistic is a special type of correlation coefficient that represents the correlation among all the items in the test. In other words, α is another way to determine inter-rater agreement, specifically, the level of agreement among questions or items in a test. Coefficient alpha can range from 0, which represents no internal consistency, to 1.00, which represents perfect agreement among all the items in the test.

For our purposes, the primary concern with this statistic is how we should interpret coefficient alpha. Unfortunately, many people misinterpret the meaning of α (Cortina, 1993; Schmitt, 1996). Some people make the mistake of assuming that a large value of α means that all the questions in a scale measure exactly the same construct. This interpretation is not necessarily correct, as Cortina and Schmitt have demonstrated. Thus, you should be careful when you read about or use α. Keep your interpretation of α straight; it indicates the extent to which responses to items within a test, or a part of a test, correlate with each other—nothing more, nothing less. The question of whether all items in a test measure exactly the same question is better answered using other statistical methods, such as factor analysis.

Improving Test Reliability

There are several things that you can do to improve the reliability of a test or measurement technique. One of the more simple methods is to increase the number of items or measurements. All other things being equal, the longer the test or the more times you measure something, the more reliable your resulting score will be. Imagine that a research team wants to examine patients' perceived quality of life after hip replacement. During a preliminary study, the researchers ask participants to rate on a scale ranging from 0 to 10 several questions related to quality of life (e.g., regarding ease of mobility, pain, need for assistance). Using these data, the researchers then determine the internal consistency of the test items. According to the results, the internal consistency is $r = .50$. The researchers can use the Spearman–Brown formula presented in Equation 10.2 to estimate the increase in reliability that will occur if they use more anagrams:

$$r' = \frac{P(r)}{1+(P-1)r} \qquad (10.2)$$

In this equation, r is the original measure of internal consistency and r' is the estimated internal consistency for a test with more items or measurements. The P in the equation represents the number of times the test is lengthened or shortened. For example, if $P = 2$, then the test will be twice as long. When $P = 0.50$, the test length is cut in half. What would happen if the researchers decided to require the participants to answer 10 rather than 2 questions? In this case, $P = 5$ because the new measure will have five times the number of measurements. According to Equation 10.2,

$$r' = \frac{5(.50)}{1+(5-1).50}$$

$$r' = \frac{2.5}{1+(4).50}$$

$$r' = \frac{2.5}{3}$$

$$r' = .83$$

Collecting more data pertaining to a single variable from each participant will increase the reliability of the measure. At a basic level, this is because the average of 10 measurements will be a more reliable estimate of an individual's standing on a construct than the average of only 2 measurements. Therefore, the researchers can use the average response to the 10 questions as their measure of quality of life for patients having their hip replaced.

More is not always better, though. Collecting more data comes at a cost. In some cases, doubling the length of a test may not be practical. For example, converting a 20-item test to a 40-item test may make the test boring and tedious. Consequently, participants may want to rush through the test, not taking the time to answer each question accurately or honestly. In this case, you will need to consider finding a more reliable test or include more participants in your study.

Apart from lengthening a scale or adding additional observations, you can also increase the reliability of a measure by using more accurate and sensitive measurement techniques. Increases in precision correspond to increases in reliability. The development of personal computing technology has also helped researchers increase the accuracy and reliability of their measurement techniques by allowing researchers to program a computer to control the experiment and data collection. The speed and consistency of the computer often allows the researcher to improve the reliability of the data collection procedures.

Using Correlation to Determine Validity

Reliability is a necessary, but insufficient requirement for validity. To determine that a test is valid, we need data that demonstrate that the test measures what we intended it to measure. This is known as construct validity. A correlation coefficient can help us to show that the test scores of a target measure correlate well with some measures that we would expect them to and not correlate with other measures with which they should not be related. In other words, researchers hope to show that their tests demonstrate evidence of convergent and discriminant validity (Campbell & Fiske, 1959). A high correlation between two measures of similar constructs is evidence of **convergent validity**. A low correlation between two measures of different constructs is evidence for **discriminant validity**.

Figure 10.7 illustrates an example of convergent validity and discriminant validity. For convergent validity, we would expect large correlations between the

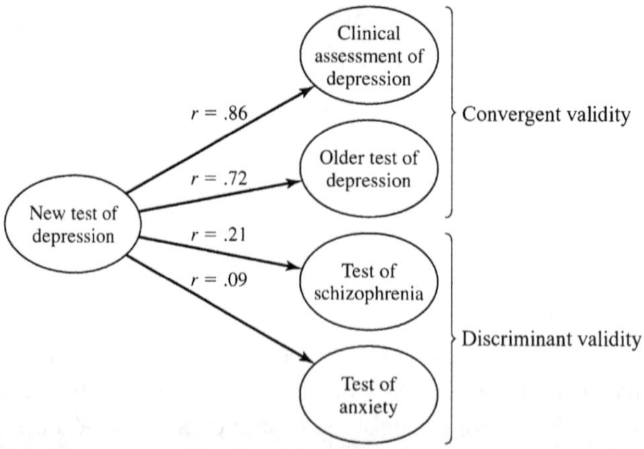

Figure 10.7 Illustration of convergent and discriminant validity.

new test of depression and other tests of depression. By contrast, the new test of depression should not correlate with scores on assessments of schizophrenia or anxiety as these are different psychiatric conditions. Evidence of convergent validity and discriminant validity indicates that the instrument measures what the author of the test intended the test to measure and not something altogether different.

For convergent validity, we would expect the test to correlate highly with other measures of the same construct. Discriminant validity indicates that the test does not correlate well with measures of other constructs. Convergent and discriminant validity is evidence that the test measures the identified construct and not other constructs.

Imagine an oncologist who wishes to develop a new test for the presence of prostate cancer. The motivation for such a project may be that current tests do not accurately test the presence of cancer cells specific to the prostate gland. The advantage of the new test would be an accurate indicator of prostate cancer that is not invasive and provides physicians an appropriate cue regarding the necessity for treatment. To test the validity of the new test, the researcher will need to compare the results of the new test against a well-established indicator of cancer such as a biopsy of cells removed from the prostate gland. If the new test is valid, we would expect a strong correlation between the results of the new screening test and the results of the biopsy.

It is also often important to validate an assessment or test in terms of its effect on meaningful outcomes (i.e., **criterion-related validity**). For criterion-related validity, we hope to show that there is a correlation between the test score and some important outcome or criterion. Specifically, researchers use criterion validity to demonstrate that one can use a test to make predictions about a person's behavior. Two forms of criterion validity are **concurrent validity** and **predictive validity**. Concurrent validity means that we can use the test to predict some condition or characteristic behavior of the person that currently exists (as in the preceding example). Predictive validity means that we can use the test to predict the person's behavior or condition in the future.

As an example, a math achievement test should be able to measure a student's current skills at mathematics. Large correlations between the test score and current math grades, for example, would be evidence of concurrent validity for the math achievement test. In contrast, an aptitude test attempts to predict whether a person will be successful at a specific task. To determine the predictive validity of an art aptitude test, we may administer the test to future employees and then determine their performance on the job.

The utility of a test is in large part determined by the extent to which a user can draw valid inferences from it regarding a specific behavior or other variable. The implication of this is that the use and interpretation of a test determines its validity and that more valid tests will have higher utility within organizations and research settings than less valid tests. This is important to remember as it means that validity is not a property of the test itself but rather of the inferences or conclusions we make based on the data from that test. Because of this, our interpretation of test results may be valid in one situation and not valid in another. Therefore, the validity of a test is much like the internal and external validity of a research project. Again, validity is not a characteristic inherent in the test scores or the data; validity is a characteristic of the interpretations we draw from the data. You can find comprehensive reviews of this issue in the work of Cronbach (1988) and Messick (1980, 1988, 1989).

REGRESSION ANALYSIS

The correlation coefficient, whether it is positive or negative, indicates the presence of a linear relationship between two variables. We can use this information to make predictions about one variable using the other. For example, in general, the longer you stay in school, the higher your annual income. As you will soon see, we can use the correlation coefficient to convert general statements like this into a specific mathematical equation. This equation defines a straight line that best describes the linear relation between the two variables. Using this equation, we can then make specific empirical predictions.

Whenever we have two variables, X and Y, we can use Equation 10.3 to define a straight line that illustrates the relationship between these two variables:

$$Y' = a + b(X) \tag{10.3}$$

In this equation, X represents the variable that we want to use to predict levels of Y. In regression analysis, we use Y' (pronounced "Y-prime") or $\hat{Y}$ (pronounced "Y-hat") to represent the predicted scores along the regression line. Do not confuse Y' and Y; Y represents the original scores and Y' represents the scores predicted by the regression equation. In most cases, Y and Y' will not be exactly the same. The other components of the equation, a and b, are constants that define the line. Specifically, a is the **intercept** and b is the **slope** of the line. The intercept (a) represents the value of Y' when $X = 0$. The slope of the line indicates how much Y' changes as X changes. Equation 10.4 represents a conceptual formula for

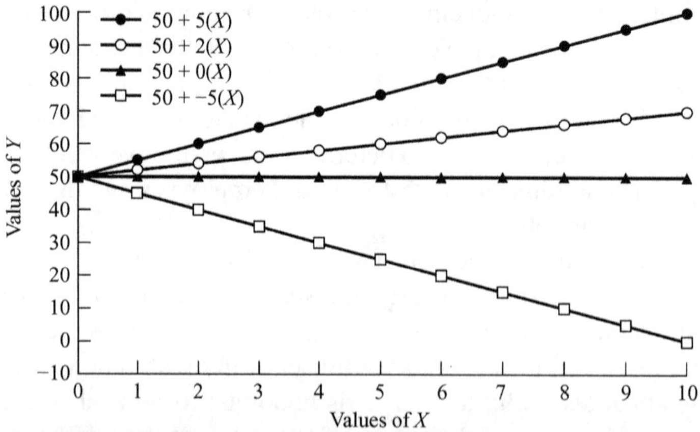

Figure 10.8 Illustrations of four regression lines with the same intercept ($a = 50$) and different slopes: $b = 5$, $b = 2$, $b = 0$, and $b = -5$.

determining the slope in a line. The b coefficient represents a ratio of the changes in Y that correspond with changes in X.

Figure 10.8 will help you see how we use Equation 10.4 to create a straight line in a simple graph. Figure 10.8 is a graph with four regression lines. For each graph, the intercept is 50, but the slopes are different. We can describe the slope as

$$b = \frac{\text{Change in } Y}{\text{Change in } X} \tag{10.4}$$

For example, when $b = 5$, for every one-point increase in X, there is a five-point increase in Y. By contrast, when the slope is $b = -5$, every increase in X corresponds to a five-point decrease in Y. When the slope is 0, the line is horizontal; there are no changes in Y that correspond with changes in X.

Characteristics of a Regression Line

The regression line has several important features we need to review. First, the regression line shares much in common with the arithmetic mean. One property of the mean is that the sum of the differences between the mean and each score in the sample is always 0. Specifically, $\Sigma(X - M) = 0$. The mean represents the balance point of the small and large scores in the sample. The regression line does the same thing; it represents the predictive equation that guarantees the smallest total discrepancy between the predicted Y' and the observed Y. In other words, with so-called least-squares regression, the sum of the differences between each value of Y and its predicted value, Y', is always 0; that is, $\Sigma(Y - Y') = 0$. The regression line represents a special type of average. Consequently, many people also call the regression line the *line of best fit*.

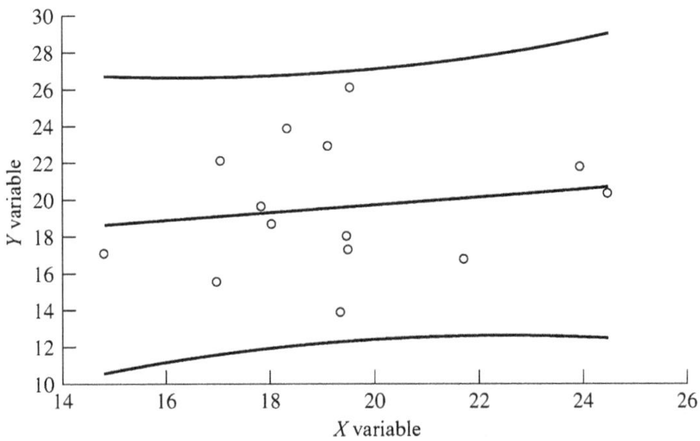

Figure 10.9 Scatter plot of the correlation between two variables.

Interpreting the Regression Equation

Unless there is a perfect correlation ($r = \pm1.00$) between the two variables, the regression line represents only an estimate of the values of Y. Therefore, we need a statistic that will indicate the accuracy of Y'. As you may recall, the standard deviation is a descriptive statistic that allows us to describe how far scores typically differ from the mean. We can do the same thing for Y' using a statistic called the **standard error of estimate**. The standard error of estimate reflects the standard deviation of observed scores around the predicted scores captured by the regression line. Using the standard error of estimate, we can then determine the confidence interval for Y'. This confidence interval is similar to the one reviewed in Chapter 7 in that it allows us to estimate the range of potential scores for a specific value of Y'. The section at the end of the chapter shows how to calculate the standard error of estimate for Y'.

Figure 10.9 presents the original scores, the regression line, and the 95% confidence interval about the regression line. The figure makes clear that values of X much less than or much greater than M_X create a broader confidence interval than values of X close to M_X. The line in the center of the data represents the regression line. The outer curved lines represent the upper and lower limits for the 95% confidence interval.

INTRODUCTION TO MEDIATION AND MODERATION

Taking our discussion of correlation and regression to the next level, it is important to realize that the simple relationship between any two variables may be impacted by one or more additional variables. These variables can be either moderator or mediator variables, depending on their relationships. According to Baron and Kenny (1986), a variable is a mediator to the extent that it accounts (partially or completely) for the relationship between the two other variables. Moderators

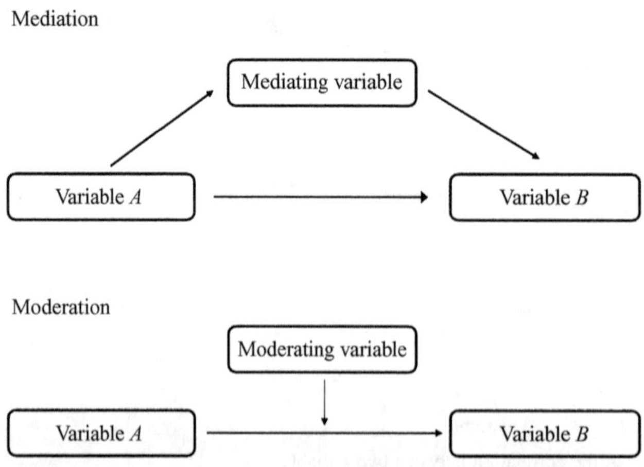

Figure 10.10 A graphical example of mediation and moderation.

specify when certain effects will hold (i.e., interactions), whereas mediators speak to how or why such effects occur.

Put another way, mediator variables explain how or why a relationship occurs. A third variable represents the mechanism through which the independent variable is able to influence the dependent variable of interest. Alternatively, moderation explains when a relationship occurs. A moderator is a third variable that affects the correlation between two other variables. In other words, the statistical relationship between two variables changes as a function of the moderator variable. Figure 10.10 presents a conceptual representation of mediation- and moderation-type relationships.

The traditional view has been that to test for mediation, it is necessary to demonstrate that (1) both the independent and mediating variables are related to the dependent variable; (2) the independent variable is related to the mediating variable; and (3) the relationship between the independent variable and the dependent variable becomes nonsignificant or is reduced significantly when controlling for the mediating variable (Baron & Kenny, 1986; Cohen, Cohen, West, & Aiken, 2003). Moderation can be assessed using either moderated regression analysis or structural equation modeling (see Aiken & West, 1991; Cohen et al., 2003).

Recent work by Preacher and Hayes (2008) and others has shown that condition 1 is sometimes irrelevant and may actually prevent researchers from identifying significant indirect effects, especially in the case of full mediation—as an example, consider the situation in which *X* predicts *Y*, but only through *Z*. In that scenario, condition 1 would never be met, and when it is not met, by traditional standards, most researchers would stop testing for mediation. As an applied example, consider the notion that intelligence (as measured by IQ test scores) is related to a person's ability to earn an advanced college degree and that this type of degree is in turn associated with higher earnings over the course of a person's career. If no evidence showed a direct link between IQ and career earnings, then it might be assumed that these two variables were not linked, even if, in reality, there is an indirect effect linking IQ to career earnings, but only through a person's completion of an advanced degree or some other form of advanced training.

These newer analysis techniques are more advanced than what you may have been exposed to yet (especially the Preacher & Hayes, 2008 approach). You can speak with your professors if you think your research interests might require these types of analytical considerations.

REGRESSION TO THE MEAN

In Chapter 4, we told you that **regression to the mean** is a potential threat to the internal validity of a study. You may also recall a promise to explain the phenomenon in this chapter. We saved the explanation of the regression-toward-the-mean phenomenon because of its relation to measurement of constructs and the reliability of our measurements.

Sir Francis Galton (1889) first recognized the regression-toward-the-mean phenomenon when he was studying various inherited traits such as height. Galton noted that tall parents tend to have taller-than-average children, who were not quite as tall as their parents. Similarly, the children of short parents tended to be taller than their parents, but shorter than average.

We can summarize the regression toward the mean phenomenon with a few simple generalizations. When the correlation between two variables is less than 1.00, the relative difference between X and its mean (i.e., $X - M_X$) will be greater than the difference between the predicted score, Y', and the mean of Y (i.e., $(Y' - M_Y)$). Consider the example of heights of parents and the heights of their children. The expression $(X - M_X)$ represents the difference between the parents' height and the average heights of parents. The expression $(Y' - M_Y)$ represents the difference between the children's height and the average height of the other children. What Galton observed was that exceptionally tall or short parents tended to have taller- or shorter-than-average children, but that the children tended to be closer to the average height than their parents.

Mathematically, regression toward the mean states that $|(X - M_X)| > |(Y' - M_Y)|$. In addition, the amount of regression-toward-the-mean effect increases as the difference between X and M_X increases. Finally, the regression-toward-the-mean phenomenon is greatest when $r = 0$ and absent when $r = 1.00$. Stated from a different perspective, as the correlation between X and Y moves closer to 0, predicted values of Y' are closer to M_Y. When $r = 0$, all predicted scores equal the mean of the Y variable. Specifically, $Y' = M_Y$.

Figure 10.11 presents an illustration of regression toward the mean. The scatter plot represents a correlation of $r = .50$. The regression line for these data is $Y' = 45.0 + .55(X)$. The other line in the graph represents the regression line had the correlation been $r = 1.00$. The shaded area between the two regression lines represents regression toward the mean. The arrows at either end of the regression lines indicate the direction of the regression. This graph clearly shows that a value

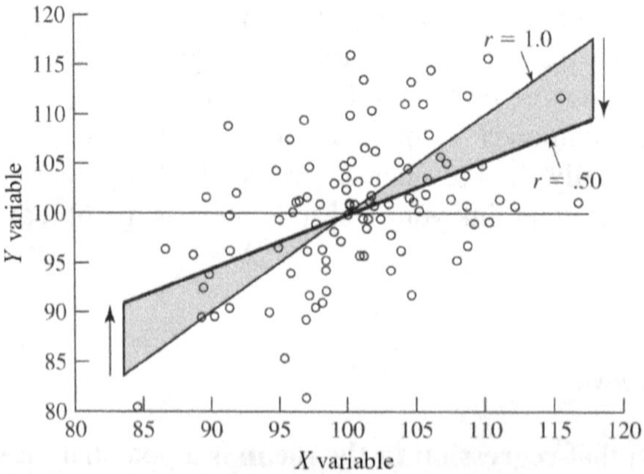

Figure 10.11 Illustration of regression to the mean.

Table 10.2 Example of Regression to the Mean

	High scores for Test 1			Low scores for Test 1		
	Test 1	Test 2	$T_2 - T_1$	Test 1	Test 2	$T_2 - T_1$
	95	33	−62	21	15	−6
	95	92	−3	21	48	27
	94	27	−67	21	17	−4
	91	42	−49	18	3	−15
	89	21	−68	16	23	7
	87	81	−6	13	21	8
	84	91	7	12	98	86
	80	8	−72	10	49	39
	79	20	−59	5	81	76
	78	26	−52	4	94	90
M	87.2	44.1	−43.1	14.1	44.9	30.8

of X above or below M_X predicts scores of Y' relatively closer to M_Y when $r = .50$. When $r = 0$, then all values of Y' equal the mean of Y.

The scatter plot represents $r = .50$. The bold line represents the regression line for the data. The other line represents the regression line if the correlation between the two variables had been $r = 1.00$. The shaded area represents the regression toward the mean. According to the regression-to-the-mean phenomenon, $|Y' - Y|$ is less than $|X - M_X|$.

Here is another example of regression toward the mean. Let us say you have two pairs of 50 random numbers ranging between 0 and 100. Think of these pairs of numbers as if they were scores on two organic chemistry exams that we gave to the same students, one on Monday and the other on Friday (hopefully not in the same week!). The first set of numbers represents test 1; the second set represents scores for test 2. Table 10.2 presents the highest 10 and the lowest 10 scores for test 1. Next to each test 1 score is the score for test 2. The regression to the mean effect is dramatic. Almost every high score on test 1 has a lower test 2 score. Similarly, almost every low test 1 score has a higher test 2 score. There is a clear regression to the mean. The amount of regression toward the mean is large in this example because we used random numbers where the correlation between the two variables is $r = 0$.

This example of regression toward the mean has practical applications for designing research projects and interpreting their results. Imagine that we wanted to test the effectiveness of a new method to reduce people's fear of public speaking. We begin the research by selecting people who are extremely anxious giving a speech. Next, we offer these people a workshop to help reduce their anxiety. After completing the workshop, we repeat our evaluation of the participants' anxiety while giving a speech.

Is a large drop in anxiety scores evidence of the effectiveness of the workshop? Not necessarily. The drop in anxiety scores could reflect regression toward the mean and nothing more. For this reason, regression toward the mean is a potential confounding variable that raises serious questions about the internal validity of the conclusion that the workshop eases people's fear of public speaking.

How would we control for regression toward the mean? One method is to have two groups: an experimental group and a control group. Participants in the experimental group would complete the workshop, whereas the control group would not. When we subsequently evaluate both groups, we would expect the experimental group to evidence greater improvement than the control group. According to the regression-toward-the-mean effect, participants in both groups should show some improvement. If the public speaking workshop is truly effective, then the participants in the experimental group should evidence greater average improvement.

RESEARCH IN ACTION: ZINC AND DEPRESSION

Zinc is a trace metal we obtain from eating such things as red meat, fish, eggs, and legumes and appears to be vital in many physiological processes. The level of zinc in our blood system may also be related to depression according to a recent study conducted by Amani, Saeidi, Nazari, and Nematpour (2010). In that study, the researchers administered Beck depression inventory to 308 university women between the ages of 20 and 25. Of these students, 23 had depression scores indicating moderate to severe depression. The blood chemistry of these 23 women was compared to another group of 23 women who showed no signs of depression but were similar in other respects to the women identified as depressed.

The research consisted of several components. First, Amani et al. (2010) asked the 46 women to complete a questionnaire regarding eating habits. This survey allowed the researchers to estimate the typical zinc consumption for each woman. The researchers then drew a blood sample from each woman and measured the serum zinc concentrations. Using these data—zinc consumption, serum zinc

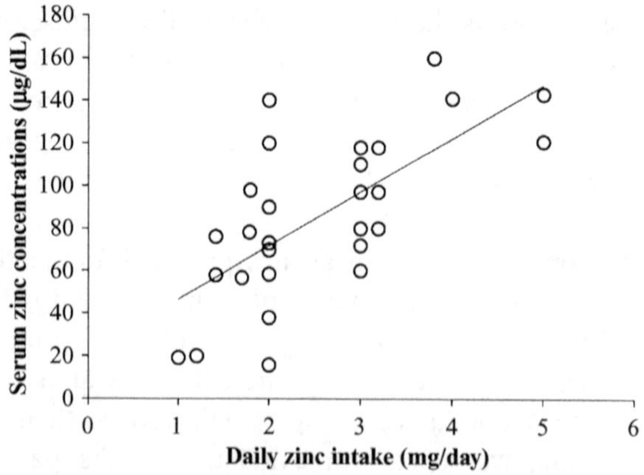

Figure 10.12 Correlation between daily zinc intake and serum zinc concentrations. *Note:* Graph based on research conducted by Amani et al. (2010).

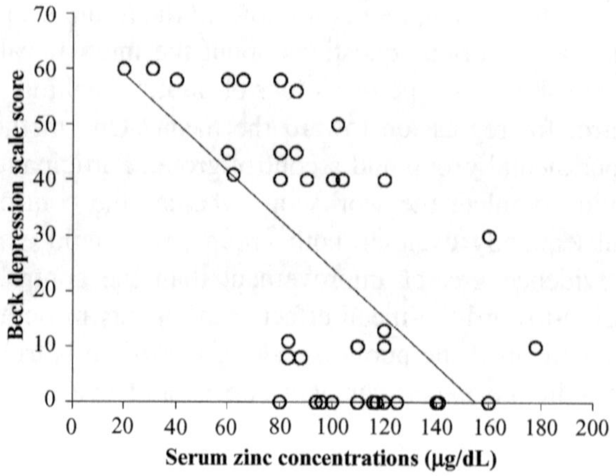

Figure 10.13 Correlation between serum zinc and level of depression. *Note:* Graph based on research conducted by Amani et al. (2010).

levels, and depression scores—the researchers were able to examine the correlation between zinc and depression.

The first analysis confirmed that eating habits influenced serum zinc levels. As can be seen in Figure 10.12, there is a strong correlation—$r = .65$—between estimated zinc consumption of serum zinc levels. These results allowed the researchers to affirm that the observed serum zinc levels were related to the participants' diet.

Figure 10.13 illustrates the relation between the Beck depression scale and zinc levels. For the Beck scale, higher scores represent greater levels of depression, whereas lower scores represent normal mood and functioning. For these data, there is a strong negative correlation, $r = -.65$, thus indicating that higher levels of zinc are related to lower levels of depression. Stated from a different perspective, as zinc levels decrease, the levels of depression increase.

SPEARMAN RANK-ORDER CORRELATION COEFFICIENT

Researchers use the Pearson product-moment correlation coefficient when the data represent interval or ratio numbers. In many cases, however, researchers will collect data best represented by the ordinal scale or when the data may be skewed, thus causing an inaccurate result. When these cases arise, researchers turn to the the Spearman rank-order correlation coefficient, or Spearman's r_S. As with the conventional correlation coefficient, the values of r_S can range between −1.00 and +1.00. Correlations close to 0 represent no linear relation between the two variables. Correlations close to ±1.00 represent large linear relations:

$$r_S = 1 - \frac{6\sum D_i^2}{n(n^2 - 1)} \tag{10.5}$$

The following data represent the relation between the per capita consumption of alcohol in various countries and the legal restrictions on access to alcohol. For this study, Brand et al. (2007) examined each country's laws regarding access to alcohol (e.g., legal drinking age, tax on alcohol sales, penalties for public intoxicated, and driving with intoxicated). The each country's control of alcohol sales, score and per capita liters of alcohol consumed were then ranked ordered from lowest to highest.

You will note that in Table 10.3 there are two values of $\text{Rank}_X = 2$ and two values of $\text{Rank}_Y = 12$. These are not typographical errors but how ranks are assigned when there are equal scores. For the sales score, both Germany and Switzerland scored 22.4. Therefore, they each share the rank of 2 and the next ranking skips to 4. When the calculations are completed, the correlation between the scales score and consumption was $r_s = -0.5884$. These results indicate that as a generality, the more restrictive a country's control of alcohol sales, the lower the overall consumption:

$$r_S = 1 - \frac{6(7{,}140)}{30(30^2 - 1)} \quad r_S = 1 - \frac{42{,}840}{26{,}970} \quad r_S = 1 - 1.5884 \quad r_s = -.5884$$

KNOWLEDGE CHECK

6. Why might the test–retest reliability of peoples' weight be greater than the test–retest reliability of their personality?
7. In what ways are coefficients κ and r similar to each other?
8. What are some ways to increase the reliability of a test?
9. Explain in your words the importance of convergent and discriminant validity for examining the validity of a test.

Table 10.3 Worked Example for Spearman Rank-Order Correlation Coefficient

Country	*X* control of alcohol sales score	$Rank_X$	*Y* per capita liters of alcohol consumed	$Rank_Y$	$D = (Rank_X - Rank_Y)$	D^2
Luxembourg	14.50	1	12.01	24	−23	529
Switzerland	22.40	2	10.56	19	−17	289
Germany	22.40	2	11.81	23	−21	441
Austria	23.00	4	12.60	27	−23	529
France	26.90	5	13.30	28	−23	529
Portugal	27.20	6	12.45	26	−20	400
Denmark	33.20	7	11.37	20	−13	169
Italy	34.20	8	8.33	8	0	0
Netherlands	34.40	9	9.55	12	−3	9
Czech Republic	35.40	10	14.97	30	−20	400
United Kingdom	35.50	11	11.67	21	−10	100
Greece	36.20	12	8.95	10	2	4
Ireland	40.80	13	13.39	29	−16	256
Spain	40.90	14	10.22	17	−3	9
Belgium	41.70	15	9.77	15	0	0
United States	43.10	16	8.44	9	7	49
Korea	43.30	17	11.80	22	−5	25
Mexico	45.10	18	5.02	2	16	256
Turkey	48.70	19	1.37	1	18	324
New Zealand	49.90	20	9.12	11	9	81
Canada	50.10	21	7.77	6	15	225
Japan	52.40	22	7.83	7	15	225
Finland	54.20	23	9.72	14	9	81
Slovakia	57.00	24	10.33	18	6	36
Hungary	57.50	25	12.27	25	0	0
Australia	62.80	26	9.89	16	10	100
Sweden	63.80	27	6.70	5	22	484
Iceland	64.50	28	5.91	3	25	625
Poland	67.00	29	9.55	12	17	289
Norway	67.30	30	6.21	4	26	676
					$\sum D^2 = 7140$	

10. Imagine that we randomly selected 25 students with extremely high systolic blood pressure. We then measure the blood pressure again later in the day. What predictions can you make about the second set of scores?

CHAPTER SUMMARY

The correlation coefficient is a descriptive statistic that indicates the degree to which changes in one variable relate to changes in a second variable. Specifically, the correlation describes a linear relation between two variables. The size of the correlation coefficient ranges between −1.00 and +1.00. When the correlation coefficient is 0, then there is no meaningful linear relation between the two variables. Correlation coefficients closer to −1.00 or +1.00 represent strong linear relations between the two variables.

Pearson's correlation coefficient is the most commonly used statistic to examine the relation between two variables. The statistic works by examining the sum of the standard score cross products ($\Sigma z_X z_Y$) of the two data sets. Using the standard scores (z-score) allows us to compare two variables even when the means and standard deviations of the two groups are extremely different.

The absolute value of the correlation coefficient indicates the strength of the relation between the two variables. One way to describe the size of the correlation coefficient is to use the coefficient of determination, r^2, which indicates the proportion of variance in one variable that the other variable can predict.

There are several errors that people often make when interpreting the size of the correlation coefficient. One common error is to use the correlation coefficient as evidence of cause and effect. Correlational research designs do not allow the researcher to control the independent variables or randomly assign participants to different levels of the independent variable. Therefore, we cannot resolve the ambiguity of cause and effect created by the third variable problem or the temporal order problem.

It is also possible to have an exceptionally low correlation because of a curvilinear relation between the two variables or because the sampling procedures did not adequately draw a representative sample from the population. These sampling errors can create a truncated range in the variances of the variables, extreme groups or outliers, and multiple unrecognized populations.

As with other descriptive statistics, we can use the concept of the standard error and the confidence interval to estimate sample size. For the correlation coefficient, the sample size needs to be large ($N = 500$) if the researcher believes that the correlation coefficient for the population is small. If the correlation is larger, however, the researcher may be able to use a smaller sample size.

The correlation coefficient has many applications for researchers. We examined how the correlation coefficient can help the researcher determine the test–retest reliability of a test or the internal consistency between two observers. A special version of the correlation coefficient, known as coefficient alpha, r_α, allows researchers to examine the internal consistency of test items. A high coefficient alpha indicates that the questions in the test are highly interrelated. Knowing the test–retest reliability or the coefficient alpha can allow the researcher to adjust the length of a test to enhance its reliability. For example, making a test longer can improve the reliability of an instrument. This increased reliability will decrease the measurement error of the dependent variable.

The correlation coefficient also allows us to examine the validity of a test. By correlating a test with tests that measure the same construct, we can determine the test's convergent validity. Tests that measure the same construct should produce high correlations. By contrast, tests that measure different constructs should not correlate with each other. Researchers call this discriminant validity.

We can use regression analysis to make a formal statement about the relation between the two variables. The regression line allows us to describe the specific linear relation between X and Y. We can also calculate a confidence interval about the regression line. Knowing the position of the regression line allows us to describe the regression-to-the-mean phenomenon. If an observed value of X is greater than the mean, then the predicted value of Y' will be closer to the mean of Y.

CHAPTER GLOSSARY FOR REVIEW

Coefficient Alpha (r_α) A common measure of internal consistency. The magnitude of r_α indicates whether the participants' responses to the individual test items correlate highly with each other, thus indicating internal agreement among the questions.

Coefficient of Determination (r^2) A descriptive statistic that indicates the percentage of shared variance between two variables.

Coefficient of Nondetermination ($1 - r^2$) A descriptive statistic that indicates the percentage of unshared variance between two variables.

Concurrent Validity A form of criterion-related validity that indicates that the test scores correlate with the current behaviors of the individuals.

Convergent Validity An index of the validity of a test that requires that different methods of measuring the same construct correlate with each other.

Correlation Coefficient A descriptive statistic that describes the linear relation between two variables. The statistic can range between $r = -1.0$ and $r = +1.0$ and indicates the degree to which changes in one variable correspond with changes in the other variable.

Criterion-Related Validity An index of the validity of a test that requires that the test score predict an individual's performance under specific conditions.

Curvilinear Relationship A systematic relation between two variables that is described by a curved rather than a straight line.

Discriminant Validity An index of the validity of a test that requires that measures of different constructs should not correlate with each other.

Intercept (a) The value of Y' when $X = 0$.

Internal Consistency A measure of the degree to which questions in a test agree or correlate with each other.

Linear Relation The correlation between two variables best described as a straight line.

Negative Correlation When $r < 0$, increases in one variable correspond with decreases in the other variable.

Positive Correlation When $r > 0$, increases in one variable correspond with increases in the other variable.

Predictive Validity A form of criterion-related validity that indicates that the test scores afford accurate predictions concerning the future behavior of the individual.

Regression to the Mean A phenomenon in regression analysis where $|(X - M_X)| > |(Y' - M_Y)|$.

Slope (b) The change in Y' that occurs when X increases by one unit.

Standard Error of Estimate The standard deviation of predicted scores about Y.

Truncated Range A condition that occurs when the range of observed scores for both variables is smaller than the range of potential scores. This condition tends to reduce the size of the correlation coefficient.

REFERENCES

AIKEN, L.S. & WEST, S.G. (1991). *Multiple regression: Testing and interpreting interactions.* Newbury Park, CA: SAGE.

AMANI, R., SAEIDI, S., NAZARI, Z., & NEMATPOUR, S. (2010). Correlation between dietary zinc intakes and its serum levels with depression scales in young female students. *Biological Trace Element Research, 137*, 150–158.

BARON, R.M. & KENNY, D.A. (1986). The moderator-mediator variable distinction in social psychological research: Conceptual, strategic, and statistical considerations. *Journal of Personality and Social Psychology*, *51*, 1173–1182.

Brand, D.A., Saisana, M., Rynn, L.A., Pennoni, F., & Lowenfels, A.B. (2007). Comparative analysis of alcohol control policies in 30 countries. *PLoS Medicine*, *4*, e151.

Campbell, D.T. & Fiske, D.W. (1959). Convergent and discriminant validation by the multitrait-multimethod matrix. *Psychological Bulletin*, *56*, 81–105.

Cohen, J. (1988). *Statistical power analysis for the behavioral sciences* (2nd ed.). Hillsdale, NJ: LEA.

Cohen, J., Cohen, P., West, S.G., & Aiken, L.S. (2003). *Applied multiple regression/correlation analysis for the behavioral sciences* (3rd ed.). Mahwah, NJ: Erlbaum.

Cortina, J.M. (1993). What is coefficient alpha? An examination of theory and applications. *The Journal of Applied Psychology*, *78*, 98–104.

Cronbach, L.J. (1951). Coefficient alpha and the internal structure of tests. *Psychometrika*, *16*, 297–334.

Cronbach, L.J. (1988). Five perspectives on the validity argument. In H. Wainer & H.I. Braun (Eds.), *Test validity* (pp. 63–79). Hillsdale, NJ: Erlbaum.

Galton, F. (1889). *Natural inheritance*. London: Macmillan.

Koren, D., Levitt Katz, L., Brar, P., Gallagher, P., Berkowitz, R., & Brooks, L. (2011). Sleep architecture and glucose and insulin homeostasis in obese adolescents. *Diabetes Care*, *34*, 2442–2447.

Messick, S. (1980). Test validity and the ethics of assessment. *The American Psychologist*, *35*, 1012–1027.

Messick, S. (1988). The once and future issues of validity: Assessing the meaning and consequences of measurement. In H. Wainer & H.I. Braun (Eds.), *Test validity* (pp. 80–95). Hillsdale, NJ: Erlbaum.

Messick, S. (1989). Validity. In R.L. Linn (Ed.), *Educational measurement* (3rd ed.). New York: American Council on Education/Macmillan.

Preacher, K.J. & Hayes, A.F. (2008). Asymptotic and resampling strategies for assessing and comparing indirect effects in multiple mediator models. *Behavior Research Methods*, *40*(3), 879–891.

Salvadori, A., Fanari, P., Giacomotti, E., Palmulli, P., Bolla, G., Tovaglieri, I., et al. (2003). Kinetics of catecholamines and potassium, and heart rate during exercise testing in obese subjects. Heart rate regulation in obesity during exercise. *European Journal of Nutrition*, *42*(4), 181–187.

Schmitt, N. (1996). Uses and abuses of coefficient alpha. *Psychological Assessment*, *8*, 350–353.

Svensson, V., Jacobsson J.A., Fredriksson, R., Danielsson, P., Sobko, T., Schioth, H.B., et al. (2011). Associations between severity of obesity in childhood and adolescence, obesity onset and parental BMI: A longitudinal cohort study. *International Journal of Obesity*, *35*, 46–52.

Zubaran, C., Foresti, K., Schumacher, M., Amoretti, A., Thorell, M., & Müller, L. (2010). The correlation between postpartum depression and health status. *Maternal and Child Health Journal*, *14*, 751–757.

Chapter 11

Between-Subjects Designs

Statistics is no substitute for judgment.

—Henry Clay

INTRODUCTION

In Chapter 9, we examined ways to design a research project that allows us to collect data relevant to the research hypothesis. Once we collect the data, we must begin the process of analysis and interpretation. Chapter 10 introduced you to correlational research and evaluating the linear relationships among variables. In many cases, however, we are interested in group differences in addition to or instead of linear relationships.

Accordingly, in the first part of this chapter, we will review the logic of inferential statistics and how they allow us to evaluate group differences. As a part of this review, we will examine the types of inferences these statistics do and do not allow us to make. Inferential statistics are important to all the behavioral sciences. Unfortunately, some researchers abuse, misapply, and misinterpret these tools. Because statistical analysis forms the basis of decision making, you need to

Understanding and Conducting Research in the Health Sciences, First Edition. Christopher J. L. Cunningham, Bart L. Weathington, and David J. Pittenger.

understand what conclusions, or inferences, you can and cannot draw from these statistical tests.

Take heed of Clay's comment opening this chapter, about statistics and judgments. Statistics are wonderful tools for helping us find patterns and trends in the data. They also help us solve problems and come to decisive conclusions. However, statistical inference requires careful consideration. Statistics may help us make judgments, but they do not, cannot, and should not replace our ability to make rational decisions.

In another section of this chapter, we will review how inferential statistics help us solve other important questions. For example, many researchers ask, "How many participants do I need in my research?" This is an important question because the sample size affects the accuracy of the results and conclusions of our research.

Throughout this chapter, we will use the two-group experiment for our examples in this chapter, not because it is the most popular form of research design, but because it is a good way to describe the important topics in this chapter. We will review more complex research designs in subsequent chapters. Fortunately, the principles you mastered in Chapters 9 and 10, and the ones you will learn in this chapter, apply to all good research designs and inferential statistics.

STUDENT'S *t*-TEST FOR INDEPENDENT GROUPS

One of the most important advances in modern inferential statistics was the development of Student's sampling distributions and the inferential statistics using those distributions. This statistic, known as **Student's *t*-ratio**, is a commonly used statistical procedure in the social and behavioral sciences. Once you master the principles used for this inferential statistic, you will be better able to use and understand other inferential statistics. The "Statistics behind the Research" section at the end of this chapter presents the *t*-ratio and its alternatives in more detail.

The *t*-ratio gets its name because it is a ratio of two measures of variability. Specifically, it is the ratio of the difference between two group means relative to the variability within the groups. Equation 11.1 presents a conceptual formula for Student's *t*-ratio:

$$t = \frac{\text{Difference between group means}}{\text{Standard error of the difference between group means}} \quad (11.1)$$

The numerator of the equation represents the difference between the two group means. The difference between these two means reflects the effect of the independent variable and the effects of random error on both of the study groups. Because the *t*-ratio compares the difference between two means, the *t*-ratio can be a negative value when $M_1 < M_2$, a positive value when $M_1 > M_2$, or 0 when $M_1 = M_2$.

The denominator in this ratio is the **standard error of the difference between means** ($\sigma_{\mu_1-\mu_2}$), which allows us to estimate the variability of scores within the groups we are comparing. Recall that the within-groups variability represents normal differences among individuals in the population and sampling error. In this way, the standard error of the difference between means is similar to the standard error of the mean, as we have discussed in previous chapters. Therefore, we can use what we have learned about the standard error of the mean and apply it to the standard error of the difference between means. According to the central limit theorem, the distribution of sample means is normally distributed with a standard deviation (often referred to as a standard error) of

$$s_M = \frac{SD}{\sqrt{n}}$$

The same occurs when we examine the difference between means drawn from separate populations. Imagine that we have two populations, such that $\mu_1 = 5.0$, $\sigma_1 = 1.0$ and $\mu_2 = 5.0$, $\sigma_2 = 1.0$. We draw a random sample of $n = 2$ from each population. For the first sample, the scores might be 3 and 5. For the second sample, the scores could be 4 and 6. Therefore, $M_1 = 4$ and $M_2 = 5$. The difference between the means is $M_1 - M_2 = -1$. If we continue this sampling procedure infinitely, we can create a sampling distribution of the difference between the means.

What would this sampling distribution look like? Figure 11.1 presents an illustration of what would happen. The sampling distribution at the bottom of the figure represents the distribution of the difference between means, $M_1 - M_2$. Because the null hypothesis states that $\mu_1 = \mu_2$, we predict that the mean of the sampling distribution will be $\mu_{\mu_1 - \mu_2} = 0$ and that values above and below 0 represent random sampling error. According to the central limit theorem, we can predict that the distribution of the difference between means will be symmetrical and bell-shaped.

According to the null hypothesis, $\mu_1 = \mu_2$. Therefore, if we were to create an infinite number of sample pairs, the difference between the pairs would fall into the sampling distribution at the bottom of the figure. The mean of the sampling distribution will be 0 and the standard deviation will equal

$$\sigma_{\mu_1 - \mu_2} = \sqrt{\frac{\sigma_1^2}{n_1} + \frac{\sigma_2^2}{n_2}}$$

We use the term $\sigma_{M_1 - M_2}$ to represent the standard error of the difference between means. When we calculate the *t*-ratio, we use the variances of the two samples to estimate $\sigma_{M_1 - M_2}$. Specifically, $\sigma_{M_1 - M_2}$ is an unbiased estimate of $\sigma_{M_1 - M_2}$, just as the *SD* is an unbiased estimate of the population standard deviation, σ.

Many statisticians refer to $\sigma_{M_1 - M_2}$ as the **error term** because we use it to estimate the random error that occurs when we take random samples from the population. Thus, sampling error and random error are synonymous and estimated by $\sigma_{M_1 - M_2}$. Using this information, we can turn our attention to hypothesis testing.

What does the *t*-ratio tell us? Using Equation 11.1, you can conclude that the larger the absolute value of the *t*-ratio, the greater the relative difference between

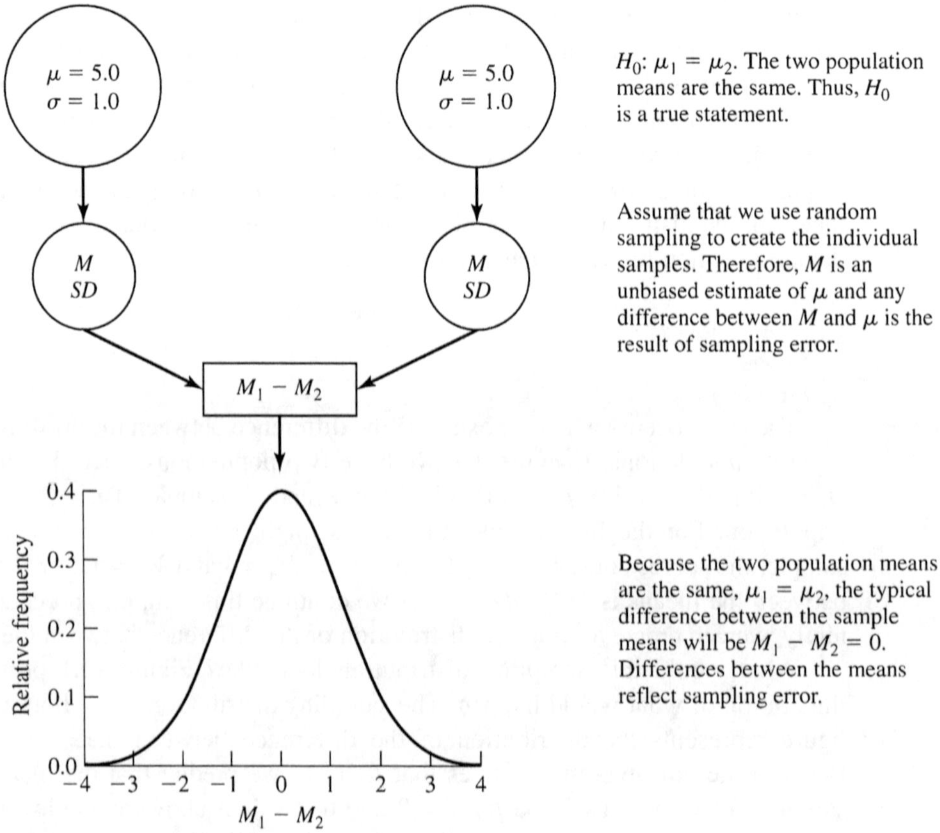

Figure 11.1 Standard error of the difference between means.

the means (taking into account the error that is present). The magnitude of the *t*-ratio, therefore, allows us to determine whether to reject the null hypothesis. The typical two-group nondirectional null hypothesis states that H_0: $\mu_1 = \mu_2$. Consequently, we predict that the average difference between the means of samples drawn from the two populations will be 0. Sampling error may cause the difference between some sample means to be greater or less than 0. However, if a *t*-ratio is sufficiently large, its magnitude signals us to reject the null hypothesis. The question before us is what we mean by *sufficiently large*.

REVIEW OF HYPOTHESIS TESTING

How do we decide whether to reject or retain the null hypothesis? We use estimates of probability. Imagine that your friend claims to be able to toss a coin and cause it to land heads up at will. To test your friend's skill, you ask that he or she toss a coin 10 times in a row. If your friend produces only 5 heads out of 10 tosses, you probably will not be impressed because your friend seems to be tossing heads at random. However, if your friend can toss 10 heads in a row, you may be impressed because the probability of a run of 10 heads is less than 1 in 1000. Could your friend have been lucky and tossed 10 heads by chance?

Yes, assuming that the probability of heads is .5. The reason for this is that when there are 10 coin tosses, the probability of having all 10 land heads up is $p = .5^{10} = .000977$. This probability is small enough that you decide that the 10 heads must represent a unique skill (either that or a weighted or two-headed coin). We use the same reasoning when we evaluate the *t*-ratio. We ask, "If the null hypothesis is a true statement, what is the probability of obtaining a *t*-ratio that differs from some critical value?" If the probability is sufficiently small, we can infer that the null hypothesis is a false statement.

As you learned in Chapter 9, we use α to represent the probability of committing a type I error (a false alarm, or rejection of the null hypothesis when it is actually correct). Most researchers use α to establish the criterion for determining when to reject the null hypothesis and set $\alpha = .05$. The α-level indicates that if the null hypothesis is correct, there is a 5% probability of committing a type I error (rejecting H_0 when we should not). In some cases, the researcher may want to be more cautious and will set $\alpha = .01$, thus lowering the risk of a type I error to 1%.

Figure 11.2 represents an example of a sampling distribution for a *t*-ratio where the nondirectional null hypothesis is H_0: $\mu_1 = \mu_2$. Using this sampling distribution, we can determine the probability of obtaining any specific *t*-ratio and thereby determine whether to reject the null hypothesis. You should recognize that the null hypothesis in Figure 11.2 is a nondirectional hypothesis and that the alternative hypothesis is H_1: $\mu_1 \neq \mu_2$. In this case, we will reject the null hypothesis if the *t*-test result is considerably less than or considerably greater than 0. As you can see, the tail ends of this figure are shaded and marked "Reject null $\alpha/2$" to denote a two-tailed or nondirectional test. If the null hypothesis is correct, then the probability that the *t*-ratio will fall within either shaded area is equal to α. Because the null hypothesis is nondirectional, we divide α equally between the two ends of the sampling distribution.

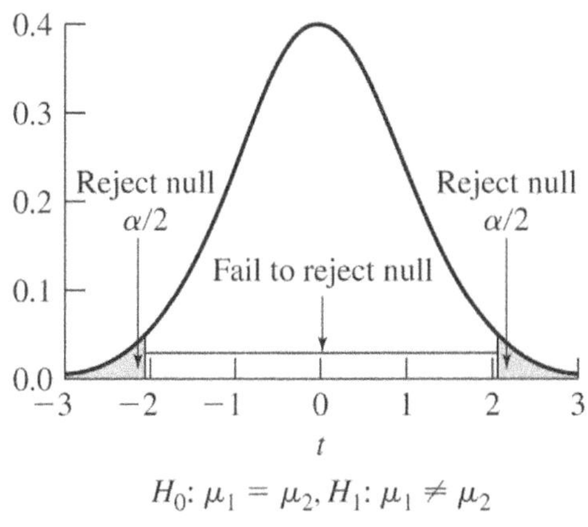

Figure 11.2 Hypothetical sampling distribution of the *t*-ratio and testing the null hypothesis H_0: $\mu_1 = \mu_2$, a nondirectional test.

The shaded areas represent the criterion for rejecting the null hypothesis. If the t-ratio falls in the shaded area, we reject H_0 in favor of H_1. If the t-ratio falls in the clear area, we fail to reject the null hypothesis.

To illustrate, we will use an example study we discussed in Chapter 9, where we wanted to compare men's and women's shopping behavior. In that example, we asked the question "Do men and women differ how they find things to buy things in a shopping mall?" This is a nondirectional hypothesis because we want to determine whether there is a difference between the two groups. In this example, the null hypothesis is that men and women will behave differently while shopping. If the null hypothesis is correct, the t-ratio for the difference between means should be close to 0, and any difference from 0 would represent the effect of random events.

Assume we decide to set $\alpha = .05$. According to Figure 11.2, the probability of obtaining a t-ratio in the upper shaded area is $p = .05/2 = .025$ or 2.5%. Similarly, the probability of obtaining a t-ratio in the lower shaded area is also $p = .025$. Added together, the two shaded areas comprise a total of .05 or 5% of the distribution. If a t-ratio falls within either shaded area, we reject the null hypothesis because the probability of obtaining such a t-ratio unless there is a real difference present between the groups is so low that we are willing to infer that null hypothesis is incorrect. In this situation, we then, by default, accept the alternative hypothesis as a better explanation of the difference between groups we have observed. In contrast, if the t-ratio falls in the clear area between the two shaded areas of our distribution, we do not reject the null hypothesis.

Figure 11.3 presents how we would proceed if we used a one-tailed or directional hypothesis. Because the directional hypothesis predicts the relation between the means (using < or >), we place the entire α region on the side of the sampling distribution representing the hypothesized difference between the means. Specifically, Figure 11.3*a* represents the test we create using the null hypothesis, H_0: $\mu_1 \geq \mu_2$; Figure 11.3*b* represents the test created using the null hypothesis, H_0: $\mu_1 \leq \mu_2$.

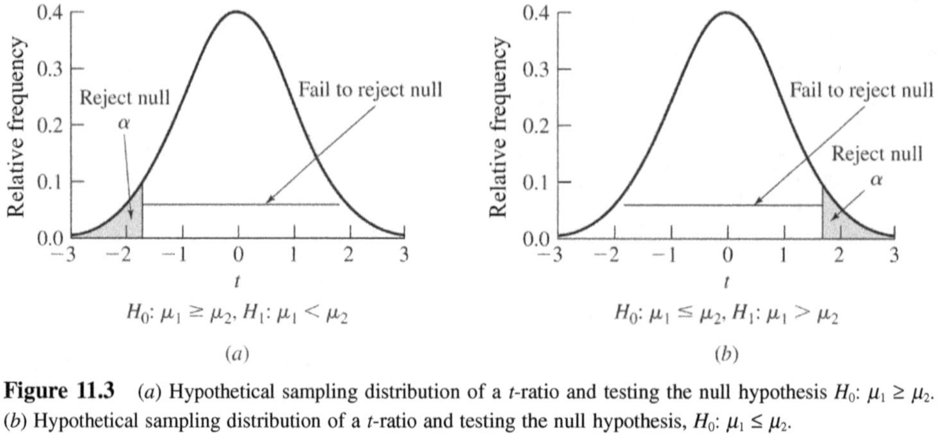

Figure 11.3 (*a*) Hypothetical sampling distribution of a t-ratio and testing the null hypothesis H_0: $\mu_1 \geq \mu_2$. (*b*) Hypothetical sampling distribution of a t-ratio and testing the null hypothesis, H_0: $\mu_1 \leq \mu_2$.

The shaded areas represent the criterion for rejecting the null hypothesis. If the Figure 11.3 *t*-ratio falls in the shaded area, we reject H_0 in favor of H_1. If the *t*-ratio falls in the clear area, we fail to reject the null hypothesis.

TESTING STATISTICAL HYPOTHESES

Statisticians and researchers invented inferential statistics during the early part of the twentieth century (Oakes, 1986). Over time, general rules and principles for using inferential statistics became common practice. For example, most researchers now routinely include a null and alternative hypothesis as a part of their statistical test. We discussed forming hypotheses in some detail in Chapter 3. With the information in this section, we build on that material and examine the specific steps involved in hypothesis testing.

One important feature of all hypothesis testing is that we form the hypothesis before we collect or examine the data. In fact, in most social and behavioral science research, the design of the research actually flows from the stated hypotheses. All of us can describe the past with accuracy; predicting the future, however, is less certain. Thus, we are more impressed by and have much greater confidence in a prediction of the future that is supported by good data from a well-designed research study. As you will see in the following steps, much of the work associated with hypothesis testing comes before we collect or analyze even a single point of data.

Steps for Hypothesis Testing

Step 1: State the Null and Alternative Hypotheses

As you learned earlier in this book, the null hypothesis, H_0, is the mathematical statement we intend to disprove with the data we collect in a study. More specifically, the typical null hypothesis states that there is no relationship between the independent and dependent variables or that two or more groups' (i.e., treatment condition) means are equal. If we can reject the null hypothesis, then we can accept the alternative hypothesis, the focus of the study. Typically, the alternative hypothesis is the central thesis of the research and states that there is a meaningful difference between the group means.

The null hypothesis can be either nondirectional or directional. The nondirectional hypothesis for a two-group study will be H_0: $\mu_1 = \mu_2$. This version of the hypothesis states that the two population means are identical and that any observed difference between the group means is due to random events unrelated to the independent variable. In contrast, the directional null hypothesis can be either H_0: $\mu_1 \geq \mu_2$ or H_0: $\mu_1 \leq \mu_2$. The directional hypothesis predicts that one population mean is less than the other mean with the exception of the effects of random sampling error.

The alternative hypothesis (often denoted H_1) is always the mathematical complement of the null hypothesis. The complementary relational signs are =

versus ≠, ≥ versus <, and ≤ versus >. As a rule, each null hypothesis has only one alternative hypothesis.

Step 2: Identify the Appropriate Statistical Test

This step may seem obvious, but it is often the downfall of many students and researchers. Part of the problem is that you can often apply any number of statistical tests to any set of data. One question you could ask is "Can I use this statistic to analyze these data?" The answer will typically be "yes" because you can often use a variety of statistical tests to analyze any set of data. A better question is "Will this statistical test allow me to answer the question I posed in my hypothesis?" Different statistical tests lead to different information from a set of data. For example, some allow you to compare groups, whereas others examine the correlation between groups.

Therefore, your null and alternative hypotheses will largely determine the type of test you use. Other factors also influence the selection of the appropriate statistical test. Many inferential statistics require that the data meet specific criteria to produce valid results. For example, the *t*-ratio for independent groups assumes that the observations for the two groups are independent, that the data are normally distributed, and that the variances of the two groups are equivalent. Failure to meet these criteria can produce spurious results. As you plan your research, you should identify the appropriate statistical tests for analyzing the data and identify the requirements for using that test. In addition, you should identify alternative strategies for analyzing the data in the case that your data do not conform to the requirements of a specific test.

Step 3: Determine the Appropriate Significance Level

Determining the appropriate significance level is another important step. As you learned in the previous chapter, researchers want to avoid making a type I error. We can control the probability of committing a type I error by selecting a suitable criterion. Because α represents the probability of committing a type I error and is determined by the researcher, you have direct control over the probability of committing a type I error. Recall that the probability of a type I error and a type II error is inversely related—as you reduce the risk of a type I error, you increase the risk of a type II error. Consequently, you should plan your research with care to ensure that the risk of both errors is at a tolerable level.

Many researchers have fallen into the habit of automatically setting $\alpha = .05$ or $\alpha = .01$ and then continuing on with their typical analysis of the data. There is nothing magical about these criteria for α, but their use is common practice. Selecting the α-level should not be reflexive. Instead, selecting α should be a deliberative process that examines the consequence of committing a type I or a type II error.

In many ways, selecting the α-level takes us back to the utilitarian principles of ethics because the process forces us to examine the consequences of our actions.

Ask yourself, "What is the cost of committing a type I versus a type II error?" For the American legal system, the proposition is that sending an innocent person to jail (a type I error) is far worse than letting a guilty person back on the streets (a type II error). Therefore, the legal criterion for finding guilt, "beyond a reasonable doubt," is equivalent to setting α to a small value, say, $\alpha = .0001$.

Rudner (1953) argued that researchers must make value judgments when deciding whether to reject the null hypothesis. He asserted that

> *[because] no scientific hypothesis is ever completely verified, in accepting a hypothesis the scientist must make the decision that the evidence is sufficiently strong or that the probability is sufficiently high to warrant the acceptance of the hypothesis . . . . How sure we need to be before we accept a hypothesis will depend on how serious a mistake would be. (p. 350)*

Rudner recognized two important components of hypothesis tests. First, inferential statistics depend on samples and probabilities. Consequently, whenever we make a statistical decision, we must recognize the legitimate risk of committing a type I or type II error. Second, either error may have grave consequences depending on the purpose of the research. Therefore, as researchers we must examine the utility of our research results and the consequences of our statistical decisions.

For some experiments, committing a type I error can have serious negative consequences. Will we, for example, use the results to justify an expensive program of research or to alter the treatment that patients currently receive? If the statistical conclusion represents a type I error, we may end up spending money on a potentially pointless set of experiments or begin to use an ineffective treatment.

In other cases, committing a type II error is the more serious mistake. A type II error may cause the researcher to ignore an important relationship between two variables or an important difference between two or more groups. For many researchers who conduct basic research, a type II error is the more serious error because it means that they have overlooked a potentially interesting finding.

Because both type I and type II errors involve potentially unpleasant consequences, we should design our research to minimize the risk of both errors. Reaching a balance between the two errors is not always easy, but it is possible with careful planning. We will return to this important topic later in the chapter. Once you determine the significance level that you wish to use, you can proceed with the analysis of the data. Using the results of your data analysis, you can then determine whether you have sufficient evidence to support your original research hypothesis.

Step 4: Determine the Significance Level for the t-Ratio

We can use Table B.3 of Appendix B to determine the critical value of the *t*-ratio required to reject the null hypothesis (or we could rely on a computer program to do this for us). As you learned previously, the table consists of a series of rows and columns. Each column represents the probability level we want to use for our hypothesis. Each row represents the degrees of freedom for the data. For the independent-groups *t*-ratio, the degrees of freedom are

BOX 11.1 ***Hypothetical Example of Time Spent Actually Reading Instructions and Background Information on Prescription Drugs***

	Time Reading	
	Women	Men
	32	20
	32	20
	12	10
	14	18
	14	12
	26	12
	24	12
	22	16
Sum	176	120
Sum of squares	4320	1912
Mean	22	15
Variance	64	16
Standard deviation	8	4
Degrees of freedom = 14		
$t = 2.214$		
$p = 0.044$		

$$df = (n_1 - 1) + (n_2 - 1) \tag{11.2}$$

To determine the critical value of t, known as t_{critical}, we need to determine the level of α, whether we wish to conduct a directional or a nondirectional test, and the degrees of freedom for the test. As an example, assume that we set $\alpha = .05$ and decide to use a nondirectional test, and that the degrees of freedom are 16. Given these facts, we see that $t_{\text{critical}} = 2.120$. What would have happened if we had decided to conduct a directional test? For $df = 16$ and $\alpha = .05$, one-tailed, $t_{\text{critical}} = 1.746$.

We can use hypothetical data for men's and women's time spent actually reading instructions regarding their prescriptions as an illustration of the steps for hypothesis testing (see Box 11.1). For the sake of illustration, we will assume that our hypothesis in this case is nondirectional.

If you used a computer program to calculate the statistic, you may have obtained an output that looks something like this:

$$t(14) = 2.214;\ p \text{ one-tailed} = .022;\ p \text{ two-tailed} = .044$$

Because the absolute value of the obtained t-ratio exceeds the critical value ($t_{\text{observed}} > t_{\text{critical}}$; $2.214 > 2.145$), we can reject H_0. If the null hypothesis were true, the probability of selecting at random two samples that differ this much or more is less than 5/100 or $p < .05$. Whenever the probability level associated with a statistical outcome (p) is less than the α-level that you establish before running your analyses, you can reject the null hypothesis. In this example, the probability associated with this t-ratio is $p = .007$ for the two-tailed or nondirectional test. Because $p < \alpha$, you can reject H_0.

Step 5: Interpreting the t-Ratio

Once we calculate the t-ratio and its probability, we can determine whether to reject H_0. Although we have completed one important task, there are still many opportunities to further analyze the results of our research. In this section, we will examine several of the errors that many people make when interpreting the results. We will also examine the conclusions that inferential statistics afford and several statistical procedures that can enhance our interpretation of the data.

Statistically Significant. Whenever the obtained p-value is less than the set α in a study, it is common to see researchers state that they found statistically significant results. This means that they feel they have sufficient statistical evidence to reject the null hypothesis. Unfortunately, many people equate "statistical significance" with "importance." Because this is clearly not the case, some researchers like to make the distinction between **statistical significance** and **clinical or practical significance**. Whereas statistical significance means that one has reason to reject the null hypothesis in favor of the alternative hypothesis, practical significance implies that the results have real implications for the target population (e.g., they may justify a specific intervention; Moyé, 2000). In other words, if our evaluation of the effectiveness of a new smoking-cessation intervention yielded practically significant findings, we would consider the size and direction of the observed effect to conclude that the intervention's benefits (or consequences) justify the cost of the treatment (or point toward a need to discontinue the treatment).

Under some circumstances, it is possible to obtain statistical significance when the effect size is trivial (e.g., when $d = .1$ or less, as discussed in Chapter 9). This is especially true when the sample size is very large. For example, we can claim that there is a statistically significant difference between men's and women's scores on math achievement tests, even when d is within the trivial effect size range. The difference between most men's and women's math skills is of little practical significance, especially for anything less than the most complex levels of mathematics. Therefore, when you hear that the results of a study are statistically significant, do not automatically assume that the results are of practical or clinical significance unless the researcher has compelling evidence to make such a claim.

Omega Squared ($\hat{\omega}^2$). The presence of statistical significance does not automatically confer the title of "important" on any particular finding. Given a sufficiently large sample size, any difference in means, no matter how trivial, can be statistically significant. One way to clarify the importance of a statistically significant t-ratio is to determine the extent to which variations in the independent variable account for variations in the dependent variable. Apart from effect size indicators such as d, one of the more popular measures of association is **omega squared ($\hat{\omega}^2$)**. In a study with two groups, we can calculate $\hat{\omega}^2$ by

$$\hat{\omega}^2 = \frac{t^2 - 1}{t^2 + n_1 + n_2 - 1} \tag{11.3}$$

Omega squared is an index of the degree to which the variance in one variable accounts for the variance in another variable. Omega squared can have positive and negative values. Specifically, when the absolute value of $t < 1.00$, $\hat{\omega}^2$ will be negative. For practical purposes, negative values of $\hat{\omega}^2$ have little meaning. Consequently, most researchers calculate $\hat{\omega}^2$ only when the t-ratio is statistically significant.

We can apply the equation for $\hat{\omega}^2$ to the sex-differences data. From that example, we know that $t = 2.21$, $n_1 = 8$, and $n_2 = 8$. Therefore,

$$\hat{\omega}^2 = \frac{2.21^2 - 1}{2.21^2 + 8 + 8 - 1}$$

$$\hat{\omega}^2 = \frac{4.8841 - 1}{4.8841 + 16 - 1}$$

$$\hat{\omega}^2 = \frac{3.8841}{19.8841}$$

$$\hat{\omega}^2 = \frac{3.8841}{19.8841}$$

$$\hat{\omega}^2 = .1953$$

We may interpret $\hat{\omega}^2$ to mean that the independent variable accounts for approximately 20% of the variance in the dependent variable. In some research contexts, accounting for 20% of the variance is a tremendous event and reflects a real advance in the science. In other contexts, 20% is barely worth considering. Consequently, you must use your knowledge of other findings in a specific area and your critical judgment to evaluate the importance of this type of effect size indicator.

It is important for us to note that you cannot judge the magnitude of an effect by merely looking at the value of the t-ratio or at the smallness of the obtained p-value. For example, imagine that you found a t-ratio of -3.17 with 70 participants in each group. The difference is clearly statistically significant. Under these conditions, the probability for $t = -3.17$ is $p = .00052$. However, $\hat{\omega}^2$ can be calculated as

$$= \frac{(-3.17)^2 - 1}{(-3.17)^2 + 70 + 70 - 1}$$

$$= .061$$

Neither the size of the t-ratio nor the size of the p-value can indicate the "importance" of the data. The t-ratio indicates only whether there is sufficient evidence to reject the null hypothesis. You must use other statistics such as $\hat{\omega}^2$ or d to further evaluate the results of the study. Then you must evaluate the findings within a broader context and existing literature before you draw your final conclusions and inferences.

Step 6: Reporting the t-Ratio

Although standards may vary by journal or professional organization, many researchers in the social and behavioral sciences use the *Publication Manual of the American Psychological Association* (American Psychological Association, 2009) as the editorial guideline when preparing their manuscripts for publication. The recommended format within this editorial style for reporting the results of the t-ratio is

$$t(df) = t_{\text{observed}},\ p = p\text{-level}$$

or as

$$t(df) = t_{\text{observed}},\ p < \alpha$$

As an example, in the results section of your research manuscript, you could write something like this:

> *Figure 1 presents the average recall by women and men. An independent-groups* t-test *confirmed that men spent more time reading than did women. $t(14) = 2.21$, $p < .05$, $\hat{\omega}^2 = .19$.*

KNOWLEDGE CHECK

1. A researcher conducts a study comparing the behavior of two groups. According to the results, the means of the two groups are $M_1 = 98$ and $M_2 = 107$.
   a. What are potential explanations for the difference between the means?
   b. Can the researcher use the difference between the means to conclude that the independent variable has a statistically significant effect on the dependent variable?
   c. How will changes in the standard error of the difference between means influence our interpretation of the difference between the two means?
2. Using the previous example, could we potentially reject the null hypothesis if it took the form H_0: $\mu_1 \leq \mu_2$? What if the null hypothesis were H_0: $\mu_1 \geq \mu_2$?
3. Why should the researcher state the null and alternative hypotheses before collecting and analyzing the data?
4. What are the factors that will determine your selection of an inferential statistic?

5. A researcher is determining the significance level of a statistical test. According to Rudner, why is this selection a value judgment? What are the consequences of selecting the significance level?
6. A researcher finds that the difference between two groups is statistically significant. Under what conditions can the researcher assume that there is a cause-and-effect relation between the independent and dependent variables?

COMMON ERRORS IN THE INTERPRETATION OF *p*

Many researchers criticize the use of statistical hypothesis testing. Much of the complaint is that some people misuse or misinterpret the results of an inferential statistic. The following is a list of the more common errors. For the most part, the greatest controversy surrounds the interpretation of the established and observed probability levels (α and p) associated with inferential statistics.

Changing α after Analyzing the Data

Imagine that a researcher set $\alpha = .05$ before collecting the data. After analyzing the data, the researcher finds that $p = .003$. Should he or she be able to revise α to $\alpha = .005$ to make the data look "more significant"? What if the researcher had set $\alpha = .01$, but the analysis of the data revealed $p = .03$? Can the researcher revise α to $\alpha = .05$ to say that there are statistically significant results? Shine (1980) argued that the answer to both questions is an emphatic *no!* The first problem with this type of revision is that it is unethical. Researchers should set α before analyzing their data and stick with that decision. Changing this probability level after the fact nullifies your ability to lock in the chance of a type I error at an acceptably low level. If you choose to modify this level after your analyses, you are, in effect, ignoring the risk of a type I error and opening yourself up to criticism for fishing through the data looking for whatever might appear to be interesting or significant, even if it was only that way because of chance.

The second reason that you should not change your interpretative standards after beginning an analysis is that the p-level is not a substitute for α. Although the complete mathematical logic behind this is beyond the scope of this text, you need to understand that p only estimates the probability of obtaining at least the observed results, if the null hypothesis is true. Because we estimate p using sample statistics and because the null hypothesis may be false, we cannot work backward and use our observed p to revise α.

Assuming That *p* Indicates Results Due to Chance

Carver (1978) calls this error the "odds-against-chance" error. Some researchers interpret p to indicate the probability that the difference between the means is due to chance. For instance, a researcher may find that for a specific t-ratio, $p = .023$ and may conclude that the probability that the results were due to chance is 2.3% or less. This interpretation is incorrect because we begin our statistical test with an H_0 that states that any and all observed differences between our study groups are due to chance, and therefore the probability of identifying a difference by chance is 1.00. Because we do not know the real parameters of the populations, we cannot predict the probability that the results are due to chance. The p-value merely indicates the probability of obtaining a specific or more extreme difference if H_0 was in fact true. In other words, if your results are associated with a $p < .05$, you can feel confident that the observed difference is not due to chance because if it were, it would happen only less than 5% of the time.

A related error in the interpretation of the p-value is to assume that p establishes the probability that we will commit a type I error. If you found that $p = .023$, you could not conclude that the probability of committing a type I error is 2.3%. The value of α determines the probability of a type I error, a probability that you as the researcher will set. As a rule, you should interpret the p-value to mean the following:

> *If H_0 is a true statement, the probability of obtaining these or more extreme results is p. If the value of p is small enough (i.e., $<\alpha$), you are willing to reject H_0 in favor of the alternative hypothesis. The probability of committing a type I error is α.*

Assuming the Size of *p* Indicates the Validity of the Results

Some researchers assume that the p-value indicates the probability that the research or alternative hypothesis is true. Again, the probability associated with a statistic does not allow us to determine the accuracy or truth of the alternative hypothesis. The ability to determine the validity of the research hypothesis is an extremely complicated issue. In general, the p-value cannot confirm the validity of the research hypothesis. The only function of the p-value is to determine whether there is sufficient evidence to reject H_0.

Some researchers infer that a small value of p indicates the degree to which there is a meaningful relationship between the independent and dependent variables. For example, if $p = .001$, the researcher may conclude that the independent variable has a large and important effect on the dependent variable, or may claim that the data are "highly significant." This inference is not true, and worse, it is misleading to those who read the results of such research. Only effect size estimates such as $\hat{\omega}^2$ appropriately indicate the strength of the relationship between the independent and dependent variables in a study.

Assuming That *p* Establishes the Probability That the Results Can Be Replicated

Some people assume that $(1 - p)$ indicates the probability that the experiment will yield the same result if repeated. If, for example, $p = 0.23$, a researcher may assume that the probability of replicating the experiment and rejecting H_0 is $1 - .023 = 0.977$. This is a false statement because the probability of correctly rejecting H_0 is not related to the value of p. The ability to replicate a given finding depends on the difference between the means, the number of participants, the amount of measurement error, and the level of α. It is possible to have a small level of p (i.e., $p = .001$) and a low probability that the results can be directly replicated. As an aside, there is some recent work on a statistic that can tell us our probability of replicating the findings from a given study. If you are interested, you may wish to check out Killeen (2005).

You are probably wondering whether you can draw any meaningful conclusion at all from a statistically significant t-ratio. If you obtain a statistically significant t-ratio at $p < \alpha$, you may conclude that there is a statistically significant difference between the mean scores on the dependent variable measure that you used in the two groups you are studying. In addition, if your comparison is part of a true experiment designed to test causality, you may also be able to conclude that the independent variable affects the dependent variable. You may also conclude the probability of a type I error is α.

THE POWER OF A TEST

Putting α and type I errors aside for a minute, you may recall that a type II error (i.e., a miss) occurs when we fail to reject the null hypothesis and it is false. Researchers want to avoid type II errors for obvious reasons. We conduct research because we want to discover interesting facts and relationships between and among variables. If our statistical tools overlook these important findings, we will have wasted much time and energy. Therefore, researchers strive to increase the power of their statistical test when they conduct research.

In Chapter 9, we reviewed several tactics that we can use to ensure that the data we collect will allow us to correctly reject the null hypothesis in favor of the alternative hypothesis. In this section, we will revisit those design tactics and show how they can increase the statistical power of an analysis.

For any statistical test, β defines the probability of making a type II error. The **power of a statistic**, which we represent as $1 - \beta$, is the probability of correctly rejecting a false null hypothesis. The four factors that influence statistical power are (1) difference between the population means, $\mu_1 - \mu_2$; (2) sample size, n; (3) variability in the population, σ^2; and (4) alpha (α) level and directionality (directional vs. nondirectional) of the test.

Difference between Population Means ($\mu_1 - \mu_2$)

If the null hypothesis is false, there will be two sampling distributions similar to the ones presented in Figure 11.4*a*. Each sampling distribution represents sample means drawn from separate populations. The sampling distribution for the null hypothesis is a distribution of sample means for which $\mu_1 - \mu_2 = 0$. The shaded

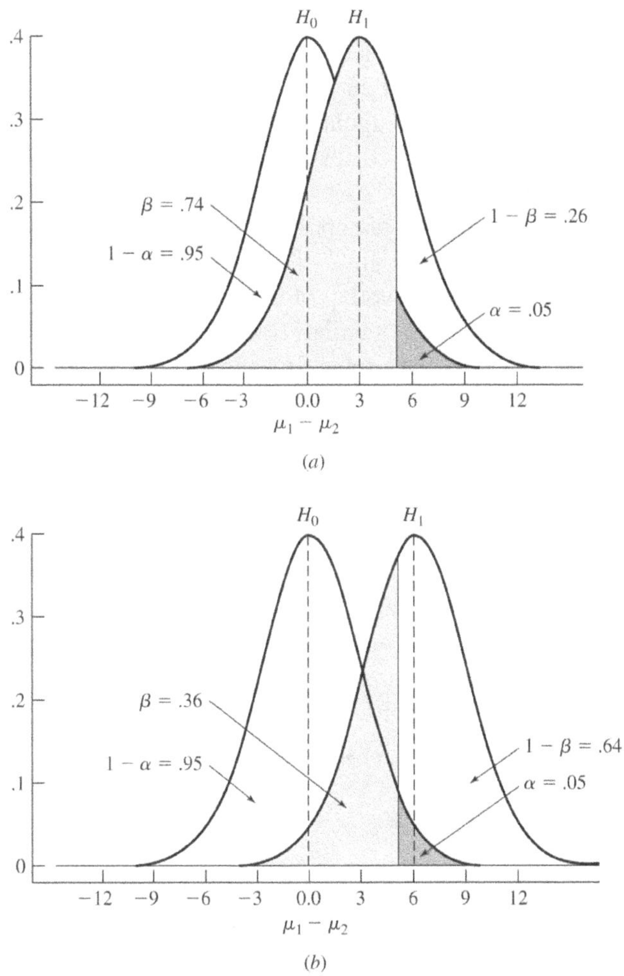

Figure 11.4 The effect on statistical power of $\mu_1 - \mu_2$.

area at the upper end of the scale represents the critical region for $\alpha = .05$ using a directional test. According to the null hypothesis, there is a 5% probability that the difference between any pair of means will be in the critical area. The second distribution is a sampling distribution of means that would occur if we draw samples of a specific size from two populations for which $\mu_1 - \mu_2 = 3$.

In the upper figure, the difference is 3.0. The difference in the lower figure is 6.0. For both pairs of distributions, the shaded areas represent the probabilities of committing a type I error (α) and a type II error (β). The power of the statistic is $1 - \beta$. The lower figure represents greater power.

As you can see, the two distributions overlap, but not perfectly. What do the different areas represent? The area shaded in light gray represents β, the probability of a type II error. Although the two population means are different, it is possible to select sample means whose differences will be less than the critical region. Those samples will not be sufficiently different to allow us to reject the null hypothesis. In this example, 74% ($\beta = .74$) of the curve is in the lightly shaded area. Thus, the probability of committing a type II error is 74%, whereas the probability of rejecting the null hypothesis is 26% ($1 - \beta = .26$). Therefore, the power of this statistic is 26%. Although the population means are different from each other, the chance that we will be able to discover this difference using sample statistics is only slightly better than 1 in 4.

Figure 11.4*b* presents a different set of conditions. Now the difference between the population means is much greater ($\mu_1 - \mu_2 = 6.0$). Notice the differences between Figure 11.4*a*,*b*. There is less overlap of the two curves. Because there is less overlap, the area representing β is smaller. Specifically, the probability of a type II error is now 36%, whereas the probability of correctly rejecting the null hypothesis is 64%. The conclusion to draw from this example is that power increases as the difference between the two population means increases.

Intuitively, this should make sense—a larger, more noticeable difference between groups should be more easily identified than a smaller difference. In Chapter 9, we reviewed tactics for increasing the differences between two study populations. The most direct method is to select levels of the independent variable that create the greatest possible difference in the dependent variable measures between the two groups. All else being equal, making the two populations as different as possible from each other will increase the statistical power of the study. There are other ways that we can increase power that we will consider next.

Sample Size

Sample size is another important consideration because it influences how well a sample statistic will estimate the corresponding population parameter and the power of a statistical test. As sample size increases, the standard error of the mean decreases. Figure 11.5*a* presents the sampling distributions that would occur if one selected samples from two populations using small samples. Figure 11.5*b* shows the two distributions obtained when we use a larger sample size. The central limit theorem explains the difference in the shape of the two sampling distributions.

For both distributions, the shaded areas represent the probability of committing a type I error (α) and a type II error (β). The power of the statistics is $1 - \beta$. The lower graph represents the greater statistical power.

As a rule, as sample size increases, the standard error of the mean decreases. Consequently, the degree of overlap of the two sampling distributions decreases and the corresponding power increases. Again, intuitively, this should make sense because as the degree of potential overlap between two groups decreases, any real difference between those two groups should be more noticeable and easily identified. Although it is true that increasing the sample size will increase your power to detect an effect if it is present, you also need to be aware of the costs associated with this design strategy.

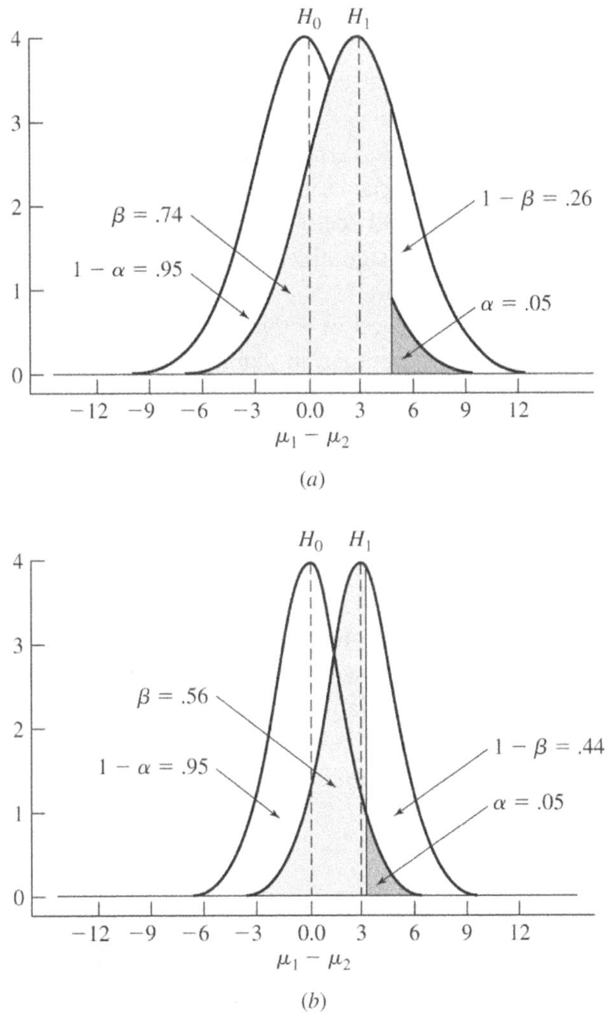

Figure 11.5 The effect on statistical power resulting from changes in sample size or changes in population variance.

Although increasing the sample size sounds easy (you just need to collect more data), the downside is that collecting data takes time and money. From the cost of purchasing materials to the time you must take out of your schedule, collecting data from each subject adds to the total cost of the research. In some cases, such as administering a set of questionnaires to students in their classes, there will be a minimal cost for collecting more data. In other cases (e.g., face-to-face interviews), collecting the data from one subject will be time consuming and expensive. Therefore, you should consider all your options for increasing power before you jump to the conclusion that you need to run more participants through your study.

Variability

If the population from which you draw the sample has considerable variability in general, the sampling distribution associated with your outcome or dependent variable scores will also have considerable variance. We can use Figure 11.5 again to represent the effects of population variance on statistical power. The distributions in Figure 11.5*a* represent what would occur when the variance in the population is large. Consequently, there is a considerable overlap of the two distributions and the power is small. Reducing the amount of variability in the population will produce a result similar to the one depicted in Figure 11.5*b*.

Because the magnitude of σ (the population standard deviation) decreased, the overlap depicted in Figure 11.5*b* decreased and the statistical power increased. The main takeaway point from all of this is that anything you can do to reduce sampling error will improve your ability to detect a statistically significant difference between the means.

Apart from making it easier to detect differences between groups, as you may recall from Chapter 9, there are also several ways that we can decrease the variability of scores within groups. For example, we can use homogeneous samples. If the participants in the individual groups are similar to each other, it is easier to detect differences among the group means. Similarly, as we use more reliable and accurate measurement techniques, the variability within the groups will tend to decrease. This, in turn, will help to minimize the spread of the data within each group, having the same effect that we described in the previous paragraph.

Alpha (α)

Alpha (α) sets the probability of a type I error. The smaller the α, the lower the probability of a type I error. Unfortunately, lowering the α-level will decrease the power of the statistical test. As the probability of the type I error decreases, the probability of the type II error increases. Look at Figure 11.6 for an illustration. In both graphs included in this figure, the difference between the means is 3.0. For the distributions in Figure 11.6*a*, $\alpha = .05$. For the distributions in Figure 11.6*b*, α is larger ($\alpha = .10$) and the power is greater. For both graphs, the probability of committing a type I error (α) and a type II error (β) are represented by shaded areas. The power of the statistic is $1 - \beta$.

Note the differences between the statistical powers for the two conditions illustrated in the preceding figure. When $\alpha = .10$, the power is $1 - \beta = .39$. Lowering the α to .05 decreases the power to $1 - \beta = .26$. All things being equal, the probability of a type I error increases as α decreases. Another way to say the same thing is that as α increases, power also increases. Think about this conceptually and it may make more sense—if you make it more difficult to identify an effect as statistically significant (i.e., linked to how low you set your α-level), then you will also make it more difficult to detect an effect when it is really there (i.e., you will reduce your statistical power). The opposite also holds true.

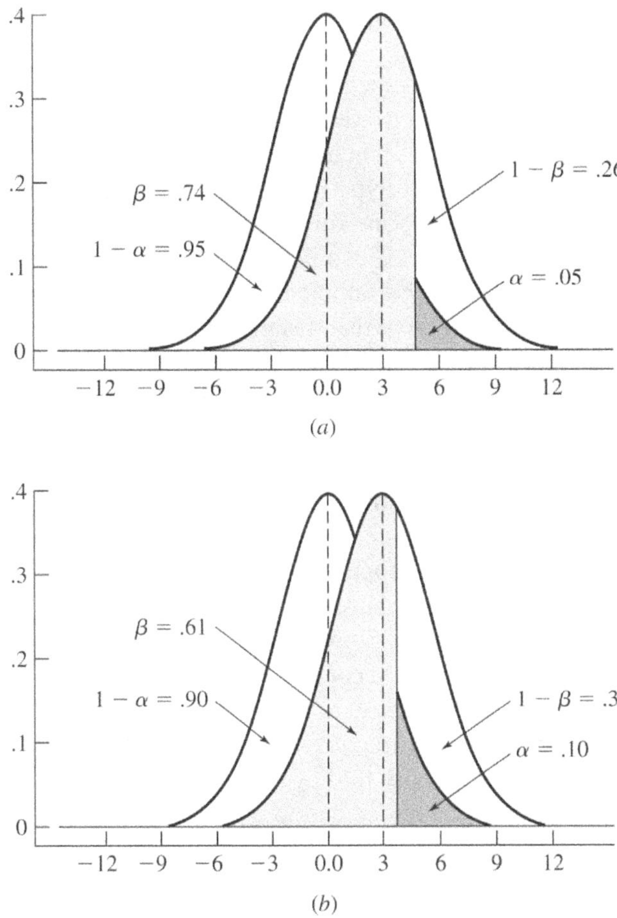

Figure 11.6 The effect of changes in α on statistical power.

Your selection of a directional and nondirectional test based on your hypotheses will also influence statistical power. In general, a directional test is more powerful than a nondirectional test. The difference lies with where we place the critical region for rejection. Consider a t-ratio with $df = 15$ and $\alpha = .05$. When we use the directional test, we place the critical region at one end of the distribution. In this example, $t_{critical}$ for the directional test is 1.753. When we conduct a nondirectional test, we split the critical regions between the two extreme ends. Consequently, the t-ratio required to reject the null is larger than it would be for a comparable directional test. For the nondirectional test where $df = 15$, $t_{critical} = 2.131$.

ESTIMATING THE SAMPLE SIZE

You know that sample size can be an important factor in ensuring sufficient statistical power, but you also know that going after too many participants might leave you broke and exhausted. How many participants should you include in your sample? We can use effect size estimates such as d to estimate the number of participants we will need for a two-group study such as those examined in this chapter.

Look at Tables B.5.1 and B.5.2 of Appendix B, which provide power estimates for two sample *t*-ratios. Along the leftmost column are sample sizes (n), the number of participants in each group. The columns represent values of t_{critical}, the critical *t*-ratio needed to reject the null hypothesis, and four levels of effect size, d. There are two versions of the table, one for directional tests and the other for nondirectional tests. To use the table, select an effect size that you believe represents your research. Then determine the sample size you will need for a specific level of power. Some researchers believe that a statistical power of $1 - \beta = .80$ is an adequate level for most research. Others may want their power to be higher or lower depending on the importance of avoiding a type II error and identifying an effect, along with the potential implications of making a type I error.

Here is an example. You want to replicate the sex differences for the time spent reading provided background information and directions regarding prescription drugs study that we discussed earlier. How many participants will you need to replicate the effect? What would happen if you used a directional versus a nondirectional test to compare the differences?

For this example, assume that $\alpha = .05$, two-tailed, and that the effect size for this phenomenon is medium, $d \approx .50$. According to Table B.5.2, you will need approximately 60 participants in each group to have your power set at $1 - \beta = .79$ for a nondirectional test. If you used a directional test, you would require approximately 50 participants in each group to have the power of approximately $1 - \beta = .82$. What would happen if you decided to collect data from 100 students and used a directional test? In that case, $1 - \beta = .98$. The larger sample size means that you will have a 98% chance of rejecting the null hypothesis if the null hypothesis is false.

Caveat Emptor: Sample Size

Please remember that increasing sample size should not be the only thing you do to increase the power of a statistical test. The decision to increase sample size should come only after you have examined the design of your research. As you learned in Chapter 9 and again here, you can increase power by increasing the reliability of the tests you use to measure the dependent variable, ensuring homogeneity of variance among the participants and increasing the difference between the two groups.

KNOWLEDGE CHECK

7. Assume that a researcher conducted a two-group study and found that $t(18) = 2.7$, $p < .01$. Determine whether the following statements are correct or incorrect and defend your answer:

a. The researcher has disproved the null hypothesis, H_0: $\mu_1 = \mu_2$.
b. The researcher has determined the probability that H_0 is true.
c. The researcher has proved that the research hypothesis is correct.
d. The researcher can determine the probability that H_1 is correct.
e. The researcher can be confident that there is an extremely high probability that he or she could replicate the findings by conducting the same experiment.
f. The probability of the data given the null hypothesis is $p < .01$.

8. Mark and Mary each conducted an experiment studying the relation between the same independent and dependent variables. Mark used 20 participants in each group and found a statistically significant effect using $p = .045$. Mary used 100 participants in each group and found a statistically significant effect using $p = .045$.
   a. Of the two experiments, which appears to have the greater power?
   b. How would you account for the differences in power between the two experiments?
9. Khalil wants to conduct an experiment and believes that the effect size for the variables selected is small, $d \approx .20$. How many participants should he place into each group to have statistical power of .80 for a two-sample t-ratio with $\alpha = .05$—two-tailed? What would be the required sample size if Khalil used $\alpha = .05$—one-tailed?
10. Khalil wants to find ways to improve the power of his study. What factors, other than increasing the number of participants in each group, should he consider in the design of his experiment that could increase power?

STATISTICS BEHIND THE RESEARCH

In this section, we will review in more depth several statistical tests that allow us to compare the differences in dependent variable scores or measures between two independent groups. The phrase *independent groups* means that there is no specific correlation in characteristics or biases between the two groups. One way to ensure the independence of the groups is to randomly assign the participants to the two research conditions, as is done in a true experiment. We can also assume independence of the groups if we use an intact group design and ensure that our selection of participants for one group has no influence on the selection of participants in the other group.

The Student's t-ratio is an inferential statistic that allows us to determine whether the difference between two sample means is statistically significant. Researchers use the test when the dependent variable represents an interval or ratio scale of measurement. The other tests are alternatives that we use when the data do not meet the mathematical assumptions of the statistic.

Student's *t*-Ratio for Independent Groups

The complete equation for Student's *t*-ratio is

$$t = \frac{M_1 - M_2}{\sqrt{\dfrac{\sum X_1^2 - \dfrac{\left(\sum X_1\right)^2}{n_1} + \sum X_2^2 - \dfrac{\left(\sum X_2\right)^2}{n_2}}{n_1 + n_2 - 2}\left(\dfrac{1}{n_1} + \dfrac{1}{n_2}\right)}} \tag{11.4}$$

The degrees of freedom are calculated as

$$df = (n_1 - 1) + (n_2 - 1) \tag{11.5}$$

Assumptions of the Test

For the statistical test to work properly, the data must meet several assumptions. The assumptions are independence of groups, normal distribution of the data, and homogeneity of variance.

Independence of Groups: The first assumption is that the groups are independent of each other. Chapter 14 reviews procedures for analyzing the data when this assumption is not met and there is a correlation between the groups.

Normal Distribution of the Data: A second assumption is that the data are normally distributed. The Mann–Whitney *U*-test is an alternative to the *t*-ratio that researchers can use when the data are not normally distributed. Other types of statistical corrections may also be appropriate. Related to this assumption is that the data represent interval or ratio data. If the dependent variable represents a count or ordinal scale, the Mann–Whitney *U*-test is often the preferred alternative test.

Homogeneity of Variance: The third assumption refers to homogeneity of variance ($\sigma_1^2 = \sigma_2^2$). As a generality, if the larger variance is 3.0 times greater than the smaller variance, then we should not assume homogeneity of variance, and we should use the Welch *t*-test or the Mann–Whitney *U*-test as an alternative. Other corrections are also available.

Forming the Hypothesis

The purpose of the *t*-ratio is to determine whether there is a meaningful difference between the two group means. To conduct the statistical test, you may select either a directional or a nondirectional test:

Nondirectional or two-tailed test	Directional or one-tailed test
H_0: $\mu_1 = \mu_2$	H_0: $\mu_1 \geq \mu_2$ or H_0: $\mu_1 \leq \mu_2$
H_1: $\mu_1 \neq \mu_2$	H_1: $\mu_1 < \mu_2$ or H_1: $\mu_1 > \mu_2$

Sampling Distribution

To determine t_{critical}, establish the level of α and type of hypothesis, and calculate the degrees of freedom. Use Table B.3 in Appendix B to find the degrees of freedom (rows) and the appropriate α-level (columns). For an example, refer to the following chart:

Nondirectional or two-tailed test	Directional or one-tailed test
$\alpha = .05$, $df = 15$	$\alpha = .05$, $df = 15$
H_0: $\mu_1 = \mu_2$	H_0: $\mu_1 \geq \mu_2$ H_0: $\mu_1 \leq \mu_2$
$t_{\text{critical}} = 2.131$	$t_{\text{critical}} = -1.753$ $t_{\text{critical}} = 1.753$

If you use a directional hypothesis, then you must convert the sign of t_{critical} to conform to the test.

As an example, consider the following scenario. A researcher decided to examine the difference between men's and women's recall of emotional events. The researcher asked randomly selected men and women to describe in two or three words positive or happy events. Another student who did not know whether the author of the comments was male or female counted the number of episodes recalled. Table 11.1 presents the data.

The researcher designed this study to test the hypothesis that women will recall more positive events than men (extending a study by Seidlitz & Diener,

Table 11.1 Hypothetical Data for Recall of Emotional Events

1		2		
Males		Females		
6	7	8	11	
12	10	10	11	
9	10	12	12	
10	9	9	12	
8	8	10	14	
$n_1 = 10$		$n_2 = 10$		$df = (10 - 1) + (10 - 1) = 18$
$\sum X_1 = 89$		$\sum X_2 = 109$		$df = 18$
$\sum X_1^2 = 819$		$\sum X_2^2 = 1215$		$t_{\text{critical}} = 1.734$
$M_1 = 8.9$		$M_2 = 10.9$		
$SD_1 = 1.7288$		$SD_2 = 1.7288$		

1998) and decided to set $\alpha = .05$. Accordingly, H_0: $\mu_W \leq \mu_M$ and H_1: $\mu_W > \mu_M$. Using Table B.3 of Appendix B, we find that $t_{\text{critical}} = 1.734$. The value of t_{critical} is positive because the researcher wants to show specifically that $\mu_W > \mu_M$:

$$t = \frac{10.9 - 8.9}{\sqrt{\dfrac{819 - \dfrac{(89)^2}{10} + 1215 - \dfrac{(109)^2}{10}}{10 + 10 - 2}\left(\dfrac{1}{10} + \dfrac{1}{10}\right)}}$$

$$t = \frac{2.0}{\sqrt{\dfrac{26.90 + 26.90}{18}(.2)}}$$

$$t = \frac{2.0}{.7732}$$

$$t = 2.5867$$

Because t_{observed} exceeds t_{critical}, we can reject the null hypothesis and conclude that there is a statistically significant difference between men's and women's recall of positive emotional events and that women recall more such events than do men.

Additional Tests

Confidence Interval of the Difference between Means

This statistic allows you to estimate the confidence interval (*CI*) for potential mean differences given sample statistics, based on the standard error of the difference between means (*SED*):

$$SED = \sqrt{\frac{\sum X_1^2 - \dfrac{\left(\sum X_1\right)^2}{n_1} + \sum X_2^2 - \dfrac{\left(\sum X_2\right)^2}{n_2}}{n_1 + n_2 - 2}} \tag{11.6}$$

$$CI = (M_1 - M_2) \pm t_{\text{critical}}(SED) \tag{11.7}$$

For this equation, t_{critical} represents the value for the two-tailed or nondirectional test using the degrees of freedom, $df = (n_1 - 1) + (n_2 - 1)$. In this example, $df = 18$ and $t_{\text{critical}} = 2.101$. Using the data from the previous example,

$$SED = 0.7732$$

$$CI = 2.0 \pm 2.101(0.7732)$$

$$CI = 2.0 \pm 1.6245$$

95% confidence interval: 0.3765 to 2.0 to 3.6245

The *CI* allows us to conclude that if we repeated the study under identical conditions, there is a 95% probability that the true difference between additional sample means will fall between 0.3765 and 3.64. In other words, there is compelling evidence that there is a difference between the two group means. The span of the *CI* also gives us some indication of the precision of our estimate (narrower = more precise).

Omega Squared ($\hat{\omega}^2$)

This statistic is a measure of association that estimates the degree to which the independent variable shares common variance with the dependent variable. The larger the value of $\hat{\omega}^2$, the greater the relationship between the variables. For this example, $\hat{\omega}^2$ is calculated as

$$
\begin{aligned}
&= \frac{t^2 - 1}{t^2 + n_1 + n_2 - 1} \\
&= \frac{-2.5867^2 - 1}{-2.5867^2 + 10 + 10 - 1} \\
&= \frac{5.6910}{25.6910} \\
&= 0.2215
\end{aligned}
$$

For this example, we can conclude that the independent variable accounts for approximately 22% of the variance in the dependent variable.

Effect Size (*d*)

Effect size is an index of the relative difference between the means. The index is similar to a *z*-score in that statistics converts the difference between the means into standard deviation units. The equation for *d* is presented as Equation 11.8. Box 11.2 presents Cohen's (1988) guide for evaluating effect sizes. Use this equation only when the sample sizes of your two groups are equal (when $n_1 = n_2$). Do not use this equation when $n_1 \neq n_2$:

BOX 11.2 ***Cohen's (1988) Guidelines for Evaluating Effect Sizes***

Small Effect Size	$d = .20$	Generally, these differences are small and difficult to detect without large sample sizes. Some researchers believe that when $d \leq .10$, then the effect is trivial.
Medium Effect Size	$d = .50$	A medium effect is sufficiently large to be obvious when graphed.
Large Effect Size	$d = .80$	A large effect size is easily seen when graphed and required few subjects to detect the effect.

$$d = \frac{|\mu_1 - \mu_2|}{\sigma}, \sigma = \frac{SD_1 + SD_2}{2}$$
$$d = \frac{|8.9 - 10.9|}{1.7288}$$
$$d = \frac{2.0}{1.7288} \tag{11.8}$$
$$d = 1.1218$$

CHAPTER SUMMARY

In this chapter, we examined the foundation of statistical inference using two independent groups and Student's t-ratio. Specifically, we examined how we compare the difference between the means to the standard error of the difference between means. As you learned, the standard error of the difference between means is an estimate of the sampling error present in the data. Therefore, if the relative difference between the sample means is sufficiently large, we can assume that the difference is in excess of sampling error and represents a meaningful difference between the population means.

There are important decisions to make when conducting an inferential statistic. First, we need to determine an acceptable α-level to control for the probability of committing a type I error. Researchers who find the cost of a type I error high use a lower α-level. Although lowering the α-level decreases the risk of a type I error, doing so increases the risk of a type II error. Consequently, researchers must balance the relative costs of type I and type II errors to determine the α-level to select. The researcher must also determine whether to use a directional or a nondirectional test. Another important consideration is the type of test to conduct. Each statistic has a special purpose and role in evaluating hypotheses. Selecting the wrong statistical test will produce misleading information.

Once one finds a statistically significant effect, the results must be interpreted with care. You learned that indices such as omega squared and effect size allow us to describe further the relation between the independent and dependent variables and the relative difference between the sample means. As a review, we can assume cause and effect if the researcher used a true experiment. In other designs, such as an intact group design, we can use the t-ratio to examine the difference between the groups but cannot assume that the statistically significant results indicate a cause-and-effect relation between the two variables.

We also examined the meaning of the p-value and that researchers often misinterpret its meaning. The p-value indicates the probability of obtaining the observed t-ratio if the null hypothesis is a correct statement. If the value of p is less than α, then we can reject the null hypothesis in favor of the alternative hypothesis. Other interpretations of p such as (1) indicating that the results are due to chance, (2) indicating the reliability of the results, or (3) indicating the "importance" of the results are incorrect interpretations.

An important component of hypothesis testing is power analysis. The aim of power analysis is decreasing the probability of committing a type II error. Researchers can increase the power of their research by attempting to increase the difference between the population means, increasing sample size, decreasing sampling and measurement error, and decreasing the α-level.

CHAPTER GLOSSARY FOR REVIEW

Clinical or Practical Significance Indicates that the results support rejecting the null hypothesis and that the results represent an effect meaningful in the context of the research.

Error Term ($\sigma_{M_1-M_2}$) Estimates the random error that occurs when taking samples from the population and is an unbiased estimate of $SD_{M_1-M_2}$.

Omega Squared ($\hat{\omega}^2$) A statistical index of the degree to which the independent variable accounts for the variance in the dependent variable.

Power of a Statistic ($1-\beta$) The probability that one will correctly reject H_0.

Standard Error of the Difference between Means ($\sigma_{\mu_1-\mu_2}$) Standard deviation of the sampling distribution of the difference between means.

Statistical Significance Indicates that the results support rejecting the null hypothesis.

Student's *t*-Ratio An inferential statistic that allows us to compare the difference between two means and to determine whether there is sufficient evidence to reject the null hypothesis.

REFERENCES

American Psychological Association (2009). *Publication manual of the American Psychological Association* (6th ed.). Washington, DC: Author.

CARVER, R.P. (1978). The case against statistical significance testing. *Harvard Educational Review, 48*, 278–399.

COHEN, J. (1988). *Statistical power analysis for the behavioral sciences* (2nd ed.). Hillsdale, NJ: Erlbaum.

KILLEEN, P.R. (2005). An alternative to null-hypothesis significance tests. *Psychological Science, 16*, 345–353.

MOYÉ, L.A. (2000). *Statistical reasoning in medicine: The intuitive p-value primer.* New York: Springer.

OAKES, M. (1986). *Statistical inference: A commentary for the social and behavioral sciences.* New York: John Wiley & Sons.

PEDHAZUR, E.J. (1997). *Multiple regression in behavioral research: Explanation and prediction* (3rd ed.). Fort Worth, TX: Harcourt Brace.

RUDNER, R. (1953). The scientist qua scientist makes value judgments. *Philosophy of Science, 20*, 349–380.

SEIDLITZ, L. & DIENER, E. (1998). Sex differences in the recall of affective experiences. *Journal of Personality and Social Psychology, 74*, 262–271.

SHINE, L.C. (1980). The fallacy of replacing an a priori significance level with an a posteriori significance level. *Educational and Psychological Measurement, 40*, 331–335.

Chapter 12

Single-Variable Between-Subjects Research

Research is to see what everybody else has seen, and to think what nobody else has thought.

—Albert Szent-Gyorgyi

INTRODUCTION

In Chapter 11, you learned about the classic research design that consists of two groups. Although that design is a model of simplicity, it has limitations. Because health is intimately linked to wondrously complex human behavior, we often must

Understanding and Conducting Research in the Health Sciences, First Edition. Christopher J. L. Cunningham, Bart L. Weathington, and David J. Pittenger.

compare or consider more than two groups. Events in nature and health sciences rarely order themselves into two neat categories (e.g., it happened vs. it did not happen).

Therefore, we often need research designs that can help us answer more complex questions. The focus of this chapter is research that involves more than two levels or conditions of a single independent variable. We will also examine an extremely useful statistical technique known as the **analysis of variance** (ANOVA). Since its development by Sir Ronald Fisher in the 1920s, the ANOVA has evolved into an elaborate collection of widely used statistical procedures in contemporary behavioral and social research. Therefore, this statistic, combined with a sound research design, can tell us much about the behavior we want to study. To help illustrate the value of the ANOVA approach to design and analysis, we will use several examples from the published research literature, including the following two central studies.

Effects of Praise on Motivation and Performance

Most of us like to be praised for our work, and we strive to receive compliments for our best efforts. Although we tend to think of praise as a reward, is it possible that some compliments could have the opposite effect and decrease one's motivation? Mueller and Dweck (1998) conducted a number of experiments in which they gave school children different types of praise. While Mueller and Dweck focused specifically on children, their study is part of a larger body of research that has been applied to the many aspects of human life and behavior including promoting healthy behavior (e.g., Cameron & Pierce, 2006; Wu, 2012).

Spotting a Liar

Everyone has lied at one time or another. Some people tell the occasional white lie to protect the feelings of others (e.g., "Gee, I *really* like what you did with your hair!"). Some lies are dishonorable and are intended to hurt others or to protect the guilty. Ekman and O'Sullivan (1991) asked, "How good are we at detecting when someone lies to us?" Specifically, the researchers wanted to know if professionals (e.g., Secret Service agents, judges, and psychiatrists) who routinely work with potential liars could detect lies better than other people could.

INDEPENDENT VARIABLE

In Chapter 11, we reviewed the importance of defining the characteristics of the independent variable because its selection influences the type of research we conduct, how we analyze the data, and the conclusions we can draw from the data. In this chapter, we will examine the *single-variable between-subjects* research design. This cumbersome name indicates that the design allows us to examine the relation across several specific levels, or forms, of the independent variable and the dependent variable. Often this type of research involves an independent variable for which there are more than two **levels** or conditions.

In Mueller and Dweck's (1998) study, for example, the independent variable was the type of praise that the children received for completing a project. Specifically, Mueller and Dweck considered three levels of this independent variable: (1) praise for intelligence (e.g., "Very good, you must be smart"), (2) praise for effort (e.g., "Very good, you must have worked hard"), and (3) no praise (this condition was the control condition).

In the Ekman and O'Sullivan (1991) study, the independent variable was the person's profession. For the Ekman and O'Sullivan study, there were seven levels or conditions to the independent variable: (1) Secret Service agents, (2) judges, (3) psychiatrists, (4) detectives, (5) college students, (6) lawyers, and (7) FBI agents.

As already mentioned, this chapter is focused on between-subjects research designs. As you should recall, this type of design is one in which we compare the behavior of separate or independent groups of participants. In a true experiment, such as the Mueller and Dweck (1998) experiment, the researcher randomly assigns participants to one of the multiple experimental conditions formed by the levels of the independent variable. In an intact group design, such as the Ekman and O'Sullivan (1991) study, the researcher uses an existing subject variable (e.g., occupation) to define the levels of the independent variable and the conditions in which participants are grouped.

Advantages of a Parametric Design

A parametric design allows for an analysis of the relationship between the independent and dependent variables. Figure 12.1 presents graphs that represent the results of four hypothetical experiments. Each graph represents a unique and complex relationship between the independent and dependent variables that a research study

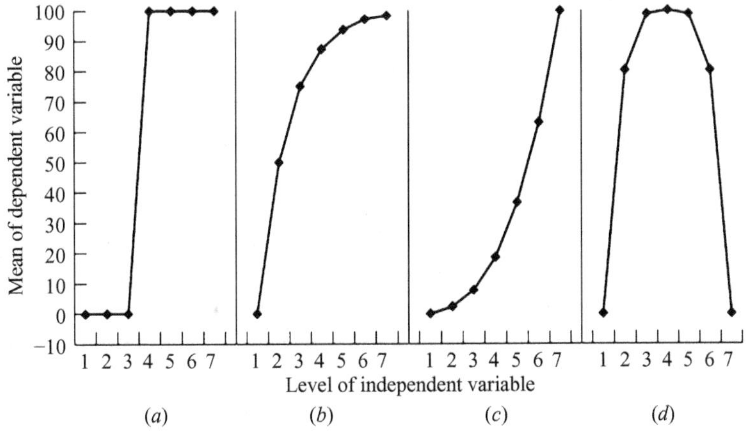

Figure 12.1 Hypothetical results of four separate studies.

with only two groups for comparison could not reveal. For instance, look at Figure 12.1*d*, which looks like an inverted U. Rotton and Cohn (2001) found an inverted U-shaped relationship, similar to the one illustrated here, between temperature and the number of aggravated assaults in a large metropolitan city.

Each experiment represents a different relation between the independent and dependent variables. A multilevel research design will help find trends like these. Specifically, Rotton and Cohn (2001) found that the number of assaults increased along with increases in temperature up to a certain point, after which the number of assaults decreased as the temperature continued to increase. *What would happen if a researcher conducted an experiment examining the relationship between temperature and aggression but used only two temperature conditions, cold (35°F) and hot (95°F)?* According to Rotton and Cohn's data, the researcher would conclude that temperature and aggression are not related, even though we can see from Figure 12.1*d* that these two variables are related, just not in a linear fashion.

Another advantage of parametric designs considering multiple levels of the same independent variable in research is its potential to increase the statistical power of the study. This, in turn, can make it easier for the researcher to detect a significant effect or relationship if it is represented by the data. The increase in power occurs if the researcher can identify additional levels of the independent variable and therefore establish additional groups of participants that are more homogeneous than would be the case if comparing groups of participants based on fewer levels of the independent variable. This, in turn, will reduce the amount of variability among participants in any one of the treatment conditions or groups, reducing the overall error term in the calculation of the resulting ANOVA test statistic (more on that later).

As an illustration, consider the Ekman and O'Sullivan (1991) study. The researchers could have compared three broad categories of professionals, including (1) detectives (Secret Service, FBI, and police detectives), (2) lawyers and judges, and (3) mental health personnel. Although this may appear to be a more simple study, it runs the risks of overlooking important differences among groups of professionals. As you will soon see, according to the results of the Ekman and O'Sullivan's study, some groups of professionals are better able to detect liars than others. By only comparing participants across these high-level occupational groupings, we would overlook these important differences. In contrast, if we grouped participants together into more specific and homogeneous occupational "conditions," then we would have a better chance of identifying differences between groups (and those differences would be more meaningful and clear-cut to interpret).

A final advantage of a parametric design emerges when the researcher can incorporate multiple control or comparison conditions into a true experiment. By having several such control conditions, the researcher can rule out alternative explanations and thereby have greater confidence in describing the effect that the independent variable has on the dependent variable. This is valuable to researchers because it can help us to more clearly identify treatment effects when they are really present.

One way to think about this is to consider the last time you tried to manage a conversation on a cell phone in a crowded room or dining hall. When everyone else in that room is talking and making noise, it is nearly impossible for you to focus in on the voices on the other end of your discussion. That one voice over your phone represents the signal you are trying to capture. The other voices and noises in the room reflect error that obscures your ability to focus on the one signal you are trying to find and maintain. When this situation occurs, you may choose to remove yourself from the crowded space to better focus on the conversation in a quieter, more private setting. When researchers add treatment conditions to a study, it is often done for a similar reason: By separating control groups and treatment groups, we can maximize our chances of identifying the signal we care about amid the noise of error that could prevent us from finding the effects we set out to study.

CAUSE AND EFFECT

Under certain circumstances, we can use the between-subjects research design to infer cause and effect between the independent and dependent variables. As a quick review, in a true experiment, the researcher randomly assigns participants to the treatment conditions, uses a manipulated independent variable, and uses control groups and procedures to account for alternative explanations of the data. Does the fact that only a highly controlled true experiment best allows us to infer cause and effect imply that other research designs are not valuable? Absolutely not!

However, there is an important difference in the researcher's ability to infer cause and effect from data collected from true experiments versus other designs that may not include the same level of control over confounding sources of error. Finding meaningful differences among intact groups can be helpful. Consider the research examining the ability to detect liars. What would happen if we found that one or two groups of professionals are able to detect lies? Although we could not conclude that being in a specific profession "causes" one to be a good lie detector, the results would indicate a need to engage in further research to identify how and why these groups of people are so good at detecting liars.

GAINING CONTROL OVER THE VARIABLES

One of the essential factors in determining cause and effect in any research is the degree of control that the researcher has over relevant variables. In general, a researcher who has greater control over the independent variable and other extraneous variables in a study will be better able to argue that there is a cause-and-effect relationship between the independent and dependent variables being studied. There are several basic strategies that a researcher can use to exercise control over the variables in a research project. Building on concepts introduced in Chapters 3 and 4, we will focus here on three broad issues: (1) control through research design, (2) control through research procedure, and (3) control through statistical analysis.

Control through Research Design

Control over variables can be enhanced by properly designing a research study. Common design strategies for this purpose include random assignment of participants to treatment conditions, use of a manipulated independent variable(s), and inclusion of control groups to highlight the relationship between independent and dependent variable(s). When designing a study, researchers may also choose to incorporate additional independent variables that may help to better explain the relationship between other focal variables in a study. We discuss this use of additional variables to add control over the relationship being studied in this and the next two chapters.

The common feature of all research designs is that they are created to help researchers determine the nature and magnitude of relationships between independent and dependent variables. The many possible designs differ in terms of the control that researchers can maintain over the variables once the study begins. Of the available designs, the true experiment provides the greatest level of control for two reasons. First, the researcher can select the levels of the independent variable. By determining the levels of the independent variable, the researcher can also identify the appropriate control group conditions. Second, the researcher can use random assignment of participants to treatment conditions to minimize the influence of any participant characteristics across the various treatment groups.

Control through Research Procedure

Once we select the design for the research, we also may need to consider techniques for controlling the influence of extraneous variables that might make it difficult for us to study the relationship between the independent and dependent variable that we are most interested in studying. Control through research procedure is based on the methods we use when collecting the data. Control is enhanced when the procedures we follow help us eliminate or at least reduce the influence of extraneous variables and thereby increase our ability to examine the relation between the independent and dependent variables. Remember that using a true experiment is not an automatic guarantee that you will be able to clearly identify a cause-and-effect relationship between two variables. No research design can overcome the effects of a sloppy research procedure.

In Chapter 10, we examined the need for control procedures such as single- and double-blind data collection techniques. Many other control-enhancing techniques exist. The primary goal of all good research procedures is to ensure that each participant experiences the same common events in the study (to the extent possible), except for the level of the independent variable, which varies depending on their assigned treatment level/condition.

Attention to detail is a valuable asset for any researcher. That said, there is a fine line between concern for good research design and becoming obsessive-compulsive regarding research minutiae. All research projects include a nearly infinite number of factors that, if identified ahead of time, the researcher could potentially control. It is important to recognize, however, that just because we *can* exert control over a variable does not mean that we necessarily *should* in every study. The goal is to focus on extraneous variables that have a legitimate chance of influencing or modifying the functioning of either the independent or dependent variable of interest, and then establishing control procedures to minimize this influence.

There are several ways you can determine the extraneous variables that you need to control. The first is to carefully read the method section of research articles about studies that are similar to your own; doing this will allow you to benefit from the collective wisdom of a long tradition of research in a particular area of study. Another tactic is to talk with other researchers who have conducted similar research. There is often a considerable amount of "laboratory lore" involved in different types of research. Many researchers consider these tricks of the trade common knowledge to those working in the field and do not describe them in the method section. Finally, you may want to consider conducting a pilot study before beginning your actual research.

A **pilot study** is a dress rehearsal for the final research project. By conducting a small-scale study with a small group of participants, you can determine (1) whether they understand your instructions, (2) whether you have effectively reduced the risk of demand characteristics and experimenter effects, and (3) whether your method of recording the data works. Armed with this information, you can make any necessary modifications to your actual study procedure before beginning your full data collection.

Control through Statistical Analysis

Control through statistical analysis depends on the mathematical techniques we use to remove or account for extraneous effects on the independent and dependent variables in a study. Statistical procedures such as ANOVA, analysis of covariance, multivariate ANOVA, and multiple regression give researchers greater control over different sources of variance acting on the data. These techniques, while useful, can be complex and require researchers to be familiar with many statistical principles that go beyond the scope of this book. *It is important to note that no statistical method, no matter how advanced, can overcome the consequences of a poorly designed and executed research project.*

THE GENERAL LINEAR MODEL

Now that we have examined the issues surrounding the design of a multigroup single-variable between-subjects design, we can turn our attention to appropriate

Table 12.1 Results for the Effects of Praise on Willingness to Continue on a Task

No praise (control group)	Praise for intelligence	Praise for effort
$n_1 = 14$	$n_2 = 14$	$n_3 = 14$
$M_1 = 5.0$	$M_2 = 4.0$	$M_3 = 6.0$
$SD_1 = 1.0$	$SD_2 = 1.0$	$SD_3 = 1.0$
$VAR_1 = 1.0$	$VAR_2 = 1.0$	$VAR_3 = 1.0$

Note: Scores range from 0 = low willingness to 10 = high willingness.

analytical strategies for data from this type of research. To help you understand the basic analytical principles, we can use data from Mueller and Dweck's (1998) praise-and-motivation study.

For this experiment, the researchers randomly assigned 42 fifth-grade students to one of three experimental groups. The researcher asked all students to complete three problem-solving tasks. After the child solved the first set of problems, the researcher told him or her, "Wow, you did very well on these problems. You got # right. That's a really high score; you did better than 80% of the other children who try to solve these problems." For the children assigned to the praise-for-intelligence group, the researcher then added, "You must be smart at these problems." By contrast, the researcher told the children in the praise-for-effort group, "You must have worked hard at these problems." The children in the control group received no additional feedback. Notice that each child received the same comments except for the attribution regarding their good score.

During the next phase of the experiment, the researcher had the children work on a set of difficult problems and then told all of them that they had done poorly. The researcher then asked each child to rate on a 10-point scale their desire to work on the third set of problems (0 = low willingness, 10 = high willingness). Table 12.1 presents the results for this experiment. Before reviewing the results, we want to draw your attention to an important ethical issue involved in this research; the researchers deceived the children. They misled the children about their performance on the first part of the study and then rigged the second portion of the study to ensure that the children would do poorly. Mueller and Dweck (1998) recognized the ethical responsibility they had to preserve the well-being of the children. Consequently, they ensured that they received informed consent for each child participating in the study. In addition, they developed extensive debriefing procedures that ensured that the children went away from the experiment feeling proud of their efforts and participation in the study.

According to the sample means in Table 12.1, children praised for intelligence expressed the least desire to continue with the task ($M = 4.0$), whereas the children praised for effort expressed the greatest desire to continue ($M = 6.0$). The mean for the children in the control condition was intermediate between the two treatment groups ($M = 5.0$). Although we know that there are real differences among the group means, is this difference meaningful? How much of the difference reflects the effects of the praise and how much reflects the effects of random events? We can use Equation 12.1 to help us understand the variation among the observed scores. The equation will also allow us to return to our discussion of between-groups and within-groups variance. Equation 12.1 is the **general linear model** for the data collected in a single-variable study:

$$X_{ij} = \mu = \alpha_j + \varepsilon_{ij} \quad (12.1)$$

In this equation, X_{ij} represents an observation for a single participant within a specific treatment condition. The i in the subscript represents the individual participant, whereas the j represents the group. For example, X_{23} represents the second subject in the third group. The next term, μ, represents the mean for the base population. The α_j represents the effect of the independent variable (via each of its levels) on the dependent variable. For the general linear model, the value of each α_j may be 0, positive, or negative, depending on the effect of the level of the independent variable. Finally, ε_{ij} represents random error. We assume that the error is a random variable with a mean effect of 0 and a standard deviation equal to σ, the population standard deviation. Figure 12.2 presents an illustration of how the general linear model applies to this experiment.

After creating a random sample from the population, the mean and standard deviation of the sample should be equivalent to the corresponding population parameter. The researcher then randomly assigns the participants into one of three groups. The participants' responses (X_{ij}) represent the combined effects of the treatment condition (α_j) and random events (ε_{ij}).

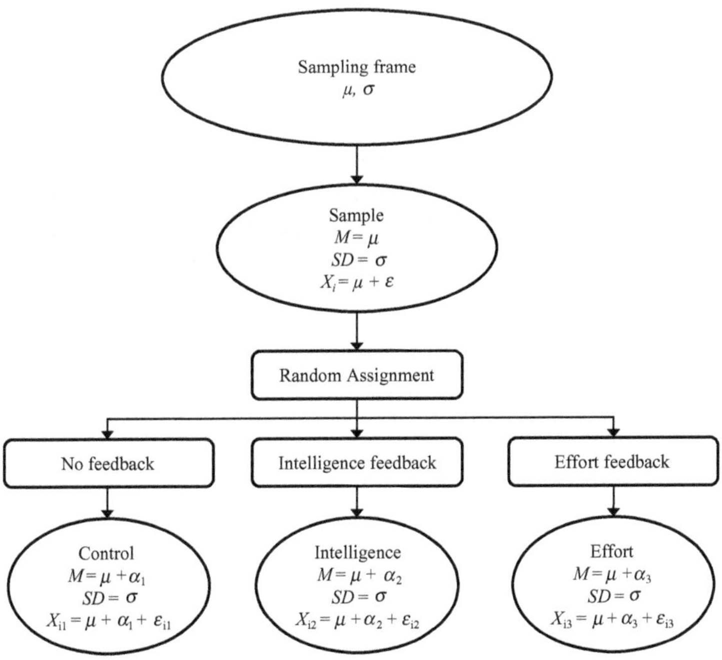

Figure 12.2 Single-variable study and the general linear model.

To accomplish the above, we begin by identifying the sampling population. In this case, the sampling population consisted of children in the fifth grade (10- to 12-year-olds) in a public elementary school located near the researchers. From this population, we draw a sample of participants. As you learned in Chapter 7, the researchers believe that the sample is representative of the sampling population and that the sampling population is representative of the target population.

The next step in the study is the random assignment of the participants to one of the three treatment conditions. As we noted previously, using random assignment for placement of participants into groups makes it possible for the groups to be equivalent regarding subject variables. Consequently, the only thing that should differentiate the three groups is the type of feedback the children receive after completing the first task.

The final stage shown in Figure 12.2 represents the data that we collect. According to the logic of the experiment, we believe that each child's performance will reflect three things. First, μ represents the typical interest that fifth-grade students have in the problem-solving task. The second influence on the dependent variable is the effect of the treatment condition, α_j. For the control group, α_1 should equal 0 because the children in this group did not receive any praise. The researchers believe that the type of feedback will influence the participants' interest in the task. Therefore, they will predict that the values of α_j for the intelligence and effort groups will not equal 0. They believe that the effect of the praise will affect each child in the group. The goal of the research is to determine how the type of praise affects the children's interest in the task. The third influence is ε_{ij}, the random error in the equation, which, in this example represents all the random events that can influence the child's interest in the task.

Whenever we conduct a study, we want to determine what proportion of the total variance in the dependent variable is due to the independent or treatment variable and what proportion is actually due to random error. Figure 12.3 may help you better understand how we can analyze the data and use comparisons to help us make these determinations. The top three lines represent the data for each of the three groups from the Mueller and Dweck (1998) study. The *M*s represent the three group means, 4.0, 6.0, and 5.0, respectively. The lines surrounding the *M*s represent one standard deviation above and below the group mean. For these

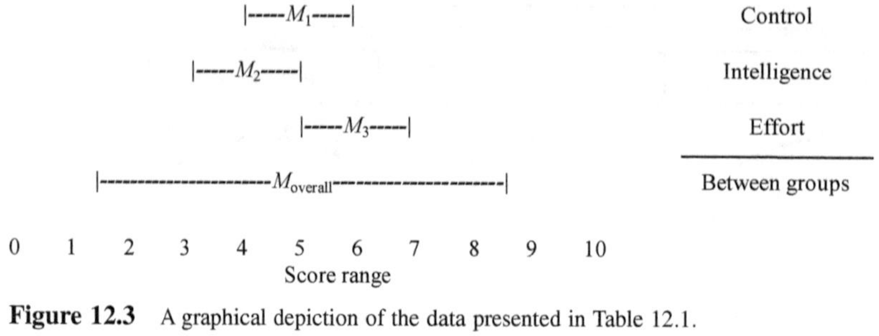

Figure 12.3 A graphical depiction of the data presented in Table 12.1.

data, within each treatment group, $SD = 1.0$. Because *VAR* (recall that $VAR = SD^2$) is an unbiased estimate of the population variance, σ^2, we can conclude that the average within-groups variance is 1.00. The bottom line represents the variance between the three group means. Specifically, $M_{overall} = 5.0$ represents the mean of all the data. The lines surrounding $M_{overall}$ represent the variance for the three group means, M_1, M_2, and M_3. As you can see, the between-groups variance is greater than the within-groups variance.

The top three lines represent the scores for the no-praise control, praise-for-intelligence, and praise-for-effort treatment conditions, respectively. The *M*s represent the mean of each treatment group. The lower line represents the data for all 42 students. The $M_{overall}$ indicates the mean across all 42 participants. For each mean, the |—| represents the limits of one standard deviation above and below the mean. The standard deviation for each group is $SD = 1.0$. The standard deviation for the three group means is $SD = 3.74$.

What affects the size of the between-groups variance? According to the general linear model, the between-groups variance represents the effect of random error, ε_{ij}, and the effect of the independent variable, α_j. The question that we need to answer is whether the between-groups variance is substantively larger than the within-groups variance. If so, then we can assume that the difference we see between two treatment groups reflects the effect of the independent variable more than random error. The ANOVA, as the name suggests, is a statistical test that allows us to compare different types of variance. For the single-independent-variable study, the ANOVA involves three types of variance.

COMPONENTS OF VARIANCE

The first form of variance is the **total variance**, which represents the variability among all the participants' scores regardless of the treatment condition they experienced. The second form of variance considered in ANOVA is the **within-groups variance**, which reflects the average amount of variability among scores for participants within each treatment group/condition. We use this measure of variability to estimate the magnitude of ε_{ij}. The third estimate of variance is the **between-groups variance**. Specifically, we examine the variance among the different treatment group means. The between-groups variance allows us to determine the joint effects of α_j and ε_{ij}. The primary statistic for the ANOVA is the ***F*-ratio**, the ratio of the between-groups variance to the within-groups variance. We use the size of the *F*-ratio to determine whether there is sufficient evidence to infer that the differences among the group means are greater than what would be due to random effects or error.

In this section, we will examine the components of the ANOVA. You will likely use a computer statistics program to actually calculate your ANOVA, but you can also perform the calculations by hand using formulas that you can find in most statistics textbooks. A convention among those who use the ANOVA is to call estimates of variance **mean squares (*MS*)**. The term implies that the variance of any set of numbers is the sum of squares of those numbers divided by the degrees of freedom. Therefore, whenever you read about a mean square or an *MS*, the author is describing an estimate of variance.

Total Variance

The total variance in a set of data represents the overall variability among all participants in the study, regardless of which treatment group or condition they were actually a part of during the data collection. As the general linear model suggests, the total variance is the sum of the effects of variance due to the independent variable, represented as α_j, and the effects of random variation among the participants, represented as ε_{ij}. When we conduct an ANOVA, we **partition**, or divide, the total variance into its smaller components—the variance within groups and the variance between groups.

Within-Groups Variance

The within-groups variance represents the differences among the observations caused by random error and factors not controlled by the researcher. The mean random error is 0 and, in the long run, it adds to or takes away nothing from the actual population mean. Within any given, smaller sample, however, this error can be problematic. Random error is a variable and thus inflates variability among participants' scores within and between groups in a study. Therefore, it causes individual scores to be greater than or less than the population mean. You can think of the within-groups variance as the average of the variance within each treatment group. For example, in Table 12.1, you can see that the variance in each group is 1.0. Therefore, we can conclude that the overall within-groups variance is 1.0.

When we use the ANOVA, we assume that the variances of each group are equal or homogeneous. Homogeneity of variance means that $\sigma_1 = \sigma_2 = \sigma_3 \ldots = \sigma_k$. This is an important assumption because we treat the within-groups variance as an unbiased estimate of the population variance, σ^2. Therefore, if the variance of one group is much larger or smaller than the other group variances, our estimate of the population variance will be inaccurate. Most computer statistics programs automatically test the homogeneity of variance assumption.

Between-Groups Variance

The between-groups variance reflects the variability present among the means of the various treatment groups in a study. Specifically, the ANOVA determines the variance between each sample mean and the mean of the entire data set. When we examine the differences among the sample means, we assume that some of the difference reflects random variation and that some of the variation reflects the independent variable. When we conduct this type of ANOVA, we assume that the groups are **independent**, or that the data observed in one treatment group have no influence or relation to the data observed in any of the other groups. There are a couple of general ways that we can ensure that the groups are independent for the single-variable between-subjects research.

First, in a true experiment, we use random assignment to send the participants to different groups. Because the assignment of participants to research conditions is random, we can conclude that the observations are independent of one another. Second, we can also conclude that the data are independent if the researcher tests each participant within only one level of the independent variable. Because the data for each group come from different people, we can assume that there is no correlation among the data from the groups.

THE *F*-RATIO

Once these three variance estimates are calculated, it is then possible to calculate an inferential statistic that helps us to determine whether the between-groups variance is sufficiently greater than the within-groups variance in a data set to be evidence of an effect rather than an error. We call this statistic the *F*-ratio in honor of Sir Ronald Fisher, the scholar who invented the ANOVA. This statistic is nothing more than the between-groups variance divided by the within-groups variance. We can write the equation for the *F*-ratio in one of several ways:

$$F = \frac{\text{Treatment variance} + \text{error variance}}{\text{Error variance}} \tag{12.2a}$$

or

$$F = \frac{\text{Between-groups variance}}{\text{Within-groups variance}} \tag{12.2b}$$

or

$$F = \frac{MS_{\text{between}}}{MS_{\text{within}}} \tag{12.2c}$$

The *F*-ratio uses the same logic as the *t*-ratio because it creates a standardized score or ratio. The *F*-ratio compares the size of the between-groups variance relative to the within-groups variance. We can use the Mueller and Dweck (1998) experiment to examine how the relation between the independent and dependent variables affects the size of the *F*-ratio.

No Treatment Effect

Assuming that the independent variable has no influence on the dependent variable, would all the means be equal to one another? This is sort of a trick question: The means across groups in this type of research would probably not be precisely

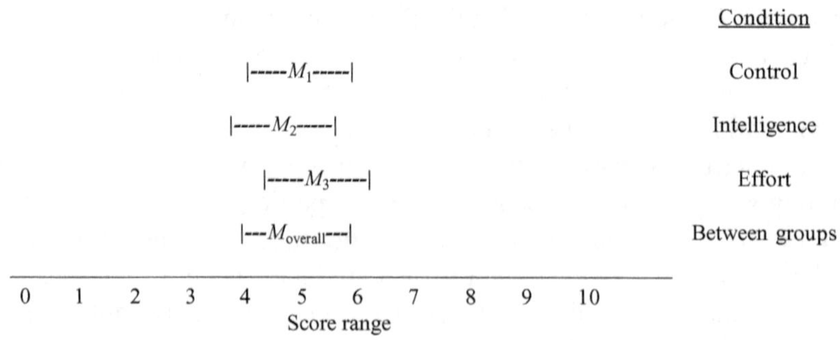

Figure 12.4 Within-groups variance is 1.0 and between-groups variance is 1.0.

equal to one another because sampling (random) error will likely cause each sample mean to be slightly different from the population mean. We would expect, however, that the means would be relatively similar. More to the point, the variance between the means of the various treatment groups should equal the variance within the groups. Figure 12.4 presents an example of such a situation. For the sake of the example, we will assume that the variance within groups is 1.0 and that the variance due to the independent variable is 0.0. Therefore,

$$F = \frac{\text{Treatment variance} + \text{error variance}}{\text{Error variance}} = \frac{0.00 + 1.00}{1.00} = 1.00$$

Therefore, the ratio of the between-groups variance to the within-groups variance is $F = 1.00$. We could interpret these data to indicate that any differences between pairs of treatment group means is random or due to error, and that there is no systematic relation between the independent and dependent variables.

In this example, the variance between groups is equivalent to the variance within groups. Therefore, whenever the F-ratio is close to 1.0, we can assume that the variance between groups is essentially due to random factors and not a meaningful treatment effect. As you can see in Figure 12.4, there is still some variance among these group means. However, because of the considerable overlap among the score distributions for these groups, we assume that the difference among the means represents chance, the effect of random error. Notice, too, that the between-groups variance is the same as the within-groups variance—when computing an F-ratio from these data, this would translate into a value close to 1.0.

Treatment Effect Present

What happens to the F-ratio when the levels of the independent variable correspond with changes in the dependent variable? In an experiment, the independent variable may cause the mean of your treatment groups (filled with members of your sample) to be greater or less than the mean of the population. Hence, the between-groups variance in your study will be larger than the within-groups variance. The between-groups variance increases because the treatment variance is

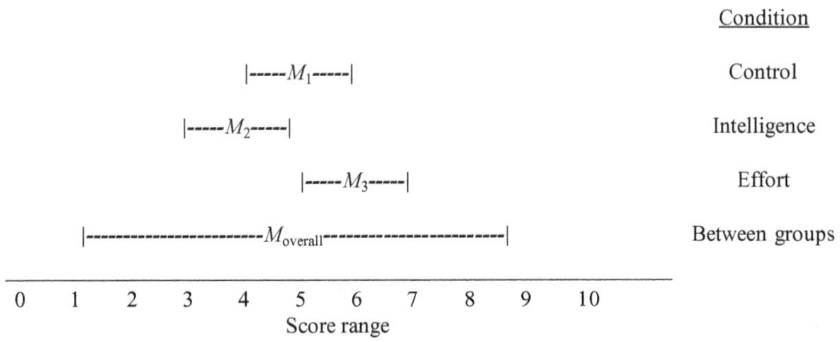

Figure 12.5 Within-groups variance is 1.0 and between-groups variance is 14.0.

greater than 0.0. Figure 12.5 represents what would occur if there were a statistically significant treatment effect. As you can see, the between-groups variance is now extremely large. You can also see that the variability among the three means has increased, whereas the variance within each group remains the same as it was for the example in Figure 12.4. We can represent the variance in Figure 12.5 using the following equation:

$$F = \frac{\text{Treatment variance} + \text{error variance}}{\text{Error variance}} = \frac{13.00 + 1.00}{1.00} = 14.00$$

For this example, $F = 14.00$ and we can infer that the difference among the group means is greater than would be expected from random error alone. We may infer that there is a systematic relation between the independent and dependent variables.

The increase in the treatment variance caused the F-ratio to be greater than 1.00. If the effect of the treatment variance is sufficiently large, we will conclude that the F-ratio is statistically significant and that there is a potentially meaningful difference between two or more of your treatment groups. Additionally, we will infer that there is a relationship between the independent and dependent variable(s) in your study. Looking to the example in Figure 12.5, we can interpret the F-ratio to indicate that the variance between the groups is 14 times larger than the variance within groups. Going a step farther, we may have statistical evidence that changes in the independent variable correspond with changes in the dependent variable.

H_0 AND H_1

The null (H_0) and alternative hypotheses (H_1) for the ANOVA are similar in form and logic to the null and alternative hypotheses used to test the t-ratio. Specifically, the null hypothesis specifies that there are no differences between treatment groups, implying that differences in the level of the independent variable are not associated with differences in the mean levels of the dependent variable. We write the null hypothesis as

$$H_0: \mu_1 = \mu_2 = \mu_3 = \mu_4$$

We used a similar null hypothesis for the *t*-ratio. The only difference is that with the ANOVA, we compare simultaneously more than two means. The interpretation of the null hypothesis is that all groups represent the same population and that any observed difference between the means is due to random factors or sampling error and not a potentially meaningful treatment effect.

Because the ANOVA is a general or omnibus test of variances, we do not make specific statements about how the means will be different from one another in the alternative hypothesis. Remember that the primary purpose of the ANOVA is to determine whether there are any systematic differences among the means. Because we do not specify the relation between the means in the ANOVA, we write the alternative hypothesis as

$$H_1: \text{Not } H_0$$

or

$$H_1: \text{All } \mu_i \text{ are not equal}$$

The alternative hypothesis for the *F*-ratio is a nondirectional hypothesis because we do not specify how the means will be different from one another, only that the between-groups variance is greater than the within-groups variance. If we can reject the null hypothesis, we must then use a special form of *t*-ratio to make specific comparisons among the treatment group means.

F-RATIO SAMPLING DISTRIBUTION

Just as Student developed a family of sampling distributions for the *t*-ratio, Fisher developed a family of sampling distributions for the *F*-ratio. The concept of the sampling distribution for the *F*-ratio is the same as that for other sampling distributions. Specifically, the sampling distribution represents the probability of various *F*-ratios when the null hypothesis is true. Two types of degrees of freedom (*df*) determine the shape of the distribution. The first *df* represents the between-groups variance. For any unbiased estimate of variance, the *df* always equals one less than the number of observations that contribute to that particular estimate of variance. For the between-groups variance, the *df* is the number of treatment groups less one. We also call this *df* the *degrees of freedom numerator* ($df_{between}$) because we use this to calculate the $MS_{between}$ in the numerator of the *F*-ratio.

In the ANOVA framework, the second *df* represents the within-groups variance. We also call this *df* the *degrees of freedom denominator* (df_{within}) because we use this to calculate the MS_{within} in the denominator of the *F*-ratio. The critical values of the *F*-ratio for various combinations of $df_{between}$ and df_{within} are presented in Appendix B. As you will see, the columns represent "degrees of freedom for numerator" and the rows represent "degrees of freedom for denominator." For the single-variable ANOVA, the degrees of freedom for the numerator are the $\boldsymbol{df_{between}}$

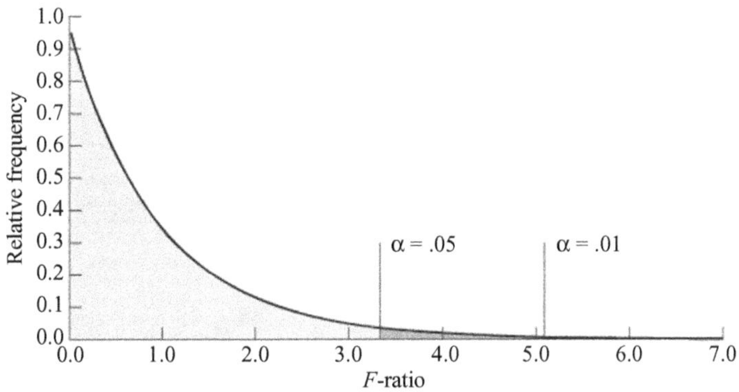

Figure 12.6 Sampling distribution for F with $df_{\text{between}} = 2$, $df_{\text{within}} = 39$.

and the degrees of freedom for the denominator are the $\boldsymbol{df_{\text{within}}}$. For degrees of freedom of $df_{\text{between}} = 2$ and $df_{\text{within}} = 39$, $F_{\text{critical}} = 3.24$.

Figure 12.6 represents the sampling distribution for the F-ratio when the degrees of freedom are 2 and 39. The distribution is positively skewed. Because we want to determine whether F is greater than 1.00, we place α on the right extreme of the distribution. When the F-ratio is 3.24, 5% of the sampling distribution is to the right of the F-ratio. Therefore, any observed F-ratio that falls in this region will allow us to reject the null hypothesis if $\alpha = .05$.

KNOWLEDGE CHECK

1. Describe in your words the advantages of using a multigroup single-variable between-subjects research design.
2. Under what conditions can the researcher use the multigroup single-variable between-subjects research design to infer cause and effect?
3. What is the difference between the between-groups variance and a variance due to the independent variable?
4. Why is the mean of $\varepsilon = 0$ for all participants in a specific treatment condition?
5. For a control group used in a true experiment, why would we state that $\alpha_j = 0$?
6. Comment on the accuracy of the following statements; explain why each statement is correct or incorrect.
   a. When the F-ratio is statistically significant, we can infer that the size of ε_{ij} is small and close to 0.
   b. All else being equal, as the size of ε_{ij} increases, the probability of rejecting the null hypothesis decreases.
   c. If the F-ratio is less than 1.00, the variance between groups, due to the independent variable, is less than 0.0.

Table 12.2 Example of an ANOVA Summary Table

Source of variance	*SS*	*df*	*MS*	*F*
Between groups	28.00	2	14.00	14.00
Within groups	39.00	39	1.00	
Total	67.00	41		

$F = \frac{14.00}{1.00}$

Variance estimates

Degrees of freedom for variance estimates

SS, sum of squares; *df*, degree of freedom; *MS*, mean square.

d. If the *F*-ratio is not statistically significant, we can assume that the independent variable had no effect on the dependent variable.

e. Even in a true experiment, the researcher cannot influence the size of α_j.

f. For a control group in an experiment, we could write the general linear model, $X_{ij} = \mu + 0 + \varepsilon_{ij}$.

g. It is impossible for the value of all α_j's to be less than 0 in the general linear model.

SUMMARIZING AND INTERPRETING ANOVA RESULTS

We can use the skills learned in Chapter 11 to determine how we will conduct an inferential statistical analysis of data similar to those summarized in Table 12.2. Before we get there, however, you should recall that in this type of hypothesis-testing analysis, we begin by identifying the null and alternative hypotheses and proceed to determining the criterion for rejecting the null hypothesis. The following considerations must be made when engaging in an ANOVA.

Null Hypothesis

If the null hypothesis, H_0: $\mu_1 = \mu_2 = \mu_3$, is true, then the average performance of the children in the three groups should be roughly equivalent. In other words, this would mean that the type of praise has no influence on the child's interest in the task. Any observed difference among the three sample means represents chance events.

Alternative Hypothesis

If the null hypothesis is false, then we by default shift to our alternative hypothesis, H_1: Not H_0. If this is the case, then we are concluding that the average performance of the children in the three treatment groups is not equivalent. In other words, the variance among the three groups is greater than what we would expect from the within-groups variance. Therefore, we conclude that the type of praise does influence the children's interest in the task. It is important to note that a significant

F-statistic does not by itself tell us which pairs of treatment group means are different from each other.

Statistical Test

We will use the single-factor independent-groups ANOVA for this type of analysis because we wish to compare the difference between the three independent treatment groups (e.g., group 1 vs. 2, group 2 vs. 3, and group 1 vs. 3).

Significance Level

Using $\alpha = .05$ and the degrees of freedom, we can determine the critical value for our F-test. If our observed statistic is larger than this critical value, then we can infer that there is at least one statistically significant difference between pairs of treatment group means. We can find this critical value by using the following information: $\alpha = .05$, $df_{\text{between}} = 2$, $df_{\text{within}} = 39$, yielding $F_{\text{critical}}\,(2, 39) = 3.235$. Again, if our F-statistic is greater than this value, then we can conclude that there is a significant difference between the means of at least one of the pairs of treatment groups.

Summary Table

Most statisticians report the sum of squares, degrees of freedom, mean squares, and the F-ratio in an **ANOVA summary table**. The summary table represents the common practice of several generations of statisticians and is now common in published research and the printouts of statistical analysis software programs. Whether you calculate the ANOVA with one of these programs or by hand, the basic format of the summary table for this type of analysis is shown previously in Table 12.2.

Because our F_{observed} is greater than the F_{critical} ($14.00 > 3.235$), we can reject the null hypothesis and conclude that the type of praise did influence the children's motivation to continue the task (our alternative hypothesis). As we mentioned in previous chapters, specific journals and publication styles vary, generally recommending a reporting format for the statistical results of an F-test that includes the following elements: $F(df_N, df_D) = F_{\text{observed}}$, $p = p$ or $F(df_{\text{between}}, df_{\text{within}}) = F_{\text{observed}}$, $p < \alpha$. For this example, we would report the F-ratio as $F(2, 39) = 14.00$, $p = .0000261$ or $F(2, 39) = 14.00$, $p < .05$.

When we reject the null hypothesis with the results of an ANOVA, we conclude that the difference between one or more pairs of treatment group means is large enough to infer that H_0 is false. We accept the alternative hypothesis, H_1: Not H_0. Although the F-ratio allows us to take a significant step toward interpretation, we need to continue the analysis of our data to learn more about the relation between the treatment and the results.

EFFECT SIZE AND POWER

All the rules you learned about interpreting the t-ratio also apply to the interpretation of the F-ratio. The size of the F-ratio and p-value indicates only whether we can reject the null hypothesis given the value selected for H_0. Estimates of the **effect size** depend upon conducting additional analyses.

To evaluate the degree to which the independent variable correlates with the dependent variable, we need to convert the F-ratio to estimated **omega squared** ($\hat{\omega}^2$):

$$\hat{\omega}^2 = \frac{df_{\text{between}}(F-1)}{df_{\text{between}}(F-1)+N} \tag{12.3}$$

The size of $\hat{\omega}^2$ indicates the degree of association between the independent and dependent variables. When we introduced you to $\hat{\omega}^2$ in Chapter 11, we told you to calculate $\hat{\omega}^2$ only if the statistic is statistically significant. The same is true for the F-ratio; it is good practice to calculate $\hat{\omega}^2$ only after you have established that the F-ratio is statistically significant.

An alternative estimate for the effect size of an F-statistic comes from Cohen (1988). This effect size estimate, represented as f, provides the same type of information as d for the t-ratio. The calculation of f is based on **eta squared**, η^2. For ANOVA, η^2 is equivalent to R^2 for regression analysis:

$$\eta^2 = \frac{df_{\text{between}}(F)}{df_{\text{between}}(F)+df_{\text{within}}} \text{ or, } \eta^2 = \frac{SS_{\text{between}}}{SS_{\text{total}}} \tag{12.4}$$

$$f = \sqrt{\frac{\eta^2}{1-\eta^2}} \tag{12.5}$$

According to Cohen, effect sizes for the ANOVA fall into one of three categories, as presented in Table 12.3 for the f estimate. The effect size for the present example experiment is what Cohen would refer to as a *large effect*.

One of the most commonly utilized measures of effect size due primarily to its ease of calculating using SPSS and other common statistics programs is partial eta-squared, *partial* η_p^2. η_p^2 calculates the proportion of total variation attributable to one factor by partialling out other factors. η_p^2 tends to be larger than η^2 and is considered to be a more appropriate and less biased measure than η^2 (see Cohen, 1973; Pedhauzer, 1997):

Table 12.3 Table of Cohen's Recommended Categories for Small, Medium, and Large Effect Sizes

Small effect size	Medium effect size	Large effect size
$f = .10$	$f = .25$	$f = .40$

$$\eta_p^2 = \frac{SS_{\text{between}}}{SS_{\text{between}} + SS_{\text{error}}} \tag{12.6}$$

Effect sizes need to be interpreted with caution, especially when in the small- to medium-effect-size range. Under some circumstances, a small effect size is an extremely important finding. For example, when studying a complex social behavior in a natural environment, one has little direct control over the environment and the treatment of the participants. Furthermore, in this situation, and many others, the measurement procedures may be prone to much error. Therefore, we need always to remember to interpret effect size within a broader context. Sometimes a "small" effect size can represent a major breakthrough in a program of research (Abelson, 1985). In other cases, a "small" effect size is insignificant from a meaning standpoint and is very easily forgotten.

According to Cohen (1988), effect sizes for most social and behavioral research are in the small-to-moderate range. Why is this the case? One major reason this is so is that there is so much random error involved when researching with humans. Most phenomena studied by social and behavioral scientists have multiple causes, and no single study can account for all of them. Second, there are always differences among people; even in a highly controlled laboratory experiment, each participant walks into the experiment with his or her own biases and predispositions. These experiences create differences among participants that the experimenter cannot control. Finally, measurement in the behavioral sciences inherently contains a good deal of random measurement error. This uncontrolled error can reduce the overall effect size of a research study by obscuring the effect that the researchers were seeking (i.e., the signal-to-noise metaphor). The successful researcher recognizes the presence of these inherent sources of error and attempts to improve the quality of the research procedure.

MULTIPLE COMPARISONS OF THE MEANS

When we reject the null hypothesis based on the results of an ANOVA, we can conclude only that the variance between the means of our treatment groups is greater than what we would expect due to sampling error alone. The F-ratio does not specify where the statistically significant differences among the groups may occur (e.g., is the significant difference between groups 1 and 2, or groups 1 and 3?). To determine which group means are statistically different from one another, we must return to the logic of the t-ratio to guide our comparisons.

You may be wondering why we would return to the t-ratio after conducting an ANOVA; why not conduct the t-ratios to begin with? The answer comes from an interesting fact related to hypothesis testing: The probability of a type I error (i.e., a false alarm) increases depending on the number of comparisons you make using the same data set (think back to Chapter 11). When we conduct a t-ratio, we assume that the groups are independent of one another. When the groups are independent, the probability of committing a type I error is α. The problem of an inflated type I error arises when we make **multiple comparisons** based on a single experiment with many groups. Consequently, if we were to compare three or more means from the sample experiment, the probability of committing a type I error is greater than α. Statisticians use Equation 12.5 to determine the probability of committing a type I error under these conditions:

$$\alpha_e = 1 - (1 - \alpha_p)^c \tag{12.7}$$

In this equation, α_e represents the **experiment-wise error**, the probability of making a type I error in any one of the potential comparisons. The other α_p term represents the **pairwise error**. The pairwise error represents the probability of a type I error for a single comparison. Finally, the superscript c in this equation represents the number of potential comparisons.

The size of the experiment-wise error increases quickly as the number of comparisons increases. For instance, if a researcher were to conduct three *t*-ratios, the α_e is approximately .143. Consequently, there is a 14.3% chance that one or more of any statistically significant *t*-ratios identified from these comparisons will be false (a type I error). Table 12.4 helps to illustrate this problem. Assume that a researcher conducts a study with three independent groups and proceeds to compare each possible combination of mean pairs. We can also assume that the researcher set $\alpha_p = .05$ for each comparison. Because $\alpha = .05$, we assume that the probability of a type I error (represented as α_p) for each comparison is 5%. In this example, there are eight potential outcomes ranging from all the comparisons being statistically significant to none of the comparisons being statistically significant.

Table 12.4 The Experiment-Wise Error Rate (α_e) for Multiple Comparisons

A vs. B	A vs. C	B vs. C	Probability of outcome	Cumulative probability	
All three comparisons statistically significant					
☑	☑	☑	$P = .000125 = \alpha^3$	.000125	
At least two comparisons statistically significant					
☑	☑	⃠	$P = .002375 = \alpha^2(1 - \alpha)$		
☑	⃠	☑	$P = .002375 = \alpha^2(1 - \alpha)$		
⃠	☑	☑	$P = .002375 = \alpha^2(1 - \alpha)$	.007250	
At least one comparison statistically significant					
☑	⃠	⃠	$P = .045125 = \alpha(1 - \alpha)^2$		
⃠	☑	⃠	$P = .045125 = \alpha(1 - \alpha)^2$		
⃠	⃠	☑	$P = .045125 = \alpha(1 - \alpha)^2$	.142625	$\alpha_e = 1 - (1 - \alpha)^c$
No comparisons statistically significant					
⃠	⃠	⃠	$P = .857375 = (1 - \alpha)^3$	1.000000	

Note: ☑ Statistically significant $p = \alpha$ $p = .05$.
⃠ Not statistically significant $p = (1 - \alpha)$ $p = .95$.

For this example, we assume that H_0: $\mu_A = \mu_B$, $\mu_A = \mu_C$, $\mu_B = \mu_C$ are true statements.

The column labeled "probability of outcome" represents the probability of results if the null hypothesis is a true statement for each of the comparisons. For instance, if H_0: $\mu = \mu$ for each of the three comparisons, then the probability that all three would be statistically significant is $p = .000125$ ($.000125 = .05 \times .05 \times .05$). In other words, the probability that we make a type I error for each comparison is .0125%. The "cumulative probability" column represents the probability of obtaining one of the potential outcomes. We will focus our attention on the next-to-last line. The cumulative probability indicates the probability of obtaining one of the seven outcomes by chance, if H_0: $\mu = \mu$ is actually true for each comparison. For this example, the cumulative probability is $p = .143$, as predicted by Equation 12.5.

You could lower α_p (e.g., from .05 to .01) to keep α_e to an acceptable level. The problem with this tactic is that the power for the individual comparisons will be so small that few, if any, of the *t*-ratios will reveal statistically significant results. For example, with three potential comparisons, we would have to reduce the α_p to approximately .01695 to maintain α_e at .05. We can calculate this using the equation

$$\alpha_e = 1 - (1 - \alpha_p)^c \qquad (12.8)$$

To maintain the probability of a type I error at .05 for the three comparisons, we would have to evaluate the statistical significance of each comparison at

$$\alpha_p = 1 - 3\sqrt{1 - .05} = .01695$$

Reducing α may reduce the risk of a type I error, but it will increase the risk of a type II error (a miss). If this is not acceptable in your particular situation, an alternative strategy is to use a post hoc (meaning after the fact) test. Post hoc tests are methods for comparing many sample means while controlling for inflated experiment-wise comparison rates. These specialized *t*-ratios allow you to compare sample means after you conduct the ANOVA and determine which pairs of means are different from one another at a statistically significant level, all while avoiding an inflated type I error. You will also find that most statistical packages offer a broad menu of post hoc statistical tests. In the following subsection, we will examine one of the more popular post hoc tests, known as **Tukey's HSD**.

Tukey's Honestly Significant Difference (HSD)

Tukey (1953) developed the HSD procedure to compare all possible pairs of treatment group means after one has rejected the null hypothesis using an ANOVA. This post hoc procedure strikes a reasonable balance between protecting against inflated type I errors and preserving statistical power (Jaccard, Becker, & Wood, 1984). We can use the data from the Mueller and Dweck (1998) experiment to

Table 12.5 Comparison of Differences between All Pairs of Means Using the Tukey HSD Test

		Intelligence	Control	Effort
Praise group	*M*	4.0	5.0	6.0
Intelligence	4.0	–	1.0*	2.0*
Control	5.0	–	–	1.0*
Effort	6.0	–	–	–

The asterisk (*) indicates the statistically significant difference between means at the .05 level.

examine the use of Tukey's HSD. Table 12.5 presents a matrix of the differences between the three means from the Mueller and Dweck experiment. The asterisk (*) indicates the statistically significant difference between means at the .05 level. If you are using a computer program to analyze the data, the program will automatically perform the necessary calculations. According to the results, all the group means are significantly different from one another. From the perspective of the purpose of the research, the HSD test confirms that children who received praise for intelligence rated the task lower than did the children in the control condition. Children in the praise-for-effort condition rated the task higher than did the children in the control condition. Therefore, it appears that the type of praise did affect the children's interest in the task.

Estimating Sample Size

An important, if not essential, step in any research project is determining the number of participants required to provide data that will allow you to test and possibly reject the null hypothesis. This is not a trivial step; we do not want to spend our time conducting a study only to find that we cannot reject the null hypothesis due to inadequate statistical power brought on by an insufficient sample. Instead, as we design the research project, we want to ensure that we do everything possible to increase the probability that we will reject a false null hypothesis. In the language of statistics, we want to increase the statistical power $(1 - \beta)$ of the study. As you should recall, one way to increase power is to increase the sample size for the research. In the following section, we will examine how to estimate the number of participants to include in a single-factor study.

Table B.13 of Appendix B presents a power estimate table that you can use to estimate the number of participants you will need for your research. Table 12.6 presents a portion of Table B.13. The table includes the two conventional α-levels (.05 and .01). The column labeled n represents the number of participants in each treatment condition, and the column marked F_c represents the critical F-ratio required to reject H_0. The four other columns represent the small ($f = .10$), medium ($f = .25$), large ($f = .40$), and very large ($f = .55$) effect sizes. We can use this table to estimate the power of an experiment and to plan future research projects.

Table 12.6 Power Table for the ANOVA

		$\alpha = .05$ $df_N = 4$					$\alpha = .05$ $df_N = 5$					$\alpha = .05$ $df_N = 6$			
		Effect Size, f					Effect Size, f					Effect Size, f			
n	**F_c**	**.10**	**.25**	**.40**	**.55**	**F_c**	**.10**	**.25**	**.40**	**.55**	**F_c**	**.10**	**.25**	**.40**	**.55**
37	2.422	.17	.77	.99	.99	2.256	.19	.82	.99	.99	2.135	.20	.85	.99	.99
38	2.420	.18	.78	.99	.99	2.255	.19	.83	.99	.99	2.134	.21	.87	.99	.99
39	2.419	.18	**.80**	.99	.99	2.254	.20	.84	.99	.99	2.133	.21	.88	.99	.99
40	2.418	.18	.81	.99	.99	2.253	.20	.85	.99	.99	2.132	.21	.89	.99	.99
45	2.413	.20	.86	.99	.99	2.248	.22	.90	.99	.99	2.128	.23	.93	.99	.99

The n column represents the number of observations in each treatment group. The F_c column represents the critical value of F for the degrees of freedom. The f columns represent the four levels of effect size.

For example, assume that you want to conduct an experiment with five levels of the independent variable ($df_N = 4$) and set $\alpha = .05$. How many participants do you need to have an 80% chance of rejecting the null hypothesis? Unless you have some idea of what effect size you should expect from previous similar studies, Cohen (1988) suggests you should plan for a maximum effect size that is moderate, $f = .25$, as it reflects the power of most behavioral research. If you wish to be a bit more conservative in your estimate, you can plan using a smaller effect size estimate (e.g., $f = .10$). Similarly, if you believe that your experiment will produce a large effect size, you can increase your estimate (e.g., $f = .40$). We can use Table 12.6 to illustrate how to use Table B.13 for the single-factor study.

For the present example, there are five groups; therefore, $df_N = 4 = 5 - 1$, and we plan to use $\alpha = .05$. How many participants will this experiment require if we assume that $f = .25$ and we want to set our power to $1 - \beta = .80$? As shown in Table 12.6, we will need at least 39 participants in each treatment group ($39 \times 5 = 195$ participants) to have sufficient statistical power to reject the null hypothesis.

RESEARCH IN ACTION

How good are individuals at detecting when someone is lying? Ekman and O'Sullivan (1991) asked that question by examining the ability of psychiatrists, Secret Service agents, FBI polygraphers, police robbery detectives, judges, lawyers, and college students to watch a series of videotapes and identify the liars. To conduct the study, the researchers developed 10 videotape presentations. Each tape consisted of a different college-aged woman in an interview. Half of the tapes presented different women lying; the other tapes were of different women telling the truth. The researchers explained to the participants that they would see 10 different women, half of whom were lying. Participants were supposed to identify

Table 12.7 Summary Statistics and ANOVA Summary Table for Ekman and O'Sullivan (1991)

SS	FBI	D	J	P	L	CS	Total
$n_1 = 10$	$n_2 = 9$	$n_3 = 10$	$n_4 = 9$	$n_5 = 10$	$n_6 = 10$	$n_7 = 10$	$N = 68$
$M_1 = 6.8$	$M_2 = 5.3$	$M_3 = 5.2$	$M_4 = 5.1$	$M_5 = 5.4$	$M_6 = 5.3$	$M_7 = 5.2$	$M_{overall} = 5.5$

Source	*SS*	*df*	*MS*	*F*
Between	20.796	6	3.466	4.388
Within	48.189	61	0.790	
Total	68.985	67		

$\hat{\omega}^2 = .23$, $\eta^2 = .30$

SS, Secret Service agents; FBI, FBI agents; D, detectives; J, judges; P, psychiatrists; L, lawyers; CS, college students.

who was lying. Table 12.7 presents the hypothetical results for this study. The data represent the number of correct identifications each person made. Ekman and O'Sullivan wanted to know whether members of some professions are better able to detect liars.

There are differences between this data set and the data collected by Mueller and Dweck (1998) regarding the relationship between praise and performance. First, in Ekman and O'Sullivan's (1991) study, the independent variable is a subject variable. This has no real bearing on the analysis of the data; we can continue to use the ANOVA. It will, however, change the way in which we interpret any results. Second, because we did not use random assignment of participants to the groups, we will not be able to determine unambiguously that the independent variable causes the dependent variable.

Table 12.7 presents the summary statistics and an ANOVA summary table for these data. Because the $F_{observed}$ is greater than the $F_{critical}$, we can reject the null hypothesis. Doing so allows us to assume that members of some professions are better able to detect liars than others. We will need to use Tukey's HSD to determine where the difference lies. Table 12.8 presents the matrix of mean differences and the results of Tukey's HSD test. From these data, we can conclude that the Secret Service agents were better able to detect liars than members of other groups. Figure 12.7 presents the results section that one might prepare for these data. In this example, we included a graph of the data that presents the mean and the standard error of the mean. This technique helps to illustrate that the Secret Service agents performed better than the other groups of participants at detecting liars.

Consider the implications of Ekman and O'Sullivan's (1991) findings. One of the authors of this textbook was recently asked to comment on a training program that was being developed to assist healthcare professionals in identifying and dealing with victims of domestic abuse. One factor that was included in this training, based partially on research like this, was the recognition that most of us are much worse at identifying lies, distortions, and falsehoods than we would like to believe.

Table 12.8 Matrix of Mean Differences for the Seven Groups Using the HSD to Determine the Statistical Significance of the Differences between Pairs of Means

		J	D	CS	L	FBI	P	SS
Profession	*M*	5.11	5.20	5.20	5.30	5.33	5.40	6.80
J	5.11	0.00	0.09	0.09	0.19	0.22	0.29	1.69*
D	5.20		0.00	0.00	0.10	0.13	0.20	1.60*
CS	5.20			0.00	0.10	0.13	0.20	1.60*
L	5.30				0.00	0.03	0.10	1.50*
FBI	5.33					0.00	0.07	1.47*
P	5.40						0.00	1.40*
SS	6.80							0.00

The asterisk (*) indicates the statistically significant difference between means at the .05 level.
J, judges; D, detectives; CS, college students; L, lawyers; FBI, FBI agents; P, psychiatrists; SS, Secret Service agents.

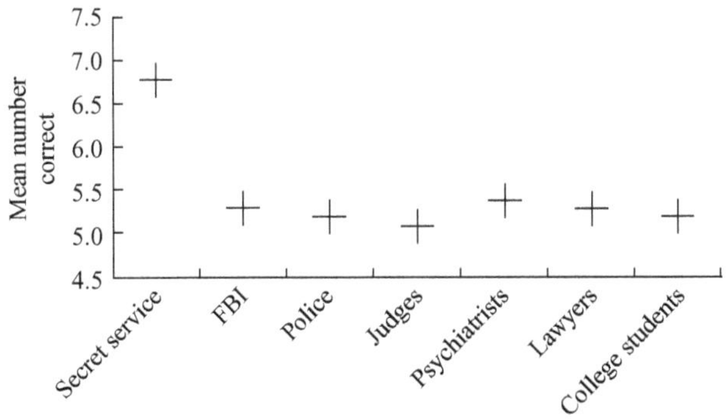

Figure 12.7 Results section describing the single-factor between-groups study presented in Table 12.7.

CHAPTER SUMMARY

This chapter is an extension of our previous review of single-variable between-subjects research designs. The important difference is that the design we examined in this chapter allows us to simultaneously examine many levels of the independent variable. The statistical test most often used to study the results of this research design is the ANOVA.

The ANOVA is an omnibus test because it allows us to examine the effect of many different levels of the same independent variable on the dependent variable. As you have learned in this chapter, the ANOVA is an extremely flexible design because we can use either quantitative or qualitative levels of the independent variable in our research. We can also use the ANOVA to analyze the data from a true experiment (for which the participants are randomly assigned to the treatment conditions) or an intact group design (the independent variable is a subject variable). In this chapter, we revisited the themes surrounding cause and effect. Specifically, the researcher must be able to control the independent variable and account for alternative explanations to infer cause and effect. We then examined several ways that the researchers can gain control over the variables in a research project.

The two that we examined are control through research design and control through research procedure.

In a true experiment, the researcher can exercise direct control over the independent variable by including control conditions and by randomly assigning the participants to the treatment conditions. Researchers exercise this type of control through the design of their study. A researcher can also exercise control over the research through good procedure. Control through procedure refers to ensuring that the experience of the participants in the different treatment conditions is identical save for the level of the independent variable.

Researchers use the general linear model to describe how the ANOVA analyzes the data. In a true experiment, for example, the participant's observed score is the total of the population mean plus the effect of the level of the independent variable plus the effects of random error. The ANOVA analyzes the variance among the scores in the study. Specifically, the ANOVA allows us to examine the within-groups and between-groups variance. The within-groups variance represents the effects of random events that influence the difference among the scores in the individual groups. The between-groups variance represents the differences among the group means. We use these variance estimates to determine the F-ratio.

The F-ratio is the between-groups variance divided by the within-groups variance. If the F-ratio is equal to 1.00, we must assume that the differences among the group means are due to random error. By contrast, if the F-ratio is statistically significant, we can assume that the differences among the means are greater than would be expected by chance if the null hypothesis were a true statement. Once we find that there is a statistically significant F-ratio, we can then calculate $\hat{\omega}^2$ or η^2; both indicate the proportion of the differences among the groups related to differences among the levels of the independent variable. Similarly, we can use the post hoc tests, such as Tukey's HSD, to determine which pairs of means are different from each other. As with any research design, we can determine the power of the study. We can estimate the number of participants required in each treatment condition to ensure that we will have sufficient statistical power to reject the null hypothesis.

CHAPTER GLOSSARY FOR REVIEW

Analysis of Variance An inferential statistical technique used to examine the difference among group means. The statistic divides the total variance among the participants into variance related to the independent variable and variance related to random error. If the ratio of the between-groups variance to the within-groups variance is sufficiently large, one may reject the null hypothesis that all group means are equal.

ANOVA Summary Table A table that presents the sum of squares, degrees of freedom, mean square, and F-ratio for an ANOVA.

Between-Groups Variance An estimate of the variance among the individual group means.

$df_{between}$ Degrees of freedom between groups, also presented as df_N.

df_{within} Degrees of freedom within groups, also presented as df_D.

Effect Size A descriptive statistic that provides a standardized measure of the relative difference among the group means. Effect sizes greater than 0 indicate a relation between the independent and dependent variables.

Eta Squared (η^2) A measure of how much of the variance in a dependent variable can be explained by a categorical independent variable.

Experiment-Wise Error (α_e) The probability of committing one or more type I errors when conducting multiple t-ratios from a single experiment.

f A measure of effect size for the ANOVA.

***F*-Ratio** The inferential statistic for the ANOVA. The statistic is the between-groups variance divided by the within-groups variance. If the *F*-ratio is sufficiently large, one may reject the null hypothesis that all group means are equal.

General Linear Model A conceptual mathematical model that describes an observed score as the sum of the population mean (μ), the treatment effect for a specific level of a factor (α_j), and random error (ε_{ij}). For the one-variable ANOVA, the general linear model is $Xij = \mu + \alpha_j + \varepsilon_{ij}$.

Independent For a multigroup research design, independence indicates that the data collected for one group have no effect or relation to data collected in the other groups.

Level The magnitude of or a condition within the independent variable. For qualitative variables, such as sex, level refers to the category for the variable (e.g., male vs. female). For quantitative variables, level refers to the amount or magnitude of the independent variable.

Mean Square (*MS*) Another term used for an estimate of variance, especially within the context of the ANOVA.

Multiple Comparisons A procedure for examining pairs of means from a multigroup research design.

Omega Squared ($\hat{\omega}^2$) A descriptive statistic that indicates the proportion of the variance between groups that is related to the independent variable.

Pairwise Error (α_p) The probability of committing a type I error for a single *t*-ratio.

Partition In statistics, refers to separating the total variance into its smaller components, the within-groups variance, and the variance due to specific variables or combinations of variables.

Pilot Study A rehearsal of the final research project that permits the researcher to ensure that the procedures will allow him or her to collect the data needed for the study.

Total Variance An estimate of the variance among all the observations collected in the study.

Tukey's HSD A test used in conjunction with the ANOVA to examine the differences between means. The test protects against inflated experiment-wise errors.

Within-Groups Variance An estimate of the average variance among scores within each group of the research.

REFERENCES

ABELSON, R.P. (1985). A variance explanation paradox: When a little is a lot. *Psychological Bulletin, 97*, 129–133.

CAMERON, J. & PIERCE, D. (2006). *Rewards and intrinsic motivation: Resolving the controversy*. Charlotte, NC: Information Age Publishing.

COHEN, J. (1973). Eta-squared and partial eta-squared in fixed factor ANOVA designs. *Educational and Psychological Measurement, 33*, 107–112.

COHEN, J. (1988). *Statistical power analysis for the behavioral sciences* (2nd ed.). Hillsdale, NJ: LEA.

EKMAN, P. & O'SULLIVAN, M. (1991). Who can catch a liar? *The American Psychologist, 46*, 913–920.

JACCARD, J., BECKER, M.A., & WOOD, G. (1984). Pairwise multiple comparison procedures: A review. *Psychological Bulletin, 96*, 589–596.

MUELLER, C.M. & DWECK, C.S. (1998). Praise for intelligence can undermine children's motivation and performance. *Journal of Personality and Social Psychology, 75*, 33–53.

PEDHAUZER, E.J. (1997). *Multiple regression in behavioral research*. Florence, KY: Wadsworth.

ROTTON, J. & COHN, E.G. (2001). Temperature, routine activities, and domestic violence: A reanalysis. *Violence & Victims, 16*, 203–215.

Tukey, J.W. (1953). *The problem of multiple comparisons*. Mimeographed monograph, Princeton University.

WU, J. (2012). Rewarding healthy behaviors—Pay patients for performance. *Annals of Family Medicine, 10*, 261–263.

Chapter 13

Between-Subjects Factorial Designs

To raise new questions, new possibilities, to regard old questions from a new angle, requires creative imagination and marks real advances in science.

—Albert Einstein

INTRODUCTION

How many times have you heard someone say something like "Eat healthy and be healthy" and thought to yourself, "It's not that simple; there are other factors involved"? As you learn more about research, you will discover that most explanations of behavior include several variables. For instance, what causes one person to help others? Researchers who study altruism suggest that many variables influence helping behavior, including the number of people in the situation, the cost of helping, and the similarity between the person needing help and the helper. There are few instances where we can use a single variable to fully explain human behavior. The implication of this is that our research needs to incorporate more than one variable to understand the phenomenon that we are studying. In this chapter, you will learn how to use the two-variable design and the two-factor analysis of variance (ANOVA). This research design has many

Understanding and Conducting Research in the Health Sciences, First Edition. Christopher J. L. Cunningham, Bart L. Weathington, and David J. Pittenger.

advantages, including the ability to (1) examine the effects of more than one independent variable at a time, (2) examine the interaction between the independent variables, and (3) conduct research that is an efficient use of time and effort. This chapter also sets the foundation for designs involving more than two variables or factors.

THE LOGIC OF THE TWO-VARIABLE DESIGN

The logic of the two-variable design is similar to the single-variable design. The primary difference is that the two-variable design allows us to examine the relationship between two independent variables and a dependent variable. Specifically, the two-variable design examines how much each independent variable, by itself, influences the dependent variable. In addition, this type of design allows us to test the joint influence of the two independent variables on the dependent variable. Figure 13.1 illustrates the logic of the one- and two-variable ANOVA. As you can see, the single-variable ANOVA partitions the total variation among scores into two general components, the between-groups variation and the within-groups variation. The between-groups variation represents in part the variance caused by systematic differences among the groups. By contrast, the two-variable design further divides the between-groups variance.

The two-variable design partitions the between-group variation into three components—effects due to variable A, variable B, and the joint effects of the two independent variables.

Specifically, the two-factor ANOVA divides the between-groups variation into the effects due to each of the independent variables and the interaction of these

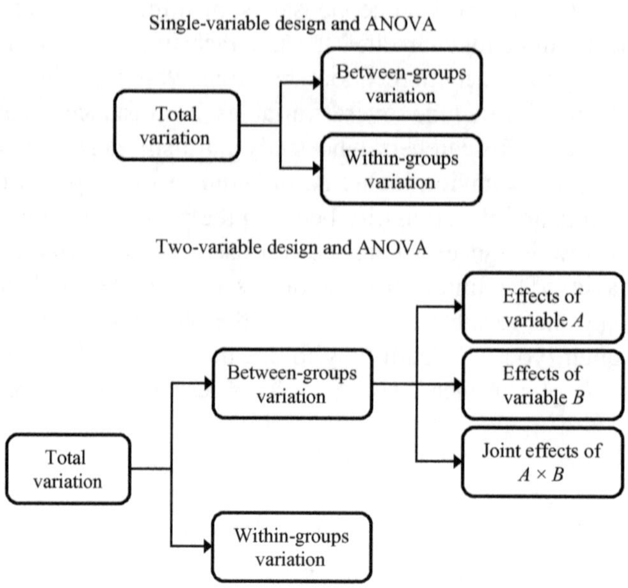

Figure 13.1 Single-variable versus two-variable design.

variables. We can use the general linear model to examine the logic of the between-subjects two-variable ANOVA:

$$X_{ijk} = \mu + \alpha_j + \beta_k + \alpha\beta_{jk} + \varepsilon_{ijk} \tag{13.1}$$

In this equation, X_{ijk} represents an observation within a specific treatment condition. The i in the subscript represents the participant in the group, the j represents the level of the first independent variable, and the k represents the level of the second independent variable. As in the single-variable model, μ represents the mean for the base population, and ε_{ijk} represents the sampling error for each observation. The other three terms represent the effects of each independent variable and their interaction. Specifically, α_j represents the effects of the first independent variable, β_k represents the effects of the second independent variable, and $\alpha\beta_{jk}$ represents the unique variation due to the interaction of the two variables.

ADVANTAGES OF THE TWO-VARIABLE DESIGN

There are several advantages of the two-variable design over the one-variable design. The two-variable design allows us to use research resources efficiently, analyze the interactive effects of two variables, and increase our statistical power.

Increased Efficiency

What would happen if we conducted two one-variable studies rather than one two-variable study? The first study could examine the effects of three levels of the first independent variable. The second study could then examine the effect of three levels of the second independent variable. Assume for a moment that a power analysis for these experiments indicated that we need 30 participants in each treatment condition to have sufficient statistical power (e.g., $1 - \beta = .80$). Therefore, we will need 90 participants for each study, or 180 participants for both studies. By comparison, the two-variable design requires only 90 participants.

As you can see in Table 13.1, the two-variable design contains the elements of the two separate one-variable designs *and* uses fewer participants. The design retains 30 participants for each level of each treatment condition. By combining the two studies, we require half as many participants as if we attempted to conduct two separate studies. Therefore, one two-variable study is more cost effective than two one-variable studies. In addition, the two-variable design provides more information concerning the combined effects of the two independent variables.

Analysis of the Interaction of Variables

Another advantage of the two-variable design is that it helps us understand how combinations of variables influence behavior. Because we can combine different levels of the two independent variables, we can observe their combined effect on the dependent variable, an effect that we would not be able to observe or test in

Table 13.1 Difference between Two Single-Variable Studies and One Two-Variable Study

Two single-variable studies

Each study uses 30 participants in each treatment condition ($n = 30$) for a total of $N = 90$ in the whole study. Conducting both studies requires $N = 180$.

Study 1

a_1	a_2	a_3	Total
$n_1 = 30$	$n_2 = 30$	$n_3 = 30$	$N = 90$

Study 2

b_1	b_2	b_3	Total
$n_1 = 30$	$n_2 = 30$	$n_3 = 30$	$N = 90$

One two-variable study

This design combines the two studies and requires only $N = 90$.

		A				
		a_1	a_2	a_3	Total	
B	b_1	$n_{11} = 10$	$n_{21} = 10$	$n_{31} = 10$	$n_{b1} = 30$	For each level of variable *B*, there are 30 participants.
	b_2	$n_{12} = 10$	$n_{22} = 10$	$n_{32} = 10$	$n_{b2} = 30$	
	b_3	$n_{13} = 10$	$n_{23} = 10$	$n_{33} = 10$	$n_{b3} = 30$	
	Total	$n_{a1} = 30$	$n_{a2} = 30$	$n_{a3} = 30$	$N = 90$	

For each level of variable *A*, there are 30 participants.

a single-variable study. As you will learn, an interaction represents a unique pattern of influence on the dependent variable that cannot be explained by the independent variables separately.

Here is an example of how studying two variables can allow us to examine an interaction. Cohen, Nisbett, Bowdle, and Schwarz (1996) conducted an experiment in which they examined the reaction of white male participants who had just been insulted. When the participants arrived for the experiment, a researcher collected a sample of the participants' saliva. Next, the participants walked through a long, narrow hall. For a random half of the men, a research associate, walking in the other direction, bumped into the participant and called him an insulting name. When the participants reached the end of the hall, another researcher greeted the participant, had him complete a short task, and then collected another sample of saliva. Cohen et al. examined the change in testosterone levels in the two saliva samples. Testosterone is a hormone that has been shown to correlate with levels of arousal (especially aggression), and it is easily measured in saliva.

Figure 13.2 represents the results for the study—specifically, the average percent increase in testosterone for participants in the control and insult groups. The results seem straightforward: Insulted men produced more testosterone after the insult than men not insulted. Do these data tell the entire story? Compare Figure 13.2 with Figure 13.3. The results presented in Figure 13.3 represent the relation between changes in testosterone and two independent variables. The first

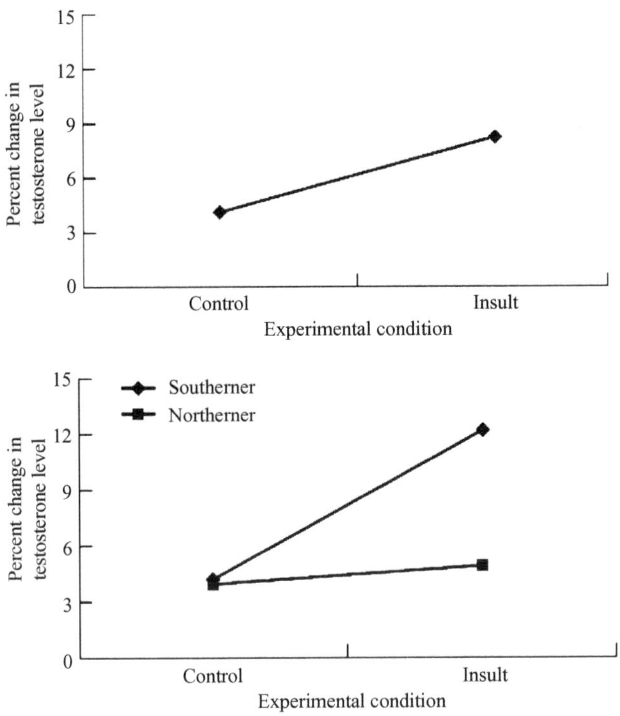

Figure 13.2 Relation between changes in testosterone levels and the control and insult conditions.

independent variable is whether the participant received an insult. The second independent variable is the regional background of the participants. Cohen et al. (1996) selected the participants for the study such that half of the men had lived most of their lives in the North. The other participants were men who had lived mainly in the South. The insult condition represents a manipulated independent variable because the researcher randomly assigned half the participants to the insult condition and the others to the control condition.

The regional background of the participant is a subject variable as it reflects a characteristic of the participant that is not controlled by the researcher. Do you notice a difference between the two figures? Did men from the North respond to the insult in the same way as men from the South? Clearly, the two groups of men reacted differently to the insult condition. More specifically, in the control condition, there is no distinguishable difference between the two groups of men in terms of their levels of testosterone. The interesting difference occurs in the insult condition. From the results of this condition, you can see that in men from the North, there was little reaction to the insult as indicated by the percent change in testosterone (in fact, their levels are almost equivalent to the levels found in the control group). By contrast, the change in testosterone level for the men from the South shows a clear spike as these men produced considerably more testosterone after the insult than any of the men in the other conditions. The data in Figure 13.3 represent an interaction because the men from the North and the South reacted differently to the same insult condition but showed no difference for the control condition.

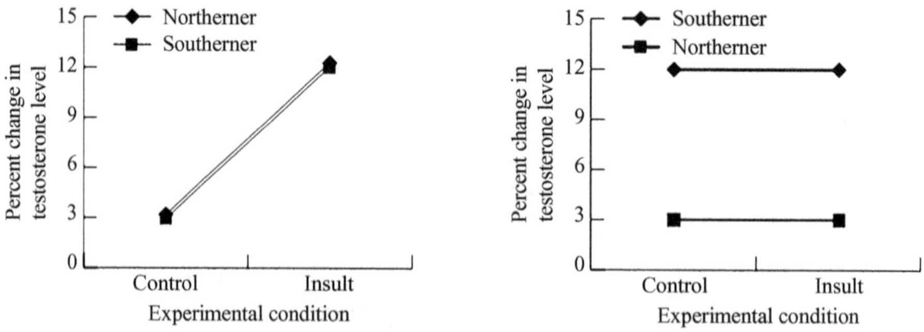

Factor A insult condition

Factor B background	**Control** a_1	**Insult** a_2	Row means
Northerner b_1	3.25	12.25	7.75
Southerner b_2	3.00	12.00	7.50
Column means	3.13	12.13	7.63

Main effect for factor A
No main effect for factor B
No interaction

(*a*)

Factor A insult condition

Factor B background	**Control** a_1	**Insult** a_2	Row means
Northerner b_1	3.0	3.0	3.0
Southerner b_2	12.0	12.0	12.0
Column means	7.5	7.5	7.5

No main effect for factor A
Main effect for factor B
No interaction

(*b*)

Figure 13.3 Two types of main effect for a 2 × 2 design.

Cohen et al. (1996) predicted that men who grew up in the South experience a "culture of honor" that requires them to protect their character when attacked, whereas men raised in the North experienced a different cultural norm. Because of these cultural differences, Cohen et al. predicted that men raised in the South would produce greater levels of testosterone, indicating greater levels of aggression and offense to the insult. Look again at Figure 13.2. Is it accurate to say that being insulted increases testosterone levels? Although the statement is by itself accurate, it is incomplete because it does not take into account the background of the participants. As you can see in Figure 13.3, men raised in the North did not react to the insult, whereas men raised in the South did react. Therefore, to offer a more complete and accurate account of the reaction to insult, we need to make full use of the data we have available—in this case then, we need to consider the interaction of the two independent variables (level of insult and regional background) and their influence on testosterone levels.

Increased Statistical Power

A third advantage of the two-variable ANOVA is an increase in statistical power. In general, a two-variable or multivariable design tends to be more powerful than a single-variable design. A main reason for this is that the within-groups variance in a two-variable or multivariable design tends to be smaller than the within-groups variance in a one-variable study. Why is this true? In the one-variable design, the within-groups variance reflects the effects of sampling error and the effects of variables that the researcher has not measured. In some research situations, this unmeasured variable may be strongly associated with the dependent variable. Consequently, measuring an additional variable in a factorial design often allows the researcher to describe the data more precisely. The benefit of this tactic is that the researcher can reduce the within-groups variance in each of the treatment groups, thereby increasing the possibility of detecting a statistically significant difference between treatment groups in terms of the dependent variable.

We can use the Cohen et al. (1996) study as an example. If the researchers had performed a single-variable study examining only the effect of the insult, as presented in Figure 13.2, they would have found a statistically significant difference between the means of the two groups. The within-groups variance would be relatively large, however, because the researchers did not take into account the potential influence of participants' regional backgrounds. If we use the same data, but now conduct a two-factor ANOVA that includes regional background, the within-groups variance will decrease because we have identified another variable that systematically affects the participants' physiological reactions (i.e., testosterone level). In essence, we have become more specific in our classification of participants: Instead of grouping them solely in terms of their exposure to insult, we are now also grouping them in terms of their respective regional backgrounds.

FACTORIAL DESIGNS: VARIABLES, LEVELS, AND CELLS

When researchers talk about the two-variable design, they often refer to concepts such as factorial designs, main effects, and the interaction. These terms refer to the design of the experiment and to the specific components of the general linear model of the two-variable ANOVA. Before we continue, you need to understand the meaning of these terms.

Treatment Combinations

One of the advantages of the two-variable ANOVA is that it allows us to examine the effects of two or more independent variables and their combined effects. When we design a two-variable study, we select the number of **levels** that we want to use for each independent variable. Because we combine the two variables into one study, we create a **factorial design**. A factorial design represents a study that includes an independent group for each possible combination of levels for the multiple independent variables. In the Cohen et al. (1996) experiment, for instance, there were two levels of the insult condition (variable A: control vs. insult) and two levels of participant background (variable B: Northerner vs. Southerner).

Table 13.2 Matrix for a 2 × 2 Factorial Research Design Using Cohen et al. (1996) as an Example

		Variable *A* Insult condition	
Variable *B* Regional background		Control a_1	Insult a_2
Northern	b_1	Control × Northern a_1b_1	Insult × Northern a_2b_1
Southern	b_2	Control × Southern a_1b_2	Insult × Southern a_2b_2

Consequently, we can say that Cohen et al. used a *2 × 2 factorial design*. This design therefore created four distinct or independent treatment groups as presented in Table 13.2.

For a 3 × 4 factorial design, there are 12 independent treatment conditions or **cells**. For example, the cell a_1b_2 represents the first level of the first independent variable (variable *A*: control condition) and the second level of the second independent variable (variable *B*: Southerner). Hence, all the participants in a_1b_2 are men from the South assigned to the control condition.

Factorial designs can vary in complexity from the simple 2 × 2 design to more complex designs such as a 3 × 5 or a 4 × 6 factorial. Factorial designs can also include more than two variables. The ANOVA allows researchers to use any number of independent variables in a study. Some researchers may use three or four independent variables in one experiment. You may, for example, read in a research report that the researcher conducted a 2 × 3 × 4 ANOVA. This information implies that the researcher examined the effects of three independent variables and the interactions among all of these variables. There were two levels for the first variable, three levels of the second variable, and four levels of the third variable. If this was a fully between-subjects design, then you should also conclude that there were 24 (2 × 3 × 4 = 24) independent treatment groups, each representing a unique combination of the three variables. Thus, a distinguishing feature of all factorial designs is that they contain all possible combinations of the treatment conditions the researcher decided to use.

In this chapter, we primarily focus on the two-variable model. If you need to use a more complex form of the ANOVA, such as one including three variables, you may want to review other sources such as Hayes (1981); Kirk (1982); or Winer, Brown, and Michels (1991) to learn how to conceptualize and analyze the results from more complex research designs. Because of complexities with multiple variable interactions and interpreting what the results actually mean, it is not common to see behavioral science research that involves more than a three-way interaction (e.g., $A \times B \times C$).

MAIN EFFECTS AND INTERACTION

The purpose of the factorial design is to examine how the two variables in the research combine and possibly interact with one another. In this section, we will examine the potential outcomes for a factorial design and describe how to interpret the results. Specifically, we will examine main effects and interactions. As a brief preview, a main effect represents results wherein one or both factors separately have a statistically significant effect on the dependent variable. An interaction represents a condition where both independent variables have a joint effect on the dependent variable. For the sake of illustration, we can use hypothetical data that could have come from the study design used by Cohen et al. (1996). As a quick reminder, in this study, variable *A* is the insult condition (control vs. insult) and variable *B* is the participant's regional background (Southern vs. Northern).

Main Effects

A **main effect** refers to the effect that one independent variable has on the dependent variable holding the effect of any other independent variable(s) constant. Specifically, a main effect represents a special form of the between-groups variance of a single independent variable. In a two-variable ANOVA, there are two main effects, one for each variable. When we examine the data using an ANOVA, each main effect can be either statistically significant or not statistically significant. Consequently, there are four potential main effect results: (1) a statistically significant main effect for variable *A*, (2) a statistically significant main effect for variable *B*, (3) statistically significant main effects for variables *A* and *B*, or (4) no statistically significant main effect for variables *A* and *B*.

The statistically significant main effect represents the **additive effect** of the independent variable. An additive effect means that when we combine two variables, the effect of each variable adds a consistent effect across all treatment conditions. By contrast, the interaction indicates that there is interplay between the two variables that is more than can be explained by either variable alone. An interaction indicates a unique treatment effect that neither independent variable can explain by itself.

Figure 13.3 presents two types of main effect. In Figure 13.3*a*, there is a main effect for variable *A* but not for variable *B*. More specifically, Figure 13.3*a* indicates that the insult condition increased the testosterone of all the participants and that there were no differences between Northerners and Southerners. In contrast, Figure 13.3*b* indicates that the Southerners had a greater change in testosterone than the Northerners, but that the insult condition had no effect on the results. There is no evidence for an interaction in either of the graphs included in Figure 13.3. An easy way to recognize this is to note that the lines in these figures are parallel, indicating that the relationship between insult condition and testosterone was identical regardless of the participant's regional background.

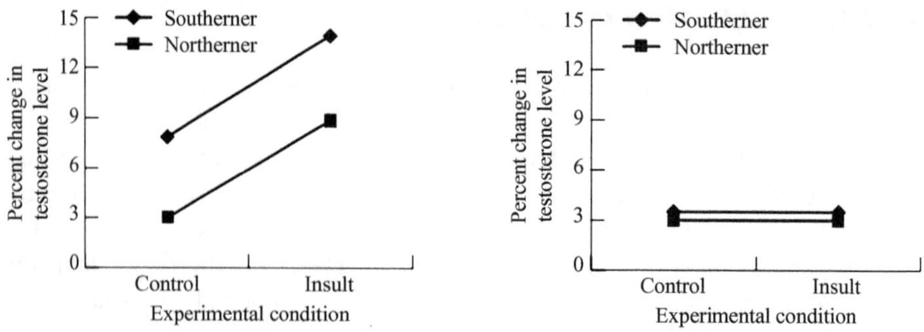

Factor B background	Factor A insult condition: Control a_1	Insult a_2	Row means
Northerner b_1	3.0	9.0	6.0
Southerner b_2	8.0	14.0	11.0
Column means	5.5	11.5	6.5

Main effect for factor A
Main effect for factor B
No interaction

(*a*)

Factor B background	Factor A insult condition: Control a_1	Insult a_2	Row means
Northerner b_1	3.5	3.5	3.5
Southerner b_2	3.0	3.0	3.0
Column means	3.25	3.25	3.25

No main effect for factor A
No main effect for factor B
No interaction

(*b*)

Figure 13.4 Two more types of main effect for a 2 × 2 design.

In Figure 13.3*a*, there is a statistically significant main effect for variable *A*, insult condition—men who were insulted had larger changes in testosterone than men who were not insulted. In Figure 13.3*b*, there is evidence for a statistically significant main effect of variable *B*, background—Southerners had a greater increase in testosterone than Northerners.

It is also possible to have results that indicate statistically significant main effects for both independent variables, without a significant interaction. This is the situation illustrated in Figure 13.4*a*. Notice that there is a statistically significant main effect for the insult condition: All the participants in the insult condition had greater increases in testosterone than all the participants in the control condition. This graph also depicts a statistically significant main effect for the participant's background: Southerners had greater increases in testosterone than Northerners in the control and insult conditions. Because the lines in this graph are parallel, there is no evidence that the influence of one variable on the dependent variable depends on the other (i.e., there is no interaction present here).

Although it can be frustrating to the researchers, there are times when research produces no statistically significant findings, as illustrated in Figure 13.4*b*. In this graph, you can see that participants' testosterone showed a slight increase for the control and insult conditions, but there are no evident systematic differences among the four group means.

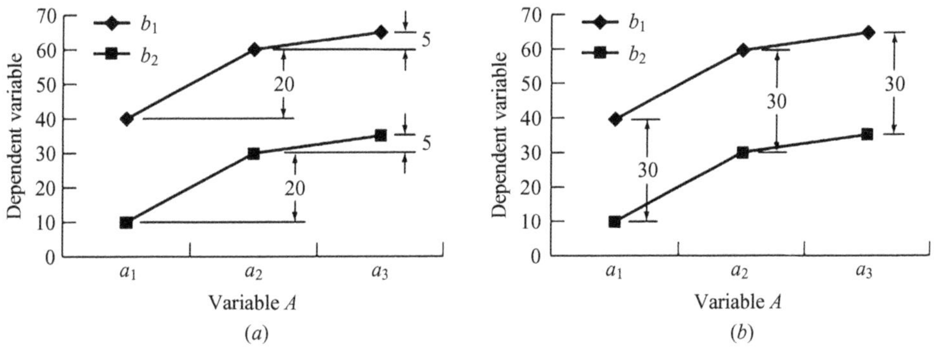

Figure 13.5 Additive main effect for a 3 × 2 ANOVA.

As a secondary and slightly more complex example, Figure 13.5 illustrates additive main effects for a 3 × 2 design with main effects for variables *A* and *B*, but no interaction. For variable *A*, each increase in level (e.g., from a_1 to a_2 and from a_2 to a_3) represents an equal change in the means. The difference between a_1 and a_2 is 20 points for both the b_1 and b_2 conditions. Similarly, the change between a_2 and a_3 is 5 points for both the b_1 and b_2 conditions. Changes in the dependent variable for all levels of variable *B* are consistent across all levels of variable *A*. We see this same consistency in the differences for variable *B*. The average of the b_2 treatment condition is always 30 points greater than the b_1 treatment condition. This consistency across all levels of *A* and *B* means that there is no interactive effect of the two independent variables on the dependent variable in this example.

The Interaction

An **interaction** indicates that the effect of one independent variable is not consistent across all levels of the other independent variable. In a two-variable design, there are four potential patterns of interaction. We will begin by examining the situation that arises when there is a statistically significant interaction and one statistically significant main effect.

Figure 13.6 presents two examples of statistically significant interactions. The most notable characteristic of the two graphs in Figure 13.6 is that the lines are not parallel. This is a hallmark of the interaction. The interaction represents variation among the treatment conditions that cannot be explained by the independent variables considered separately. When there is an interaction, one must examine how the two variables combine to produce unique results. In either panel of Figure 13.6, the insult condition, by itself, cannot account for all the differences among the four group means. Instead, it is more accurate and to the point to highlight that the effect of the insult condition interacted with participants' backgrounds. In this example, Northern men showed little reaction to the insult, as measured by the level of testosterone. However, Southern participants showed a considerable

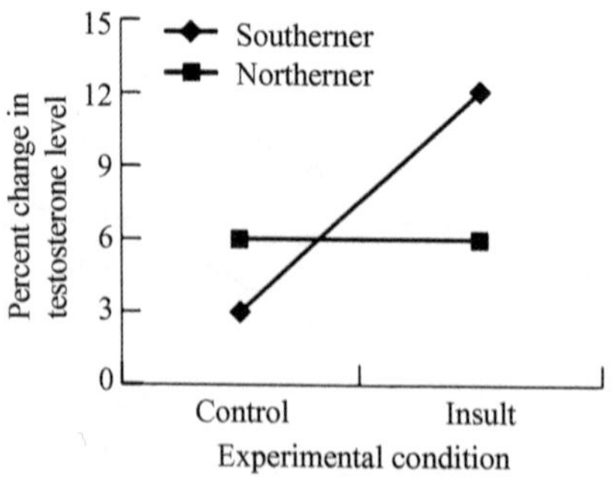

Factor B background	Factor A insult condition: Control a_1	Insult a_2	Row means
Northerner b_1	6.0	6.0	6.0
Southerner b_2	3.0	12.0	7.5
Column means	4.5	9.0	6.75

Main effect for factor A
No main effect for factor B
Interaction

(*a*)

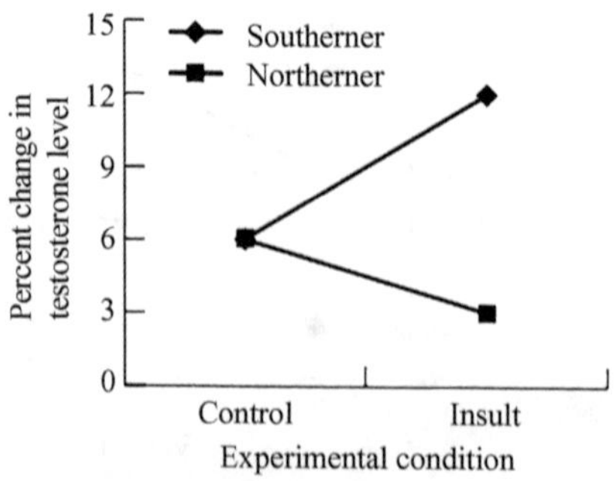

Factor B background	Factor A insult condition: Control a_1	Insult a_2	Row means
Northerner b_1	6.0	3.0	4.5
Southerner b_2	6.0	12.0	9.0
Column means	6.0	7.5	6.75

No main effect for factor A
Main effect for factor B
Interaction

(*b*)

Figure 13.6 Two types of interaction for a 2×2 design.

reaction to the same insult condition. This is evidence for an interaction of insult × regional background.

In Figure 13.6*a*, there is also evidence for a statistically significant interaction of insult × regional background and a main effect for variable *A*, the insult condition. This main effect indicates that, all else being equal, participants in the insult condition had greater testosterone increases than participants in the control condition. As you can see in the graph, this statement does not take into account the effect of the interaction between the insult condition and the participant's background. Looking at Figure 13.6*a*, it is clear that the most important finding to highlight is that Northern participants' testosterone levels did not change across the two insult conditions, but Southern participants' testosterone levels did rise for Southern participants who experienced the insult condition (this type of interaction is often referred to as a disordinal interaction).

Figure 13.6*b* also indicates a statistically significant interaction and main effect. In this example, the insult had opposite effects on the Southern and Northern participants. The insult caused the Southern participants' testosterone to increase but the Northern participants' testosterone to decrease (this type of interaction is often referred to as an ordinal interaction as the lines do not cross within the bounds of measurement for the variables involved).

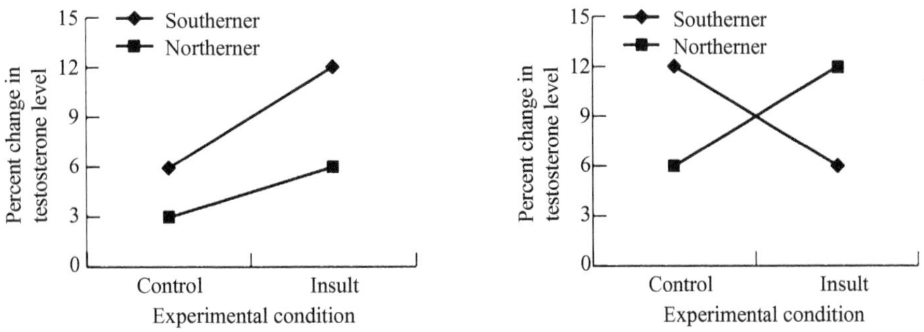

(a)

Factor B background	Factor A insult condition Control a_1	Insult a_2	Row means
Northerner b_1	3.0	6.0	4.5
Southerner b_2	6.0	12.0	9.0
Column means	4.5	9.0	6.75

Main effect for factor A
Main effect for factor B
Interaction

(*a*)

(b)

Factor B background	Factor A insult condition Control a_1	Insult a_2	Row means
Northerner b_1	6.0	12.0	9.0
Southerner b_2	12.0	6.0	9.0
Column means	9.0	9.0	9.0

No main effect for factor A
No main effect for factor B
Interaction

(*b*)

Figure 13.7 Two types of interaction for a 2 × 2 design.

Figure 13.7 presents the last set of potential patterns for interactions and main effects in a two-variable design. Figure 13.7*a* represents the condition where the interaction and both main effects are statistically significant. As you can see in the means, the insult condition raised the testosterone levels more than the control condition. In addition, Southern men had, overall, greater increases in testosterone levels. Although both of these statements are correct, they do not fully describe all the components of the results. From the observed interaction, however, we can more completely conclude that Southern participants reacted more to insults, as measured by the change in testosterone, than did Northern participants.

The data presented in Figure 13.7*b* represent a scenario in which the interaction of the two independent variables is significant, but neither of the main effects is significant. Do these results mean that the independent variable had no effect on the results? No! Both independent variables influence the dependent variable, but their effects can be understood only within the context of the interaction. This is why it is so important to consider and interpret interaction effects first and then to follow that with a consideration of any possible main effects in this type of design. In the present example, the main effects cancel each other because of the pattern of means.

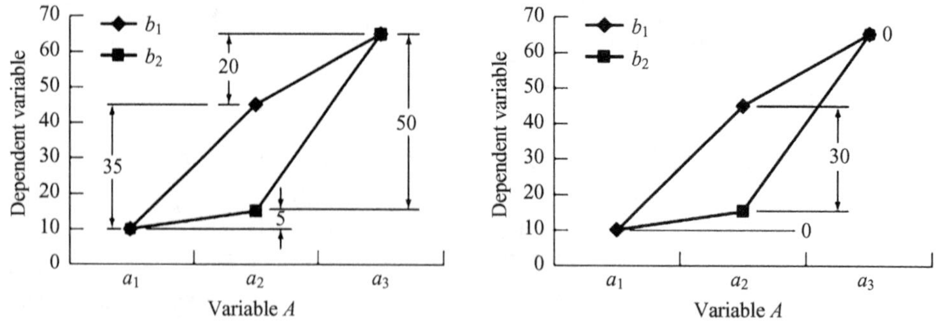

Figure 13.8 Significant interaction and significant main effects for variables A and B.

Figure 13.8 presents another look at a significant interaction. Looking closely at the data, you can see that the changes in the dependent variable are not consistent for the different combinations of the variables. For example, the difference between a_1b_1 and a_2b_1 is 35 points, whereas the difference between a_1b_2 and a_2b_2 is 5 points. In addition, the difference between a_1b_1 and a_1b_2 is 0, whereas the difference between a_2b_1 and a_2b_2 is 30 points.

This inconsistency across treatment conditions is evidence of an interaction. You should interpret any differences among treatment group means with care when there is a statistically significant interaction. This is why you should always consider the interaction effect(s) first when reviewing the results of your ANOVA based on such data. According to the general linear model, each treatment group mean reflects the effects of each independent variable *and* their interaction. Therefore, when you see that the mean of $a_2b_1 = 45$, you must recognize that its value represents the effect of variable A, variable B, and the interaction of $A \times B$.

KNOWLEDGE CHECK

Each of the following represents the cell totals and sample sizes for 2×2 factorial studies. For each study, determine the means for the individual treatment conditions as well as the row and column means. Then prepare a graph of the means and determine whether there is evidence of main effects for variables A and B and an interaction.

1.

	a_1	a_2			a_1	a_2	
b_1	$\Sigma X_{11} = 30.00$ $n_{11} = 15$	$\Sigma X_{21} = 45.00$ $n_{21} = 15$	$\Sigma X_{b1} = 75.00$ $n_{b1} = 30$	b_1	$M_{11} =$	$M_{21} =$	$M_{b1} =$
b_2	$\Sigma X_{11} = 45.00$ $n_{12} = 15$	$\Sigma X_{22} = 60.00$ $n_{22} = 15$	$\Sigma X_{b2} = 105.00$ $n_{b2} = 30$	b_2	$M_{12} =$	$M_{22} =$	$M_{b2} =$
	$\Sigma X_{a1} = 75.00$ $n_{a1} = 30$	$\Sigma X_{a2} = 105.00$ $n_{a2} = 30$	$\Sigma X_{ijk} = 180.00$ $N = 60$		$M_{a1} =$	$M_{a2} =$	$M_{overall} =$

2.

	a_1	a_2			a_1	a_2	
b_1	$\Sigma X_{11} = 100.00$	$\Sigma X_{21} = 150.00$	$\Sigma X_{b1} =$	b_1	$M_{11} =$	$M_{21} =$	$M_{b1} =$
	$n_{11} = 25$	$n_{21} = 25$	$n_{b1} =$				
b_2	$\Sigma X_{12} = 125.00$	$\Sigma X_{22} = 75.00$	$\Sigma X_{b2} =$	b_2	$M_{12} =$	$M_{22} =$	$M_{b2} =$
	$n_{12} = 25$	$n_{22} = 25$	$n_{b2} =$				
	$\Sigma X_{a1} =$	$\Sigma X_{a2} =$	$\Sigma X_{ijk} =$		$M_{a1} =$	$M_{a2} =$	$M_{\text{overall}} =$
	$n_{a1} =$	$n_{a2} =$	$N =$				

3.

	a_1	a_2			a_1	a_2	
b_1	$\Sigma X_{11} = 100.00$	$\Sigma X_{21} = 50.00$	$\Sigma X_{b1} =$	b_1	$M_{11} =$	$M_{21} =$	$M_{b1} =$
	$n_{11} = 25$	$n_{21} = 25$	$n_{b1} =$				
b_2	$\Sigma X_{12} = 50.00$	$\Sigma X_{22} = 100.00$	$\Sigma X_{b2} =$	b_2	$M_{12} =$	$M_{22} =$	$M_{b2} =$
	$n_{12} = 25$	$n_{22} = 25$	$n_{b2} =$				
	$\Sigma X_{a1} =$	$\Sigma X_{a2} =$	$\Sigma X_{ijk} =$		$M_{a1} =$	$M_{a2} =$	$M_{\text{overall}} =$
	$n_{a1} =$	$n_{a2} =$	$N =$				

4.

	a_1	a_2			a_1	a_2	
b_1	$\Sigma X_{11} = 125.00$	$\Sigma X_{21} = 225.00$	$\Sigma X_{b1} =$	b_1	$M_{11} =$	$M_{21} =$	$M_{b1} =$
	$n_{11} = 25$	$n_{21} = 25$	$n_{b1} =$				
b_2	$\Sigma X_{12} = 75.00$	$\Sigma X_{22} = 125.00$	$\Sigma X_{b2} =$	b_2	$M_{12} =$	$M_{22} =$	$M_{b2} =$
	$n_{12} = 15$	$n_{22} = 25$	$n_{b2} =$				
	$\Sigma X_{a1} =$	$\Sigma X_{a2} =$	$\Sigma X_{ijk} =$		$M_{a1} =$	$M_{a2} =$	$M_{\text{overall}} =$
	$n_{a1} =$	$n_{a2} =$	$N =$				

5. Nisbett (1968) conducted a study to examine the relation between food taste and food consumption for obese and underweight people. Nisbett classified participants as obese or underweight by comparing their actual weight to their ideal weights. He then asked people to eat vanilla ice cream. Half the participants ate good-tasting ice cream; the others ate bitter-tasting ice cream. The dependent variable was the number of grams of ice cream consumed. The following chart presents the average amount of ice cream consumed (assume equal sample sizes).

		Variable *A* Weight of participant		
Variable *B* Taste of food		Obese a_1	Underweight a_2	Row means
Good	b_1	200	120	
Bitter	b_2	45	95	
Column means				

a. Draw a graph of the data and write a short essay describing the results.

b. Given these results, why was it important for Nisbett to include the weight of the person as a variable in the study?

6. A researcher wanted to study state-dependent learning. To examine the phenomenon, she had half the participants learn a list of 30 unrelated words

while listening to a tape of Grateful Dead songs. The other participants learned the list in a quiet room. Two days later, all participants returned to the laboratory to recite the list of words. Half the participants in the music group and half the participants in the quiet group recited the list while listening to the Grateful Dead tape. The other participants recited the list in a quiet room.

a. Draw a graph of the data and write a short essay describing the results.

b. Could the researcher have studied the state-dependent learning phenomenon using two single-variable studies? Describe the specific advantages to using the factorial design for this research project.

		Variable *A* Learned list condition		
Variable *B* Recite list condition		Music a_1	Quiet a_2	Row means
Music	b_1	18	9	
Quiet	b_2	11	15	
Column means				

DESIGNING A FACTORIAL STUDY

Designing a factorial study is relatively straightforward in that all the principles you have already learned about good research design still apply. In the following section, we will briefly review these issues as they relate to the independent and dependent variables and the necessary sample size.

Independent Variables

We have already discussed several of the distinguishing features of the factorial design. The factorial design allows us to simultaneously examine the relation between two or more independent variables and the dependent variable. Using two or more independent variables allows us to examine how combinations of treatment conditions relate to different outcomes in the study. Of particular interest, the factorial design allows us to examine the presence of an interaction, something that we could not observe in a single-variable study. Also, we can use manipulated as well as subject variables in the study. This feature allows us to examine the relationships among a number of predictor variables and the dependent variable.

Between-Subjects Variables

The specific research design we examine in this chapter requires that all the variables be between-subjects variables. This requirement means that we test each participant in only one treatment group or combination of treatment conditions. For examples of this, consider some of the studies we have examined so far. In the Cohen et al. (1996) study, both the insult condition and the participant background are between-subjects variables. For the insult condition, Cohen randomly assigned the participants to either the control or the insult treatment condition. Participant background is also a between-subjects variable in that the participants grew up either in the South or the North. Each participant is therefore found in one combination treatment condition of insult and regional background. The dependent variable in this study was percent change in testosterone level, and this was used in each of the four treatment conditions.

When both independent variables are between-subjects variables, we can conclude that the individual cells or treatment groups are **independent groups**. In this case, independence means that the data collected in one treatment group are not correlated with the data collected in any of the other groups. This independence of the treatment conditions helps to determine the type of statistical analysis we can use to examine the data. In a following section, we will review how to analyze the data for the two-variable between-subjects design. In the next two chapters, we will discuss techniques for analyzing data from studies in which at least one of the variables is a within-subjects variable.

Levels of the Variables

An important design consideration is the number of levels to use for each independent variable or factor. At a minimum, you must have at least two levels of each variable. You can use as many levels for both variables as you need. Could Cohen et al. (1996) have added more levels to their study? For the participant background condition, they could have identified other regions of the country (e.g., West Coast or Rocky Mountain states) or other countries (e.g., men from Asian or African countries). In addition, Cohen et al. could have used three treatment conditions by adding a condition where the men received a compliment rather than an insult. Adding more treatment conditions may produce results that further enhance our understanding of the phenomenon we are studying. For example, if Cohen et al. (1996) had included men from the West Coast states, they could have determined whether the Southerners' reaction to the insult is unique to men raised in the South. Similarly, the addition of a compliment condition (i.e., an "anti-insult") would help determine whether the increase in the testosterone occurs only in the insult condition. Such a study would be an interesting follow-up project to the original work.

In the previous chapter, you learned that adding more levels of the independent variable has the advantage of improving our understanding of the relation between the independent and dependent variables. Adding more levels of the variable can also improve the power of a study if adding treatment conditions increases homogeneity of the participants with each condition and increases the range of effect for the independent variable. These same advantages occur in the factorial design.

There is a cost, however, to adding more levels of a variable in this type of design. Each level creates additional treatment groups, requiring you as the researcher to identify more participants.

Imagine that a researcher begins by planning for a 2 × 2 factorial design and determines that there must be at least 10 participants in each research condition. For this design, the researcher will require 40 participants to conduct the study. After thinking more about the purpose of the study, the researcher decides to switch to a 3 × 3 factorial design. This design will require 90 participants, more than twice as many as the 2 × 2 design requires. Even switching to a 2 × 3 factorial design will require 60 participants. Selecting the number of levels for each independent variable requires a balance between the intent of the research and the cost of conducting the study.

Dependent Variables

Several characteristics of the dependent variable also require attention when conducting a factorial design. The first characteristic refers to the dependent variable's scale of measurement. In general, an ANOVA works best when the dependent variable data are measured using an interval or ratio scale. It is possible to conduct a factorial design for which the data represent a nominal ordinal scale, but in these situations, a different statistical test is often required (as we discuss in Chapter 16).

IDENTIFYING SAMPLES AND ESTIMATING SAMPLE SIZE

Apart from the variables, it is also necessary to consider the sample size of your treatment groups when planning to use an ANOVA to examine data from a factorial design. An ANOVA works best when the number of observations in each treatment group (typically denoted as n) is equivalent. Although minor differences in sample sizes have little effect on the validity of the ANOVA, large differences in sample sizes among the cells can cause an ANOVA to produce spurious results. Plan your study with sufficient care to ensure equal sample sizes across all treatment conditions. If circumstances beyond your control cause unequal sample sizes, you should refer to an advanced statistics textbook for recommendations for analyzing the data that you are able to collect.

When designing a factorial-type study, it is common to perform a statistical power analysis to determine the number of participants necessary for each treatment group to ensure the best likelihood of detecting statistically significant effects if they are present. The steps that we follow for estimating the sample size in a factorial design are essentially the same that we use for the single-factor experiment.

First, to determine power for the two-factor ANOVA, we again turn to Table B.13 in Appendix B. For this example, assume that you want to conduct a 3 × 4

Table 13.3 Equations for Determining the Numerator Degrees of Freedom for a Factorial Design

Equations	Example for a 3 × 4 factorial design
Main effect for factor *A*	$df_A = (3 - 1) = 2$
$df_A = (j - 1)$, where j = # of levels of *A*	
Main effect for factor *B*	$df_B = (4 - 1) = 3$
$df_B = (k - 1)$, where k = # of levels of *B*	
Interaction for $A \times B$	$df_{A \times B} = (3 - 1)(4 - 1) = 6$
$df_{A \times B} = (j - 1)(k - 1)$	

factorial design. Following Cohen's (1988) recommendation, we will assume that and also that the effect size for the main effects and interaction is moderate, $f = .25$, and that we want the power of our statistic to be $1 - \beta = .80$. At this point, it is important to recognize that a two-variable between-subjects ANOVA will generate three *F*-ratios: (1) for the main effect of variable *A*, (2) for the main effect of variable *B*, and (3) for the interaction of variables *A* and *B*.

This being the case, you should plan your *n* based on the smallest effect size that you expect to observe (usually the interaction of $A \times B$).

The next step will be to calculate an adjusted sample size. To complete this step, you will need the degrees of freedom for the ANOVA. Table 13.3 presents the method for determining the numerator degrees of freedom for each *F*-ratio. Using these degrees of freedom, we can use Equation 13.2 to estimate power using specific sample sizes.

We will begin by examining the power of the two-factor ANOVA with 10 participants in each cell. Table 13.4 presents the application of Equation 13.2 assuming $n_{ij} = 10$, $\alpha = .05$, and $f = .25$ for each *F*-ratio:

$$n'_{\text{effect}} = \frac{jk(n_{ij} - 1)}{df_{\text{effect}} + 1} + 1 \tag{13.2}$$

According to Table 13.4, the probability of obtaining a statistically significant interaction is less than 50% ($1 - \beta \approx .45$). The probability of obtaining statistically significant main effects is also on the slim side: $1 - \beta \approx .68$ for variable *A* and $1 - \beta \approx .61$ for variable *B*. One way to increase power is to increase sample size. Table 13.5 presents the predicted power if we increase the sample size to 20. Doubling the sample size does increase the power of the interaction, but at a cost. You will need a total sample size across all treatment groups of $N = 240$ ($3 \times 4 \times 20 = 240$) participants to conduct the study. This is where you would have some important decisions to make:

- Do you have the time and energy to collect this much data?
- Are there other things that you can do to increase the effect size of the study?
- Have you used the best possible measure of the dependent variable?

Table 13.4 Estimated Power for a 3 × 4 Factorial ANOVA with $n_{ij} = 10$, $\alpha = .05$, and $f = .25$

	df_{effect}		Adjusted n	Rounded n	Estimated power
Variable A	2	$n' = \frac{12(10-1)}{2+1} + 1$	$n' = 37$	$n' = 40$	$1 - \beta \approx .68$
Variable B	3	$n' = \frac{12(10-1)}{3+1} + 1$	$n' = 28$	$n' = 30$	$1 - \beta \approx .61$
Variable $A \times B$	6	$n' = \frac{12(10-1)}{6+1} + 1$	$n' = 16.4$	$n' = 16$	$1 - \beta \approx .45$

Note: The adjusted sample size has been rounded to conform to the values in the power tables. These power estimates are sufficient for general estimates of the power of the design. If you need exact power estimates, refer to Cohen (1988) or software that performs such calculations.

Table 13.5 Estimated Power for an Interaction in a 3 × 4 Factorial ANOVA with $n_{ij} = 10$, $\alpha = .05$, and $f = .25$

	df_{effect}		Adjusted n	Rounded n	Estimated power
Variable $A \times B$	6	$n' = \frac{12(20-1)}{6+1} + 1$	$n' = 33.57$	$n' = 30$	$1 - \beta \approx .76$

- Consider the independent variables. Do you need that many treatment levels?
- Can you create greater differences among the levels of the variables?

Attending to these considerations may help you increase the power without having to use this many research participants.

CHAPTER SUMMARY

This chapter is a further extension of the between-subjects research design as we examined how we can conduct a study that includes more than one independent variable. The two-variable design allows us to simultaneously examine the effect of two variables and their interaction on the dependent variable.

The two-variable design is extremely useful to researchers because it makes better use of research resources, has greater statistical power, and allows us to examine the complex interplay between variables. This interplay between the independent variables is the interaction. The interaction is an extremely important finding in any research project because it represents an outcome that cannot be explained by either of the independent variables. The interaction indicates results beyond the simple addition of two variables.

In this chapter, we reviewed the many possible outcomes for a two-factor research design. As a generality, the results can indicate that one or more main effects are statistically significant and that the interaction is statistically significant. We also reviewed how to examine a graph and to interpret the effects of the independent variables and the interaction. A large portion of the chapter reviewed the ANOVA for the two-variable design. As with the single-variable ANOVA, the two-variable ANOVA uses the *F*-ratio to determine whether there is evidence for a statistically significant effect of the independent variables and their interaction.

An important lesson to learn is that a statistically significant interaction means that the data cannot be explained simply by describing the simple effects of the independent variables.

As with other inferential statistics, we can use power tables to estimate the number of participants that we require to obtain statistically significant effects.

CHAPTER GLOSSARY FOR REVIEW

Additive Effect In a factorial design, the effect of one variable has a consistent effect on the other variables.

Cell Within a factorial design, the cell represents a unique combination of the levels of the independent variables.

Factorial Design A research design that uses two or more independent variables and all possible combinations of treatment levels the researcher selects.

Independent Groups A characteristic of the treatment conditions when both variables are between-subjects variables. If the groups are independent, we assume that there is no correlation.

Interaction In a factorial design, an interaction indicates that the effects of the independent variables are not consistent across all treatment combinations. The interaction represents the differences among the sample means that neither main effect can explain.

Level A unique condition or magnitude of the independent variable.

Main Effect Within a factorial design, refers to the primary effect that an independent variable has on the dependent variable, holding the other effects constant.

REFERENCES

Cohen, D., Nisbett, R.E., Bowdle, B.F., & Schwarz, N. (1996). Insult, aggression, and the southern culture of honor: An "experimental ethnography." *Journal of Personality and Social Psychology, 70*, 945–960.

Cohen, J. (1988). *Statistical power analysis for the behavioral sciences* (2nd ed.). Hillsdale, NJ: LEA.

Hayes, L.V. (1981). *Statistics*. New York: Holt, Rinehart & Winston.

Kirk, R.E. (1982). *Experimental design* (2nd ed.). Monterey, CA: Brooks/Cole.

Nisbett, R.E. (1968). Taste, deprivation, and weight determinates of eating behavior. *Journal of Personality and Social Psychology, 10*, 107–116.

Winer, B.J., Brown, D.R., & Michels, K.M. (1991). *Statistical principles in experimental design* (3rd ed.). New York: McGraw-Hill.

Chapter 14

Correlated-Groups Designs

INTRODUCTION

As you have learned in the previous chapters, we can be creative with our use of the independent-groups or between-subjects design. The primary feature of the independent-groups or between-subjects design is its ability to examine the differences among different groups of people. Using analysis of variance (ANOVA), we can determine whether the variance between groups is larger than the variance within groups and thereby infer that there is some form of relationship between the independent and dependent variables.

A **between-subjects variable** is an independent variable to which the researcher randomly assigns the participants. For each level of a between-participants variable, there is a different and independent group of participants. The essential characteristic of a between-subjects variable is that participants experience only one level of the variable. Consequently, if participants are exposed to only one level of the variable, it is a between-subjects variable. If the researcher uses a subject variable (e.g., men vs. women, depressed vs. not depressed), then the variable is also a between-subjects variable. The term *between-subjects* refers to the fact that the ANOVA compares the differences between independent groups of different participants. This is the case with most traditional single-variable and factorial ANOVA designs that utilize different participants in each treatment condition or group.

In the initial four research designs we discussed, the treatment groups are independent as they are based on between-subjects independent variables. In an

Understanding and Conducting Research in the Health Sciences, First Edition. Christopher J. L. Cunningham, Bart L. Weathington, and David J. Pittenger.

ideal experimental scenario, we achieve this independence by randomly assigning different participants to each of the treatment groups, each of which represents a distinct level of the independent variable(s). Because these different treatment groups consist of different people receiving different levels or conditions of the independent variable, there should be no correlation in outcomes between pairs of participants from different groups. This basic ANOVA framework can also be easily modified to fit other research designs. As you learned in Chapter 13, a factorial design allows us to examine the interaction between two or more variables. We can extend this capability by using ANOVA when studying the relationship between independent and dependent variables in a correlated-groups design.

In this chapter, we examine research designs in which we treatment groups are created in such a way that there is an intentional dependency or correlation among pairs of scores from participants across the different conditions. This correlation is the result of one or more of the independent variables being a **within-subjects variable**. Correlated-groups or within-subjects research typically involves collecting data from the same participants at multiple time points or under multiple different levels of the within-subjects independent variable. The resulting ANOVA and comparisons, then, are not solely contrasting separate groups of participants but rather examining how change occurred within an individual person or group of individuals who shared the same experiences in a research context.

The two primary design techniques used for this type of research are the **repeated-measures design** and the **matched-groups design**. Although these designs are different, it is possible to use the same statistical techniques to analyze the data generated by these research designs. You will learn in this chapter that research designs such as the repeated-measures design and the matched-participants design offer you the opportunity to conduct research with considerable statistical power. These designs also allow the researcher to exercise greater control over the variables that influence the dependent variable. The distinguishing characteristic of the research designs discussed in this chapter is that there is a known or intentional correlation among the groups being studied.

LOGIC OF THE CORRELATED-GROUPS RESEARCH DESIGN

Many factors contribute to random variation in participant behavior during a research project. Individual differences among participants often play a major role in determining this random variation within treatment groups. This is important because if the within-groups variance is too large, it will be difficult to detect differences between separate treatment groups that are due to the effect of the independent variable. Stated differently, if the variance between groups is small relative to the error variance within groups, then the corresponding F-ratio will be too small to allow us to reject the null hypothesis (think back to how F is calculated).

There are several research design decisions we can make to reduce the within-groups variance. For example, we can try to select participants who are similar to

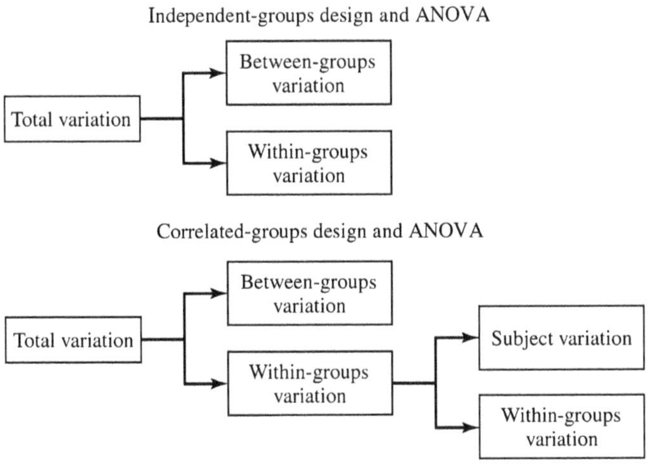

Figure 14.1 Difference between an independent-groups and a correlated-groups research design.

each other because the more homogeneous the sample, the lower the within-groups variance. Another tactic is to treat significant subject variables, such as sex, as one of the independent variables in a factorial design. If the subject variable systematically contributes to the total variance in the outcome(s), identifying it as a variable will reduce the within-groups variance. This tactic works best if the subject variable is a categorical variable (e.g., sex or marital status) or if you can cluster the participants into logical categories (e.g., age groups).

Another way to reduce the within-groups variance is to use a correlated-groups design. This procedure is an exceptionally cost-effective method for increasing statistical power in a study. Like the factorial design, the correlated-groups design allows the researcher to reduce the within-groups variance. Figure 14.1 helps to illustrate the difference between the between-subjects and the correlated-groups research designs. For the between-subjects design, the variance among the participants is treated as part of the within-groups variation. The within-groups variation, therefore, reflects variance due to random error *and* individual differences among participants in each treatment or study condition. Using a correlated-groups design and corresponding ANOVA, we can partition out the portion of within-groups variance that is due to individual differences and treat it as a separate source of variance. Doing this reduces the size of the ultimate MS_{within} term, thereby increasing the resulting F-value. Consequently, a correlated-samples design typically has greater statistical power to detect significant differences among treatment groups than an equivalent between-subjects design.

In a between-subjects design, there is no way to identify variability in participants' behaviors or responses due to factors that are not relevant to the study's focus. Because of this, the error term in an ANOVA based on these designs will include random differences in participant characteristics and tendencies. The correlated-groups design, in contrast, makes it possible to partial out within-groups variance that is due to participant differences, thereby reducing the size of the overall within-groups variance term and increasing the chances of identifying a significant F-value when conducting an ANOVA based on the data from this type of design.

While there are many types of correlated-groups designs out there, two basic forms are most widely used: repeated-measures and matched-participants designs. Although these forms entail different research design considerations, both allow the researcher to exercise greater control over the variables that influence the independent variable than would be possible in a between-subjects design. The benefit to the researcher is tremendous, in terms of the increased statistical power that results.

REPEATED-MEASURES DESIGN

In a repeated-measures design, data are collected from the same participants over a series of occasions or under several different levels of the independent variable to determine whether there are systematic changes in each participant's behavior. There are two general types of repeated-measures designs.

The first type of repeated-measures design allows us to test the same participant using several levels of the same independent variable. For example, a health sciences researcher might be interested in testing the physical functioning of a patient against several different levels of resistance. To conduct the study, the researcher may use no resistance, light resistance, and heavy resistance (possibly even more levels of the independent variable that is resistance). Each patient participant in this study would then be exposed to each of the various levels of resistance over a period of time and their physical functioning would be assessed in this fashion. The advantage to this procedure is that it requires fewer participants because each participant is tested under each treatment condition rather than assigning separate groups of participants to each of the three treatment conditions. Another advantage to this design is that we can use each participant as his or her own control or comparison benchmark, reducing the amount of variance within each treatment condition in the eventual ANOVA (more on this later).

The second general type of repeated-measures design allows us to use time as an independent variable. For this type of research, a study is designed so that data are gathered from participants at multiple points over a specified period of time. Using this procedure, we can then determine how the participants' behaviors or attitudes change over time. This technique is useful when time is a critical variable of interest, such as when we are interested in studying how the effects of a health promotion intervention may develop in the short term but may or may not last over an extended period of time. Thus, we can use this type of design when we want to observe a behavior or other type of outcome before and after a critical event (e.g., intervention) or when we want to observe natural changes in a participant that occur over time.

Figure 14.2 illustrates the difference between an independent-groups design and a correlated-groups design. Both designs begin with a sampling population from which we draw our sample. A primary difference is found in the number of

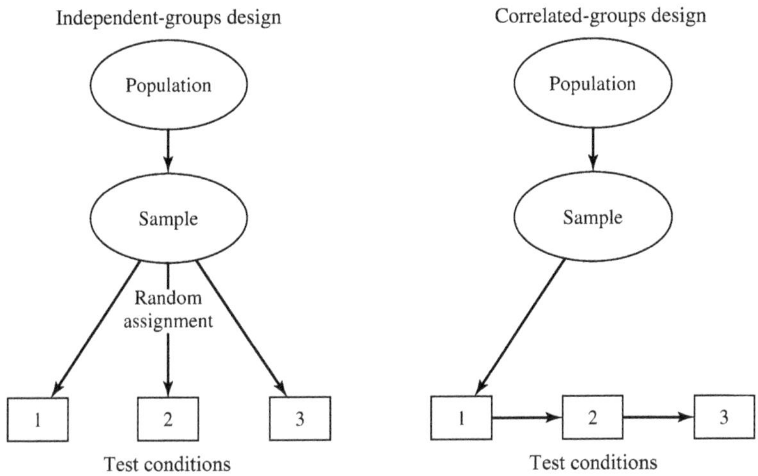

Figure 14.2 Difference between an independent-groups and a correlated-groups design.

Table 14.1 Hypothetical Data for a Study of the Serial-Position Effect Using Four Participants

	Position of word							
Participant	1	2	3	4	5	6	7	Participant means
1	4	4	3	2	2	2	3	2.86
2	5	3	2	3	2	2	5	3.14
3	6	4	5	5	3	3	4	4.29
4	7	6	6	5	5	3	5	5.29
Average number of words recalled	5.5	4.25	4	3.75	3	2.5	4.25	

samples that are required to conduct each type of study. For the independent-groups design, we create a separate sample for each treatment condition using random assignment of different participants into each condition. For the correlated-groups design, we create a single sample and then test each participant in that sample under multiple treatment conditions.

Example of a Repeated-Measures Design

As an example, imagine that we want to replicate a study that has many implications for written communication, the serial-position effect. To conduct our study, we ask four participants to listen to seven lists of seven abstract, three-syllable words. Immediately after hearing the last word of each list, the participants must repeat the list. If our results replicate the serial-position effect, words at the start and end of a list have the greatest probability of recall, whereas words toward the middle of a list have a lower probability of recall. The data in Table 14.1 and

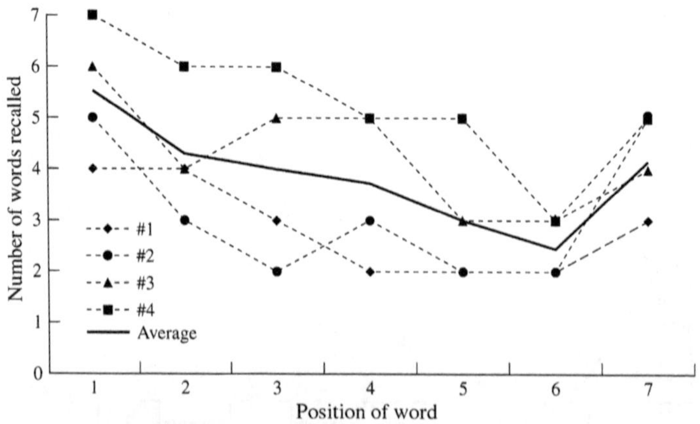

Figure 14.3 Number of words each participant recalled. *Note:* The solid black line represents the average performance of the four participants.

Table 14.2 Summary Table for an Independent-Groups ANOVA Using Data in Table 14.1

	SS	*df*	*MS*	*F*	*p*
Between	22.4286	6	3.7381	2.166	0.088
Within	36.2500	21	1.7262		
Total	58.6786	27			

Figure 14.3 represent hypothetical data for such a study (this example comes from research conducted by Walker & Hulme, 1999).

There is considerable variability among the four participants in this study; some were good at recalling the words, whereas others did not do as well. If you look carefully at each participant's performance, you can see some evidence of the serial-position effect. Participants recalled the most words that came at the start or at the end of the lists. The solid black line in Figure 14.3 represents the average performance of the four participants for each serial position. This line clearly illustrates the serial-position effect. As the position of the word in the order of presentation increases, the probability that a participant will remember the word deceases up to the sixth position. Participants recall the word in the seventh position more than the word in the sixth position, as the serial-position effect predicts.

What would happen if we analyzed the data from this study using a conventional, between-subjects ANOVA? Although this is solely a hypothetical scenario designed to highlight the differences for you between these two types of designs, Table 14.2 presents the ANOVA summary table for a conventional independent-groups ANOVA. For this analysis, we treated the seven word positions as if they were independent conditions each involving four different people. In other words, we treated the data as reflecting the responses of 28 different people who participated in the study and were randomly assigned to one of the seven conditions. The conclusion to draw from this summary table is that we cannot reject the null hypothesis if $\alpha = .05$.

There is a major problem with analyzing these data using a between-subjects ANOVA—the treatment groups (and therefore subjects) are not truly independent from each other. If you reread the previous paragraph, you can see that the description of the "assumptions" had no bearing on how the researcher conducted the study. In the actual study, the data for all seven word positions come from the same four participants. Thus, we should assume that each participant's recall of a word in one position will influence that individual's scores in the other position conditions. Participant 1 does not seem to have done well on the memory test as the scores for this person are generally low. In contrast, participant 4 generally recalled the most words. If you look closely at the data, although there is quite a difference between participants 1 and 4, both show evidence of the serial-position effect.

Given this situation, we would be better off analyzing these data using a within-subjects ANOVA that allows us to examine the differences among groups that consist of the same participants and therefore have a systematic or consistent influence on their responses or outcomes. Table 14.3 presents a summary table for a correlated-groups ANOVA applied to these data. There are some differences in the summary table for an independent-groups versus a correlated-groups ANOVA. You can see that with the correlated-groups ANOVA, we can reject the null hypothesis and conclude that there are statistically significant differences in terms of recall among the seven word positions, after taking into account the variability within each participant.

According to Table 14.3, the F-ratio increased from $F(6, 21) = 2.17$ to $F(6, 18) = 6.63$. How did this change occur? For both ANOVA summary tables, the $MS_{\text{between}} = 3.74$. This number remains a constant for both forms of the ANOVA because we used MS_{between} as an estimate of the variance among the seven serial-position means. Because each ANOVA determines the between-groups variance using the same procedures, this value will not change for the two tests. What does change for the two forms of the ANOVA is the value of MS_{within}.

Earlier, we noted that the importance of the correlated-groups ANOVA is that it divides the variance within groups (remember this is a reflection of error) into two components: variance due to consistent differences within participants and

Table 14.3 Summary Table for a Correlated-Groups ANOVA Using Data in Table 14.1

	SS	*df*	*MS*	*F*	*p*
Between	22.43	6	3.74	6.63	0.00
Participants	26.11	3	8.70		
Within	10.14	18	0.56		
Total	58.68	27			

variance due to random error. In Table 14.3, you can see that there is a sum of squares (*SS*) for participants and a *SS* for within-groups variance. Adding $SS_{\text{participants}}$ and SS_{within} in Table 14.3 equals the SS_{within} of the independent-groups ANOVA reported in Table 14.2 (36.25 = 26.11 + 10.14). In other words, a sizable proportion of variance within each group reflects systematic differences among the participants. More concretely, some people are just better at recalling lists of words than others. When we divide the MS_{between} by the revised MS_{within}, we obtain a much larger *F*-ratio (6.63 = 3.7481/0.5635).

In this way, then, the correlated-groups ANOVA allows us to detect the systematic differences among the serial position of the words that the independent-groups ANOVA could not detect. This success occurred because the correlated-groups ANOVA removes from the calculation of the *F*-ratio the variability within groups due to consistent differences among participants. Hence, we can conclude that the correlated-groups ANOVA is more statistically powerful than the independent-groups design when there are consistent differences among the participants.

We do not need to calculate an *F*-ratio for the variance due to participants because our primary interest is the differences in the mean number of words recalled across the serial positions, not in the differences among the participants. You could, if you wanted, calculate the *F*-ratio by dividing the $MS_{\text{participants}}$ by the MS_{within}. The result would tell you what you had anticipated, that there are statistically significant differences among the participants. Because we want to treat the variance among participants as a nuisance variable that we control for, such findings are inconsequential.

Advantages and Disadvantages of a Repeated-Measures Design

The primary advantage of the repeated-measures design is the increase in statistical power we obtain by identifying the systematic variance caused by consistent differences within the participants. In effect, this increase in our ability to detect differences in participant responses or behaviors across different treatment conditions comes from allowing each participant to serve as his or her own control condition. We do this by testing the same participant under several treatment conditions or across time.

A related advantage of a repeated-measures design is that it requires a smaller overall number of participants because the same participants are evaluated under more than one treatment condition. This is primarily a resource and cost advantage, but when combined with the increased statistical power, it helps to make this type of research design extremely attractive when the research question permits.

Despite these general benefits, repeated measurement of the same participant can introduce unwanted effects in some cases. *Can you think of how repeated testing of the same participants might make it difficult for a researcher to draw clear and unbiased conclusions?* Consider a study for which we require the participants to complete the same test after receiving three separate treatments (each

Table 14.4 Common Carryover Effects That Can Confound a Repeated-Measures Design

Contrast	In some cases, the change in level of the independent variable creates a contrast effect. For example, a participant switched from 50 to 25 reward points will respond differently than a participant switched from 25 to 50 points or a participant maintained at 25 points.
Fatigue	The participants become tired from or bored by repeating the same task on multiple trials. The effects of fatigue and boredom occur when the individual has little opportunity to rest between trials.
Habituation/adaptation	Habituation and adaptation refer to participants becoming used to the effects of the independent variable. In some cases, the participants may develop a tolerance for the independent variable. For example, a stimulus that produces fear the first time will lose its effectiveness if presented repeatedly.
Learning	The participants learn how to perform the task during the first test. Consequently, their performance on subsequent tests reflects the effect of the independent variable and the effect of learning. This form of carryover effect is prevalent whenever one uses the same test or skill-based task.
Pretest sensitization	A pretest, given before the treatment, may alter the participant's reaction to the eventual treatment. For example, asking participants questions about racism or sexism may change how they respond to a treatment condition that causes them to confront these issues.

treatment being a level of the independent variable). Over time, the participants' performance on this test will change, though this change may not be entirely due to the effects of the independent variable; simply taking the same test multiple times is bound to improve the participants' test scores.

In a more general sense, the term **carryover effect** is used to describe situations in which experience in one treatment condition affects participants' performance in one or more subsequent conditions. Because carryover effects are typically not the focus of the research, they can confound any eventual interpretation of the results. Table 14.4 lists several of the more common carryover effects that can confound the interpretation of the results.

Reducing Carryover Effects

There are several research designs that adequately address the threat of a carryover effect. The first is a between-subjects design. In many cases, this may be your only real alternative to the within-subjects design that is the focus of this chapter, especially if your treatment conditions are likely to lead to irreversible changes in the participants. For example, if you were comparing the effects of invasive brain surgery on patients' ability to focus, you could not likely use a within-subjects design to compare the efficacy of different types of brain surgery. Once a patient experienced one of the surgery conditions, the effects of any further or additional surgeries would be confounded by the effects of the initial surgery.

A second option for reducing carryover effects is to use a technique known as the **Solomon four-group design**. This procedure is a creative use of multiple control groups that allows you to judge the extent to which there is a carryover effect. This procedure is effective for accounting for the effects described in Table 14.4. When the changes in a participant due to an independent variable *are* temporary or reversible, we can shuffle the order in which participants experience different levels or conditions of the independent variable. This shuffling of events will distribute any effects that are due to the order of condition exposure across all the participants. The goal of shuffling is to make the order of condition experience random so that it has no consistent effect on the participants' behavior.

Two other techniques for addressing carryover effects are **counterbalancing** and **Latin square designs**. A final option is to recognize that changes in the participant are an important independent variable to examine. From this perspective, an observed carryover effect is not a confounding variable but a central focus of the research question we are asking. In the following sections, we will examine each of these techniques for addressing carryover effects in more detail.

Solomon Four-Group Design

Solomon (1949) developed a creative research design (presented in Table 14.5) that incorporates three different types of control groups of participants to account for possible carryover or sequence events. This research method design is viewed as a superior one for addressing potential carryover effects (e.g., Campbell & Stanley, 1963).

For the Solomon design, only groups 1 and 3 actually experience the treatment condition. The difference between these groups is that group 1 receives a pretest and a posttest; group 3 receives only the posttest. Groups 2 and 4 represent special control conditions that do not experience the treatment condition. Group 2 acts as a control condition for group 1; the participants in both groups complete both the pretest and posttest, but only participants in group 1 experience the treatment. Similarly, group 4 acts as a control group for group 3; the participants in both groups complete the posttest, but only participants in group 3 receive the treatment.

Figure 14.4 presents an idealized graph of results from a study that employs a Solomon four-group design. As illustrated in this figure, the participants who

Table 14.5 The Solomon Four-Group Design

	Pretest		Treatment		Posttest
Group 1	O_1	⟶	Treatment	⟶	O_2
Group 2	O_3	⟶	⟶	⟶	O_4
Group 3			Treatment	⟶	O_5
Group 4					O_6

Note: O_x represents an observation or test score.

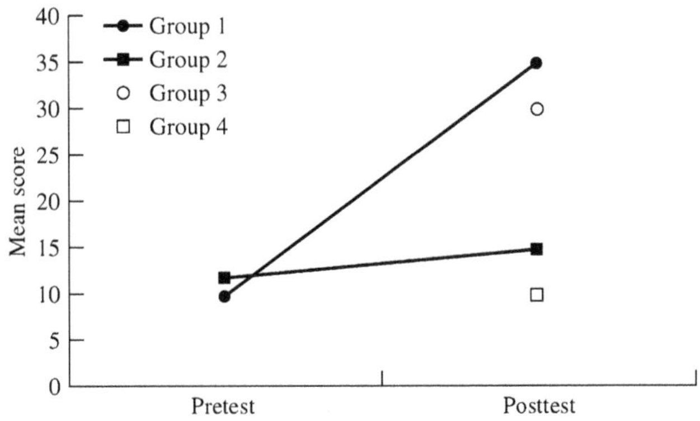

Figure 14.4 Idealized results of a Solomon four-group design.

Table 14.6 The 2 × 2 Factorial Analysis of the Solomon Four-Group Design

	Factor A Treatment	
	a_1 Treatment	a_2 No treatment
b_1 Pretest	O_2	O_4
Factor B: Pretesting		
b_2 No pretest	O_5	O_6

experienced the treatment (those in groups 1 and 3) show much higher average posttest scores than the participants in the control groups (groups 2 and 4). There is also evidence of a carryover effect in that the posttest scores for groups 1 and 2 are higher than the posttest scores of groups 3 and 4 (the groups that did not have a pretest). The strength of this type of design, however, is that it allows the researcher to identify whether a possible carryover effect exists, but also whether there is evidence of a significant effect of the independent variable even if a carryover effect exists—in the present illustration, for example, the conclusion would be that the independent variable is impacting the dependent variable.

Solomon (1949) recommended that researchers use a simple 2 × 2 ANOVA factorial design to analyze the data from this type of design. Table 14.6 presents the layout of such a design. Braver and Braver (1988) noted that the Solomon four-group design is underused and frequently misinterpreted. Contrary to popular opinion, Braver and Braver demonstrated that the Solomon design does not require twice the number of participants even though it has more control groups. They demonstrated that the Solomon design can be a cost-effective research technique if used with the proper statistical analyses. In summary, the Solomon four-group design is a sophisticated research tool that offers considerable methodological and statistical control over carryover effects. It is a research tool often overlooked for little good reason.

Counterbalancing

We can also reduce the presence of carryover effects by randomly changing the sequence of testing for each participant. Researchers call this shuffling procedure counterbalancing. For example, if you wanted to test the effects of three possible orders of treatments for a particular type of cancer on a single set of patients, you could shuffle the order in which each treatment is administered to each patient. More specifically, one patient would receive treatment 1 followed by treatment 2, and then treatment 3. Another patient might receive the treatments in the 3, 2, then 1 order, and so on, randomly mixing the order of treatment administration.

The number of potential patterns for complete counterbalancing is $k!$, where k represents the levels of the independent variable to be tested (! represents the factorial of a number, so, e.g., $2! = 2 \times 1 = 2$). For three levels of the independent variable, there are 3! or $3 \times 2 \times 1 = 6$ patterns. Because there are six patterns, you would need enough participants for the study to ensure that you could use each treatment pattern the same number of times. In other words, if there are three treatment conditions, the number of participants in the experiment must be some multiple of six to ensure equal representation of each counterbalance sequence in the final set of data.

Latin Square Design

Complete counterbalancing can be logistically challenging and expensive to conduct because of the number of participants required. Imagine conducting an experiment that used five levels of the independent variable. A completely counterbalanced version of this study would require at least 120 (i.e., 5!) participants to fill out all the sequence conditions with one participant per condition (and only having one person in each condition is not generally considered a very strong sampling strategy).

An alternative is to use a Latin square design, which is a collection of counterbalanced treatment sequences that ensures each treatment condition occurs once within each possible position of the ordering or sequence of conditions. The Latin square also ensures that the sequence of treatments is random, thus controlling for unwanted sequence events.

Perhaps seeing how one constructs a Latin square will illustrate the value of the procedure. Table 14.7 is a summary of the steps needed to generate a Latin square plan for a research study involving a single independent variable with any number of levels. Once you have designed a Latin square for your research study, you can then randomly assign equal numbers of participants to each treatment order condition specified by the rows of the Latin square. The Latin square is a

Table 14.7 Steps to Create a Latin Square for Any Number of Levels of the Independent Variable

	1	2	3	4	5	
1						Step 1: Create a square of *k* rows and *k* columns, where *k* represents the number of levels of the independent variable. For this example, there are five levels of the independent variable.
2						
3						
4						
5						
	1	2	3	4	5	Step 2: Starting with row 1, write the numbers 1–5. For each subsequent row, start the line with the number of the line and then continue the sequence of numbers until you reach *k*, and then restart the numbering at 1.
1	1	2	3	4	5	
2	2	3	4	5	1	
3	3	4	5	1	2	
4	4	5	1	2	3	
5	5	1	2	3	4	
	4	1	5	3	2	Step 3: Using a random number table, randomly shuffle the order of the columns.
1	4	1	5	3	2	
2	5	2	1	4	3	
3	1	3	2	5	4	
4	2	4	3	1	5	
5	3	5	4	2	1	
	4	1	5	3	2	Step 4: Using a random number table, randomly shuffle the order of the rows. The numbers within the gird represent the Latin square. Randomly assign equal numbers of participants to each of the five sequences you generated.
3	1	3	2	5	4	
1	4	1	5	3	2	
4	2	4	3	1	5	
5	3	5	4	2	1	
2	5	2	1	4	3	

useful design tool to overcome most carryover effects. The analysis of data from this type of design is, however, complex and beyond the scope of this textbook. Fortunately, advanced textbooks on statistics, such as those by Hayes (1981), Kirk (1982), and Winer, Brown, and Michels (1991), all review the statistical procedures needed for analyzing data gathered with a Latin square design.

Examining the Effects of a Sequence

Sometimes a carryover, order, or sequence effect is of interest and not simply a confounding variable. Consider, for example, Hoeffer et al.'s (2006) effort to improve the person-centered caregiving of certified nursing assistants (CNAs) assisting nursing home residents with bathing. To control for possible sequence or order effects (which could be caused by the order in which the intervention segments were presented), CNAs were randomly assigned to one of three conditions: (1) *group 1* participants (control condition) received no education but were

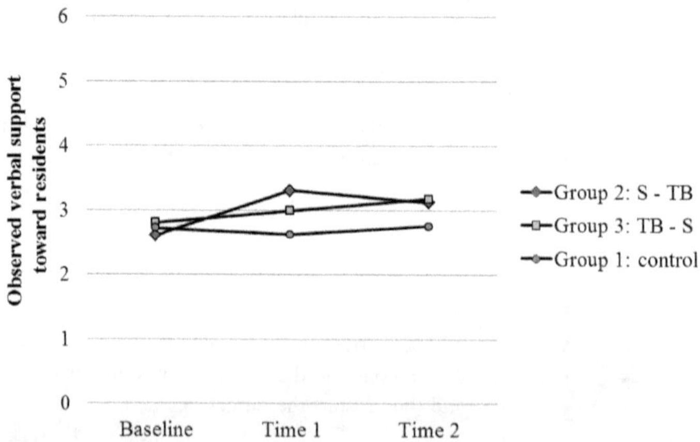

Figure 14.5 Change in verbal support toward nursing home residents for CNAs receiving person-centered training.

assessed at the same time points as the participants who were receiving person-centered care education; (2) *group 2* participants were taught about person-centered showering care during weeks 1–6 and then person-centered towel bathing care during weeks 7–12; and (3) *group 3* participants were taught about person-centered towel bathing care during weeks 1–6 and then person-centered showering care during weeks 7–12. The outcome variable in this study was quality of caregiving, which was operationalized in a variety of ways. For the present example, we will focus on the results pertaining to trained research assistants' observation-based ratings of verbal support provided by CNAs to residents when providing bathing assistance at before (baseline), during (time 1), and at the end (time 2) of the study.

Hoeffer et al. (2006) demonstrated that their educational intervention led to improvements in the quality of caregiving behaviors toward nursing home residents during bath time. Figure 14.5 illustrates the results of this study with respect to one of the multiple caregiving outcomes, demonstrated verbal support.

As you can see in Figure 14.5, participating CNAs demonstrated improved levels of verbal support at time 1 and time 2 compared with baseline, and as compared with the control group of CNAs who did not receive any person-centered training. In this particular instance, the effects of the training differed slightly depending on whether the CNAs were educated about showering first and towel bathing second (group 2) or about towel bathing first and showering second (group 3). This is evidence of a sequence effect, although the impact of this effect in the present example is minimal because the overall implication is that regardless of the order in which CNAs were trained, the person-centered training did leave to an improvement in the observe quality of their verbal support toward residents. *What would the implications of this type of research be, however, if you noticed that CNAs who received towel bath training before showering training demonstrated significantly higher levels of verbal support toward residents than CNAs who received showering training before towel bath training?*

KNOWLEDGE CHECK

1. Describe in your words the meaning and purpose of the matched-groups and repeated-measures designs.
2. What are the similarities and differences between a matched-participants design and a repeated-measures design?
3. What are common carryover effects?
4. Describe in your words how the Solomon four-group design helps us contend with carryover effects.
5. In a repeated-measures research design, how does each participant act as his or her own control group?
6. Why does a correlated-groups design increase the power of the ANOVA?
7. When is the use of counterbalancing and the Latin square design necessary?

CROSS-SECTIONAL SEQUENTIAL DESIGNS

Many researchers study the transitions that people make as they age. As the general population in America and other countries continues to grow older and more diverse, the effect of retirement on quality of life during one's later years is an increasingly important public health issue. As an example, a researcher may want to examine changes in perceived quality of life among retirees as they move from the moment of retirement into later years postretirement. Using a conventional longitudinal design would require years to collect the data for the study, a considerable test of one's patience and persistence. The **cross-sectional sequential design** is an alternative method that allows the researcher to examine developmental transitions but over a shorter period of time.

The researcher can select cross sections of the age span that he or she wants to study. Starting in year 1, for example, the researcher could select participants age 50 through 60. Over the course of the next 5 years, the researcher can continue to monitor changes in the reported quality of life experienced by these participants. By the end of year 5, the researcher will have data that indicate differences in quality of life across a wide range of ages and how those changes relate to the passage of time since the point of retirement. The cross-sectional sequential design provides the researcher with much information.

First, the researcher can use the data to compare the differences among blocks of ages. Analyzing these data will indicate whether there are systematic differences between these age groups. Second, the cross-sectional sequential design allows the researcher to examine the changes observed in workers as they grow older.

MATCHED-GROUPS DESIGN

The repeated-measures design allows researchers to examine or test participants' behavior under more than one condition. Although this procedure has its advantages, there are some shortcomings related to carryover effects, as we have already discussed. Perhaps the most serious problem with carryover effects arises when the exposure to the treatment (or independent variable of any form) creates an irreversible change in the participants or their behavior. In this section, we present another form of correlated-groups research design that retains the statistical power of the repeated-measures design but allows researchers to test each participant under only one treatment condition. This design is known as a matched-groups design.

Researchers use the matched-groups design when they believe that an important subject variable may correlate with the dependent variable. The goal of the matched-groups design is to ensure that the participants assigned to the different treatment groups are equivalent with respect to this subject variable. Because each group in the study contains a separate group of participants, standard repeated-measures procedures cannot be used. Nevertheless, the matched-groups design allows us to assume that there is a meaningful correlation among pairs of participants across the different treatment conditions. This sampling-based design tactic makes it possible for researchers to identify and remove variance in the dependent variable that is due to underlying subject variable(s) that might influence the participant's behaviors or reactions to the independent variable that is the focus of the study.

Table 14.8 presents the steps for creating a matched-groups design. First, test each participant using a measure of some subject variable that you believe or know to be related with the dependent variable (e.g., underlying ability, intelligence, personality traits). Second, using these pretest scores, rank the participants

Table 14.8 Steps Used to Create a Matched-Groups Design Study

Step 1		Step 2		Step 3				Step 4		
Pretest all participants using a measure correlated with the dependent variable.		Rank order participants from highest to lowest based on test score.		Group participants in clusters equal to the number of treatment groups.				Randomly assign participants, by cluster, to treatment groups.		
Participant	Score	Participant	Score							
A	114	A	114						**Group**	
B	101	G	111					**1**	**2**	**3**
C	104	F	109							
D	105	I	106	**1**	A	G	F	A	F	G
E	92	D	105							
F	109	C	104	**2**	I	D	C	C	I	D
G	111	H	103							
H	103	B	101	**3**	H	B	E	E	H	B
I	106	E	92							

from highest to lowest score. Third, cluster the participants into groups based on their pretest score ranking. The size of the cluster must equal the number of treatment conditions used in the study. Finally, randomly assign the participants within each cluster to one of the treatment conditions. What is the result of all this effort?

In the example presented in Table 14.8, the first participant in each group earned one of the highest three scores on the test. Similarly, the second participant in each group earned the next highest level of test scores. This pattern continues for all the participants within each group. Therefore, we can assume that ordered pairs of participants (e.g., the fifth participant in each group) are similar to each other inasmuch as they each received similar scores on the original test.

Some researchers call the matched-groups design a *randomized block design*. The term *block* refers to agricultural experiments conducted during the early 1900s. A block was a specific plot of land that the researcher randomly selected to receive a specific treatment. Although behavioral researchers often use both of these phrases to describe this type of design, we prefer *matched-groups design* as it highlights the researcher's intention to equate the participants across different treatment groups in terms of some underlying subject variable.

We can use a hypothetical example of an experiment designed to test the efficacy of a nutritional intervention for middle-aged men at risk for developing diabetes. For this example, assume that the researcher will examine the effects of a brief printed handout, a group-based nutrition consultation, and a one-on-one nutritional coaching session with participants. Because the researcher believes that socioeconomic status (SES) will influence the dependent variable or outcome for this study (i.e., intention to adhere to the nutritional guidance), the researcher gathers information about each potential participant's annual household income in advance of the study and matches these individuals across the three nutrition "treatment" conditions according to this proxy indicator of SES. This is accomplished by identifying the three participants with the highest household incomes and randomly assigning each to one of the three treatment groups. Repeating this procedure for all participants, moving down the rank-ordered list based on annual household incomes ensures the equivalence among the groups with respect to the underlying SES.

There is no reason to limit participant matching to one subject variable with this type of design. In some research projects, researchers find it necessary to match participants across a number of subject variables, including age, sex, level of education, income, and other quantitative and qualitative variables that differentiate individuals and correlate with the dependent variable. A caveat is in order, however: Using more than one matching variable can greatly increase the difficulty of achieving a true balance among the participants.

In this type of design, the matching of participants in terms of underlying characteristics that are associated with the dependent variable means that the groups are not fully independent. As you know by now, *independent* in this case means that the behavior of one group of participants has no influence on the behavior of another group of participants. In a typical between-subjects design, participants are randomly assigned to independent treatment groups, which are then later compared to one another. In that type of design, the behavior of one group of participants is unlikely to be related in any systematic way to the behavior of participants in another treatment condition. In a matched-groups design, however, the researcher intentionally creates a correlation among different treatment groups by matching participants on underlying subject variables and then partialling the influence of these variables on the dependent variable out when the subsequent data analysis is performed. This is why this type of design is a correlated-groups design similar to repeated-measures designs.

Advantages and Disadvantages of the Matched-Groups Design

There is an advantage to the matched-groups design: Any difference among the different treatment groups is easier to detect than would be the case in a standard between-subjects design. As you have already learned in this chapter, within-subjects and correlated-groups designs result in reduced MS_{within} terms in the resulting ANOVA. However, a potential disadvantage to using the matched-groups design is low statistical power. The correlated-groups ANOVA treats participants' scores across the treatment conditions as if they came from the same participants. The result is that the ANOVA based on data from this type of design assumes there are fewer participants in the study than is actually the case. In other words, the degrees of freedom for a correlated-groups design are smaller than for the comparable independent-groups design. This is evident in Table 14.2 and Table 14.3. The degrees of freedom are (6, 21) for the independent-groups design and (6, 18) for the correlated-groups design. In general, ANOVAs based on smaller degrees of freedom require larger F-values to reject the null hypothesis. Consequently, if the matching technique does not effectively identify a relevant subject variable, the statistical power of a matched-groups analysis test can actually be lower than the same analysis for a comparable between-subjects design. Therefore, the advantage of a matched-groups design occurs only when the subject variable(s) used for participant matching are at least moderately correlated with the dependent variable.

KNOWLEDGE CHECK

8. How will using a matched-groups design increase the power of the ANOVA?
9. In what ways are the matched-groups design and the **yoked-control design** similar and different from each other?
10. Describe how a yoked-control design works and how it ensures that participants in the experimental and control groups receive the same treatment except for the independent variable.
11. How can you distinguish between an independent-groups and a correlated-groups design?

RESEARCH IN ACTION: MIXED-MODEL DESIGN

In addition to the purely within-subjects, correlated-groups type of designs detailed so far in this chapter, it is also possible to mix independent variables that are between- and within-subjects in nature. In this "Research in Action" extension, we present an example of a **mixed-model design**, which gets its name from the combination of at least one between-subjects and one within-subjects independent variable.

Consider a simple experiment to demonstrate how to use a mixed-model ANOVA. A student wants to replicate a classic experiment conducted by Wickens, Born, and Allen (1963) that examined the effects of proactive interference. Proactive interference occurs when you attempt to learn new information and previously learned information interferes with the new material. As you read through the example presented next, consider the implications of proactive interference for patient recognition of heavily marketed and advertised prescription medications. Research has found that repetition improves brand name recall. However, when competitive advertising is presented (especially in the same product category), it has been found that repetition provided no improvement in brand name recall over a single exposure and recall of brand relevant information is lower overall (Unnava & Sirdeshmukh, 1994).

In this hypothetical student's replication of the Wickens et al. (1963) study, participants complete a series of trials. On each trial, the researcher presents the participant with a short list of words related to a marketing campaign. After seeing the words, the participant must then recall the words in the list. For the experimental and control groups, to which participants are randomly assigned, the first three trials use the same category of words. On the fourth trial, the researcher changes the meaning of the words for the experimental group (e.g., present an ad for an unrelated product) but leaves the meaning of the words for the control group unchanged. Table 14.9 presents the percentage correct for the two groups across the four trials and Figure 14.6 presents a graph of the group means.

As a review, let us examine several components of this experiment. First, the researcher is using a mixed-model design. The researcher randomly assigned participants to either the experimental or the control group, making that first independent variable a between-subjects variable (i.e., participants are tested under one of the two conditions or levels of that variable). The second independent variable, however, is the sequence of trials. Because the researcher tested each participant over multiple trials, this independent variable is a within-subjects variable. As such, the researcher assumes that there will be a correlation between the treatment groups that are formed based on this independent variable because each participant is tested across all of the different trial conditions.

During the first three trials, the participants' performance in both groups decreased, as is predicted by the proactive interference phenomenon. The data for the fourth trial are interesting. The performance of the participants in the control group continued to decline, while the performance of the participants in the experimental group improved.

Table 14.9 Hypothetical Data for an Experiment Examining Release from the Proactive Interference Effect

Experimental	1	82	53	48	63	
	2	92	51	39	60	
	3	63	41	10	52	
	4	75	57	51	72	
	5	78	49	43	77	
		$M_{11} = 78.0$	$M_{12} = 50.2$	$M_{13} = 38.2$	$M_{14} = 64.8$	$M_{a1} = 57.80$
Control	6	84	60	47	33	
	7	85	44	42	46	
	8	78	71	34	45	
	9	76	49	38	25	
	10	63	69	55	30	
		$M_{21} = 77.2$	$M_{22} = 58.6$	$M_{23} = 43.2$	$M_{24} = 35.8$	$M_{a2} = 53.60$
		$M_{b1} = 77.6$	$M_{b2} = 54.4$	$M_{b3} = 40.7$	$M_{b4} = 50.3$	$\bar{M} = 55.75$

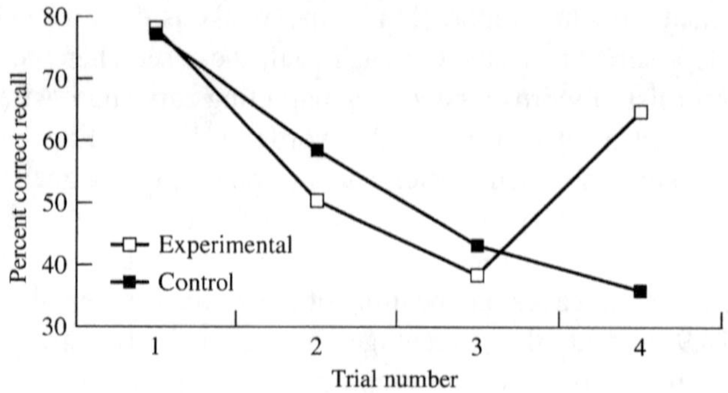

Figure 14.6 Graph of the data presented in Table 14.9.

Table 14.10 presents the ANOVA summary table for the data presented in Table 14.8. There is a statistically significant interaction between the two independent variables ($A \times B$) and significant main effects for the between-subjects factor (experimental vs. control) and the within-subjects factor (trial).

As with the other forms of the ANOVA, we can follow the statistically significant *F*-ratio with additional tests, including post hoc tests of the main effects and interaction and measure of effect size. These follow-up tests will make it easier to provide a clearer interpretation of the findings. For example, a closer inspection of the pattern of means within the groups across the four trials indicates that participants in the experimental group performed just like the participants in the control group until trial 4. Switching the meaning of the words in the fourth trial undid the effects of the proactive interference.

Table 14.10 The ANOVA Summary Table for a Mixed-Model ANOVA

Source	*SS*	*df*	*MS*	*F*
A (experimental vs. control)	168.10	1	168.10	0.84**
Error *A*	1,597.40	8	199.68	
B (trial)	7,354.50	3	2,451.50	30.13**
$A \times B$	2,174.90	3	724.97	8.91**
Error (trial)	1,952.60	24		
Total	13,247.50	39		

$^{**}p < .01$ (reject H_0).

CHAPTER SUMMARY

In this chapter, we examined a type of research design that allows us to increase the power of our research by identifying and removing variance due to differences among the participants. Specifically, we examined several forms of correlated-groups designs that leverage the power of within-subjects independent variables.

In one of these forms, the matched-groups design, the researcher identifies a subject variable that he or she believes influences the dependent variable of interest and then assigns the participants to the treatment conditions in such a way that each group has the same proportion of people who are high and low on the subject variable.

In another type of correlated-groups design, the repeated-measures design, the researcher tests the same participants under a series of different treatment conditions. As a generality, the matched-groups and repeated-measures designs can increase the power of the statistical test because the corresponding ANOVA identifies systematic variance among the subjects and removes this variance from the common error term. Therefore, the designs make it easier to detect the differences among the various groups.

The repeated-measures design does have a potential disadvantage, however. In some cases, there may be a significant carryover effect that will bias or confound our interpretation of the data. In some cases, the carryover effect can be overcome using a Latin square design. In other cases, the carryover effect may be the focus of the study.

The ANOVA for the correlated-groups design is similar to the ANOVA for between-subjects designs. The notable difference is that the correlated-groups ANOVA estimates the variance due to the differences among the participants. This strategy can reduce the common error term and then increase the probability that the differences among the groups will be detected.

Another version of the within-subjects design allows us to mix between-subjects and within-groups variables. A between-subjects variable is one where the participants in one treatment condition are different from the participants in the other treatment conditions. A within-subjects variable is one where there is an intended correlation among the groups. This correlation occurs when we use either a matched-groups or a repeated-measures design.

CHAPTER GLOSSARY FOR REVIEW

Between-Subjects Variable An independent variable that represents different groups of participants. Each level of the variable consists of a different group of participants.

Carryover Effect A form of confounding variable that can occur when using a repeated-measures design. Exposing the participant to one level of the independent variable can affect the participant's performance when tested under other conditions.

Cohort Effect A systematic difference between different cohorts due to unique experiences of the cohorts.

Counterbalancing A technique to reduce the carryover or sequencing effects in a repeated-measures design. The researcher tests the participants under the identified treatment conditions; however, the order of the exposure is different. Specialized tactics for counterbalancing include Latin square designs.

Cross-Sectional Sequential Design A type of longitudinal design for which the researcher selects participants that represent a cross section of ages. The researcher then monitors the relevant variables across a number of subsequent times.

Latin Square Design A special form of counterbalancing. The procedure balances the sequence of treatment conditions such that the order of treatments is balanced across all participants.

Longitudinal Design A procedure that allows the researcher to monitor the behavior of participants across a long period.

Matched-Groups Design A form of correlated-groups research design in which the researcher matches participants for common characteristics and then assigns them to one of the treatment conditions. The procedure ensures that the groups will be similar regarding the relevant subject variables.

Mixed-Model Design A research design that contains both a between-subjects variable and a within-subjects variable.

Repeated-Measures Design A form of correlated-groups research design that allows the researcher to test the same participants under different treatment conditions or on different occasions.

Solomon Four-Group Design This procedure is a creative use of control groups that allows you to judge the extent to which there is a carryover effect.

Within-Subjects Variable An independent variable that represents either a repeated-measures condition or a matched-groups design.

Yoked-Control Design A form of control procedure for which the researcher randomly pairs participants; one member of the pair is assigned to the experimental group and the other to the control condition. The participant in the control condition experiences the exact same sequence of events as the experimental partner but does not experience the critical feature of the experiment.

REFERENCES

Braver, M.C.W. & Braver, S.L. (1988). Statistical treatment of the Solomon four-group design: A meta-analytic approach. *Psychological Bulletin, 104*, 150–154.

Campbell, D.T. & Stanley, J.C. (1963). *Experimental and quasi-experimental designs for research.* Chicago: Rand McNally.

Flynn, J.R. (1998). IQ gains over time: Toward finding the causes. In U. Neisser (Ed.), *The rising curve: Long-term gains in IQ and related measures* (pp. 25–66). Washington, DC: American Psychological Association.

Hayes, L.V. (1981). *Statistics*. New York: Holt, Rinehart & Winston.

Hoeffer, B., Talerico, K.A., Rasin, J., Mitchell, C.M., Stewart, B.J., McKenzie, D., et al. (2006). Assisting cognitively impaired nursing home residents with bathing: Effects of two bathing interventions on caregiving. *The Gerontologist, 46*(4), 524–532.

Kirk, R.E. (1982). *Experimental design* (2nd ed.). Monterey, CA: Brooks/Cole.

Lewinsohn, P.M., Rohde, P., Seeley, J.R., & Fischer, S.A. (1993). Age-cohort changes in the lifetime occurrence of depression and other mental disorders. *Journal of Abnormal Psychology, 102*, 110–120.

Solomon, R.L. (1949). An extension of control group design. *Psychological Bulletin, 46*, 137–150.

Stice, E., Killen, J.D., Hayward, C., & Taylor, C.B. (1998). Age of onset for binge eating and purging during late adolescence: A 4-year analysis. *Journal of Abnormal Psychology, 4*, 671–675.

Unnava, H.R. & Sirdeshmukh, D. (1994). Reducing competitive add interference. *Journal of Marketing Research, 31*(3), 403–411.

WALKER, I. & HULME, C. (1999). Concrete words are easier to recall than abstract words: Evidence for a semantic contribution to short-term serial recall. *Journal of Experimental Psychology. Learning, Memory, and Cognition*, *25*, 1256–1271.

WICKENS, D.D., BORN, D.G., & ALLEN, C.K. (1963). Proactive inhibition and item similarity in short-term memory. *Journal of Verbal Learning and Verbal Behavior*, *2*, 440–445.

WINER, B.J., BROWN, D.R., & MICHELS, K.M. (1991). *Statistical principles in experimental design* (3rd ed.). New York: McGraw-Hill.

Part Four

Special Research Designs

Chapter 15

Single-Participant Experiments, Longitudinal Studies, and Quasi-Experimental Designs

If we knew what we were doing, it wouldn't be called research.

—Albert Einstein

INTRODUCTION

After a cursory study of standard research guidelines for health sciences research, it is easy to conclude that the results of a research project can be valid only if the researcher used legions of participants randomly selected from the target population and randomly assigned to treatment conditions in a highly controlled experiment. Although it is true that much health sciences research relies on random assignment of large groups of participants to a set of well-defined treatment groups or conditions, it is also true that researchers in the health sciences conduct many fine and valid experiments using single or very small groups of participants. Similarly, many researchers conduct revealing studies for which random assignment of participants to treatment conditions is impossible or unethical. Alternatives to

Understanding and Conducting Research in the Health Sciences, First Edition. Christopher J. L. Cunningham, Bart L. Weathington, and David J. Pittenger.

the typical true experiment design offer researchers valuable tools that allow them to address interesting, albeit complex, questions about human behavior. In this chapter, we review three important research paradigms: (1) single-participant experiments, (2) longitudinal designs, and (3) quasi experiments.

These methods for conducting research are extremely useful tools that all health science researchers can use depending on their specific research questions. Used correctly, these research techniques allow one to examine the relation among variables. Furthermore, these techniques adhere to the same rigorous standards of objectivity and validity as the other research designs we have reviewed in this book. Thus, each of these paradigms offers researchers analytical tools to help them understand various behavioral phenomena.

Single-Participant Experiments

The features that set these methods apart from other research designs are the methods used to collect and analyze the data. For instance, the single-participant experimental paradigm, as the name implies, is a collection of research techniques used to conduct controlled experiments that might, for instance, focus on changes in the health of individuals. Whereas much research examines the average condition of groups of participants, the single-participant research examines how the individual responds to the treatment conditions.

Longitudinal Designs

A longitudinal design also allows researchers to examine complex research questions from a different perspective by making it possible to track a group of individuals over time. Longitudinal designs are a special case of the repeated-measures procedure as the researcher tracks changes in the participants' health over time. There is an important difference between longitudinal designs and repeated-measures research. In most cases, longitudinal designs use a time frame that covers months and even years. By contrast, most repeated-measures research examines behavior over a much shorter period. So, the difference between the repeated-measures and longitudinal design is the emphasis and time frame of the research.

In the repeated-measures design, we typically want to know how individuals perform under a series of experimental conditions that we control. We may test the same participants using several different dosages of a drug to determine its effect on a behavior. In another experiment, we may expose one group of participants to different sequences of treatments to find how the pattern of treatment improves the participants' health. We may also be interested in observing an individual's behavior over time.

Indeed, for a longitudinal study, the researcher is more interested in developments or changes that occur over time. Consider the Nun Study, a longitudinal study of aging started in 1986 to follow a group of nuns over time. The women in this program have been routinely tested for the effects of aging on mental conditions such as Alzheimer's disease (Tyas, Snowdon, Desrosiers, Riley, & Markesbery, 2007).

Quasi Experiments

The quasi-experiment paradigm is unique because it does not use random assignment of participants. Quasi-experimental methods have evolved as a way to examine behavior under conditions where meeting the requirements of a true experiment is impossible or impractical. The name *quasi experiment* represents that the researcher may have partial control over the independent variable. For most quasi-experimental research, the researcher cannot randomly assign participants to the research conditions. However, quasi-experimental designs can yield interesting insights into human behavior.

Before we dig into each of these alternative research designs more deeply, it is important to emphasize that there is no such thing as a *best* method for studying any phenomenon. As you learn more about research, you will discover that health science researchers use many different techniques to study the same issues and questions. Their selection of techniques is the result of their training, experiences, and perspectives regarding the best way to conduct a specific research project. Knowing about these three types of designs can help to add more possible tools to your research method toolbox, giving you more options for your own use in the future as a researcher or consumer of research.

SINGLE-PARTICIPANT EXPERIMENTS

As the name implies, single-participant experiments (often known as *case studies*) examine how individuals respond to changes in the independent variable. Another term you will often find used interchangeably with *single-participant research* is *small n research* (i.e., research that includes only a small number of participants). While we will use the term *single-participant* research most often during this discussion, keep in mind that the information and techniques we will discuss in this chapter are equally applicable to studies utilizing only one participant or a small group of participants.

Many of the single-participant research designs were initiated by psychological researchers looking for ways to use the scientific method to examine the behavior of individual participants. As a set, these techniques are often referred to as methods for the **experimental analysis of behavior**.

When to Use Single-Participant Procedures

Those who routinely use single-participant procedures believe that single-participant designs are most useful when the researcher wishes to understand the changes in an individual that correspond to changes in the individual's environment (Morgan & Morgan, 2001). This definition surely covers many research situations.

The primary goal of the single-participant design is to determine how specific changes in the environment or treatment of the individual affect that participant's behavior. For example, a psychiatrist might want to determine whether exposing a person to a particular therapy will reduce undesirable behaviors and increase desirable behaviors. In such an example, the researcher wants to show that the therapy causes the person's behavior to change, hopefully for the better. In one of the following examples, we will review how a team of researchers used a single-participant research design to study the effectiveness of a method of treating panic attacks.

There are three design features that enhance the likelihood of a researcher conducting a successful single-participant design. First, the researcher needs to have direct control over the independent variable. Thus, the single-participant design is like a true experiment in that the researcher can control significant elements of the participant's environment and determine when to add and remove the independent variable. Second, the researcher must be able to regularly measure or observe the participant's behavior. Monitoring the participant's behavior allows the researcher to determine how changes in the independent variable correspond to changes in the dependent behavior. Third, the researcher must be able to observe the participant across a significant span of time. For many single-participant studies, the researcher will examine the participant's behavior across many days, even weeks and months.

These conditions are not always attainable, however, so there are research situations in which the single-participant design is not appropriate. Most of these cases involve situations in which researchers are interested in defining a population or examining the differences between populations. For example, a researcher may be interested in the emotional development of children who attended preschool versus children who did not. For this type of research, we want to know whether there are systematic differences among the populations representing the two groups of children. In addition, the predictor variable (preschool vs. no preschool) is a variable that the researcher does not control.

Types of Single-Participant Experiments

There are two broad categories of single-participant experiments: **baseline studies** and **discrete trial studies**. As a generality, researchers use baseline studies to examine changes in an ongoing or continuous behavior. For example, a researcher may want to examine how a specific treatment technique changes the social skills of an autistic child. The researcher would perform a baseline study to track changes in the child's behavior before, during, and after the treatment. This research would allow the researcher to determine whether the treatment influences the child's social skills.

By contrast, a discrete trial study allows us to examine how a participant reacts to specific test conditions. A simple example would be to examine how a person responds to a specific medication or different dosages of a medication. By systematically testing different prescription treatments, the researcher can determine the best amount of that medication for treating a particular underlying condition.

Baseline studies examine changes in the frequency of the ongoing health of an individual. Consequently, all baseline studies begin by developing operational definitions of the behaviors to be measured. We want to develop a reliable method for observing the behavior or behaviors that we want to modify. Once we decide on the appropriate behavior to observe and the data recording strategy, we can begin by establishing a baseline.

The general format of a baseline design is first to establish a baseline of the individual's health and then to introduce the change or intervention into the individual's environment. If the intervention is effective, there should be a commensurate and consistent change in the individual's behavior. Specifically, we expose the same participant to different treatment conditions to examine how the independent variable influences the dependent variable. The baseline is, therefore, the control condition for the experiment as it represents the participant's condition in the absence of the treatment condition.

Establishing Cause and Effect

The rules for determining cause and effect are straightforward: Use a control condition, demonstrate that there is a relationship between the independent and dependent variables, and rule out alternative explanations. There are a number of specific techniques that we can use to achieve these goals using the single-participant experiment.

Establish a Baseline

The first step is to establish a baseline that we can use for a frame of reference. The goal of this step is to determine the steady state of the person's health, specifically the symptoms we want to modify. For example, if we wanted to examine the effectiveness of a social skills training technique for children diagnosed with autism, we would want to examine how often the child engages in specific types of social interactions with other children.

In some cases, the baseline represents a small sample of the individual's symptoms. A child with autism may spend little time playing with other children. Because the behavior is extremely consistent, the baseline may be short. For the autistic child, we may observe his or her behavior for three or four sessions to establish the baseline.

If the target condition is variable, then the baseline will need to be longer to establish a suitable pattern of the variability. Sometimes symptoms are cyclical as the condition increases and decreases. For such cases, the baseline will need to be longer to establish the normal pattern for the behavior. Having the longer baseline will help the researcher determine the extent to which changes in the behavior reflect the independent variable or the cyclical nature of the behavior. The baseline should contain at least three observations (Barlow & Hersen, 1984). There is no hard-and-fast rule to establish the length of the baseline, as highlighted by Sidman (1960), who advised his readers that the "rule to follow for the criterion will depend upon the phenomenon being investigated and upon the level of experimental control that can be maintained" (p. 258). Therefore, the researcher needs to exercise common sense to determine whether the baseline is a fair and stable representation of the individual's behavior.

Direct Replication

A single-participant experiment does not mean that there is only one participant in the study, merely that the researcher does not aggregate the behavior of many participants into a single index such as the arithmetic mean. Instead, the hallmark of the single-participant research design is that the researcher reports the behavior of each individual in the study. **Direct replication** means the researcher will use the same treatment procedures to examine the pattern of behavior for each of several single participants. If the treatment is effective, then there should be similar patterns of behavior change for each participant who is exposed to the treatment. Therefore, the single-participant experiment is like other true experiments in that the researcher examines the data of several individuals under highly controlled conditions. The difference, however, is that in most experiments, the researcher examines and reports the *average* performance of the participants. In contrast, the researcher using a single-participant design will report the individual behavior of each participant.

One of the benefits of the single-participant design is that researchers can use it to demonstrate the relationship between independent and dependent variables with both inter- and intraparticipant replication to demonstrate the relationship between the independent and dependent variables. The intraparticipant replication occurs when one uses ABAB or other reversal designs. By adding and removing the independent variable, we can establish the link between changes in the independent variable and changes in the dependent variable. If we use the same procedure on several participants, we will generate interparticipant replication. The interparticipant replication is, therefore, another way of describing direct replication.

Examine the Effects of the Intervention

There are a number of general techniques used in an experiment to study the effects of the intervention. The prototypes for these techniques are the **ABA** and the **ABAB designs**. Researchers call these designs **reversal designs** because the researchers use several phases in the research during which they repeatedly add and remove a treatment condition. As a generality, the "A" phase represents the baseline or no-treatment condition and the "B" represents the intervention or treatment-present condition. The goal of the reversal design is to demonstrate that as the research conditions change (A to B and B to A), there is a corresponding shift in the individual's condition.

Potential Problems with Reversal Designs

In many cases, once we change a person's condition, it may be impossible to undo the change with the removal of the treatment. We must also consider the ethics of removing an intervention once we show that it is effective. Why would we want to remove something that we now believe improves the individual's quality of life? In most clinical research, once the researcher finds that the treatment is effective, the researcher will begin to administer the treatment to all the participants receiving the placebo or in the baseline phase of the study. When these conditions arise, it is hard for the researcher to have absolute proof of a cause-and-effect relation between the treatment and the outcome. As you might imagine, there are techniques to help the researcher demonstrate the underlying effectiveness of an intervention. The following is a brief review of several research techniques that overcome the problems created in a reversal design.

Multiple-Baseline Design

A suitable alternative to the reversal design is the **multiple-baseline design**. The advantage of this procedure is that it allows us to demonstrate the effectiveness of an intervention when the reversal design is not appropriate. In addition, the procedure allows us to examine the effectiveness of an intervention across behaviors, individuals, or settings.

The primary characteristic of the multiple-baseline design is that we monitor the ongoing behavior of the participants and then systematically introduce the independent variable at different times. One of the more popular of these procedures is the *multiple baseline across participants*. To use this procedure, we begin by monitoring the behavior of a group of participants in a study. Once we establish the baseline, we introduce the treatment to one of the participants while the others remain in the baseline condition. After some time has elapsed, we introduce the treatment to the second participant while keeping the remaining participants in the baseline condition. The goal of this procedure is to introduce the independent variable to each participant on a staggered basis to demonstrate that the participant's behavior changes only after the researcher introduces the independent variable.

Figure 15.1 illustrates the application of the multiple-baseline-across-participants procedure. There are three participants. For each participant, we begin to track the behavior that we want to modify. Participant 1 is the first to experience the independent variable while the other participants remain in the baseline condition. Later, we introduce the independent variable into the environment of participant 2. Finally, we add the independent variable to the environment of participant 3. Using this procedure, we can be sure that the introduction of the independent variable, and not some confounding variable, caused the change in the participant's behavior.

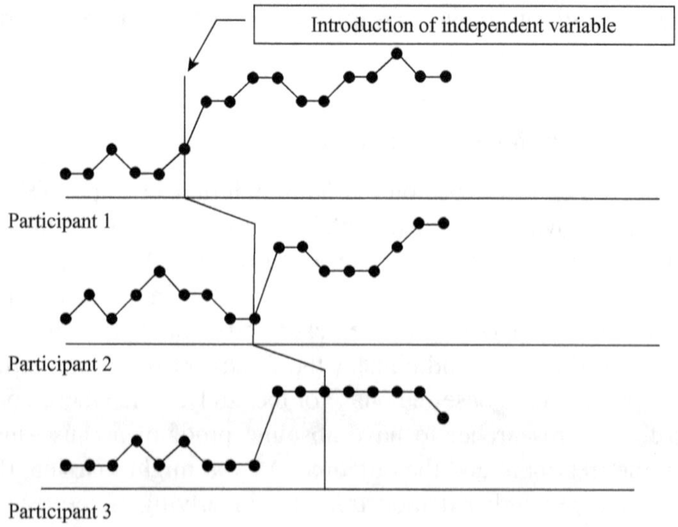

Figure 15.1 Multiple-baseline-across-participants procedure.

We can also apply the multiple-baseline procedure across situations and across behaviors. For the *multiple-baseline-across-behaviors* procedure, the researcher may identify several behaviors that he or she wants to modify with the same intervention. To use this technique, the researcher needs to ensure that the behaviors are independent of each other.

As an example, a researcher examining the effectiveness of an intervention with children diagnosed with autism may examine how the intervention works with three behaviors: (1) talking to other children, (2) participating in group activities, and (3) following the instructions of the teacher.

As with the multiple-baseline-across-individuals procedure, the researcher would collect baseline information for all three behaviors and then stagger the start of the intervention for each behavior. In theory, the intervention should influence the target behavior but not the other behaviors. The specific behavior will not change until the intervention begins.

RESEARCH IN ACTION: TREATMENT FOR PANIC DISORDER

Because we will review three research paradigms in this chapter, three separate "Research in Action" sections are also included, to illustrate the application of each paradigm. First, we will examine an application of the single-participant design.

Figure 15.2 presents an example of a single-participant experiment conducted by Laberge, Gauthier, Côté, Plamondon, and Cormier (1993). The researchers wanted to demonstrate the effectiveness of cognitive-behavior therapy for the treatment of panic disorder. The researchers examined the frequency of panic attacks of seven individuals using a multiple-baseline procedure.

The dependent variable is the frequency of panic attacks experienced by seven participants over a 35-week period. The vertical dashed lines represent the start

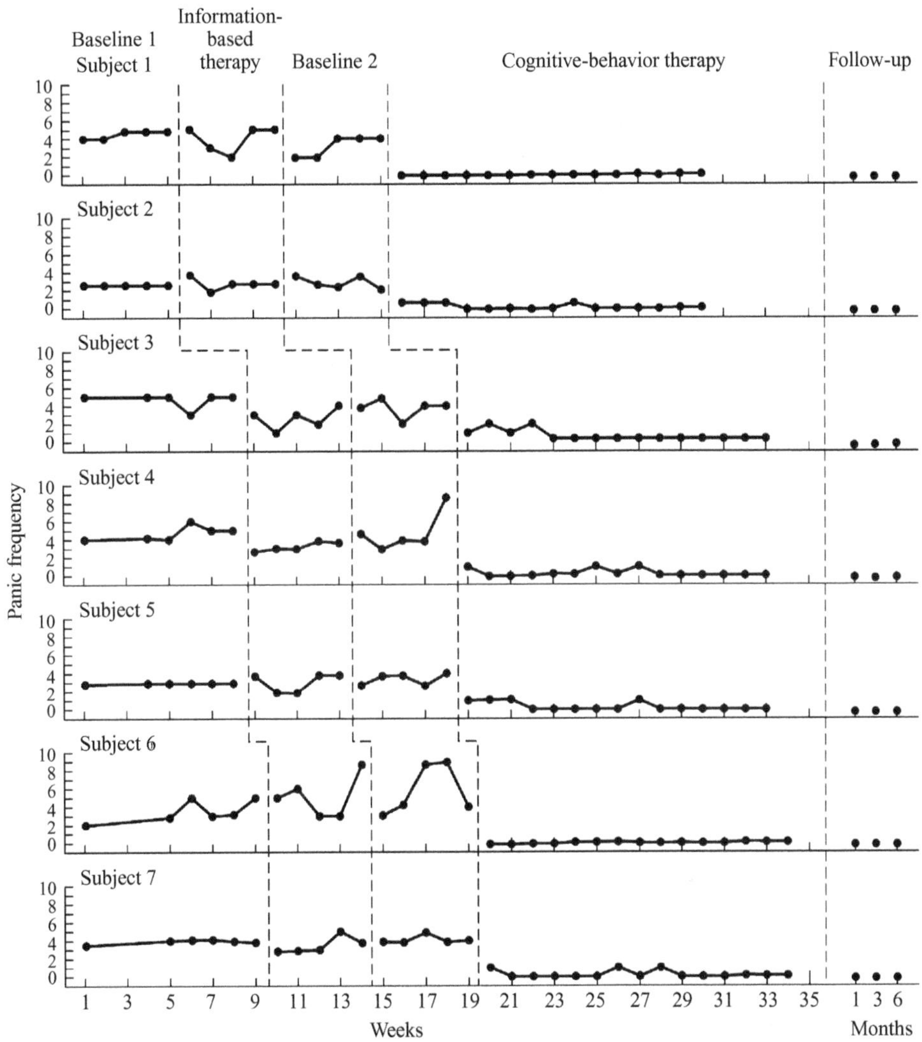

Figure 15.2 Multiple-baseline-across-participants design. *Note:* Data from Laberge et al. (1993).

of each phase of the experiment for the individual participants. During the first three phases of the experiment, the participants' panic attacks were assessed during three baseline phases. The fourth phase of the experiment included the cognitive-behavior modification treatment. The last three points represent a long-term follow-up of the patients.

This experiment also contained a slight methodological twist that helps demonstrate the effectiveness of cognitive-behavior therapy. During the first phase, the researchers monitored the frequency of panic attacks to establish a baseline. Then the researchers started a session called "information-based therapy." During this intervention, the participants received detailed information about panic attacks and learned how their thinking leads to panic attacks. Laberge et al. (1993) treated this phase of the experiment as a placebo treatment as they believed that it did not include the critical components of a legitimate treatment. Following the information-based therapy phase was another baseline period during which the researchers continued to monitor the frequency of the panic attacks. The first and second baseline conditions allowed the researchers to estimate the frequency of panic attacks while the person was under observation. The second control condition was the information-based therapy session. This condition allowed the researchers to determine whether the placebo treatment had any effect on the frequency of panic attacks.

The critical part of the experiment began when Laberge et al. (1993) initiated cognitive-behavior therapy to treat the participants for their panic attacks. The researchers staggered the start of each phase of the experiment after the initial baseline period across the individuals. By staggering the onset of the different intervention periods, the researchers demonstrated that the change in the panic attack coincided with the start of the legitimate treatment intervention (and not as a result of other factors that could have changed in the environments of the individuals). In this example, the researchers used direct replication within the multiple-baseline-across-participants procedure to demonstrate that the intervention brings about a change in the individual's behavior. The researchers also demonstrated the effectiveness of the cognitive behavior by including the placebo treatment. There were no discernible changes in the frequency of the panic attacks during the first three phases (baseline 1, information-based therapy, and baseline 2) of the experiment. The panic attacks stopped, however, when the participants received the cognitive-behavior therapy. Looking at the graph, it appears that the effectiveness of the intervention lasted well into the follow-up period.

One thing you may note about single-participant designs is the lack of formal statistical analysis. Many of the early pioneers of single-participant research designs believed that inferential statistical analysis was not appropriate for the analysis of individual participants and believed that the effect of the independent variable should be clear and unambiguous from visual inspection of the data (Morgan & Morgan, 2001). Therefore, most single-participant designs use simple figures or graphs and basic descriptive statistics to present the data and facilitate interpretation.

KNOWLEDGE CHECK

1. Explain why a reversal design, such as the ABAB design, produces results that support conclusions regarding cause and effect.
2. Imagine that two researchers examined the effectiveness of a new treatment intervention for depression. One researcher used a multiple-baseline-across-participants research design to examine the effect of the treatment on 12 participants. The other researcher randomly assigned 25 participants to a placebo treatment group and another 25 participants to the treatment condition and then compared the average level of depression between the

two groups. Assume that the researchers work at the same institution and sampled from the same subject pool.

a. Explain why the two studies are comparable with regard to internal validity.

b. Explain why the two studies are comparable with regard to external validity.

3. Using the previous example, what are the ethical considerations that would make one research design preferable to the other research design?

4. Imagine that your professor wants to demonstrate the effectiveness of homework for helping students learn research methods. How could he or she use the multiple-baseline-across-participants or across-situations procedures to conduct the research?

LONGITUDINAL DESIGNS

The primary goal of **longitudinal designs** is to examine how behavior, attitudes, or other outcomes change over time. Health science researchers have long used these techniques to examine the systematic changes in people's health.

The fundamental feature of a longitudinal design is relatively straightforward. The researcher identifies a sample of participants and then arranges to monitor some aspect or behavior of these participants over a predetermined period of time. For example, a cardiologist may be interested in the development of various ailments related to the heart by studying a diverse group of adults over an extended period. During that time, the researcher would periodically assess the particular aspect of heart function along with various lifestyle conditions that may affect heart health.

During the past two decades, health science researchers have become extremely interested in longitudinal research methods due to the development and application of new and advanced statistical methods for analyzing data from these types of studies. As you will soon learn, these new statistical tests make the longitudinal research design an extremely useful research tool. Before we describe these new statistics, we should first examine the foundations of the longitudinal design.

The conventional longitudinal design requires that the researcher track one or more groups of individuals over a specified time. The groups can represent an important subject variable (e.g., sex, intelligence, parent's marital status, or other categorical condition) or a treatment condition randomly assigned by the researcher. The researcher then arranges to periodically evaluate the members of the sample.

Although the conventional longitudinal design is popular and produces useful information, it does suffer from some shortcomings. First, the data may take a long time to obtain. Often this is an unavoidable problem. Tracking the progress of a group of patients who have just completed a program of treatment may require a significant follow-up period to evaluate the long-term effectiveness of the treatment.

Another problem related to the length of this type of study is the problem of participant dropout or attrition. Over time, it is common for individuals to leave the sample when they stop participating. These people may become bored with the study, move to another city or state, or otherwise become incapable of participating in the study. This problem of attrition can play havoc with many standard statistical tests such as analysis of variance (ANOVA). Fortunately, alternative statistical tools can overcome the problems created by participant dropout.

There are other problems related to the longitudinal research that are more difficult to resolve. One of these problems is the **cohort effect**. The cohort effect refers to differences among groups of people that share a common characteristic. For example, a cohort may be a group of people born at the same time and who had a common experience. A cohort effect occurs when members of one group are consistently different from people in a different group who have had different experiences.

There are more serious examples of a cohort effect. For example, Lewinsohn, Rohde, Seeley, and Fischer (1993) reported a large cohort effect for the prevalence of depression and other psychological disorders. Lewinsohn et al. found higher rates of depression among people born in more recent decades than in the past. The presence of a cohort effect forces researchers to examine the external validity of conclusions based on data from one cohort before they are generalized to others.

Survival Analysis

For many years, researchers used the repeated-measures ANOVA to analyze the data from longitudinal studies. Although this tactic yields useful information, it does have its limitations. First, the correlated-groups ANOVA requires data to meet specific statistical assumptions that are often difficult to achieve with real-world data. Failure to meet these requirements can produce invalid F-ratios. Second, the longitudinal design is subject to high levels of participant attrition. Losing large numbers of participants from a longitudinal research project can wreak havoc on the data analysis, especially with repeated measures ANOVA techniques, which are based on changes within persons over time. Fortunately, statisticians have developed a class of statistical tests that provide researchers an alternative method for analyzing the data generated by longitudinal designs (Greenhouse, Stangl, & Bromberg, 1989; Gruber, 1999; Lewinsohn et al., 1993; Luke & Homan, 1998; Singer & Willett, 1991; Willett & Singer, 1993). These statistical tests overcome the problems created by participants dropping out of the research and the failure of the data to meet the mathematical requirements of the ANOVA.

One example of these alternative analytic methods for longitudinal data is survival analysis. Within the health sciences, survival analysis is a common and popular statistical tool when the goal is to determine the amount of time it takes for an event to occur. For example, a physician may want to know which of two surgical procedures is the best for protecting the long-term health of patients. To collect the data, the researcher will track the patients after the surgery to determine which surgical technique is more effective at preventing a second heart attack or other significant problems. One characteristic that makes survival analysis unique is that it treats time as a dependent variable rather than as an independent variable.

This is in contrast to more conventional longitudinal analyses in which time is typically treated as an independent variable. When treated as such, the passage of time is often considered as a possible predictor of some outcome. In survival analysis, however, we treat the *time* between several events as the dependent variable. Perhaps an example will help illustrate how we can use survival analysis and the longitudinal design.

Stice, Killen, Hayward, and Taylor (1998) examined the onset of the binge-and-purge eating disorder in adolescent girls. The researchers identified a sample of 543 young women aged 13–17 and then used a comprehensive set of surveys and structured interviews to assess the participants' health and binge eating and purging behaviors. Stice et al. then reassessed the young women each year for the next 4 years. As the researchers noted, the average annual attrition rate was 15%.

Using the survival analysis procedures, Stice et al. found that the risk of the onset for an eating disorder begins after age 14, peaks at approximately 16, and then decreases. Other researchers have used the same techniques to examine the difference between men and women for the onset of depressive disorders and also for the risk of relapse among cocaine abusers following treatment (Willett & Singer, 1993). In all these examples, the critical questions are "How much time typically elapses between two critical events (e.g., treatment and relapse)?" and "If an event has not occurred after a specific interval, what is the probability that it will occur?"

The statistical analysis rules that underlie this approach are beyond the scope of this book and require intensive computational procedures best done by a computer. Fortunately, most professional statistical packages used by psychologists (e.g., BMDP®, SAS®, and SPSS®) include survival analysis routines. The primary message that we want you to appreciate is that these statistical tests are well suited for longitudinal data you might collect within a health sciences research project. They correct for dwindling sample size without compromising the validity of the interpretations of the results.

RESEARCH IN ACTION: ONSET OF BULIMIA

For this example of research in action, we can return to the study of the risk of bulimia among young women conducted byStice et al. (1998). The researchers began with a sample of 543 students from three high schools in California. The average age of the women was 14.9 years and ranged from 13 to 17. At the start of the study, none of the participants showed signs of bulimia.

The researchers then reassessed the participants once a year for the subsequent 3 years. The assessments consisted of a structured interview conducted by trained researchers who looked for symptoms of eating disorders, especially bulimia. As with most research of this type, some of the participants did not complete the study. Approximately 10% of the original sample moved to another school, and others were absent during the assessment period or chose not to complete the remaining interviews.

Stice et al. (1998) then examined the risk of the women developing bulimia between the ages of 13 (the youngest age of the participants) through 19 (the oldest age by the end of the 3 years of the study). For most statistical tests, the researchers would have to throw out all the data of the participants who did not complete the study because many statistical tests such as correlation, regression, and ANOVA require a complete record of observations for each participant. Survival analysis does not make this requirement, however, as the statistic uses the available data for each wave of measurement to determine the proportion of participants who show evidence of the syndrome.

Even though some women dropped out of the study, we can use their data to make inferences about the presence of bulimia. Consider, for example, a participant who is 16 at the start of the study and who is available for the first and second measurement sessions but drops out before the third session. Although she dropped out of the study, she still provided the researchers with 5 years of data. Recall that at the start of the study, none of the participants showed evidence of bulimia. Therefore, we can assume that the young woman was symptom-free when she was 13, 14, 15, and 16.

Using this information, Stice et al. (1998) were able to determine the age when the greatest proportion of women began the binge-and-purge eating pattern associated with bulimia. Because the survival analysis uses *proportions* (proportion = number of people with condition/total in current sample), the statistic is not affected by loss of participants from the sample. Figure 15.3 presents an example of the data reported by Stice et al.

As Stice et al. (1998) noted, their project was the first to examine the onset of bulimia in a population of young women. Having these data allow researchers to understand the forces associated with the onset of bulimia. Perhaps more importantly, the data can help counselors and others working with adolescents to find

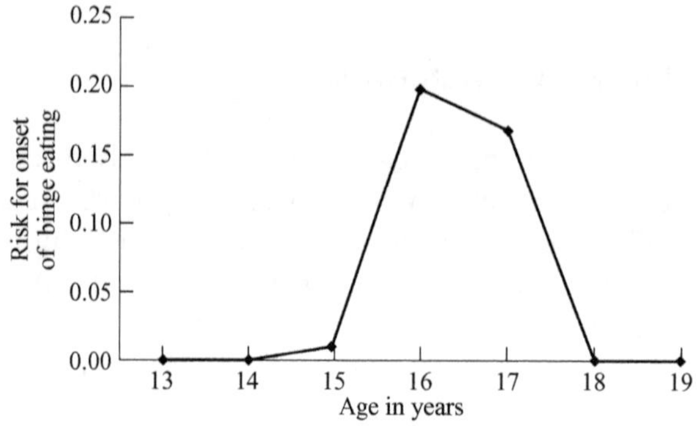

Figure 15.3 Risk of onset of binge eating.

the optimal time to introduce prevention programs and to look for risk factors that predict the onset of bulimia and other eating disorders.

QUASI EXPERIMENTS

In some circumstances, a true experiment is either impractical or unethical. We may not be able to randomly assign participants to different treatment conditions, although we can randomly select one group of participants to receive a treatment while the other group acts as a control group. As Harris and his colleagues (2006) noted, quasi experiments offer health science researchers powerful tools to examine various health science phenomena when the application of traditional experimental techniques is not appropriate. In this section of the chapter, we will introduce you to two of the more common quasi-experimental techniques, the nonequivalent control-group and the interrupted time series designs.

The term *quasi experiment* literally means that the research resembles an experiment. In a quasi experiment, there is an independent variable and a dependent variable, and the researcher wants to determine whether the introduction of the independent variable produces a change in the dependent variable. In addition, there is a control group of individuals who do not experience the independent variable. Consequently, the researcher can compare the differences in outcome between the treatment and control groups. The missing component from this research is the random assignment of participants to the treatment conditions.

The primary disadvantage of the quasi experiment is that the lack of random assignment of participants to treatment conditions means that we can never fully rule out alternative explanations for any effects of the independent variable(s) that we observe. The primary advantage of the quasi experiment is that we can collect useful information that will allow us to examine the effects of the independent variable on the dependent variable. The quasi experiment is another useful tool that the researcher can select when the circumstances of the research prohibit the conventional true experiment.

Nonequivalent Control-Group Design

The nonequivalent control-group design is a common example of the quasi-experimental design. The researcher begins with two groups created by situations beyond his or her control. As an example, a researcher believes that a new triage procedure will increase the quality of treatment for patients coming to one of two hospitals in a large city. The researcher could receive the cooperation of one hospital to implement the new triage procedure while the other hospital follows its standard procedures. The researcher can then review the consequence of the new routine by comparing the overall quality of care offered by the two hospitals. In a quasi experiment, we must consider the groups to be nonequivalent because we cannot be sure whether factors unrelated to the study create systematic differences between the participants.

Although the researcher cannot randomly assign the participants to the treatment conditions, he or she can randomly determine which group will receive the treatment. The main feature of the design includes two observations for both groups. The researcher will assess both groups at the same time, introduce the intervention for the experimental group, and then observe the relevant dependent variable a second time.

Although the nonequivalent control-group design is a popular design that controls for some threats to internal validity (Campbell & Stanley, 1966), there are other threats to internal validity that it cannot directly withstand (West, Biesanz, & Pitts, 2000). The design is subject to threats due to history, regression to the mean, and instrumentation.

History

It is possible that one of the groups will experience an event or change in the environment, unrelated to the purpose of the study, which confounds the analysis and interpretation of the data. For example, a researcher may wish to study the effects of a binge-drinking prevention campaign for college students. After finding two colleges equivalent in size and student body, the researcher randomly selects one of the colleges as the experimental group and the other as a control group. During the course of the study, the researcher cannot control for outside factors that may change one of the campuses. A student at one of the colleges may die due to binge drinking. This tragic and uncontrolled event may affect the other students' attitudes toward binge drinking. The event also confounds the results as we do not know whether changes in student attitude represent the treatment program or the student's death.

Regression to the Mean

Regression to the mean may be a confounding factor if the researcher selected the experimental and control groups using a test or measure that had low reliability. Consider again the researcher studying drinking behavior at colleges. It is also possible that the time during the semester when the researcher collected the data could confound the results. For example, collecting the data just before or just after midterms may cause the researcher to underestimate or overestimate the amount of drinking at the college. Any changes observed may merely reflect regression to the mean.

Instrumentation

Another threat to the internal validity of a nonequivalent control-group design is instrumentation. During the course of the study, it is possible that there will be a change in the methods or measures used to collect data from one group, but not the other. Consider the binge-drinking study. The researcher may require that the student life staff keep track of the number of episodes or problems related to drinking. It is possible that the staff of the colleges become more selective in what they are willing to ignore and what they document. Thus, the changes in drinking patterns may reflect changes in the way the staff at one of the schools collects and reports the data.

Interrupted Time Series Design

The interrupted time series is much like an AB design used in single-participant research. The researcher makes repeated measures of the frequency or rate of behavior in a sample of participants before and after a critical event. These data can come from many sources. As with most quasi experiments, the researcher may not be able to control when the critical event occurs but can use its presence to examine interesting aspects of human behavior.

Hawton et al. (1999) reported the results of an interrupted time series that examined the effects of a television program on suicide attempts. The researchers learned that the medical television drama *Casualty* (a popular television show in England) would include an episode in which one of the characters dies of an overdose of the drug paracetamol. Hawton and his associates contacted the emergency rooms of 49 hospitals in England and asked them to monitor the number of cases of self-administered overdoses of paracetamol and other substances. The hospitals tracked the frequency of drug overdoses for 3 weeks before the episode aired and for 3 weeks after the broadcast.

Figure 15.4 presents the percent change in the number of reported overdoses for paracetamol after the broadcast of the episode of *Casualty*. Hawton et al. (1999) found that there was a statistically significant increase in all forms of self-administered overdoses. The cases of paracetamol overdoses were greatly elevated and continued at higher rates for 2 weeks. Data such as these might lead one to conclude that the television show led some of the viewers to mimic the characters in the drama.

The simple time series study suffers from the same problems of internal validity we faced with the simple AB research design. In the present example, we cannot be sure whether the television show produced the changes in drug

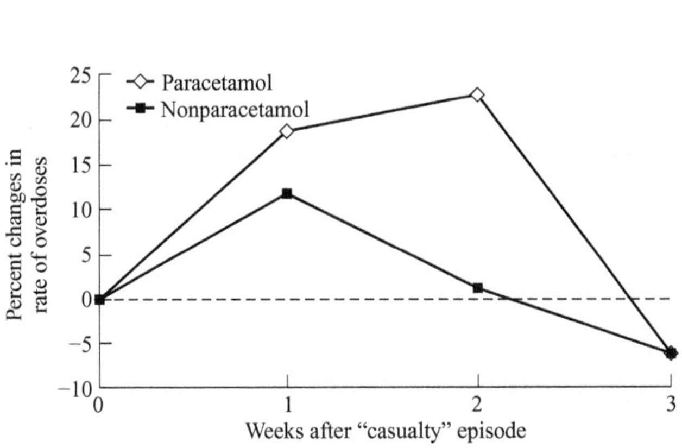

Figure 15.4 Percent change in the rate of overdoses presented in hospitals after a particular episode of the British television show *Casualty*.
Note: Based on data presented by Hawton et al. (1999).

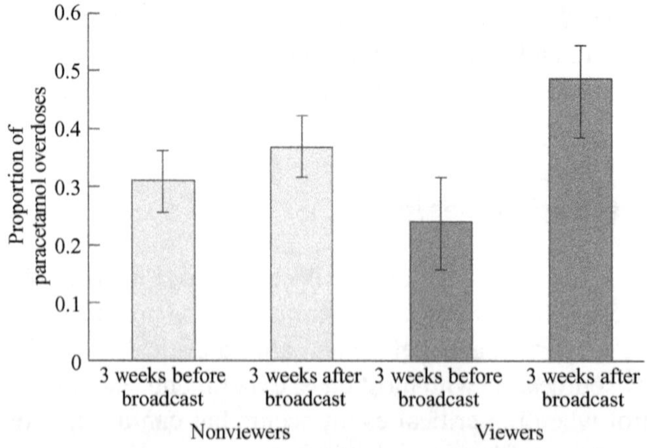

Figure 15.5 Proportion of self-administered overdoses of paracetamol that occurred 3 weeks before and 3 weeks after the broadcast of an episode of the television drama *Casualty*.

overdoses or if other events, unrelated to the show, contributed to the effect. Therefore, researchers hope to find instances where nature provides the equivalent to the controlled experiment. *Can you think of some alternative design options?*

One option would be to find two groups for which only one experiences the critical event. Hawton et al. (1999) were able to include this control in their research. They designed their questionnaire to include questions asking patients whether they had seen the particular episode of *Casualty*. Using this information, the researchers were able to compare the rates of self-poisonings for nonviewers versus viewers of *Casualty*. The results of this analysis help to reinforce the hypothesis that the television episode contributed to the increase in drug overdoses. Figure 15.5 presents the results of this analysis. Among nonviewers of the program, there was no statistically significant difference in the proportion of paracetamol overdoses over time following the broadcasting of the show. Among viewers of the program, however, there was a marked increase in paracetamol overdoses in the weeks following the *Casualty* episode. Evidence such as this reinforces the conclusion that a television show may influence rates of drug overdoses.

For each bar in the figure, the vertical lines represent the 95% confidence interval associated with the proportion of viewers and nonviewers who overdosed before and after the airing of the *Casualty* episode.

As with the single-participant design, the creative researcher can use many variants of the interrupted time series to examine the relation between two variables. Cook and Campbell (1979) provide a comprehensive review of the more subtle and specialized methods along with a review of the statistical procedures used to study interrupted time series data.

RESEARCH IN ACTION: TRAFFIC LAWS AND SAFETY

For this example of applied research, we will examine a study conducted by West, Hepworth, McCall, and Reich (1989). The researchers wanted to study the effects

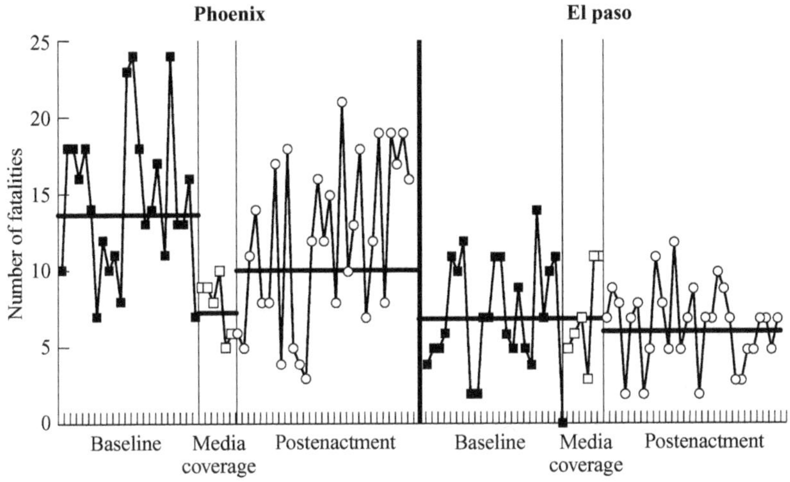

Figure 15.6 Number of traffic fatalities reported in Phoenix, Arizona, and El Paso, Texas.

of a drunk-driving law that Arizona had adopted during the 1980s. In 1982, Arizona instituted a drunk-driving law that required those convicted of driving while intoxicated to receive jail sentences. The law required first-time offenders to be jailed for at least 24 hours, fined $250, and have their driver's license temporarily suspended. Those who supported the law argued that the benefit of the law would be, among other things, a decrease in traffic fatalities (West et al., 1989).

To examine the effects of the new law, West et al. (1989) reviewed the monthly police reports for the city of Phoenix, Arizona. Recognizing that there are cyclical patterns to traffic fatalities, they reviewed the records 30 months prior to the enactment of the law and 30 months after the enactment of the law. West and his colleagues also collected the same information from the El Paso police department. El Paso is a city in Texas that has much in common with Phoenix. The primary difference between the two cities is that Texas had not recently passed any drunk-driving legislation. Thus, this study combines the features of an interrupted time series and a quasi-control group.

Figure 15.6 presents the number of reported fatalities for the two cities. Two characteristics of this graph require explanation. First, both graphs contain a segment labeled "media coverage." These months represent the time before the enactment of the law that the Phoenix newspaper *Arizona Republic* reported extensively about the pending law. Second, the horizontal lines within the graph represent the average reported fatalities during the period.

The baseline and media coverage phases represent the time before the enactment of the new and stricter drunk-driving law in Arizona. The media coverage phase represents the time during which there was intense media coverage in a Phoenix newspaper. The thick horizontal lines represent the average monthly fatalities for each period (baseline, media coverage, and postenactment).

Looking at the graphs, it is apparent that the new law did affect Arizona drivers. Prior to enforcement of the law, there were, on average, 14 traffic fatalities each month. During the months prior to the enforcement of the new law, when the media coverage was at its highest, the fatality rate decreased to seven deaths per month. Over time, the average number of fatalities per month in Phoenix actually increased slightly to 10. In contrast, the number of traffic fatalities for El Paso remained relatively constant during the changes in Phoenix, at just under seven deaths per month.

KNOWLEDGE CHECK

5. In what ways are repeated-measures designs and longitudinal designs similar to and different from each other?
6. Explain how time can be a dependent variable.
7. Describe the similarities between an interrupted time series study and a multiple-baseline study.

CHAPTER SUMMARY

The research designs considered in this chapter are alternatives to the conventional experiment. Although single-participant, longitudinal, and quasi experiments are empirical forms of analysis, they differ from the research designs considered in the earlier chapters.

The distinguishing feature of the single-participant experiment is the attention paid to the individual participant. Rather than average the responses of all the participants for subsequent statistical analysis, researchers who use single-participant research designs focus their analysis on the consistency of behavior observed across individuals. We revisited the issue of internal and external validity for the single-participant research design. Single-participant research designs can afford substantial internal validity and external validity that compares favorably to other research projects.

The typical research designs for the single-participant baseline experiment are the ABA and ABAB reversal designs and the multiple-baseline experiment. In the reversal design, the researcher observes the participant's behavior during a baseline period and then examines how additions to and removal of the intervention influence behavior. The multiple-baseline experiment allows the researcher to stagger the onset of the intervention across participants, behaviors, or situations.

Longitudinal research designs allow researchers to track changes in behavior over a long period. An alternative to the longitudinal design that avoids the cohort effect is the cross-sectional sequence procedure. This design allows the researcher to examine the differences among participants of different ages, potential cohort effects, and the effects of growing older.

One of the more recent developments in the analysis of longitudinal research is survival analysis. Survival analysis is a collection of statistical techniques that allows the researcher to examine changes in the participant's behavior over time. The statistical techniques allow researchers to overcome problems with participant dropout and other events that may make statistical tests such as the ANOVA inappropriate.

The quasi experiment represents a method of conducting research used when the researcher cannot randomly assign participants to the treatment conditions. Two of the more popular quasi experiments are the nonequivalent control group and the interrupted time series analysis.

CHAPTER GLOSSARY FOR REVIEW

Baseline Study A form of single-participant research design in which the researcher monitors the changes in an ongoing behavior of the participant during changes in the participant's environment.

Cohort Effect A systematic difference between different cohorts due to unique experiences of the cohorts.

Direct Replication A tactic used in single-participant experiments to demonstrate that the intervention brought about changes in the behavior. Specifically, the researcher exposes several individuals to the same treatment conditions to determine whether the intervention has the same general effect on all the participants.

Discrete Trial Study A form of single-participant research design in which the researcher assesses the participant's reaction to a specific event or stimulus.

Experimental Analysis of Behavior A specific form of single-participant research that emphasizes the analysis of changes in an individual's behavior that result from controlled changes in the environment.

Longitudinal Design A procedure that allows the researcher to monitor the behavior of participants across a long period.

Multiple-Baseline Design A research tactic used in single-participant baseline studies wherein reversal procedures (e.g., the ABAB) design is not appropriate. The researcher staggers the start of the intervention across participants, behaviors, or situations to determine whether the change in behavior occurs with the change in the environment.

Reversal Design (ABA and ABAB Designs) A form of baseline study for which the researcher examines the participant's behavior when the intervention is added to and removed from the environment.

REFERENCES

BARLOW, D.H. & HERSEN, M. (1984). *Single case experimental designs: Strategies for studying behavior change* (2nd ed.). Boston: Allyn & Bacon.

CAMPBELL, D.T. & STANLEY, J.C. (1966). *Experimental and quasi-experimental designs for research.* Chicago: Rand McNally.

COOK, T.D., & CAMPBELL, D.T. (1979). *Quasi-experimentation: design and analysis issues for field settings.* Rand McNally.

GREENHOUSE, J.B., STANGL, D., & BROMBERG, J. (1989). An introduction to survival analysis: Statistical methods for analysis of clinical trial data. *Journal of Consulting and Clinical Psychology, 57,* 536–544.

GRUBER, F.A. (1999). Tutorial: Survival analysis—A statistic for clinical, efficacy, and theoretical applications. *Journal of Speech, Language, and Hearing Research, 42,* 432–447.

HARRIS, A.D., MCGREGOR, J.C., PERENCEVICH, E.N., FURUNO, J.P., ZHU, J., PETERSON, D.E., et al. (2006). The use and interpretation of quasi-experimental studies in medical informatics. *Journal of the American Medical Informatics Association, 13*(1), 16–23.

HAWTON, K., SIMKIN, S., DEEKS, J.J., O'CONNOR, S., KEEN, A., ALTMAN, D.G., et al. (1999). Effects of a drug overdose in a television drama on presentations to hospital for self-poisoning: Time series and questionnaire study. *British Medical Journal, 318,* 972–977.

LABERGE, B., GAUTHIER, J.G., CÔTÉ, G., PLAMONDON, J., & CORMIER, H.J. (1993). Cognitive behavioral therapy of panic disorder with secondary major depression: A preliminary investigation. *Journal of Consulting and Clinical Psychology, 61,* 1028–1037.

LEWINSOHN, P.M., ROHDE, P., SEELEY, J.R., & FISCHER, S.A. (1993). Age-cohort changes in the lifetime occurrence of depression and other mental disorders. *Journal of Abnormal Psychology, 102,* 110–120.

LUKE, D.A. & HOMAN, S.M. (1998). Time and change: Using survival analysis in clinical assessment and treatment evaluation. *Psychological Assessment, 10,* 360–378.

MORGAN, D.L. & MORGAN, R.K. (2001). Single-participant research design bringing science to managed care. *American Psychologist, 56*, 119–127.

SIDMAN, M. (1960). *Tactics of scientific research.* New York: Basic Books.

SINGER, J.D. & WILLETT, J.B. (1991). Modeling the days of our lives: Using survival analysis when designing and analyzing longitudinal studies of duration and timing of events. *Psychological Bulletin, 110*, 268–290.

STICE, E., KILLEN, J.D., HAYWARD, C., & TAYLOR, C.B. (1998). Age of onset for binge eating and purging during late adolescence: A 4-year analysis. *Journal of Abnormal Psychology, 4*, 671–675.

TYAS, S.L., SNOWDON, D.A., DESROSIERS, M.F., RILEY, K.P., & MARKESBERY, W.R. (2007). Healthy ageing in the Nun Study: Definition and neuropathologic correlates. *Age and Ageing, 36*(6), 650–655.

WEST, S.G., BIESANZ, J.C., & PITTS, S.C. (2000). Causal inference and generalization in field settings. In H.T. Reis & C.M. Judd (Eds.), *Handbook of research methods in social and personality psychology* (pp. 40–84). Cambridge, UK: Cambridge University Press.

WEST, S.G., HEPWORTH, J.T., MCCALL, M.A., & REICH, J.W. (1989). An evaluation of Arizona's July 1982 drunk driving law: Effect on the city of Phoenix. *Journal of Applied Social Psychology, 19*, 1212–1237.

WILLETT, J.B. & SINGER, J.D. (1993). Investigating onset, cessation, relapse, and recovery: Why you should, and how you can, use discrete-time survival analysis to examine event occurrence. *Journal of Consulting and Clinical Psychology, 61*, 952–965.

Chapter 16

Research with Categorical Data

INTRODUCTION

Throughout this book, we have examined research designs that are appropriate when the dependent variable is measured in terms of an interval or ratio scale. Although much behavioral research relies on interval and ratio scales, there are also many important applications of nominal and ordinal scales in social and behavioral research, especially within the health sciences. For example, it is often the case that we may want to know how many people fit into specific categories or groupings. Research along these lines is focused on the frequency of different events rather than the average responses or ratings of a group of participants. Because data that reflects frequencies (e.g., counts) do not conform to standard parameters associated with most other traditional statistics, we will need a different set of nonparametric research design and analysis tools. This chapter focuses on research designs that use categorical or nominal data such as frequencies. Before getting too far into this, some background information is needed.

Remember that nominal and ordinal measurement scales consist of a set of mutually exclusive classes. **Mutually exclusive** means that a person or a measurement fits only one classification when these types of measurement scales are used. For example, biological sex is a mutually exclusive classification system because any person is either male or female. Political party is also a mutually exclusive

Understanding and Conducting Research in the Health Sciences, First Edition. Christopher J. L. Cunningham, Bart L. Weathington, and David J. Pittenger.

system—most states require you to register as a Democrat, Republican, or some other political party.

The independent variable for research involving categorical data is often represented in terms of a nominal or an ordinal scale. A nominal scale is typically used to represent a construct that consists of readily identified groups. As indicated above, one's sex and political party are examples of nominal scales in which there is no implied order in the different scale options. An ordinal scale, however, suggests that there is a continuum underlying the scale options. For example, most colleges use the terms *freshman*, *sophomore*, *junior*, and *senior* to identify students' class standing or seniority. These groups are mutually exclusive in that a student can be in only one level. The groups also represent an ordinal scale in that we presume that the classification scheme represents the number of courses or credit hours the student completed (i.e., seniors are expected to have more education completed than freshmen).

The dependent variable for categorical research is typically the number of people who (or number of observations that) fit within each of the predefined independent variable categories. For some research, we may use a single independent variable with many classification levels. As an example, the nursing coordinator for a rural hospital that is part of a larger health system may want to understand whether the staff model throughout the system is appropriate for the needs at this rural hospital. To figure this out, this coordinator can compare the observed number of emergency department (ED) admissions in this hospital over the last 12 months against what is seen on average through the overall health system on a month-to-month basis. If there is a noticeable difference between these two values, then it is possible that a modified nurse staffing model would be needed for this rural hospital. The independent variable in this applied research example is time (measured in terms of month), while the dependent variable is the number of ED admissions within each month.

It is also often necessary to study the relationships among multiple independent variables that are categorical in nature and dependent variables that are frequency based. As an example of this type of research, consider an industry political lobbyist who needs to understand whether there is a relationship between a politician's political party (Democrat vs. Republican) and his/her support for new legislation to mandate healthcare coverage for all citizens (favor legislation vs. oppose legislation). Such a relationship is summarized in Table 16.1. Both

Table 16.1 Example of Categorical Research Examining Two Variables

		Party affiliation	
		Democrat	Republican
Attitude toward legislation	Favor	103	98
	Oppose	97	102

variables, political party and attitude toward the legislation, are mutually exclusive categories. The dependent variable is the number of politicians who fall into each of the four quadrants of this table.

As with all research, the reliability and validity of the measurement tools used by researchers are essential to the quality of the research project. The same rule applies to research involving categorical data. Researchers who use categorical data place considerable emphasis on creating clear and useful definitions of the categories used in the research. As you recall from Chapter 8, there are many ways to prepare clear definitions of our terms and to verify the reliability of those terms. Research with categorical data has two distinguishing features: the data we collect and the statistical tests we use to analyze the data.

In the previous chapters, we studied research projects that examine the typical or average performance of people within specific treatment conditions. Comparing averages makes sense when the data collected from each participant reflect relatively continuous information that can range from low to high on a given scale. When the data are categorical in nature, however, averaging is not a legitimate statistical option. As reviewed in the following sections of this chapter, a much more appropriate statistical treatment of categorical data typically involves the use of some form of **chi-square (χ^2)** analysis. Karl Pearson, the statistician who popularized the correlation coefficient (r), also is credited with developing the χ^2 statistic. The χ^2 is an inferential statistic that allows us to determine whether the number of observations within mutually exclusive classes differs significantly from hypothetical conditions which typically reflect our null hypothesis expectations in a particular research situation. That definition may not make sense to you yet, but we think it becomes clearer once you work through the following sections and examples.

GOODNESS-OF-FIT TEST

The purpose of the χ^2 **goodness-of-fit test** is to determine whether the frequencies of different mutually exclusive categories match or fit a hypothesized population. Returning to an example mentioned in the introduction, imagine you are a nursing coordinator at a rural hospital and you believe that the staffing model you have been asked to use for coordinating nurse schedules is inappropriate for the reality of your situation. More specifically, you are concerned that more nurses are needed during certain months than others but that the current model does not accurately address this need. You know that a critical factor in determining nurse staffing levels within the health system that operates your hospital is the number of ED admissions each month. You decide to compare the actual number of ED admissions at your hospital against the average number of admissions per month at hospitals throughout the rest of your health system. You pull admission records for a sample of 600 unique ED admissions (i.e., all different people) over the past 12 months and you record the month of ED admission for each case. Figure 16.1 summarizes these data.

Month		Observed frequency
January	O_1	26
February	O_2	41
March	O_3	36
April	O_4	41
May	O_5	62
June	O_6	75
July	O_7	60
August	O_8	67
September	O_9	58
October	O_{10}	52
November	O_{11}	41
December	O_{12}	41
TOTAL (T) =		600

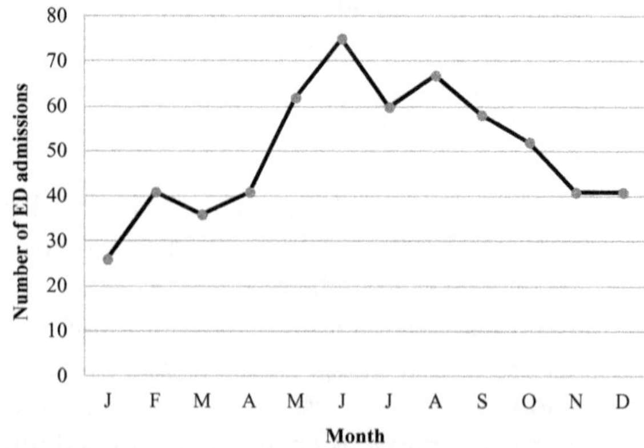

Figure 16.1 Hypothetical number of ED admissions per month over the last year.

The main research question here is whether the number of observed ED admissions in each month within your hospital is similar to what is expected by the health system that owns your hospital. This is an important question for you as a hospital administrator, as the answer may impact your staffing strategies going forward. To test this hypothesis, a χ^2 statistic can be used. To calculate this statistic, we use O_i to indicate the observed frequencies for ED admissions each month within your hospital. The symbol T represents the total number of all observations. Looking at the data in this example, it is clear that June stands out as a high ED admissions month, whereas January does not.

A χ^2 analysis also requires you to have knowledge of a set of expected frequencies (E_i), against which you will compare your observed frequencies. Table 16.2 presents the O_i along with these E_i. The column listed "expected proportions" represents hypothetical national ED admission rates throughout the health system of which your hospital is a part. According to this information, approximately 4% of ED admissions occur in January, whereas 12% of ED admissions occur in June. Using these proportions, we can estimate how many ED admissions should be expected in your hospital, if the pattern of ED admissions in your hospital matches or fits with what is seen throughout the health system. These calculated E_i are summarized in Table 16.2.

According to these data, we would expect 24 ED admissions in January and 72 ED admissions in June for any random sample of 600 marriages. Looking at Table 16.2, you can see that there are slight differences between the O_i from within your hospital and the E_i based on the national norms. For example, your hospital saw more ED admissions than expected in June and July and fewer ED admissions than expected in November. When these types of differences are identified, the question that arises is whether they represent random variation or are indicative of something that is not likely due to chance (i.e., differences that are statistically significant)? To answer that question, we would use a χ^2 goodness-of-fit test, which involves using this formula:

Table 16.2 Extension of the Data Presented in Figure 16.1

Month	Observed frequency	Expected proportions	Expected frequency = $p \times T$
January	$O_1 = 26$	0.04	$E_1 = .04 \times 600 = 24$
February	$O_2 = 41$	0.07	$E_2 = .07 \times 600 = 42$
March	$O_3 = 36$	0.06	$E_3 = .06 \times 600 = 36$
April	$O_4 = 41$	0.07	$E_4 = .07 \times 600 = 42$
May	$O_5 = 62$	0.10	$E_5 = .10 \times 600 = 60$
June	$O_6 = 75$	0.12	$E_6 = .12 \times 600 = 72$
July	$O_7 = 60$	0.10	$E_7 = .10 \times 600 = 60$
August	$O_8 = 67$	0.11	$E_8 = .11 \times 600 = 66$
September	$O_9 = 58$	0.10	$E_9 = .10 \times 600 = 60$
October	$O_{10} = 52$	0.09	$E_{10} = .09 \times 600 = 54$
November	$O_{11} = 41$	0.08	$E_{11} = .08 \times 600 = 48$
December	$O_{12} = 41$	0.06	$E_{12} = .06 \times 600 = 36$
Totals	600	1.00	600

Note: The proportions represent data based on national data. The expected values (E_i) equal the hypothetical proportion (p) multiplied by the total number of observations (T).

$$\chi^2 = \sum \frac{(O_i - E_i)^2}{E_i} \tag{16.1}$$

Equation 16.1 defines the basic χ^2 statistic. The numerator is the squared difference between the O_i and E_i for each category or group. The denominator is the E_i for each category. If random or chance factors were responsible for a difference between these observed and expected scores, then the χ^2 would be relatively small. In contrast, χ^2 will be large if there is a nonrandom difference between the observed and expected scores (i.e., if the difference is sufficiently large that it reflects a poor fit or match between the observed and expected frequencies).

As with the other inferential statistics you have learned to use, the χ^2 statistic has its own family of sampling distributions. The degrees of freedom for the χ^2 statistic is one less than the number of groups. Stated mathematically, the degrees of freedom for the goodness-of-fit test is

$$df = k - 1 \tag{16.2}$$

In this example, there are 12 months; therefore, the degrees of freedom are 11. Table B.14 of Appendix B lists the critical values for the χ^2 statistic. As you can see, the critical value to reject the null hypothesis in this example (at $\alpha = .05$) is 19.675. Like the analysis of variance (ANOVA), the χ^2 is an omnibus statistic. This means that the χ^2 will indicate whether the pattern of observed frequencies deviates significantly from the expected frequencies in some way. This statistic does not, however, indicate which specific observed frequencies differ from the expected frequencies; post hoc tests can help in this regard and we will discuss these options later in this chapter.

Assumptions of the Goodness-of-Fit Test

The goodness-of-fit χ^2 test requires the data to meet several assumptions. These assumptions include (1) mutually exclusive categories, (2) exhaustive conditions, (3) independence of observations, and (4) sufficient sample size.

We have already examined the requirement of *mutually exclusive categories*—each subject fits into only one category. Examples of mutually exclusive events include one's sex, political party, and marital status. It is critical to ensure that we use procedures that prohibit double counting the individuals. If we allow some people or observations to count in several categories, this type of statistical test will produce biased results. Because the data for this running example were based on unique ED admissions, we can be confident each person is being counted as an admission in only one of the 12 months being considered in this study.

The issue of an inflated T, or total number of observations (i.e., a very large sample size or large number of measurements), can be a serious problem for the accuracy of the χ^2 test. Counting the same participant several times biases the data. Therefore, researchers who conduct categorical data research spend much time ensuring that their definitions of the categories and the procedures they use to collect the data ensure that the data conform to the mutually exclusive requirement. The *exhaustive-conditions* assumption indicates that there is a category available for each person in the sample. In our present example, we can consider the conditions exhaustive because the researcher included a category for each month of the year.

Turning our attention to the other previous example, what if a lobbyist wanted to examine the relationship between political party and attitude toward new healthcare legislation? To meet the exhaustive-conditions requirement, the researcher would need to include a category for each political party (Communist, Democrat, Green, Libertarian, Republican, etc.) or a category for each of the major parties and a category of "other" party affiliation. The purpose of the exhaustive-conditions requirement is to ensure that the researcher classifies each person in the sample. Dropping people from the sample because they do not meet the classification scheme will bias the results.

The *independence of observations* assumption is met when the classification of one or more participant(s) into one category has no effect on the classification of other participants into the other categories. To meet the independence requirement, there must be equivalent criteria for entry into each category. For this example, we can assume that the decision of one couple to wed in February had no effect on the plans of the other couples. The final assumption, *sufficient sample size*, is met by meeting rather rough general guidelines. Typically, none of the E_i should be less than 5; otherwise, the χ^2 will be more likely to produce erroneous results.

In the present example, we can see that we have met all the requirements of the χ^2 test. The groups are mutually exclusive, the conditions are exhaustive, the classification of the people is independent, and the sample size appears to be sufficiently large. Therefore, we can proceed with the analysis of these data. Table

Table 16.3 Extension of the Data Presented in Tables 16.2 and 16.3

Month	O	E	$O_i - E_i$	$(O_i - E_i)^2$	$\frac{(O_i - E_i)^2}{E_i}$
January	$O_1 = 26$	$E_1 = 24$	2	4	0.1667
February	$O_2 = 41$	$E_2 = 42$	−1	1	0.0238
March	$O_3 = 36$	$E_3 = 36$	0	0	0.0000
April	$O_4 = 41$	$E_4 = 42$	−1	1	0.0238
May	$O_5 = 62$	$E_5 = 60$	2	4	0.0667
June	$O_6 = 75$	$E_6 = 72$	3	9	0.1250
July	$O_7 = 60$	$E_7 = 60$	0	0	0.0000
August	$O_8 = 67$	$E_8 = 66$	1	1	0.0152
September	$O_9 = 58$	$E_9 = 60$	−2	4	0.0667
October	$O_{10} = 52$	$E_{10} = 54$	−2	4	0.0741
November	$O_{11} = 41$	$E_{11} = 48$	−7	49	1.0208
December	$O_{12} = 41$	$E_{12} = 36$	5	25	0.6944
Totals	600	600			$\chi^2 = 2.2771$

Note: This table shows how one calculates the goodness of fit, χ^2.

16.3 presents the χ^2 test for the number of ED admissions by month. For these data, the null hypothesis is that the observed frequencies will match or "fit with" the expected frequencies (hence the goodness-of-fit label for this type of analysis).

The value of χ^2 in this example is less than χ^2_{critical} $(2.28 < 19.68)$, so we cannot reject the null hypothesis. The implication of this conclusion for this type of statistical test is that the observed distribution of ED admission rates in your hospital matches those identified by the broader health system in which you work. We must conclude, therefore, that we have no reason to assume that the small differences we see between the observed and expected frequencies is due to anything other than chance. In practical terms, these ED admission rate data do not support a need for adjusting the staffing levels at your ED from what was originally recommended by the health system.

In the preceding example, we used empirical data from national norms to determine the values of the E_i. There are other situations in which we want to determine whether the data we have observed are evenly distributed across a set of categories. Consider the following example. A safety expert working with the state department of transportation wants to know whether there is a relationship between the time of year and the number of traffic-related fatalities on state highways and interstates. Using state records as a source of data, this safety expert counts the number of fatalities on these roads over the past 2 years and then subdivides this total into four groups based on the four seasons of the year (i.e., winter, spring, summer, and fall). The resulting O_i and E_i data are summarized in Table 16.4, along with the basic calculations for this example.

This safety practitioner (and researcher) is interested in knowing whether the number of traffic-related fatalities varies depending on the season of the year. The

Table 16.4 Hypothetical Data Representing the Number of Traffic Fatalities within Your State across the Four Seasons (over a 2-Year Interval)

Season	O	E	$O_i - E_i$	$(O_i - E_i)^2$	$\frac{(O_i - E_i)^2}{E_i}$
Spring	495	517.5	−22.5	506.25	0.9783
Summer	503	517.5	−14.5	210.25	0.4063
Autumn	491	517.5	−26.5	702.25	1.3570
Winter	581	517.5	63.5	4032.25	7.7918
Total	2070	2070	0.0		$\chi^2 = 10.5334$

simplest null hypothesis in this scenario is that the rate of fatalities is equal across all four seasons of the year. Therefore, the appropriate E_i in this scenario is that the number of patients exhibiting depressive symptoms should be equivalent across each of the four season groups. Thus, the E_i for each season is $517.5 = 2070/4$. The remainder of the test calculations follows the same procedures already demonstrated. With $df = 3$ and $\alpha = .05$, $\chi^2_{\text{critical}} = 7.815$. Because the observed value of χ^2 is greater than the critical value ($10.5334 > 7.815$), we may assume in this example that the number of traffic fatalities does deviate from the E_i. This suggests that this state's department of transportation may wish to work on tailoring the deployment of state police and traffic enforcement strategies during the higher risk months of the year.

χ^2 TEST OF INDEPENDENCE

We can extend the logic of χ^2 to research scenarios that involve two categorical independent variables. The question that we want to address in this type of situation is whether these variables jointly explain the dependent variable or whether each independent variable operates separately from the other. A very serious application of this type of **χ^2 test of independence** analysis is illustrated by the following example. What are the long-term effects of childhood sexual abuse? Are children who experience this type of abuse more likely to suffer severe emotional disorders later in their life?

Gladstone, Parker, Wilhelm, Mitchell, and Austin (1999) addressed that question by conducting an extensive review of 171 women receiving treatment for a major depressive episode. Of these women, 40 reported experiencing sexual abuse as children. The researchers then examined the case history of all the women and looked for evidence of self-injurious behavior. The data in Table 16.5 represent the number of patients who attempted to commit suicide (i.e., the observed frequencies, or O_{ij}).

In this example, the χ^2 test of independence allows us to examine the relationship between history of childhood sexual abuse and rates of attempted suicide. If the two variables are independent of each other, then the rate of suicide attempts

Table 16.5 Contingency Table Representing the Number of Women Reporting Childhood Sexual Abuse and One or More Suicide Attempts

	Childhood sexual abuse		
	Abused	Not abused	Row total
Attempted suicide	16	23	39
No suicide attempts	24	108	132
Column total	40	131	171

will have nothing to do with the abuse experience. If, however, a history of sexual abuse does affect risk of suicide, then there should be a nonrandom pattern evident in the data.

As with the goodness-of-fit test, we need to first determine the expected values (E_{ij}) for each cell in the matrix that is created by crossing the levels of one of the categorical independent variables with the levels of the other categorical independent variable. The expected value represents the frequency within a cell that we would expect to occur given the independence of the two variables. According to the data presented in Table 16.5, we know that 40 women in Gladstone et al.'s (1999) sample reported experiencing childhood sexual abuse. Converting this number to a percentage, we can say that 23.39% χ^2 of the women experienced such abuse. We also know that 39 women in this sample attempted to commit suicide. If the rate of suicide attempts and history of abuse are independent of each other, then we can conclude that 23.39% of the women who attempted suicide also have a history of abuse. We would expect, therefore, that 9.12 (i.e., 39 × .2339) women would be in the cell, "attempted suicide"/"abused".

We can also use the following equation for determining the expected frequencies for each cell:

$$E_{ij} = \frac{R_i C_j}{T} \tag{16.3}$$

In this equation, R_i and C_j represent the row-and-column totals for each category, and T represents the total number of observations. Therefore, E_{ij} represents the expected frequency for each cell if there is no correspondence between the two variables. Table 16.6 presents the expected values for the data.

If there is no relationship between the two independent variables in this type of study, then the overall difference between the O_{ij} and E_{ij} should be minimal. In this case, each E_{ij} should equal the corresponding O_{ij}, and if it does, χ^2 will equal 0. If, however, the differences between the O_{ij} and E_{ij} are sufficiently large, we can infer that there is a relationship between the two variables and that the two variables are therefore not independent of each other in terms of their influence on the dependent variable. In the present example, a statistically significant χ^2 would lead us to conclude that there is a correspondence between a history

Table 16.6 Extension of the Data Presented in Table 16.5 and Illustration of How to Calculate E_{ij}

	Childhood sexual abuse		
	Abused	Not abused	Row total
Attempted suicide	$O_{11} = 16$	$O_{12} = 23$	$R_1 = 39$
	$E_{11} = \frac{40 \times 39}{171} = 9.12$	$E_{12} = \frac{131 \times 39}{171} = 29.88$	
No suicide attempts	$O_{21} = 24$	$O_{22} = 108$	$R_2 = 132$
	$E_{21} = \frac{40 \times 132}{171} = 30.88$	$E_{22} = \frac{131 \times 132}{171} = 101.12$	
Column total	$C_1 = 40$	$C_2 = 131$	$T = 171$

of childhood sexual abuse and suicide attempts (i.e., that these two factors do not operate independently of each other).

The formal mathematical statement for the χ^2 test of independence is

$$\chi^2 = \sum_{i=1}^{r} \sum_{j=1}^{c} \frac{(O_{ij} - E_{ij})}{E_{ij}} \tag{16.4}$$

The equation is much like the goodness-of-fit test. We need to calculate a ratio for each cell in the matrix. The number of rows, r, and the number of columns, c, determine the degrees of freedom. Specifically,

$$df = (r-1)(c-1) \tag{16.5}$$

For Equation 16.5, r represents the number of rows and c represents the number of columns in the matrix of data. With this information, we can now proceed to our calculations. Carrying out the steps in Equation 16.4, we find

$$\chi^2 = \frac{(16-9.12)^2}{9.12} + \frac{(23-29.88)^2}{29.88} + \frac{(24-30.88)^2}{30.88} + \frac{(108-101.12)^2}{101.12} = 8.77$$

Because the calculated χ^2 is much greater than the critical χ^2 for the degrees of freedom in this example, we can reject the null hypothesis. In conclusion, the data presented in this example would lead the researcher to conclude that the risk of suicide attempt corresponds with the experiencing of sexual abuse as a child. Looking at the differences between the observed and expected values, you can see the rather obvious effect: Women who experienced sexual abuse as children are more likely to attempt suicide than women who did not experience the abuse.

Interpreting χ^2

As with most other statistics, we need to interpret a statistically significant χ^2 test for independence with caution. Specifically, we need to be mindful of whether a statistically significant χ^2 allows us to infer cause and effect. The primary purpose of the χ^2 test of independence is to determine whether the variables involved function independently from one another. If we fail to reject the null hypothesis for this test, we must conclude that we do not have evidence for the interdependence between the variables. We must then assume that the data represent the random distribution of observed scores. If we *do* reject the null hypothesis, we can then conclude that there is a correspondence between the variables and that they are not independent from each other.

However, as we discussed in Chapter 3, cause and effect is not easy to demonstrate. Just as with the Pearson product-moment correlation (r), a relationship between variables in this type of analysis is also not, by itself, guaranteed evidence of a cause-and-effect relationship between the independent and dependent variables. We can use the present example to illustrate this important consideration. What were the researcher's independent variables? Both variables in this example are subject variables as they represent conditions associated with the participants that are beyond the control of the researcher. The researchers could not randomly assign the participants to the different categories. These observations mean that we cannot easily rule out alternative explanations for the data we have observed.

We also need to examine how the researchers created the sample. Is the sample representative of the general population? The participants in this research were women already seeking therapy for depression. Consequently, this sample may not represent all the women who suffered childhood abuse; some may not have sought treatment from a psychologist. Thus, we cannot automatically assume that all people who experience sexual abuse are at risk for suicide. In determining the long-term effects of childhood sexual abuse, we would need to sample from the general population and work to find those with a history of childhood sexual abuse, independent of their current treatment for a psychological disorder.

χ^2 TEST OF HOMOGENEITY

Are there conditions when we can assume a cause-and-effect relationship for the χ^2 test? The short answer to this question is yes, if we conduct the study using a slightly different approach to sampling and also establish a measureable dependent variable in advance. Imagine that a hospital is trying to improve the use of personal protective equipment (PPE) by its nurses. This hospital randomly selects groups of nurses from different departments and assigns them to one of four intervention conditions: (1) control/no intervention, (2) PPE for personal safety, (3) PPE for patient safety, and (4) PPE as the moral choice. Now that the trainings are over, the training coordinator at this hospital needs to figure out if any of these trainings is working and, if so, which one is most effective at improving nurse adherence to PPE usage requirements. The dependent variable in this example can be nurses' reported personal use of PPE after the intervention versus before the intervention, categorized into the following levels: (1) more frequent, (2) no different, or (3) less frequent than before the intervention.

Table 16.7 Hypothetical Data for an Example of a χ^2 Test of Homogeneity

	Control		PPE for personal safety		PPE for patient safety		PPE as the moral choice		Row total (O_i)
More frequent	O_{11}	19	O_{12}	15	O_{13}	7	O_{14}	15	56
	E_{11}	14	E_{12}	14	E_{13}	14	O_{14}	14	
No different	O_{21}	21	O_{22}	22	O_{23}	9	O_{24}	16	68
	E_{21}	17	E_{22}	17	E_{23}	17	O_{24}	17	
Less frequent	O_{31}	20	O_{32}	23	O_{33}	44	O_{34}	29	116
	E_{31}	29	E_{32}	29	E_{33}	29	O_{34}	29	
Column total (O_j)		60		60		60		60	$T = 240$

$df = (3 - 1)(4 - 1) = 6$; $\chi^2_{\text{critical}} = 12.592$, $\alpha = 0.5$; $\chi^2 = 23.46$ reject H_0 because $\chi^2 > \chi^2_{\text{critical}}$.

The χ^2 **test of homogeneity** allows us to determine whether the pattern of differences among O_{ij} deviates from what we would expect due to chance. If a statistically significant χ^2 results from this type of study, then we can then assume that a cause-and-effect relationship is present. The data in Table 16.7 represent hypothetical results for such a study.

In this example, we can treat the type of PPE intervention as a manipulated independent variable because the researcher randomly assigned nurses to one of the conditions. The second variable, the outcome measure, is the reported change in PPE use following the intervention administration. Therefore, this study has all the hallmarks of a true experiment: (1) manipulated independent variable, (2) a control group, and (3) random assignment of flight crews to the treatment conditions. Data such as these could lead the researcher to conclude that the type of PPE intervention does influence the actual usage of nurses. Follow-up analyses could then help to identify which of the intervention conditions was most effective.

The procedures for calculating the χ^2 in this type of analysis are identical to those presented for the test of independence. Therefore, the only differences between the χ^2 test of independence and the χ^2 test of homogeneity are the sampling methods and the interpretations we can draw from the test. To review, for the test of independence, we create a sample and then assess the participants for the variables in the study. In the Gladstone et al. (1999) study, the researchers created a sample of women seeking treatment and then determined the number of individuals who had attempted suicide and the number who had been sexually abused as children. For the test for homogeneity, however, we created a sample of equal numbers of participants within the different levels of one variable and then assessed the dependent variable using an outcome measure.

Reporting χ^2

Although statistics reporting guidelines may differ somewhat from journal to journal and discipline to discipline, most editorial styles require something similar to what the American Psychological Association (2009) recommends. For this type of statistic, a standard reporting format would look something like this: $\chi^2(df, N = n) = \chi^2, p < \alpha$, or, using data from the present example, $\chi^2(6, N = 240) = 23.46$, $p < .05$. Most editorial styles require a reporting format that includes the key elements found in this illustration, including the degrees of freedom and the number of subjects used in the study. We also indicate whether the probability level associated with the statistic is greater than or less than the α-level selected for the study.

Assumptions of the χ^2 Test

Valid interpretations from this type of statistic are only possible when the data that are analyzed meet three specific assumptions. Meeting these assumptions requires a well-designed study in the first place. These assumptions are similar to the ones associated with all forms of χ^2 tests, including the goodness-of-fit test, which we already reviewed. Specifically, (1) the categories must be mutually exclusive, exhaustive, and independent; (2) the data must represent frequencies; and (3) the sample size must be sufficiently large. Regarding sample size, many authors offer general guidelines for the χ^2 test of independence. For example, Siegel and Castellan (1988) noted that when either the number of rows or the number of columns is greater than two, "no more than 20% of the cells should have an expected frequency of less than 5, and no cell should have an expected frequency less than 1" (p. 199).

Some researchers have found that the χ^2 test maintains the same rate of type I errors even with severe violations of this assumption. For example, Good, Grover, and Mitchell (1970) suggested that expected values could be as low as 0.33 with no change in the probability of committing a type I error. Although the χ^2 test appears to be robust and able to withstand the effect of small frequencies, we should not forget the issue of statistical power. As with all inferential statistics, the probability of correctly rejecting the null hypothesis increases as the sample size increases. Therefore, although χ^2 may be computationally valid when the sample size is small, your power to identify a significant effect may be insufficient.

FOLLOW-UP ANALYSES

The χ^2 test allows us to determine whether to reject the null hypothesis; however, the results provide no information for determining the strength of the relationship between the independent variables. In the next two sections, you will learn how to determine the degree of correspondence between the variables and how to more closely examine the differences observed within each cell.

Effect Size: Cramér's Coefficient χ^2

In Chapter 10, you learned that Pearson's product-moment correlation coefficient is a statistic used to indicate the degree of association between two variables. The larger the absolute value of r, the greater the association between the two variables. **Cramér's coefficient Φ** (phi) serves the same purpose as r; it indicates the degree of association between two variables analyzed using the χ^2 statistic:

$$\Phi = \sqrt{\frac{\chi^2}{T(S-1)}} \tag{16.6}$$

Equation 16.6 presents Cramér's Φ. In the equation, T represents the total number of observations and S represents the smaller value of either the number of rows or number of columns. Because the size of Φ depends on the value of χ^2, we can infer that if χ^2 is statistically significant, the same is true of Φ.

There are important differences between r and Φ, however. Whereas r can range between −1 and 1, Φ can have values only between 0 and 1. Nevertheless, we can use the magnitude of Φ as an index of the relation between the two variables. Another important difference between r and Φ is that r is a measure of the linear relationship between two variables. In Chapter 10 you learned that we use r when we assume that a straight line describes the relationship between two variables. There are, however, instances when there are systematic relationships between two variables best described by curved lines. The Φ does not assume a linear relation between the two variables. The Φ is a general statistic that allows us to estimate how one variable predicts the other variable.

We can use the data from Table 16.6 to illustrate how to calculate Cramér's Φ:

$$\Phi = \sqrt{\frac{8.77}{171(2-1)}} = \sqrt{\frac{8.77}{171}} = \sqrt{0.05129} = .226$$

Therefore, we can conclude that the relationship between experiences of childhood sexual abuse and attempted suicides is of moderate strength.

Post hoc Interpretation

You may recall from our discussion of the ANOVA that it is a general test because it allows us to conclude that there is a relationship between at least two means among others in a larger set, but it does not tell us which specific treatment groups or conditions differ significantly from the others. For the ANOVA, Tukey's honestly significant difference (HSD) is a commonly used post hoc test to compare the differences among specific means. A similar situation arises for the χ^2 test. We can use the χ^2 test to reject the hypothesis that the two conditions are independent of one another, but we cannot determine from the test which condition or conditions contributed to the statistically significant result.

One technique that many researchers use is to convert the differences between the observed and expected values to a statistic called the **standardized residual** (***e***). The standardized residual allows us to determine the relative difference between the observed and expected frequencies. Using the standardized residual, we can then determine which cells in the data table represent statistically significant differences and which represent chance findings. Equation 16.7 presents the method of calculating e:

$$e_{ij} = \frac{O_{ij} - E_{ij}}{\sqrt{E_{ij}}} \tag{16.7}$$

Although Equation 16.7 provides an estimate of the standardized residual, many researchers prefer to use a more refined estimate of the residual. Equation 16.8 (Delucchi, 1993; Haberman, 1973) presents the method of calculating the variance for each cell:

$$v_{ij} = \left(1 - \frac{C_i}{T}\right)\left(1 - \frac{R_j}{T}\right) \tag{16.8}$$

Using Equations 16.7 and 16.8, we can now calculate what statisticians call the **adjusted residual** ($\hat{e}$):

$$\hat{e}_{ij} = \frac{e_{ij}}{\sqrt{v_{ij}}} \tag{16.9}$$

The advantage of calculating $\hat{e}$ is that it is normally distributed with a mean of 0 and a standard deviation of 1. Therefore, we can treat $\hat{e}$ as if it were a z-score. If the absolute value of $\hat{e}$ is sufficiently large, we can assume that the difference between the observed and expected values are statistically significant. For example, $z = \pm 1.96$ represents the critical value for $\alpha = .05$, two-tailed. In other words, if $\hat{e} \leq -1.96$ or $\hat{e} \geq 1.96$, we can assume that the difference between O and E is statistically significant at $\alpha = .05$. Table 16.8 presents the calculations for e, v, and $\hat{e}$. In this running example, we can conclude that the difference between the observed and expected frequencies for each cell is statistically significant.

McNEMAR TEST

The McNemar test is a special form of the χ^2 test that we can use to compare correlated samples (i.e., samples that share something in common, making the responses of one group not fully independent from responses in the other group). We can use the McNemar test to compare the proportion of participants who fit within a specific category before and after some event or under different conditions. The McNemar test works specifically with a 2 × 2 table, as presented in Table 16.9.

Equation 16.10 presents the basic form of the McNemar test:

$$\chi^2 = \frac{(|B - C| - 1)^2}{B + C}, \; df = 1 \tag{16.10}$$

Table 16.8 Extension of the Data Presented in Table 16.6

	Childhood sexual abuse		
	Abused	Not abused	Row total
Attempted suicide	$O_{11} = 16 \quad E_{11} = 9.12$	$O_{12} = 23 \quad E_{12} = 29.88$	
	$e_{11} = \frac{16 - 9.12}{\sqrt{9.12}}$	$e_{12} = \frac{23 - 29.88}{\sqrt{29.88}}$	$R_1 = 39$
	$e_{11} = 2.278$	$e_{12} = -1.258$	
	$v_{11} = 766 \times .772 = .591$	$v_{12} = 234 \times .772 = .181$	$.772 = 1 - \frac{39}{171}$
	$\hat{e}_{11} = \frac{2.278}{\sqrt{.591}} = 2.96$	$\hat{e}_{12} = \frac{-1.258}{\sqrt{.181}} = -2.96$	
No suicide attempts	$O_{21} = 24 \quad E_{21} = 30.88$	$O_{22} = 108 \quad E_{22} = 101.12$	
	$e_{21} = \frac{24 - 30.88}{\sqrt{30.88}}$	$e_{22} = \frac{108 - 101.12}{\sqrt{101.12}}$	$R_2 = 132$
	$e_{21} = -1.238$	$e_{22} = 0.684$	
	$v_{21} = 766 \times .228 = .175$	$v_{22} = 234 \times .228 = .053$	$.228 = 1 - \frac{132}{171}$
	$\hat{e}_{21} = \frac{-1.238}{\sqrt{.175}} = -2.96$	$\hat{e}_{22} = \frac{0.684}{\sqrt{.053}} = 2.96$	
Column total	$C_1 = 40$	$C_2 = 131$	$T = 171$
	$.766 = 1 - \frac{40}{171}$	$.234 = 1 - \frac{131}{171}$	

Table 16.9 Layout for the McNemar Test

		Condition II	
		1	2
Condition I	1	A	B
	2	C	D

Once you calculate the χ^2 test, you can use Table B.14 of Appendix B to determine whether to reject the null hypothesis. An alternative method is to take the square root of the χ^2, which produces a z-score. This transformation works only when the degrees of freedom of the χ^2 tests are $df = 1$:

$$z = \sqrt{\chi^2} \tag{16.11}$$

The null hypothesis of this test is that the difference between the B and C cells is due to chance. A significantly large value of χ^2 indicates that the observed frequencies in the B and C cells do not represent chance factors.

We can use an example from research conducted by Cairns et al. (1989) to illustrate the utility of the McNemar test. Cairns et al. were interested in the developmental changes in aggressive tendencies of children. More specifically,

Table 16.10 Procedure for Calculating the McNemar Test

Boys' reports of conflicts: Grade 4

	Physical aggression	No physical aggression	Total
Conflict with males	48	56	104
Conflict with females	26	78	104

$$\chi^2 = \frac{(|56-26|-1)^2}{56+26} = \frac{(|30|-1)}{82} = \frac{841}{82} = 10.2561$$

$$z = \sqrt{10.2561} = 3.20$$

$$\Phi = \sqrt{\frac{10.2561}{104(2-1)}} = 0.314$$

Boys' reports of conflicts: Grade 7

	Physical aggression	No physical aggression	Total
Conflict with males	49	55	104
Conflict with females	12	92	104

$$\chi^2 = \frac{(|55-12|-1)^2}{55+12} = \frac{(|43|-1)}{67} = \frac{1764}{67} = 26.3284$$

$$z = \sqrt{26.3284} = 5.13$$

$$\Phi = \sqrt{\frac{26.3284}{104(2-1)}} = 0.5034$$

they examined the number of times boys and girls described aggressive tendencies toward other boys and girls. To collect the data, research associates interviewed the children, first in the fourth grade and then again in the seventh grade. During the interviews, the researcher asked the children to describe two recent conflicts with other children. Another research associate, blind to the research questions and nature of the children responding, read and scored verbatim transcripts of the interviews. Part of the coding indicated whether the child described physical aggression toward another child. The data in Table 16.10 present an example of Cairns et al.'s results for boy participants.

In this table, the numbers represent frequency of events for two conditions. Look at the data for Grade 4. These data represent the results for the 104 boys interviewed. The question that the researchers wanted to address is the number of times the boys described physical conflicts with other boys and the number of physical conflicts with girls. According to the results, 48 of the boys reported a physical conflict with another boy. These same 104 boys reported 26 physically aggressive conflicts with girls. The research question is whether boys are more likely to be physically aggressive with boys than with girls.

According to Table B.14 of Appendix B, the critical value of χ^2 for $df = 1$ and $\alpha = .05$ is $\chi^2 = 3.841$; therefore, we can reject the null hypothesis for both the fourth-grade data and the seventh-grade data. We can conclude that the pattern of reported physical conflicts is not random. Specifically, these data suggest that boys are more likely to report themes of physical aggression toward other boys than toward girls and that this pattern appears to become stronger as boys grow older.

KNOWLEDGE CHECK

1. Describe in your words the uses of the χ^2 statistic. How are these uses of the χ^2 test similar to and different from each other?
2. A student conducted a study in which three groups of rats (five per group) were reinforced under three different schedules of reinforcement (100%, 50%, and 25%). The number of bar-pressing responses obtained during extinction is as follows: 100%, 615; 50%, 843; and 25%, 545. Criticize the use of the chi-square as the appropriate statistical technique.
3. Neil asked 50 participants to think about their earliest memory of their father and their earliest one of their mother. Then he asked them to evaluate each memory as to whether it is positive or negative. He obtained the results listed in the accompanying table.

	Memory of mother	Memory of father
Positive	35	31
Negative	15	19

Neil plans to analyze his data using χ^2. Do you think such an analysis is appropriate? Explain.

4. Bill believes that the current generation of students is much different from students in the 1960s. He asks a sample of students at his college the following question: Which of the following best describes your reason for attending college (pick only one alternative)?
   a. develop a philosophy of life
   b. learn more about the world around me
   c. obtain a good-paying job
   d. unsure of reason.

   Researchers asked the same question of students at the college in 1965. Here are the results of the two surveys:

	1965	2001
Develop a philosophy of life	15	8
Learn more about the world around me	53	48
Obtain a good-paying job	25	57
Unsure of reason	27	47

Is Bill justified in concluding that there is a difference between the current class of students and students in 1965?

5. Erin is the chair of a department of psychology. She sent a questionnaire to all students who graduated from the program during the past 10 years. One question asked whether the student attended a graduate program and the type of program attended. Of the students returning questionnaires, 88 indicated that they were in or had completed a PhD program:

Program	Number of students
Clinical	12
Counseling	23
Developmental	8
Experimental	7
Industrial-organizational	15
Physiological	5
Social	8

Can Erin assume that graduates of the department equally attend each of the seven programs?

6. Jackie conducted a study in which she compared the helping behavior of passersby under two conditions. What can you conclude from the results shown in the accompanying table?

	Condition A	Condition B
Helped	75	45
Did not help	40	80

Set up this study in formal statistical terms and draw appropriate conclusions.

7. Robert wanted to determine the effects of a televised debate on voters' preferences. The day before a televised debate, Robert asked 100 randomly selected people their current preference. After the debate, Robert asked the same people to identify their candidate of choice. Use the following data to describe how voters' preference changed after watching the debate:

		After the debate	
		Candidate A	Candidate B
Before the debate	Candidate A	20	25
	Candidate B	30	25

RESEARCH IN ACTION: IMPROVING COLONOSCOPY SCREENING RATES AMONG AFRICAN AMERICAN MEN

Among the many health challenges faced by aging men, colorectal cancer (CRC) is perhaps one of the most common and yet preventable. The rates of CRC are not uniform across different subgroups of men, however, and in the United States, the incidence of CRC is especially high among African Americans. Xirasagar, Hurley, Burch, Mansaray, and Hébert (2011) set out to do something about this disparity. Specifically, they believed that training primary care physicians who serve primarily African American populations in how to properly conduct and interpret CRC screenings might be an effective strategy for improving the rates of CRC screening within this historically difficult population.

To test this hypothesis, Xirasagar et al. (2011) conducted a quasi-experimental study in which primary care physicians, who were themselves African American, were provided intensive training by a specialist and were given assistance by trained technical staff. The hypothesis at the center of this study was that "patients of colonoscopy-grained African American PCPs have higher colonoscopy rates than those of untrained PCPs" (p. 5152). The logic behind this hypothesis is fairly straightforward in that research has shown that patients are more likely to get specific health screenings if and when their primary source of care can provide such screenings. In addition, because of the demonstrated challenges within the African American community, it was believed that the connection between African American patients and their primary care physician would be stronger when that physician was also African American and also because African American physicians are more likely to care for patients who are themselves minorities (Xirasagar et al., p. 5152, citing also Komaromy et al., 1996 and Moy & Bartman, 1995).

Several different types of statistical analyses were used to fully test the hypotheses and offer practical recommendations, but a chi-square test was appropriately at the heart of the analyses based on the data collected by Xirasagar et al. (2011). In total, seven African American physicians were trained in CRC screening techniques and five physicians were not trained. The dependent data reflect the number of CRC screenings among established patients for each of the 12 participating primary care physicians. Table 16.11 provides a summary of just a portion of the data collected by Xirasagar et al. to illustrate how the analyses described in this chapter can be used to at least partially test the overarching hypothesis in this

Table 16.11 Data Representing Post-Study CRC Screening Rates for African American Patients of Trained versus Untrained Physicians

		Trained PCPs		Untrained PCPs	Row total (O_i)
Patients getting CRC screening	O_{11}	485	O_{12}	195	680
	E_{11}	454	E_{12}	226	
	$\hat{e}_{11}$	3.07	$\hat{e}_{12}$	−3.07	
Patients not getting CRC screening	O_{21}	919	O_{22}	504	1423
	E_{21}	950	E_{22}	473	
	$\hat{e}_{21}$	−3.07	$\hat{e}_{22}$	3.07	
Column total (O_i)		1404		699	$T = 2103$

study. The subset of data reported in this example pertains to African American participants and their CRC screening rates following the physician training intervention (from Xirasagar et al., p. 5156).

The overall χ^2 based on these data is statistically significant, χ^2 (1, $N = 2103) = 9.41, p < .01$. As also summarized in Table 16.11, the adjusted residual ($\hat{e}$) indicates clearly that a significantly greater proportion of patients of trained primary care physicians received CRC screenings after the completion of this training intervention than patients of untrained primary care physicians. Finally, the Cramér's Φ for this example, 0.07, indicates that this effect is a small effect. It is important to note, however, that even a small effect is important given the life-threatening consequences to high-risk patients who fail to get regular CRC screenings. Interested readers are encouraged to read the full article by Xirasagar et al. (2011).

KNOWLEDGE CHECK

8. A student conducts a study of binge drinking at her college. She randomly selects 300 students who attend a small liberal arts college and asks them to complete a questionnaire that examines their drinking behavior. The results of the questionnaire allow the researcher to identify students who engage in binge drinking. One question that she wants to address is whether there is a relation between sex and binge drinking. Here are the data from the study:

		Sex	
		Male	Female
Binge drinker	Yes	28	30
	No	98	144

Use these data to determine whether there is a relation between sex and binge drinking.

9. A social scientist wanted to find whether there are ways to increase the return rate of questionnaires sent in the mail. She randomly selected 300 people listed as registered voters and prepared to mail them a short questionnaire with a stamped return envelope. For a random subsample of 100 people, the researcher first sent a separate letter explaining that the recipient would receive a survey and that it was essential to return the questionnaire. Another randomly selected 100 people received a telephone call explaining the importance of compliance. The other 100 people received no presurvey contact. The following data indicate the number who returned the survey (compliance) or did not return the survey.

	Presurvey contact		
	Letter	Phone call	Nothing
Compliance	44	68	21
No compliance	56	32	79

a. Is there evidence that there is a relation between the type of contact and the rate of compliance?

b. Which techniques appear to be the most successful in creating compliance?

CHAPTER SUMMARY

Many research projects use a nominal scale to assess the dependent or outcome variable. These data require a different statistical method for analysis. Therefore, the purpose of this chapter was to examine the analysis of categorical data using varieties of the analyses and appropriate follow-up tests. The goodness-of-fit test is a special version of the χ^2 test that allows us to determine whether the observed frequencies differ from the expected values. As with most statistical tests, the χ^2 is based on specific assumptions in order for its results to be validly interpreted. Specifically, the χ^2 requires that the data the categories represent be mutually exclusive categories, that the observations be independent, and that the sample size be greater than five per category.

The χ^2 test of independence allows us to determine whether two groups are independent. As with the χ^2 goodness-of-fit test, the test of independence compares the difference between the observed and expected frequencies. As with other statistical tests, the χ^2 has a number of post hoc tests that allow us to analyze the data further. For example, Cramér's coefficient Φ allows us to determine the extent to which one variable predicts the other variable. The adjusted standardized residual allows us to convert the difference between the observed and expected frequencies to a *z*-score.

The McNemar test is another version of the χ^2 test. It allows us to observe the change in frequencies that occur when monitoring participants on two occasions or when the participants are matched.

CHAPTER GLOSSARY FOR REVIEW

Adjusted Residual ($\hat{e}$) An enhancement of *e* that fits the normal distribution.

χ^2 Goodness-of-Fit Test A version of the χ^2 test that determines whether a statistically significant difference exists between observed frequencies and expected frequencies within mutually exclusive classes.

Chi-square (χ^2) A form of inferential statistic used to examine the frequencies of categorical, mutually exclusive classes. The statistic determines whether the observed frequencies equal hypothesized frequencies.

χ^2 Test of Homogeneity A version of the χ^2 test that determines whether the distribution of frequencies for independent populations is independent.

χ^2 Test of Independence A version of the χ^2 test that determines whether the distribution of frequencies for two categorical variables is independent.

Cramér's Coefficient (Φ) A descriptive statistic used with the χ^2 test of independence to quantify the degree of association between the two categorical variables.

Mutually Exclusive A classification scheme for which a person or observation may be placed into only one of the available categories (e.g., sex, political party, and religion).

Standardized Residual (*e*) A descriptive statistic that indicates the relative difference between the observed and expected scores for a single condition in the χ^2 test of independence.

REFERENCES

American Psychological Association (2009). *Publication manual of the American Psychological Association* (6th ed.). Washington, DC: Author.

CAIRNS, R.B., CAIRNS, B.D., NECKERMAN, H.J., FERGUSON, L.L. & GARIÉPY, J.L. (1989). Growth and aggression: I. Childhood to early adolescence. *Developmental Psychology*, *25*, 320–330.

DELUCCHI, K.L. (1993). On the use and misuse of the chi-square. In G. Keren & C. Lewis (Eds.), *A handbook for data analysis in the behavioral sciences: Statistical issues* (pp. 295–320). Hillsdale, NJ: LEA.

GLADSTONE, G., PARKER, G., WILHELM, K., MITCHELL, P., & AUSTIN, M. (1999). Characteristics of depressed patients who report childhood sexual abuse. *The American Journal of Psychiatry*, *156*, 431–437.

GOOD, I.J., GROVER, T.N., & MITCHELL, G.J. (1970). Exact distributions for chi-squared and for the likelihood-ratio statistic for the equiprobable multinomial distribution. *Journal of the American Statistical Association*, *32*, 525–548.

HABERMAN, S.J. (1973). The analysis of residuals in cross-classified tables. *Biometrics*, *29*, 205–220.

SIEGEL, S. & CASTELLAN, N.J. (1988). *Nonparametric statistics for the behavioral sciences* (2nd ed.). New York: McGraw-Hill.

XIRASAGAR, S., HURLEY, T.G., BURCH, J.B., MANSARAY, A., & HÉBERT, J.R. (2011). Colonoscopy screening rates among patients of colonoscopy-trained African American primary care physicians. *Cancer*, *117*(22), 5151–5160.

Chapter 17

Qualitative and Mixed-Methods Research

You may have heard the world is made up of atoms and molecules, but it's really made up of stories. When you sit with an individual that's been here, you can give quantitative data a qualitative overlay.

—William Turner

INTRODUCTION

Up to this point, we have discussed research methods and techniques that are primarily quantitatively focused. In other words, they require you to collect some sort of numerical information ranging from observations of frequency and intensity of behaviors to participant responses to scale or survey questions. As we have illustrated in the preceding chapters, these types of quantitative observations are tremendously useful in all science disciplines. There are times, however, especially when conducting behavioral and social science research, when the goal is not so much to quantify an effect or difference between groups but rather to deeply understand and more descriptively explain or qualify a phenomenon.

Understanding and Conducting Research in the Health Sciences, First Edition. Christopher J. L. Cunningham, Bart L. Weathington, and David J. Pittenger.

Our purpose in this chapter is to introduce you to a slightly different set of research tools that target improved understanding and to remind you one final time that it is your research question that should drive the method you use, not the other way around. This being the case, there will likely be times in your future career (whatever that may be) when you will want to ask a question that requires less quantification and more qualification, fewer numbers and statistics, and more detail and description. In these cases, a qualitative and/or combined qualitative and quantitative method for data collection and analysis may be useful to you. The present chapter is designed to provide you with an overview of this process. Many other excellent resources provide more detailed treatments of the information presented here. If you are interested in learning more than what we can provide in this brief chapter, please begin by checking out the works we have cited in this chapter and included in the chapter references.

QUALITATIVE VERSUS QUANTITATIVE RESEARCH

Whereas quantitative research is designed to empirically identify the presence and magnitude of differences between individuals and/or groups of individuals, **qualitative research** is typically more focused on sense-making in a purer sense. Quantitative research is also typically designed to test predetermined hypotheses that are formed from existing theory (a deductive process), while qualitative research often functions to develop theory from the data that are collected (an inductive process). With these two distinctions in mind, it is also often suggested that qualitative research tends to focus more on the rich description of a phenomenon than on its quantification. Thus, instead of relying on numbers, counts, and frequency-type data, qualitative research will often involve the collection and analysis of detailed observations, stories or narrative histories, sounds, pictures, or video.

Qualitative methods often bring a new or fresh perspective to existing research in areas that have been dominated by quantitative methods. When combined with quantitative techniques, qualitative strategies can often help researchers to more strongly support their research design choices and final inferences (Shaw, 2003). For example, in the case of healthcare-related interventions, which are implemented in highly complex environments, answering the question of why some interventions work while others fail is not easily done within a quantitative framework. Such a question can be more comprehensively addressed with the use of a qualitative or a combination of qualitative and quantitative (i.e., **mixed-methods research**) strategy (e.g., O'Cathain, 2009).

Despite the prevalence of qualitative methods in some social science areas (e.g., ethnography, sociology), there has not been widespread acceptance of qualitatively influenced research within applied research settings. The reasons for this are varied and not fully understood, although one major contributing factor has been the lack of education regarding these research methods (Tashakkori & Teddlie, 2003; Teddlie & Tashakkori, 2003). It has also been suggested that qualitative researchers are somehow inherently different from quantitative researchers in terms of their basic research-related goals. As such, these types of individuals may tend to gravitate toward areas of study that are more or less removed from direct practice or application than others.

Despite the existence of these and other possible distinctions, we believe there is no good rationale to support the continued separation of quantitative and qualitative research methods, particularly in research that is targeting social and behavioral science issues. There are many reasons for this, not the least of which being that the benefits of combining these two general approaches to research often outweigh the challenges, as we discuss later in this chapter. It is interesting to note that Corbin and Strauss (2008) suggest that the best qualitative researchers may share a set of personal characteristics that at first look seem antithetical to what is conveyed during standard quantitative method instruction. Specifically, they suggest that good qualitative researchers are those who, among other things, are (1) interested in the humanism present in everyday life; (2) inherently curious, creative, and imaginative; (3) capable of guiding their actions in a logical fashion; (4) able to live with and appreciate ambiguity; and (5) capable of accepting themselves as a research instrument (p. 13).

Careful consideration of these characteristics outlined byCorbin and Strauss (2008), however, should make it clear to you that these are the same qualities that ensure success as a researcher in any field, using any combination of methods. The only difference that immediately emerges is that while qualitative researchers do typically acknowledge themselves as a major part of the research instrument and process, quantitative researchers are often interested in removing themselves as much as possible from the research process so as to minimize the influence of potential sources of systematic error or bias.

THEORY AND PERSPECTIVES GUIDING QUALITATIVE RESEARCH

At the heart of most qualitative research is a belief that understanding about a phenomenon can be induced from the data as they are collected. This emergent perspective on theory development has been criticized and modified many times in the recent decades, but it remains in some way, shape, or form an underlying distinguishing feature of this type of research from the more deductive, theory-based reasoning applied in most quantitative research. Perhaps the most well-known discussion of this inductivist approach to qualitative research is contained within explanations of the **grounded theory** methodology, which remains a common guiding perspective for many qualitative researchers. By definition, this orientation toward research guides researchers to develop relevant theory for a study from the data as they are collected and analyzed in a qualitative fashion (Glaser & Strauss, 1967; Locke, 2001).

The primary challenge to pure or "simplistic inductivism" (Silverman, 2005) is the fact that no matter how detached a researcher may wish to be from existing theory and preconceptions, when we enter into a research situation, we will have an impact on that system. In previous chapters, we have referred to this impact using words such as *reactivity*, or citing Bacon's idols to remind us that when we carry preconceptions into the data collection or analysis process, we may influence our findings and interpretations. All of these risks also exist when we conduct qualitative research, and for this reason, there is no way to ensure that the development of understanding can occur completely from the data alone, without being influenced in some way by the researcher's own experience or knowledge.

Indeed, it is increasingly common to see modern qualitative research guided by some (at least loosely) preconceived hypotheses. This is also a reflection of the trend in qualitative research away from a singular model or paradigm of study toward multiple approaches to this type of research (Teddlie & Tashakkori, 2003). A primary reason for this is that just as in quantitative research, the body of scientific knowledge is cumulative. This makes it more difficult (and not desirable) to ignore the existing knowledge base completely when initiating any sort of research study. As we have emphasized and discussed in multiple ways in this textbook, your challenge when doing research (qualitative or otherwise) is not to reinvent the wheel but rather to continue the development and modification of the entire vehicle to which that wheel belongs (i.e., our collective understanding of some phenomenon).

MIXING METHODS: QUANTITATIVE AND QUALITATIVE COMBINED

In many cases, qualitative research is based on a different set of theories regarding the research process. In some forms of qualitative work, the research questions emerge from the data rather than the data emerging from an effort to address specific hypotheses that are established in advance. In addition to a different perspective on the research process itself, qualitative research often requires a different set of data collection tools. Some of these we have already discussed, but in their quantitative forms. The following sections will detail more fully some of the most common prevailing theories and methods used in qualitative research.

Despite the differences in these two general approaches to research, there are times when it is appropriate to combine these methods. Such combination can take place during any of several stages of the research process, from identification of research questions and hypotheses through sampling and selection of participants, through the actual collection, analysis, and interpretation of the data. In many cases, these two approaches may be mixed at more than one stage of the process. Mixed-methods research can, therefore, be seen as

> *[t]he type of research in which a researcher or team of researchers combines elements of qualitative and quantitative research approaches (e.g., use of qualitative and quantitative viewpoints, data collection, analysis, inference techniques) for the broad purposes of breadth and depth of understanding and corroboration. (Johnson, Onwuegbuzie, & Turner, 2007, p. 123)*

What does mixed-methods research look like? It really depends on how quantitative or qualitative the research is in its underlying nature. Bryman (2006) offers several insights regarding these issues in his qualitative content analysis of 232 social science articles based on mixed-methods research. In general, Bryman's findings suggest that the most common mixed-methods approaches have, to this point, included either a self-administered survey or questionnaire, or a structured or semi-structured interview. The vast majority of studies in this review were also of a cross-sectional variety, meaning all data were collected concurrently at one point in time.

After coding the reasons for mixed-methods work given by the authors of the studies that Bryman (2006) reviewed, a list of the common reasons or justifications for combining quantitative and qualitative methods was generated. Because these reasons sometimes differed from what was actually practiced by the researchers and shared in the article, Bryman also coded the articles for what was actually practiced. The top five most common explanations for mixed-methods research actually demonstrated in these published articles were (adapted from p. 108)

1. **Enhancement:** building on findings from one method using evidence gathered from the other method
2. **Triangulation:** using findings from one method to verify or corroborate findings collected using the other method
3. **Completeness:** using both methods to more comprehensively examine an area or issue of interest than would be possible with one method used alone
4. **Illustration:** using qualitative data to explain quantitative findings or vice versa
5. **Sampling:** using one method to help with the sampling of participants or cases in a targeted and focused fashion

Although Bryman's (2006) findings suggest there may be a disconnect between what researchers say they are doing and what they actually accomplish when attempting mixed-methods research, a more important implication is that the combination of quantitative and qualitative methods benefits researchers in multiple ways, providing all of us with improved information and richness of detail that is not obtained when singular methods are utilized.

So, when should you consider qualitative or mixed-methods research methods? In general, when your research questions are more in line with developing understanding than identifying differences, some form of qualitative inquiry may be helpful. Perhaps the set of considerations listed in Box 17.1 will be even more useful to you when you develop your research design.

QUALITATIVE AND MIXED-METHODS DATA COLLECTION AND ANALYSIS

There are numerous qualitative research techniques in use today. The one feature they all share in common is that they are designed to allow the researcher to gather

BOX 17.1 ***Qualitative or Mixed-Methods May Be a Good Idea If. . .***

- . . . you are not totally sure what the question is that you are trying to answer, but you know where you might start looking for relevant information
- . . . your goal is to study a particular phenomenon within its natural context, in detail
- . . . other researchers who have studied this topic have relied on qualitative or mixed methods (and you agree with this approach)
- . . . other researchers have relied on quantitative methods to study this issue (and you disagree with this approach and want to try something else)
- . . . it is feasible/practical for you to engage in qualitative inquiry given your available resources (i.e., you have a fair amount of time and interest)
- . . . you feel you will learn more about the topic using these methods than quantitative techniques alone
- . . . you feel sufficiently comfortable with qualitative or mixed-methods research techniques to use them.

Note: Adapted from Silverman (2005).

a very rich and complex data set. These methods are also designed to assist the researcher in making sense of the rich data collected. To give you a taste for what qualitative research techniques and analyses look like, we have selected a few very common ones to briefly present to you here.

Data Collection Techniques

In many ways, the same rules for good design of quantitative research studies still apply when qualitative data are involved. There are a few areas, however, where these two general types of methods diverge. One major area has to do with sampling, in that much qualitative research is directed at a particular sample and, as such, may not lend itself to the use of random sampling as easily as standard quantitative studies. Instead, it is more common to see purposive and/or stratified sampling techniques utilized within qualitative studies to ensure that the phenomenon or process of interest will be observed. As an example, a qualitative researcher (perhaps a medical anthropologist) interested in studying the role that culture plays in the perception of illness in one region of the world versus another would focus on observing and studying individuals with and without medical conditions, who are living in these different regions. The data collection process might involve interviewing, observing, and studying the ways in which the concept of illness is described, discussed, and ultimately treated.

Field Notes

A second major area of difference between typical quantitative and qualitative methods involves the critical role of field notes. In general, **field notes** are detailed observations made by the researcher throughout the research process. These types of notes often include observations made about the people and places in which data are being collected. Also included will be personal reflections on the data and research process as well as initial attempts to begin piecing together the story or puzzle that the data are helping to reveal. Although the taking of comprehensive field notes is more common among qualitative researchers, we strongly recommend to our students that keeping field notes is good practice for all types of research (quantitative and qualitative). It is absolutely necessary for qualitative research that is based on observations within a specific context because such notes often provide the only reliable source of information about these contexts. This information is then used by the researcher later to interpret and make sense of various participants' responses within those specific contexts.

Wolfinger (2002) indicates that the generation of field notes is in itself both a skill and an art. As such, the most effective strategy for field notes will depend on the research. From a practical perspective, he notes that there are three general considerations that may influence a researcher's field note-taking strategy:

1. the researcher's ability to take notes in the field (i.e., can observations be recorded in real time, or do they need to be summarized at the end of a period of more secretive or participant-type involvement?)
2. the degree to which the research has changed as the researcher narrows from a broad concept to more focused data collection needs (i.e., this will change the types of information the researcher can and will attend to)
3. the intended audience of this research (i.e., depending on where this work will be presented or published, different types of information may need to be emphasized).

In terms of how to actually condense and summarize your field notes, some guidance can be gleaned from Emerson, Fretz, and Shaw (1995) and from Wolfinger's (2002) extension of Emerson et al.'s work. One technique is to focus on the observations you have made during the data collection that seem most unique, special, or interesting (i.e., salient episodes). To most quantitatively trained researchers, this approach seems less than desirable, primarily because the identification of such salient episodes is based on each researcher's highly subjective judgment. However, as Wolfinger points out, the identification of these special events can be guided by at least two general rules of thumb: typically that the information may "stick out" or distinguish itself (1) from the other responses collected by the researcher or (2) from what the research was expecting to observe (p. 89).

A second general technique for field note taking and organization is to be as comprehensive or thorough as possible, describing all details that you can capture about your observations during a set time period. For this technique to be most beneficial, the notes should be generated in a systematic fashion, meaning that the researcher should follow a note-taking guide of sorts, which helps the researcher capture all relevant details. As highlighted byWolfinger (2002), such a list can include the basics of *who*, *what*, *when*, *where*, and *why*, expanded to fit the situation, or it can be focused on components of the observable situation, including such features as the space, actor, activity, time, goal, and feelings (for more details on these and other approaches to field note taking, see also Lofland, Snow, Anderson,

& Lofland, 2005). It is also possible to organize comprehensive field notes in some form of chronological order, which can often help researchers remember details of their observations that they may otherwise miss (Wolfinger).

One of the most difficult things for field researchers of any discipline to learn is to separate the act of information gathering from interpretation. We like to think of it in rather simple terms: When you are in the field collecting data (whether it is quantitative or qualitative by nature), you should act like a sponge—nothing more, nothing less. Then, once you are finished gathering data, you can wring yourself out and attempt to make sense of what you have just learned. The reason for this is that if you are constantly allowing yourself to interpret data as you perceive it (especially when you are gathering observations), then you are more likely to confine yourself to only information that connects with your preconceived notions (also known as *tacit beliefs*, as noted by Wolfinger, 2002). If instead you allow yourself to take in all information that you possibly can, your later interpretation will be based on a more complete set of details. Although we suggest this as an important strategy for data gathering, it should be noted that many qualitative researchers do prefer to begin interpretation and analysis concurrently with the collection of data, especially in inductivist research. This is one area where there is no one "right" approach—each of us has to adopt a philosophy of data collection that we can live and work with.

Other Techniques

Apart from differences in sampling and the critical role of field notes, many other qualitative or mixed-methods data collection techniques have actually already been presented in earlier portions of this textbook. Table 17.1 highlights a few of these methods, illustrating how they can be used for primarily quantitative or qualitative research means.

Obviously, this is an extremely brief summary and illustration of how similar methods can be used in both quantitative and qualitative ways. It is important to emphasize that these methods can also be used in mixed-methods fashion, incorporating elements of quantitative and qualitative focus to help the researcher achieve his or her aims. There are many other approaches to data collection within qualitative research, but most of these ways involve elements of the techniques summarized in Table 17.1. More than anything, the distinction between quantitative and qualitative data collection is often that qualitative data involve more than finite numbers and often include details regarding behavioral processes and contextual factors that may be linked to and helpful in explaining what the participant or respondent does or says.

Analytical Approaches

Although we have said that qualitative research is often more about sense-making than hypothesis testing, this does not mean that you are allowed to forget

Table 17.1 Application of Similar Methods for Quantitative and Qualitative Research

	Research method	
	Quantitative examples	Qualitative examples
Behavioral observations	• Counting frequency • Rating intensity	• Describing process or actions • Highlighting perceived causes and effects
Interviews	• Rating specific responses of the interviewee to specific questions for quality or accuracy	• Recording and transcribing the entirety of the interview experience, including interviewer questions, interviewee responses, and gaps in conversation, viewing all as meaningful information and reflections of interviewees' own reality
Surveys	• Gathering self-ratings of agreement or satisfaction with descriptive items and scales	• Eliciting writing responses to open-ended questions or comment-request boxes
Case study	• Small-*n* or single-participant study of changes in behaviors, attitudes, or other measureable variables	• More in-depth review of individualized cases, with goal of providing detail regarding context and process rather than generalizable quantified effects

Note: Portions adapted from Patton (2002), Rapley (2001), Shaw (2003), and Silverman (2005).

everything else you have learned up to this point about appropriate analytical techniques. On the contrary, the systematic research process is at least as important in qualitative research as it is in the quantitative forms we have already discussed. The reason for this should be rather clear to you, if you remember why we strive for systematic processes of data collection and analysis in the first place. Remember Bacon's idols. We know it has probably been a while since you read Chapter 1, but try to remember that while inferences based on quantitative research can be valid only when based on good design and systematic measurement with high-quality measures, inferences from qualitative research can also only be as valid as the process that the researcher follows when gathering and interpreting the data.

This being said, it is very common for qualitative researchers to view their analyses as a combination of science and art (Corbin & Strauss, 2008). This view may seem to contradict the emphasis we have placed on the scientific method throughout this textbook. In reality, though, your efforts to adhere to the basic scientific method will at times test you and require you to adapt and respond in creative and imaginative ways. In this way, it can be argued that all forms of research involve, at some level, a mixture of science and art.

Indeed, there may be more room for flexibility and imagination within qualitative research than in quantitative analyses because the objectives are typically rather different. More specifically, in quantitative research, the objective of an analysis is typically to quantify a difference between groups, a change over time, or the existence of a measurable phenomenon. In qualitative research, the emphasis is on telling a story or piecing together a puzzle that can help to explain a particular phenomenon and its relationship to other factors in the people and environment(s) involved.

Because of this, there is often a distinction to be made regarding the role of participant responses when the data collection method involves direct communication with respondents (e.g., interviews, case studies, content analysis). This distinction pertains to whether the responses are viewed as representative of that person's external reality or internal experience. A related approach to analysis partially avoids this distinction, choosing to highlight instead the possibility that participant responses are really constructed narratives that capture the respondents' perceptions within a specified context (Silverman, 2005). In effect, this narrative approach to analysis of responses may enhance the realism and eventual generalizability of the qualitative researchers' conclusions as it facilitates a more comprehensive consideration of the context in which the responses were generated.

Another distinction between qualitative and quantitative research with respect to analysis strategies, is that qualitative researchers are often taught to begin their analysis of the data as soon as the first piece of information is collected (Corbin & Strauss, 2008). This is in direct contrast to most quantitative research, in which analyses are often delayed until the full set of data can be collected. There are several reasons for this rather obvious difference between these two approaches to research. A primary one is that the theory that will be used to explain findings is often derived from the data in qualitative research, whereas it is established beforehand in quantitative research, as already mentioned. This being the case, it is often necessary for qualitative researchers to begin developing the explanatory framework, story, or theory immediately so that it can be as comprehensive as possible upon project completion.

While experimental methods are necessary to most effectively demonstrate causality, a qualitative perspective can aid in identifying causal mechanisms, identifying the ordering of events across time, and facilitating the interpretation of complex relationships among variables (Miles & Huberman, 1994). The reason for this is that qualitative techniques may facilitate more complete consideration of contextual factors and potential third variables (or confounds, as discussed earlier in this textbook) than typical quantitative methods, making it possible to better understand the true nature of an underlying causal relationship.

With respect to mixed-methods analytical approaches, one very commonly used technique involves the coding of data into categories (Duriau, Reger, & Pfarrer, 2007). These categories can be created for the purposes of quantification or qualification, making this technique especially useful to the beginning researcher interested in mixed-methods research. This analysis approach is often referred to as **content analysis**, and the goal with this approach to qualitative data analysis is often to quantify or at least categorize otherwise qualitative data. Although this may sound confusing or somehow contradictory to the goal of qualitative research,

Table 17.2 Common Content Analysis Steps

1. Define the recording unit (word, phrase, sentence, paragraph).
2. Define the coding categories.
3. Test the coding by applying it to a sample of your collected data.
4. Assess the accuracy and reliability of this sample coding (perhaps using multiple coders to check for intercoder consistency).
5. Revise the coding rules and/or instructions to improve consistency and accuracy in coding.
6. Return to step 3 and run the cycle until you achieve sufficient consistency and accuracy in coding.
7. Code all data.
8. Assess overall consistency and accuracy of the codings.

Note: Adapted from Weber (1990, pp. 21–25), and Duriau et al. (2007, p. 19).

it really is not. Such organization is often applied to help identify the qualitative content areas that are of highest importance or greatest prevalence.

The qualitative aspect of this analytical approach comes into play during the formation of content categories. In addition, once the categories are formed and data are sorted, a new door may be opened to more discussion or additional data collection to try to explain the initial categorization findings. Content categories may be defined beforehand, based on existing theory and/or relevant literature, or they may be developed from the data themselves once they are collected. This latter form is often referred to as *thematic coding*, although it is common to see *content* and *thematic coding* used interchangeably.

As summarized in Duriau et al. (2007), there are multiple sets of guidelines in the literature regarding the proper steps in a content analysis. One of the most common ones is associated with Weber's (1990) discussion of the content-coding process. Table 17.2 summarizes these steps, also adapting the simplified presentation made by Duriau et al.

Thematic coding differs from other, more quantitatively oriented forms of content coding. The goal of thematic coding is typically to extract representative or meaningful themes from a set of data. Typically, these data are verbal in nature, either collected via open-ended survey or interview questions, or extracted from communication records (e.g., e-mail or Internet chat logs). Thematic coding also can be used with records of speeches or recorded dialogues between people. Conceptually, the same strategies used when extracting themes from any of these forms of data could be used with a video of a person's movements or nonverbal behaviors as well, although there are not many good examples of this analytical technique being applied to these types of data.

In thematic coding, the first task of the researcher is typically to familiarize oneself with the data that have been collected. By reviewing the material closely and comparing one person's responses to another's; for example, the researcher will likely start to see patterns emerge in the responses. For example, when doing organizational surveys regarding employee satisfaction or morale, it is common to ask respondents to provide "any other comments" they might have regarding what is more or less satisfying to them at work. As one begins to sift through these responses, it typically becomes clear that most issues will fall into a smaller subset of categories, often involving pay, supervisors or management, coworkers, work environment concerns, and aspects of the job itself. In addition, if the job is a service job, it is not uncommon to see a thread of responses all related to interactions with difficult customers.

By sorting and labeling responses into thematic categories, the qualitative research can begin to make sense of the rich information that has been provided. The goal in thematic coding is not necessarily to quantify the number of responses per category (although that is always an option) but rather to identify which categories are most prevalent and/or important to respondents and then to dig into the responses in those categories to start to make sense of whatever phenomenon is under study. As an example, consider a study by Santos, Ortiz, Morales, and Rosales (2007) in which the meaning of ethnic identity was explored, with an emphasis on how this was associated with students' adjustment to college in ethnically diverse college settings.

Although some studies regarding these issues had been performed with quantitative surveys, Santos et al. decided to lean more on qualitative semistructured interviews with each participant as their data collection strategy. The semistructured nature of these interviews means that while all students were asked the same 13 general questions, additional probe or follow-up questions were also used if necessary to elicit sufficient detail from respondents regarding the five core issues Santos et al. were interested in examining: (1) ethnic identification, (2) personal meaning of ethnicity, (3) expressions of ethnicity, (4) influences of interethnic interactions on ethnicity, and (5) sociohistorical forces that have impinged on ethnic identity.

The data from these interviews were then content analyzed to identify representative themes. To minimize the influence of the two primary researchers' experience in this area of study, they used the skills of two student assistants (the last two authors on the paper) to assist in the coding. Nevertheless, despite this precaution, Santos et al. (2007) clearly note that because they are all of Latino heritage, "bias may have unintentionally entered the data analysis process" (p. 106). The steps in their content analysis followed very closely those outlined by Weber (1990) and summarized in Table 17.2. At multiple points, the consistency among codings was checked with inter-rater or intercoder reliability estimates. Defined themes served as the core categories into which the interview data were separated.

From this careful approach to qualitative content analysis and thematic coding, Santos et al. (2007) identified nine themes associated with ethnic diversity and the student participants' experience on college campuses. These themes could be further grouped into positive and negative groupings. The positive themes included sense of belonging, multicultural competence, evolving ethnic identity, interethnic connectedness, and politicized ethnicity. The negative themes were ethnic discomfort, perceived discrimination, interethnic tension, and ethnic segregation (p. 107).

Instead of merely reporting the prevalence of these types of themes in a quantitative summary table (what would be a very nonqualitative analysis of qualitative

Table 17.3 Themes and Supporting Example Interview Responses from Santos et al. (2007)

Themes	Illustration/example
Positive themes	
Sense of belonging	There is so much diversity [here], it helps me to feel comfortable and if people do look at me differently, I do not notice it. I do not think people look at me and say, "Look there is a Korean girl." No, it's just another . . . student. (from a Korean American participant)
Multicultural competence	I like the diversity, there's a good blend here. . . You don't understand how people are unless you're in contact with them and talk to them. I think being here has helped me understand the plight of different ethnic groups.
Interethnic connectedness	Everyone here is an [campus mascot name]. We are all the same age basically . . . so it's easy to find people that have the same interest. When people have closer interests, it's easier to accept someone into your group. There's still that common ground. (from a white participant)
Negative themes	
Feelings of ethnic discomfort	I am not sure what it means to be White really. There are so many functions, like Black History Month, the Cinco de Mayo celebration, and there's not really a White celebration. I don't really see anything on this campus for my culture. I kind of feel left out. (from a white participant)
Perceptions of discrimination	I just feel that being an Asian minority is limiting. Like, you know. If you're supposedly going [here] there's a lot of Asians there. You might have a better GPA than another person of a different ethnic group that you hardly see at that school; they'll get in before you because of their ethnic background and race. I thought that was pretty unfair.
Interethnic tension	Caucasians move back a little to let you through. They do this to prevent any type of conflict between African Americans and Caucasians. I guess I can understand that [whites] just want to avoid a conflict, but they act as though, gosh you know, [African Americans] might try and beat me up or something.

data), Santos et al. (2007) used the qualitative data from the interviews to further describe and illustrate each of these nine themes. There is insufficient space here to represent all of this, but to give you a sense for these illustrations, Table 17.3 captures a few examples of the positive and negative themes. You will find much more detail in Santos et al.'s (2007) article.

BENEFITS AND CHALLENGES OF THE MIXED-METHODS RESEARCH

In this chapter, we have tried to present you with a general overview of perspectives on qualitative and mixed-methods research theory and analytical methods.

In case you have not already identified our bias in this chapter, we believe that it is rarely the case that a research study is best served by *either* qualitative or quantitative methods. In many, dare we say *most*, cases, the researcher can learn the most about a phenomenon by using a combined, mixed-methods approach.

This mixing of methods can be done at the same time or sequentially, with some data collected quantitatively and other data collected qualitatively, for instance. The mixing can apply to the collection and analysis of data. Combining qualitative and quantitative methods is also a fantastic way that you as a researcher can successfully apply principles of data triangulation to ensure you are gaining the clearest possible understanding of the phenomenon you are studying. As you will recall from earlier in this text, *triangulation* refers to the process of operationalizing and measuring constructs or variables in multiple ways to converge upon a more accurate observation or assessment (Webb, Campbell, Schwartz, & Sechrest, 1966/2000). This can include multiple approaches within either quantitative or qualitative methodologies, or a combination of the two.

The mixing of these two more general forms of research has been around for quite some time, but it has recently been gaining in popularity (Teddlie & Tashakkori, 2003). Underlying most mixed-methods research efforts is the guiding philosophy that quantitative and qualitative methods are compatible and complementary (Teddlie & Tashakkori, 2003). When are mixed methods most appropriate? A mixed-methods approach is most appropriate if you think a combination of quantitative and qualitative methods will help you to answer your research questions as clearly as possible (Teddlie & Tashakkori, 2003). In a related fashion, the use of mixed methods makes it possible for you to engage in both theory testing and theory building within the same research project.

These benefits are all that most researchers need to at least consider mixed-methods in their own work. There are additional more practical, yet related reasons to also consider. Although most physical sciences (e.g., chemistry, biology, physics) lend themselves to quantification of observations and measurements, health science researchers are often challenged with the goal of understanding phenomena that exist within and between people and groups of people. As noted throughout this text, our efforts to operationalize and quantify psychological, behavioral, and social constructs are always limited by the presence of error. Although we have highlighted many strategies for minimizing error through careful design and planning of research throughout this textbook, there are also elements of error that cannot be fully avoided due to the idiosyncrasies of study participants and the situations in which research is conducted. In many of these research situations, it is valuable to collect data with a variety of methods.

At times, this can be done within the confines of either a quantitative or qualitative research framework, but it is often the case that a mixed-methods approach will provide more comprehensive information. This can have value for your ability to draw accurate and rational inferences from your data. It can also have positive implications for the external validity or generalizability of your findings in that a combination of quantitative and qualitative data is likely to provide you with a clearer picture of the phenomenon as it functions within a detailed and specific

Table 17.4 Tips for Successfully Combining Quantitative and Qualitative Methods

1. Keep good field notes. This will be especially useful to you if/when some aspect of your quantitative analyses does not fully make sense.
2. When measuring specific constructs, include an opportunity for participants to provide their own description or definition of what you are trying to measure, perhaps with an open-ended response option or comment box.
3. Collect observations while collecting survey responses; focus on the environment and the nonverbals in the participants (similar to field notes but perhaps more specific).
4. When planning for the use of multiple methods (triangulation), utilize at least one more qualitative technique such as interviews, focus groups, or textual analysis.
5. When developing your discussion of your findings, look for ways to encourage both quantitative and qualitative forms of follow-up research based on your findings.

context. Assuming valid inferences are made from the data you collect, the use of mixed methods should enhance your ability to retain the transferability of inferences across situations, populations, and time periods (Teddlie & Tashakkori, 2003).

As a concluding summary statement, we thought it might be helpful to list for you in Table 17.4 a few clear examples of ways in which you can meaningfully connect quantitative and qualitative research methods into your own research.

Challenges in Qualitative and Mixed-Methods Research

There are many challenges associated with the effort to integrate quantitative and qualitative methods. It is not as simple as deciding to make it happen. Bryman (2007) undertook his own mixed-methods study of these issues and found that there are several core explanations for why quantitative and qualitative methods may not "play nice" and integrate well within the same study. Among the many reasons that emerged from his interviews with researchers were

1. a tendency to report only the qualitative or the quantitative data, but not both
2. a tendency to report findings for one method, followed by findings from the other method, without any attempt at integration
3. the perceived need to focus the presentation of results toward either a quantitatively or qualitatively focused audience
4. a personal preference or comfort with one method over the other
5. a research design that inhibits the integration of quantitative and qualitative data (e.g., if the quantitative structure overly limits the qualitative follow-up)
6. the pressure to publish some results as quickly as possible, which might push the researcher to lean on one method more than the other
7. skill deficits regarding one method or the other, which might force the researcher to use one technique more

8. the tendency for some journals to favor one type of data more than the other, influencing the researchers' leanings.

It is our hope that by making you aware of these common barriers to successful integration of quantitative and qualitative methods, you will be better prepared to avoid or resolve these issues when they arise in your own research.

Apart from actual barriers to the successful use of mixed methods in research, there are also lingering questions among many researchers and practitioners regarding the validity of qualitative research. Cho and Trent (2006) point out that this "attack" on validity is increasing and is in large part due to the insistence that scientific research is used to support practice, and that this science should have evidence of replicability, testable hypotheses, and objective measurement procedures, all of which are techniques foreign to most qualitative researchers (p. 319). In contrast to these elements of good quantitative research, the standard perspective on validity among qualitative researchers is that it is linked to the correspondence between what the researcher observed or studied and what is true in reality (Cho & Trent, 2006).

Two newer perspectives regarding the validity of qualitative research are referred to as *transactional* and *transformational* validity. Cho and Trent (2006) define transactional validity as "an interactive process between the researcher, the researched, and the collected data that is aimed at achieving a relatively higher level of accuracy and consensus by means of revisiting facts, feelings, experiences, and values or beliefs collected and interpreted" (p. 322). In many ways, the techniques used by the researcher are seen as the means to achieving high levels of this type of validity. This is similar to the use of highly reliable scales in quantitative research as being necessary to lead to valid inferences. Among the widespread techniques for enhancing this form of validity in qualitative research is **member checking**, or reviewing data/observations with the research participants or respondent to ensure the information was collected accurately and to gather further reactions from the respondent (Cho & Trent, 2006).

Evidence of transformational validity develops through the research process and is not contingent on the particular techniques used by the researcher (Cho & Trent, 2006). The real support for the transformational validity of a study comes from its influence on the actions of others (i.e., the degree to which the study motivates some sort of action or change in others; Cho & Trent, 2006). As Cho and Trent are careful to note, however, there is no way to guarantee that any one set of techniques will lead to valid conclusions. Each qualitative study presents researchers with similar challenges to quantitative studies: Validity is a goal that all forms of research must strive to achieve.

SOURCES OF PUBLISHED QUALITATIVE AND MIXED-METHODS RESEARCH

So, where can you turn for good examples of solid qualitative and mixed-methods research? As mentioned early in this chapter, it is still not yet very common to see much qualitative research published within most health science, applied psychological, or social science outlets. This is due to the lack of consistent or readily available guidance regarding how such research should be presented in writing (Bryman, 2007). This is, however, changing as researchers increasingly see the benefits. In recent years, several very good examples of qualitative and mixed-methods research have been published in solid research outlets. Among the most common peer-reviewed journals publishing qualitative research are *Journal of Mixed Methods Research*, *Qualitative Research*, and *Qualitative Inquiry*.

Regarding how to go about publishing mixed-methods research, there is an increasing amount of attention paid to this issue, especially by those interested in improving education regarding these methods. If you find yourself in a position to be publishing qualitative or mixed-methods research, we strongly recommend that you first read Kidd (2002) for some practical guidance regarding qualitative research. Kidd found that there was increasing interest in reading qualitative research but that there are not consistent guidelines or standards for evaluating it in many well-established psychology journals. This being the case, it is of the utmost importance for you to clearly, concisely, but thoroughly, describe your qualitatively oriented research so that others actually can understand you.

For mixed-methods research, Creswell and Tashakkori (2007) suggest in an editorial in the *Journal of Mixed Methods Research* that articles summarizing the results of mixed-methods research should "report both qualitative and quantitative research and include both approaches in the data collection, analysis, integration, and the inferences drawn from the results" (p. 108). Creswell and Tashakkori also summarize some core features of high-quality mixed-methods journal articles, highlighting the following features:

- Well-developed quantitative *and* qualitative components (the writing should not emphatically favor one method over the other and both components should be evident through most, if not all, phases of the research process).
- An effort should be made to integrate the quantitative and qualitative components rather than to treat them as separate elements (this should facilitate more complete understanding of the findings).
- The methods reported should in some way build on existing methods (i.e., authors should emphasize the contribution of their research methodology to the broader field).

We acknowledge that these are criteria for publication in one specific journal, but these criteria are good guiding rules for any journal in which you may hope to publish your own mixed-methods research. We hope these, combined with our earlier discussion of writing up your research, will help you effectively present all your hard work.

RESEARCH IN ACTION: SPIRITUAL ASPECTS OF NURSING CARE

Our final "Research in Action" section presents you with an example of a qualitative study within the field of nursing (van Leeuwen, Tiesinga, Post, & Jochemsen,

2006). This was a very focused qualitative study, designed to shed light on the issues of (1) how nurses pay attention to the spirituality of their patients and (2) what the nurses' role is when it comes to the spiritual care of the patient. This study was conducted with three samples: experienced Dutch nurses (n = 30), patients (n = 25), and hospital chaplains (n = 12). Qualitative data were collected through a series of focus group-style interviews. As emphasized in van Leeuwen et al.'s article, the use of group interviews made it possible for individual participants to think more broadly about their responses and to consider other perspectives when forming their own responses. This is a very common qualitative data-gathering technique, which builds on the notion that the reality of the situation may not be fully understood and/or expressed by any single participant. In each of the focus group sessions, interviews were audio-recorded and a written transcript was also generated to record salient nonverbal responses.

Given the nature of these data (i.e., verbal and nonverbal responses to interview questions from three separate samples of participants), the data analysis involved analyzing the responses from nurses, patients, and hospital chaplains separately and then comparing the responses across the groups. van Leeuwen et al. (2006) utilized a content analysis process that involved the following steps: (1) coding the transcripts, (2) identifying themes, (3) categorizing the themes, and (4) matching characteristics of response fragments (i.e., interpretable phrases, words, or nonverbal signs) from participants within each group (p. 878). They also used a qualitative data analysis program to help in the identification of themes within the response fragments.

To facilitate meaningful interpretation of these qualitative data, van Leeuwen et al. (2006) then grouped the response themes into a table, aligning them as best as possible with an existing set of nursing competencies for spiritual care (citing van Leeuwen & Cusveller, 2004). This is an example of the art as well as cumulative nature of science, as this organization of identified themes from one set of data was facilitated by an existing structure of competencies identified in separate research and a creative interpretation of the identified themes by the researchers involved with this study (interested readers are directed to table 2 in the van Leeuwen et al. article for a full summary of this alignment). In the end, van Leeuwen et al. drew several valuable conclusions regarding the role of nurses in providing spiritual care for patients. Among their findings was evidence for (1) differing opinions (from nurse, patient, and chaplain perspectives) regarding the role of nurses in this respect, (2) a strong link between spirituality and emotional displays by nurses, (3) a connection between spirituality and nurses' attitudes toward and communication with patients, and (4) nurse time and care limitations being barriers to discussion of spirituality with patients.

These types of insights hold value not just for understanding the phenomenon of spirituality as it affects nurses or patients in isolation. Instead, the data gathered by van Leeuwen et al. (2006) provides rich insights into the complex role that spirituality plays at all levels of the nurse–patient relationship. The findings from this research also hold implications for the training of nurses in patient sensitivity and, more generally, compassionate care. Traditional quantitative methods would not generate understanding of this variety, and this study is a good illustration of how qualitative methods can provide meaningful insight into complicated and important issues that affect healthcare professionals and patients within the health science domain.

KNOWLEDGE CHECK

1. How would you explain the distinction between qualitative and quantitative research to your mother? What about mixed-methods research?
2. For each of the following general research objectives, outline a basic research plan that incorporates qualitative and quantitative methods:
   a. an evaluation of patient satisfaction with physician care during office visits
   b. assessing employee job satisfaction at an oil refinery
   c. examining the impact of work on children in families where both parents (if present) work full-time jobs
   d. consideration of the benefits and costs associated with religious involvement on marital satisfaction.
3. What are some of the most important pros and cons associated with using quantitative and qualitative methods? Are there any new technologies that could help you enhance these pros and minimize these cons?
4. Generate an example to fit with each of the possible reasons for doing qualitative or mixed-methods research highlighted in Box 17.1. These examples should come from your own interests in your primary field of study.
5. With a partner, locate the text (transcript) of a recent speech by some politician (easily available on the Internet) and follow the steps for content coding summarized in Table 17.2. What primary themes did you observe? Provide illustrative snippets from the transcript of the speech.
6. Read the full text of the van Leeuwen et al. (2006) article summarized in the "Research in Action: Spiritual Aspects of Nursing Care" section of this chapter. What are some alternative qualitative and mixed-methods techniques that these researchers could have integrated into this study that would add something to what they already accomplished?

CHAPTER SUMMARY

There are times when the goal of research is to describe or qualify an object of study. Accordingly, this chapter began with an overview of qualitative research and a comparison of the differences between qualitative and quantitative research. Qualitative research is an inductive process that utilizes data to drive theory creation, whereas quantitative research is typically deductive in nature and is based on an existing body of knowledge. Our contention in this chapter is that both qualitative and quantitative techniques are valid and useful methodologies. The research question being addressed should drive the methodologies, not the other way around.

Mixed-methods research is growing in popularity and represents using the "best of both worlds" by combining qualitative and quantitative techniques to more fully answer a research question. Many techniques, such as interviews, used in traditional quantitative research can be adapted for qualitative purposes. Increasing training and publication outlets for qualitative research is leading to an increase in the use of alternative techniques. Whereas fields such as ethnography have used these techniques for some time, the other social sciences are just now catching up and recognizing the benefit of mixed-methods designs.

CHAPTER GLOSSARY FOR REVIEW

Content Analysis Empirical method used to quantify qualitative information.

Field Notes Detailed observations made by the researcher based on observation or other qualitative methods.

Grounded Theory Research orientation that involves the identification and development of theory from the data as they are collected and analyzed.

Member Checking Reviewing data with respondents to ensure information was collected and coded accurately.

Mixed-Methods Research Research involving the use of both quantitative and qualitative methods.

Qualitative Research An inductive method of conducting research typified by purposive sampling, rich descriptions, detailed observations, and exploratory hypotheses.

REFERENCES

BRYMAN, A. (2006). Integrating quantitative and qualitative research: How is it done? *Qualitative Research, 6*(1), 97–113.

BRYMAN, A. (2007). Barriers to integrating quantitative and qualitative research. *Journal of Mixed Methods Research, 1*(1), 8–22.

CHO, J. & TRENT, A. (2006). Validity in qualitative research revisited. *Qualitative Research, 6*(3), 319–340.

CORBIN, J. & STRAUSS, A. (2008). *Basics of qualitative research* (3rd ed.). Los Angeles: SAGE.

CRESWELL, J.W. & TASHAKKORI, A. (2007). Editorial: Developing publishable mixed methods manuscripts. *Journal of Mixed Methods Research, 1*(2), 107–111.

DURIAU, V.J., REGER, R.K., & PFARRER, M.D. (2007). A content analysis of the content analysis literature in organization studies: Research themes, data sources, and methodological refinements. *Organizational Research Methods, 10*(1), 5–34.

EMERSON, R.M., FRETZ, R.I., & SHAW, L.L. (1995). *Writing ethnographic fieldnotes*. Chicago: University of Chicago Press.

GLASER, B.G. & STRAUSS, A.L. (1967). *The discovery of grounded theory: Strategies of qualitative research.* London: Wiedenfeld and Nicholson.

JOHNSON, R.B., ONWUEGBUZIE, A.J., & TURNER, L.A. (2007). Toward a definition of mixed methods research. *Journal of Mixed Methods Research, 1*(2), 112–133.

KIDD, S.A. (2002). The role of qualitative research in psychological journals. *Psychological Methods, 7*(1), 126–138.

LOCKE, K. (2001). *Grounded theory in management research.* Los Angeles: SAGE.

LOFLAND, J., SNOW, D.A., ANDERSON, L., & LOFLAND, L.H. (2005). *Analyzing social settings: A guide to qualitative observation and analysis* (4th ed.). Belmont, CA: Wadsworth.

MILES, M. & HUBERMAN, A. (1994). *Qualitative data analysis: An expanded sourcebook.* Thousand Oaks, CA: SAGE.

O'CATHAIN, A. (2009). Editorial: Mixed methods research in the health sciences: A quiet revolution. *Journal of Mixed Methods Research, 3*(1), 3–6.

PATTON, M.Q. (2002). *Qualitative research and evaluation methods*. Thousand Oaks, CA: SAGE.

RAPLEY, T.J. (2001). The art(fulness) of open-ended interviewing: Some considerations on analysing interviews. *Qualitative Research*, *1*(3), 303–323.

SANTOS, S.J., ORTIZ, A.M., MORALES, A., & ROSALES, M. (2007). The relationship between campus diversity, students' ethnic identity and college adjustment: A qualitative study. *Cultural Diversity and Ethnic Minority Psychology*, *13*(2), 104–114.

SHAW, I. (2003). Qualitative research and outcomes in health, social work and education. *Qualitative Research*, *3*(1), 57–77.

SILVERMAN, D. (2005). *Doing qualitative research* (2nd ed.). Thousand Oaks, CA: SAGE.

TASHAKKORI, A. & TEDDLIE, C. (2003). Issues and dilemmas in teaching research methods courses in social and behavioural sciences: U.S. perspective. *International Journal of Social Research Methodology*, *6*(1), 61–77.

TEDDLIE, C. & TASHAKKORI, A. (2003). Major issues and controversies in the use of mixed methods in the social and behavioral sciences. In A. Tashakkori & C. Teddlie (Eds.), *Handbook of mixed methods in social & behavioral research* (pp. 3–50). Thousand Oaks, CA: SAGE.

van LEEUWEN, R., TIESINGA, L.J., POST, D., & JOCHEMSEN, H. (2006). Spiritual care: Implications for nurses' professional responsibility. *Journal of Clinical Nursing*, *15*, 875–884.

WEBB, E.J., CAMPBELL, D.T., SCHWARTZ, R.D., & SECHREST, L. (1966/2000). *Unobtrusive measures* (rev. ed.). Thousand Oaks, CA: SAGE.

WEBER, R. (1990). *Basic content analysis* (2nd ed.). Thousand Oaks, CA: SAGE.

WOLFINGER, N.H. (2002). On writing fieldnotes: Collection strategies and background expectancies. *Qualitative Research*, *2*(1), 85–95.

Appendix A

Reviewing the Statistics behind the Research

INTRODUCTION

We assume that you have already had a course in statistics and have learned the basics of descriptive and inferential statistics. We also assume that you, like many students, can benefit from an opportunity to review several of the most basic and central statistical concepts. This appendix will provide you with a brief review of commonly used statistics within the behavioral and social sciences. If you believe that you understand concepts such as measures of central tendency, measures of dispersion, and *z*-scores, then you may skip this appendix. If, however, you want a quick review of these concepts, this appendix provides an overview of the critical concepts you need to know.

Many students say they hate math and statistics, though it often seems that these strong emotions are linked more to discomfort with numbers in general rather than to any real underlying animosity. It is true that calculating statistics can require a significant investment of time and cognitive resources, but many of the elements of statistics that were time consuming in the past are now easily handled by computers. The problem this has created, however, is that many students do not understand how the computer generates a particular statistic. This, in

Understanding and Conducting Research in the Health Sciences, First Edition. Christopher J. L. Cunningham, Bart L. Weathington, and David J. Pittenger.

turn, makes it difficult for students (and professionals) to accurately interpret any resulting output. As you learn about the procedures for designing and conducting high-quality behavioral and social science research, you should also strive to understand not just how to calculate the necessary statistics for your study but also to properly interpret and explain these statistics to others. We hope that reviewing the material in this appendix will allow you to understand how to interpret and use basic statistics. This appendix should also help you handle the material presented in the other chapters of this text.

X, Y, N, n

For all equations in this book, the letters *X*, *Y*, *N*, and *n* have a consistent meaning. In all cases, *X* and *Y* represent the variables being measured. More specifically, *X* and *Y* represent sets of data. In correlational research, for example, we might use *X* and *Y* to represent the two tests we administer to participants in the study. For example, we might collect data for 10 people using two measures. In this case, we use *X* to represent the data for the first variable and *Y* to represent the data for the second variable. Here are two sets of data:

$$X\{18, 10, 11, 13, 8, 11, 6, 10, 3, 10\}$$
$$Y\{18, 17, 13, 6, 6, 7, 4, 9, 3, 10\}$$

We use the letters *N* and *n* to indicate the number of observations we have in a particular study. The distinction between *N* and *n* is one of scope: *N* represents all the observations used in a study, whereas *n* represents the number of observations from a particular subset or group within a broader study. For this example, we would write $N = 20$ to indicate that there are 20 total individual data points. In each group, however, there are 10 observations. Therefore, we would write $n_X = 10$ and $n_Y = 10$.

SUBSCRIPTS (X_1) AND SUPERSCRIPTS (X^2)

Mathematicians use subscripts and superscripts to indicate types of information and specific mathematical operations, respectively. Whenever you see a subscript, such as X_1 or Y_1, the subscript represents the individual score within a specific group. Using the two data sets we have already presented, we can identify the individual scores as

$$\begin{array}{llll} X_1 = 18 & X_6 = 11 & Y_1 = 18 & Y_6 = 7 \\ X_2 = 10 & X_7 = 6 & Y_2 = 17 & Y_7 = 4 \\ X_3 = 11 & X_8 = 10 & Y_3 = 13 & Y_8 = 9 \\ X_4 = 13 & X_9 = 3 & Y_4 = 6 & Y_9 = 3 \\ X_5 = 8 & X_{10} = 10 & Y_5 = 6 & Y_{10} = 10 \end{array}$$

We can also use letters in the subscripts to help identify the groups. For example, n_X and n_Y represent the sample sizes for the groups X and Y. In a true experiment, we might use X_{a1} and X_{a2} to represent two groups of participants. The symbol X_{a1} represents the first group, whereas X_{a2} represents the second group. In several of the more advanced chapters in this book, we introduce you to more complicated designs that may involve multiple subscripts. It will be easier for you to understand that material if you remember that we use subscripts to identify different variables and observations within sets of data.

In contrast to subscripts, a superscript, such as X^2 or Y^2, represents a mathematical operation. The most commonly used superscript in statistics is squaring, or X^2. Finally, subscripts and superscripts can be combined—thus, if $X_2 = 10$, then $X_2^2 = 100$.

POPULATION PARAMETERS

In statistics, Greek letters (μ, σ, ρ) represent population parameters. A parameter is a statistic that describes a characteristic of a population. We use Roman letters (e.g., *M*, *SD*, and *r*) to represent sample statistics. The more commonly measured population parameters and the equivalent sample statistics are

Measure	Population parameter	Sample statistic
Mean	*M*	*M*
Variance	σ^2	*VAR*
Standard deviation	Σ	*SD*
Correlation	*P*	*r*

SUMMATION (Σ)

Sum of Scores, Σ*X*

The Greek letter Σ (sigma) is a mathematical symbol that tells us to find the sum of a group of numbers. Specifically, it indicates that all values in a set should be added together. For example, Σ*X* represents the sum of all scores in set *X*; Σ*Y* represents the sum of all scores in set *Y*. For example, if

$$X\{18, 10, 11, 13, 8, 11, 6, 10, 3, 10\}$$
$$Y\{18, 17, 13, 6, 6, 7, 4, 9, 3, 10\}$$

then

$$\Sigma X = 18 + 10 + 11 + 13 + 8 + 11 + 6 + 10 + 3 + 10 = 100$$
$$\Sigma Y = 18 + 17 + 13 + 6 + 6 + 7 + 4 + 9 + 3 + 10 = 93$$

Sum of Squared Scores, ΣX^2

In many cases, you will see a summation sign in front of some other statistical operation, such as ΣX^2 or ΣY^2. Whenever you see more than one operation linked together, it is important to remember the appropriate order of operations for basic mathematical processing (i.e., exponentiation before addition). In these examples, you should first square each score in the set and then add all the squared scores together. For example,

$$\Sigma X^2 = 18^2 + 10^2 + 11^2 + 13^2 + 8^2 + 11^2 + 6^2 + 10^2 + 3^2 + 10^2$$
$$\Sigma X^2 = 324 + 100 + 121 + 169 + 64 + 121 + 36 + 100 + 9 + 100 = 1144$$
$$\Sigma Y^2 = 18^2 + 17^2 + 13^2 + 6^2 + 6^2 + 7^2 + 4^2 + 9^2 + 3^2 + 10^2$$
$$\Sigma Y^2 = 324 + 289 + 169 + 36 + 36 + 49 + 16 + 81 + 9 + 100 = 1109$$

Sum of Scores Squared, $(\Sigma X)^2$

You will also often see a summation sign embedded within another statistical operation, perhaps $(\Sigma X)^2$ or $(\Sigma Y)^2$. In this type of situation, you should first add the scores together and then square this summated value. The reason here, again, is that this follows the proper mathematical order of operations (i.e., parentheses before all else). For example,

$$\Sigma X = 100 \qquad \Sigma Y = 93$$
$$(\Sigma X)^2 = 100^2 = 10{,}000 \quad (\Sigma Y)^2 = 93^2 = 8649$$

Sum of Cross Products, ΣXY

Another common summation operation is ΣXY, which we call the sum of cross products. ΣXY is not the same thing as $(\Sigma X)(\Sigma Y)$. To calculate the cross products, we first multiply each number in one set by its pair in the other set and then add the individual cross products together. For example,

X		Y		Cross product XY
18	×	18	=	324
10	×	17	=	170
11	×	13	=	143
13	×	6	=	78
8	×	6	=	48
11	×	7	=	77
6	×	4	=	24
10	×	9	=	90
3	×	3	=	9
10	×	10	=	100
$\Sigma X = 100$		$\Sigma Y = 93$		$\Sigma XY = 1063$

For these data, $(\Sigma X)(\Sigma Y) = 100 \times 93 = 9300$, whereas $\Sigma XY = 1063$. This important difference cannot be overlooked.

MEASURES OF CENTRAL TENDENCY

Measures of central tendency are descriptive statistics that summarize the data with a single number that represents the most typical score within a set of data. We use measures of central tendency because we typically believe that most scores on a scale or responses to an item cluster around the typical or most common score or response in the data set. Consequently, a measure of central tendency is a convenient statistical tool for indicating the most typical score in a set of data. Selecting an appropriate measure of central tendency depends on the measurement scale used to represent a particular variable, the symmetry of the distribution of data you have collected, and the inferences to be made from the measure of central tendency. In the following subsections, we will review the three most commonly used basic measures of central tendency, the mode, median, and arithmetic mean.

Mode, M_0

Of the many measures of central tendency, the mode is the easiest to calculate. It is also the least precise, but it still has value to us when we want to describe certain types of data (more on that later). By definition, the mode is

$$M_0 = \text{Most frequently occurring score or scores in the data} \tag{A.1}$$

Here is an example of how to calculate the mode using the data set X:

$$X\{18, 10, 11, 13, 8, 11, 6, 10, 3, 10\}$$

Step 1: Rank the numbers from lowest to highest:

$$X\{3, 6, 8, 10, 10, 10, 11, 11, 13, 18\}$$

Step 2: Find the most frequently occurring score:

$$M_0 = 10$$

As you can see, the mode is easy to determine and it can be a useful tool for indicating the location of a cluster of frequently occurring scores in your data set. This central tendency statistic is also useful when there are multiple *peaks* or equally *common scores* in the data set. In such cases, we may describe the data as *bimodal* (two modes) or *multimodal*. The mode is most often used to describe the central tendency of data gathered with a nominal scale. The mode can also be useful when the data are discrete values that have no intermediate values, such as when the data represent the number of children in a family.

Median, *Mdn* or Q_2

The median is a second and slightly more precise measure of central tendency. The definition of the median is

$$Mdn \text{ or } Q_2 = \text{Score that divides a ranked data set in half, or the midpoint of a set of scores} \quad (A.2)$$

Although there is no single formula for calculating the median, there are easier and more difficult techniques. Here is a relatively straightforward approach for calculating a median using the two example data sets from this appendix. For this example, $n_X = 10$ and $n_Y = 9$:

$$X\{18, 10, 11, 13, 8, 11, 6, 10, 3, 10\}$$
$$Y\{18, 17, 4, 6, 6, 13, 7, 9, 10\}$$

Step 1: Rank the numbers from lowest to highest:

$$X\{3, 6, 8, 10, 10, 10, 11, 11, 13, 18\}$$
$$Y\{4, 6, 6, 7, 9, 10, 13, 17, 18\}$$

Step 2: Add 1 to *n* within each data set and then divide by 2:

$$\text{For } X: (10+1)/2 = 5.5$$
$$\text{For } Y: (9+1)/2 = 5$$

Step 3: Starting with the lowest score, count up to the value from step 2 within each data set (counting in from both ends of the rank-ordered data):

$$X(3, 6, 8, 10, \mathbf{10}, \mathbf{10}, 11, 11, 13, 18)$$
$$Y(4, 6, 6, 7, \mathbf{9}, 10, 13, 17, 18)$$

Step 4: If the midpoint falls between the two numbers, add them and divide by 2:

$$\text{For } X, Mdn = (10+10)/2 = 10$$
$$\text{For } Y, Mdn = 9.0$$

The median is a useful descriptive statistic especially for skewed or nonsymmetrically distributed sets of scores or data. Such data are very common when research is being conducted from observations of people in real-world environments that are not controlled by the researcher. The median, unlike our final central tendency statistic, the arithmetic mean, is not affected by the presence of outlying or extremely high or low scores.

Arithmetic Mean, *M*

Perhaps the most common measure of central tendency in social and behavioral science research is the arithmetic mean, M. When the data or set of scores we have collected roughly follows a normal or bell-curved shape, then this central tendency statistic is the preferred one because of its value to many of the other statistics we commonly rely on when testing our hypotheses. For the sake of definition, the mean is the sum of all observed scores divided by the number of observations in the data set. We define the arithmetic mean as

$$M = \frac{\sum X}{n} \tag{A.3}$$

Here is an example of how to calculate the arithmetic mean using two data sets, X and Y:

$$X\{18, 10, 11, 13, 8, 11, 6, 10, 3, 10\}$$
$$Y\{18, 17, 4, 6, 6, 13, 7, 9, 10\}$$

Step 1: Calculate the sum of scores for each set:

$$\Sigma X = 18 + 10 + 11 + 13 + 8 + 11 + 6 + 10 + 3 + 10 = 100.0$$
$$n_X = 10$$
$$\Sigma Y = 18 + 17 + 4 + 6 + 6 + 13 + 7 + 9 + 10 = 90.0$$
$$n_Y = 9$$

Step 2: Divide the sum of scores (ΣX) by the number of scores that were summated in step 1 (in this case, the number of scores is equivalent to the sample size of each group, n_x or n_Y):

$$M_X = \frac{100}{10} = 10$$
$$M_Y = \frac{90}{9} = 10$$

The mean has a number of unique features. Conceptually, the mean represents the fulcrum or tipping point upon which a set of scores is balanced. The statistical implication of this is that deviations of each score in a data set will deviate less from the mean than from any other value. In other words, $\Sigma(X - M) = 0$. When this value is squared, that is, $\Sigma(X - M)^2$, the resulting value is considered to be an index of the smallest possible deviation or variability within a given data set (this second feature is known as the *least squares criterion*). In addition to these two properties, the arithmetic mean is also viewed as an unbiased estimate of the population mean, μ.

Table A.1 Summary of Frequently Used Measures of Central Tendency

Statistic	Definition	Appropriate for these scales of measurement	Features/properties
Mode, M_0	Most common score in a data set	Nominal, ordinal interval, ratio	Imprecise and potentially misleading for ordinal, interval, or ratio data
Median, *Mdn*	Midpoint of a ranked distribution of scores	Ordinal, interval, ratio	Good to use when extreme scores skew the distribution of data
Arithmetic mean, *M*	Total of a set of scores divided by the number of scores	Interval, ratio	Commonly used measure; $\Sigma(X - M) = 0$ and $\Sigma(X - M)^2$ = minimal value; *M* is an unbiased estimate of μ

Despite these strengths, this central tendency statistic is sensitive to outliers and, therefore, to skewed distributions of data. When the data are positively skewed, the mean will be greater than the median (e.g., $M > Mdn$). When the data are negatively skewed, the mean will be less than the median (e.g., $M < Mdn$). More advanced statistical techniques are increasingly available to help researchers who are dealing with data that do not conform to core distributional assumptions. These techniques are beyond the scope of this text, but interested readers are encouraged to learn more about these options by consulting texts such as Wilcox, R.R. (2005). *Introduction to robust estimation and hypothesis testing* (2nd ed.). Burlington, MA: Elsevier Academic Press.

Table A.1 provides a summary of the main measures of central tendency presented here.

MEASURES OF VARIABILITY

A central tendency statistic indicates only one characteristic of a set of data: its typical or most common score. Although this is a useful bit of information, we also really need additional information to fully describe a set of data. One of the more important characteristics of a data set is the amount of variability present among its scores. Variability reflects differences between data points in the data set and the typical or most common score (i.e., the spread or dispersion of the data points around the central tendency of the overall set of data). For researchers to draw valid inferences from their data, it is very important to know whether the data they have collected are clustered close to the measure of central tendency or spread widely apart (e.g., indicating lack of consistency in responses or scores). Much like the measures of central tendency, selecting an appropriate measure of variability depends on the measurement scale used to represent the variable, the symmetry of the data, and the inferences to be made from the measure of variability.

Simple Range

The range is the easiest measure of variability to calculate. The definition of the range is

$$\text{Range} = X_{\text{highest}} - X_{\text{lowest}} \tag{A.4}$$

Like the mode, the range is easy to calculate; however, it is greatly affected by outliers. Therefore, most researchers use it only as a general descriptive tool. Here is an example of how to calculate the range:

$$X\{18, 10, 11, 13, 8, 11, 6, 10, 3, 10\}$$
$$Y\{18, 17, 4, 6, 6, 13, 7, 9, 10\}$$

Step 1: Rank the scores from lowest to highest:

$$X\{3, 6, 8, 10, 10, 10, 11, 11, 13, 18\}$$
$$Y\{4, 6, 6, 7, 9, 10, 13, 17, 18\}$$

Step 2: Determine the difference between the highest and lowest scores:

$$\text{Range}_X = 15.0 = 18 - 3$$
$$\text{Range}_Y = 14.0 = 18 - 4$$

Semi-Interquartile Range

Another measure of score variability is the semi-interquartile range (*SIR*). The semi-interquartile range is defined as half of the difference between the 75th percentile and the 25th percentile in a set of scores. This probably makes more sense when you look at the formula

$$SIR = \frac{Q_3 - Q_1}{2} \tag{A.5}$$

In this equation, Q_3 represents the 75th percentile and Q_1 represents the 25th percentile. The 75th percentile indicates that 75% of the scores in the data set are at or below the value for Q_3, while 25% of the scores in the set are at or below the value of Q_1. To help you keep things in perspective, remember that the median (*Mdn*) is also sometimes called Q_2, and it represents the midpoint of a set of scores or the value below which 50% of the scores in a set would exist in the distribution of scores. In this way, the *SIR* can be thought of as an appropriate index of variability around a median, ignoring the upper 25% and lower 25% of scores (which might be extreme, relative to the middle 50% of scores). The following example illustrates how to calculate the *SIR*:

$$X\{18, 10, 11, 13, 8, 11, 6, 10, 3, 10\}$$
$$Y\{18, 17, 4, 6, 6, 13, 7, 9, 10\}$$

Step 1: Rank the scores from lowest to highest:

$$X\{3, 6, 8, 10, 10, 10, 11, 11, 13, 18\}$$
$$Y\{4, 6, 6, 7, 9, 10, 13, 17, 18\}$$

Step 2: Calculate the median (add 1 to the sample size and divide by 2):

$$\text{For } X, Q_2 \text{ or } Mdn = (10+1)/2 = 5.5$$
$$\text{For } Y, Q_2 \text{ or } Mdn = (9+1)/2 = 5$$

Step 3: Starting with the lowest score, count up to the value from step 2:

$$X\{3, 6, 8, 10, 10, 10, 11, 11, 13, 18\}$$
$$\uparrow$$
$$Y\{4, 6, 6, 7, 9, 10, 13, 17, 18\}$$
$$\uparrow$$

Step 4: Determine the midpoint of the upper and lower halves using the same procedure described in step 2. Specifically, add 1 to the number of scores in each half and divide by 2:

$$\text{Position of } Q_3 \text{ and } Q_1 \text{ for } X = (5+1)/2 = 3$$
$$\text{Position of } Q_3 \text{ and } Q_1 \text{ for } Y = (4+1)/2 = 2.5$$

Step 5: Using numbers from the previous step, determine Q_1 and Q_3. If the location of Q_1 or Q_3 falls between two numbers, take the average of the two:

$$X\{3, 6, 8, 10, 10, 10, 11, 11, 13, 18\}$$
$$\uparrow \qquad \uparrow$$
$$\text{For } X, Q_1 = 8.0 \text{ and } Q_3 = 11.0$$

$$Y\{4, 6, 6, 7, 9, 10, 13, 17, 18\}$$
$$\uparrow \qquad \uparrow$$
$$\text{For } Y, Q_1 = (6+6)/2 = 6 \text{ and } Q_3 = (13+17)/2 = 15$$

Step 6: Calculate the *SIR*:

$$SIR_X = (11-8)/2 = 1.5$$
$$SIR_Y = (15-6)/2 = 4.5$$

As with the median, extreme scores do not affect the semi-interquartile range. Therefore, researchers often use the statistic to describe a distribution of data that is skewed or somehow deviant from a normal bell-shaped form.

Variance and Standard Deviation

The variance and standard deviation are two of the most commonly used measures of score variability within behavioral and social science research. There are several ways to calculate these statistics; we present a definitional and a computational form of their equations. Statisticians use the *definitional equation* to help describe and explain the purpose of these two important variability statistics. *Computational equations*, as the name implies, are equations people use when they have to calculate these statistics by hand. Even if you never perform one of these calculations yourself, understanding these formulas can help you interpret these statistics when they are generated by your preferred statistical analysis software program.

As you can see in the following summary table, the sample statistic form yields the variance and standard deviation when the data are the population. The second form, the unbiased estimate of the population, allows us to estimate the population parameter using data from a sample. In most cases, you will likely use the unbiased estimate of the population form of the equations, especially if your ultimate goal is to generalize your findings beyond the sample you studied.

Sample statistic	Unbiased estimate of population value
Variance	
Definitional equations	
$s^2 = \frac{\sum (X-M)^2}{N}$	$VAR = \frac{\sum (X-M)^2}{n-1}$
Computational equations	
$s^2 = \frac{\sum X^2 - \frac{\left(\sum X\right)^2}{N}}{N}$	$VAR = \frac{\sum X^2 - \frac{\left(\sum X\right)^2}{n}}{n-1}$
Standard deviation	
Definitional equations	
$s = \sqrt{\frac{\sum (X-M)^2}{N}}$	$SD = \sqrt{\frac{\sum (X-M)^2}{n-1}}$
Computational equations	
$s = \sqrt{\frac{\sum X^2 - \frac{\left(\sum x\right)^2}{N}}{N}}$	$SD = \sqrt{\frac{\sum X^2 - \frac{\left(\sum X\right)^2}{n}}{n-1}}$

Considering the definitional equations for the variance and standard deviations can further help to illustrate what these statistics actually mean. If you look at the numerator of these equations, you will see the expression $\Sigma(X - M)^2$. This element of the equation is also referred to as the *sum of squares*. More specifically, the sum of squares is the sum of the squared deviation scores, which are calculated by subtracting the *M* of a set of scores from each of the scores in that set (i.e., each of the $X - M$). As such, these deviation scores reflect the difference between the mean and an observed score. As you can see, a larger difference between the mean and the observed score yields a larger deviation score. Because $\Sigma(X - M) = 0$, we square each deviation score. Squaring the deviation scores ensures that there are no negative deviation scores and that $\Sigma(X - M)^2$ will always be greater than 0 (unless all the scores in a data set are the same number).

Once we determine the sum of squares, we divide this number by another number, the denominator. For the unbiased estimate of the population equation, the denominator is $n - 1$. Thus, the variance is an estimate of the typical squared deviation between the scores and the mean. The final step is to calculate the square root of the variance, otherwise known as the *standard deviation*. Doing so returns the value of this variability statistic to the scale that was used to measure variable *X* in the first place. This means that the standard deviation is an index of the typical deviation of scores within a collected data set gathered from a sample of participants.

The following is a step-by-step example of how to determine the variance and the standard deviation of a set of data. We can begin by using a sample of data:

$$X\{18, 10, 11, 13, 8, 11, 6, 10, 3, 10\}$$

Step 1: Calculate the sum of scores:

$$\Sigma X = 18 + 10 + 11 + 13 + 8 + 11 + 6 + 10 + 3 + 10 = 100$$

Step 2: Calculate the sum of squared scores:

$$\Sigma X^2 = 18^2 + 10^2 + 11^2 + 13^2 + 8^2 + 11^2 + 6^2 + 10^2 + 3^2 + 10^2$$

$$\Sigma X^2 = 324 + 1100 + 121 + 169 + 64 + 121 + 36 + 100 + 9 + 100 = 1144$$

Step 3: Complete the equations.

Sample statistic	Unbiased estimate of population value
Variance	
$s^2 = \dfrac{\sum X^2 - \dfrac{\left(\sum X\right)^2}{N}}{N}$	$VAR = \dfrac{\sum X^2 - \dfrac{\left(\sum X\right)^2}{n}}{n-1}$
$s^2 = \dfrac{1144 - \dfrac{100^2}{10}}{10} = \dfrac{144}{10} = 14.4$	$VAR = \dfrac{1144 - \dfrac{(100)^2}{10}}{10-1} = \dfrac{144}{9} = 16$
Standard deviation	
$s = \sqrt{s^2}$	$SD = \sqrt{VAR}$
$s = \sqrt{14.4} = 3.79$	$SD = \sqrt{16} = 4$

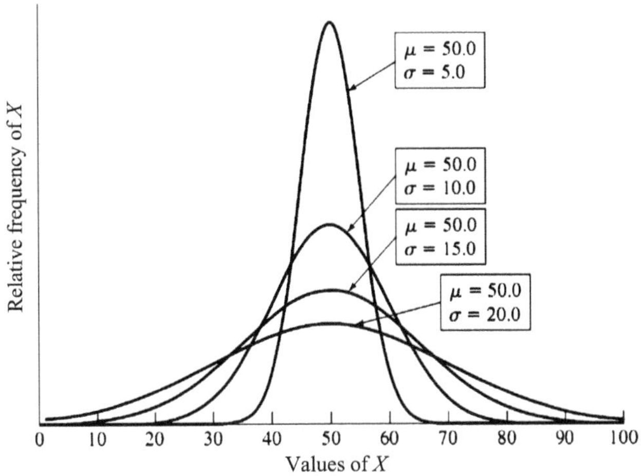

Figure A.1 Four distributions of data, each with a mean of 50 and a standard deviation of 5, 10, 15, or 20. *Note:* As the standard deviation increases, the spread of the scores around the mean becomes much wider.

Figure A.1 presents four hypothetical distributions with population means of 50 and standard deviations of 5, 10, 15, and 20. As the standard deviation increases, the spread of data becomes much wider. For the distribution with a standard deviation of 5, the majority of the scores are close to the mean of 50, with close to 95% of these scores falling between 40 and 60. In contrast, the distribution with the standard deviation of 20 has scores spread across the entire response scale from 10 to 90. Table A.2 summarizes the frequently used measures of variability we have discussed.

STANDARDIZED OR TRANSFORMED SCORES

Sometimes we need to compare scores that come from multiple different assessments or measurement periods. Because these scores may not share exactly the same measurement scale or properties (including variance and central tendency), it is often necessary to transform or standardize them first so that a direct, apples-to-apples comparison is possible. The z-score is the most widely used form of this type of transformed or standardized score. These standardized scores are also the basis for calculating correlation coefficients between two variables. Another advantage of the z-score is that it can help us interpret or make inferences about specific observations. If we assume that the normal distribution represents our data, we can use the z-score to convert observed scores to percentiles.

There are two ways to calculate the z-score. The first method assumes that the data represent an intact group. Therefore, we use s to calculate the standard deviation of the data. The equation for this version of the z-score is

$$z = \frac{(X - M)}{s} \quad \text{(A.6)}$$

In the second method, if the data are collected from a sample and used to estimate population parameters (a more common scenario in most social science

Table A.2 Summary of Frequently Used Measures of Variability

Statistic	Definition	Appropriate for these measurement scales	Features/properties
Range	$\text{Range} = X_{\text{highest}} - X_{\text{lowest}}$		
Semi-interquartile range, *SIR*	$SIR = \frac{Q_3 - Q_1}{2}$	Ordinal, interval, or ratio	Imprecise and potentially misleading when the data contain outliers
Variance of a sample, s^2	$s^2 = \frac{\sum (X - M)^2}{N}$	Ordinal, interval, or ratio	Good to use when extreme scores skew the distribution of data
Standard deviation of a sample, *s*	$s = \sqrt{\frac{\sum (X - M)^2}{N}}$	Interval or ratio	Used when the data represent the population of scores
Unbiased estimate of standard variance, *VAR*	$VAR = \frac{\sum (X - M)^2}{N - 1}$	Interval or ratio	Used when data represent a sample, but the desire is to draw inferences about the larger population (more common in behavioral and social science research)
Unbiased estimate of standard deviation, *SD*	$SD = \sqrt{\frac{\sum (X - M)^2}{n - 1}}$		

research), we may use *SD* (calculated with $n - 1$ in the denominator) as a less-biased estimate of the variability element of this equation. Thus, in this case, we would use the following equation to calculate the *z*-score:

$$z = \frac{(X - M)}{SD} \tag{A.7}$$

To illustrate the use of these formulas, consider the following example. Imagine that you give 10 people two tests, test A and test B. We will assume that the data from these tests represent an intact group. Therefore, we will calculate *s* to determine the standard deviation. Looking at Table A.3, you can see the tests have different means and standard deviations. For test A, $M_A = 100$ and $s_A = 75$. For test B, $M_B = 3.0$ and $s_B = 1.0$.

Table A.3 also presents the *z*-scores for the tests. The means and standard deviations of the *z*-scores for test A and test B are $M = 0.0$ and $s = 1.0$. This is always the case for *z*-scores. Thus, when you calculate *z*-scores using *s* for the standard deviation, then

$$\Sigma z = 0.0 \quad \Sigma z^2 = n \quad M_z = 0.0 \quad s_z = 1.0$$

When you use *SD* to calculate *z*-scores, then

$$\Sigma z = 0.0 \quad \Sigma z^2 = n - 1 \quad M_z = 0.0 \quad s_z = 1.0$$

Table A.3 Example of z-Scores within a Study in which Each Person Completed Two Tests

Person	Test A	Test A z-Scores	Test B	Test B z-Scores
A	440	(440 – 500)/75 = **–0.8**	3.4	(3.4 – 3.0)/1 = **0.4**
B	560	(560 – 500)/75 = **0.8**	2.4	(2.4 – 3.0)/1 = **–0.6**
C	515	(515 – 500)/75 = **0.2**	2.8	(2.8 – 3.0)/1 = **–0.2**
D	395	(395 – 500)/75 = **–1.4**	4.2	(4.2 – 3.0)/1 = **1.2**
E	500	(500 – 500)/75 = **0.0**	3.0	(3.0 – 3.0)/1 = **0.0**
F	545	(545 – 500)/75 = **0.6**	2.6	(2.6 – 3.0)/1 = **–0.4**
G	605	(605 – 500)/75 = **1.4**	1.2	(1.2 – 3.0)/1 = **–1.8**
H	425	(425 – 500)/75 = **–1.0**	3.6	(3.6 – 3.0)/1 = **0.6**
I	605	(605 – 500)/75 = **1.4**	4.8	(4.8 – 3.0)/1 = **1.8**
J	410	(410 – 500)/75 = **–1.2**	2.0	(2.0 – 3.0)/1 = **–1.0**
$\Sigma X_A = 5{,}000$		$\Sigma z_A = 0.0$	$\Sigma X_B = 30$	$\Sigma z_B = 0.0$
$\Sigma X_A^2 = 2{,}556{,}250$		$\Sigma X_A^2 = 10$	$\sum z_B^2 = 100$	$\sum z_B^2 = 10$
$M_A = 500$		$M_{A_z} = 0.0$	$M_B = 3.0$	$M_{B_z} = 0.0$
$s_A = 75$		$s_{A_z} = 1.0$	$s_B = 1.0$	$s_{B_z} = 1.0$

Note: Next to each test score is the corresponding z-score, which represents the relative difference between the observed score and the group mean score.

Once we convert the observed scores to z-scores, we can directly compare pairs of scores from different groups. As an example, consider person B's scores on the two tests. Person B earned a score of 560 on the first test and a score of 2.4 on the second test. We can say that person B scored better than average on the first test because that person's z-score for that test is $z = 0.80$. This score indicates that person B scored 0.8 standard deviations above the group mean score on that test. However, person B scored 0.6 standard deviations below the group average on the second test, as reflected by the z-score of $z = -0.6$.

Thus, using the z-score, we can quickly determine whether a person's score is above or below the mean and the relative magnitude of the difference. For example, z-scores of –0.2 and 0.1 indicate that the scores are not too far below or above the mean. In contrast, z-scores of –1.95 or 1.64 indicate scores far below and far above the mean.

Determining Percentiles Using the *z*-Score

Once we have converted an observed score (X) to a z-score, we can then use this z-score to determine a percentile for each score. A percentile is a number that represents the percentage of individuals that received the same or a lower/higher score. For example, if you took a test and learned that you scored at the 90th percentile, then you can conclude that you did as well as or better than 90% of the people who took the test. If, however, your test score was at the 45th percentile, you should conclude that your score was lower than average, being only as high as or better than 45% of the test takers.

Converting z-scores to percentiles is rather easy, especially if we use Table B.1 of Appendix B. If you look at that table, you will see that there are three columns. Column A represents the z-scores. Because the normal distribution is symmetrical, we can use this column to examine positive as well as negative z-scores. Column B of the table represents the area of the normal distribution that exists between the mean and the z-score. Column C represents the area of the curve beyond the z-score. Using this table, we can convert the z-score to a percentile. The following instructions show you how to convert negative and positive z-scores to percentiles.

Negative *z*-Scores

If the z-score is negative or equal to 0 ($z \leq 0$), then use the following procedure.

Step 1: Convert the observed score to a z-score:

$$X = 440 \quad M = 500 \quad s = 75$$

$$z = (440 - 500)/75 = -0.8$$

Step 2: Locate the value of the z-score, ignoring the sign, using column A from Table B.1 of Appendix B, and then note the value listed in column C. This value represents the proportion of the distribution beyond the z-score:

$$z = 0.80 \quad \text{Proportion beyond } z - \text{score} = .2119$$

Step 3: Multiply the proportion beyond the z-score by 100 and round to the nearest whole number. The resulting product is the percentile that you are seeking:

$$.2119 \times 100 = 21.19 \rightarrow 21\%$$

In this example then, we can conclude that only 21% of people taking the test will earn a score of 440 or less (remember that the original z-score was negative). In other words, if you gave this test to 100 people, you would expect that only 21 of them would have a score of 440 or lower. You could also conclude that 79% (100% – 21%) of the people taking the test would have a score greater than 440.

Positive *z*-Scores

If the z-score is positive ($z > 0$), then use the following procedure.

Step 1: Convert the observed score to a z-score:

$$X = 515 \quad M = 500 \quad s = 75$$

$$z = (515 - 500)/75 = 0.20$$

Step 2: Locate the value of the z-score using column A from Table B.1 of Appendix B, and then note the value listed in column B. This value represents the proportion of the distribution of scores that falls between the mean and this individual z-score:

$$z = 0.20 \quad \text{Proportion between mean and } z - \text{score} = .0793$$

Step 3: Add the value found in step 2 and .5000. This step accounts for the half of the normal curve that falls below the mean:

$$.5793 = .5000 + .0793$$

Step 4: Multiply the result of step 3 by 100 and round to the nearest whole number. The product is the percentile:

$$.5793 \times 100 = 57.93 \rightarrow 58\%$$

Therefore, a score of 515 represents the 58th percentile, indicating that 58% of the people taking this test would be expected to score at or below 515. What percentage of people taking the test will score greater than 515? Given the mean and standard deviation of the test, we would expect that 42% (100% – 58%) will receive a score of 515 or greater. Figure A.2 and Table A.4 illustrate how to determine the percentile score for each of the 10 scores recorded for test A. When you convert z-scores to percentiles, it is a good idea to draw a sketch of the normal distribution and label the mean and relative location of the z-score.

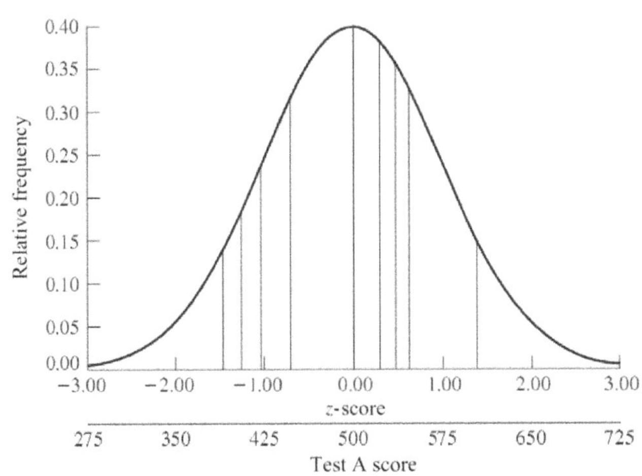

Figure A.2 Normal curve.
Note: The vertical lines represent the location of values of X presented in Table A.4.

Table A.4 Complete Illustration of Converting Observed Scores to Percentiles

X	z-Scores	Area below z-scores	Rounded percentile
395	(395 − 500)/75 = **−1.40**	.0808	.0808 × 100 = 8
410	(410 − 500)/75 = **−1.20**	.1151	.1151 × 100 = 12
425	(425 − 500)/75 = **−1.00**	.1587	.1586 × 100 = 16
440	(440 − 500)/75 = **−0.80**	.2119	.2119 × 100 = 21
500	(500 − 500)/75 = **0.00**	.5000	.5000 × 100 = 50
515	(515 − 500)/75 = **0.20**	.5000 + .0793 = .5793	.5793 × 100 = 58
545	(545 − 500)/75 = **0.60**	.5000 + .2257 = .7257	.7257 × 100 = 73
560	(560 − 500)/75 = **0.80**	.5000 + .2881 = .7881	.7881 × 100 = 79
605	(605 − 500)/75 = **1.40**	.5000 + .4192 = .9192	.9192 × 100 = 92
605	(605 − 500)/75 = **1.40**	.5000 + .4192 = .9192	.9192 × 100 = 92

Note: Areas found in Table B.1 in Appendix B.

KNOWLEDGE CHECK

XA_1	ZA_1	Percentile	XA_2	XA_2	Percentile
26			11		
23			29		
17			17		
29			41		
22	1.200	88	27	−1.200	12
19			25		
11			5		
20			23		
18			19		
15			33		

Using the data in the preceding table, calculate the following:

a. $\sum X_{a1}$ $\sum X_{a2}$

b. $\Sigma_1 + \Sigma_2$

c. $(\Sigma X_{a1})^2$ $(\Sigma X_{a2})^2$

d. Mdn_{a1} Mdn_{a2}

e. M_{a1} M_{a2}

f. S^2_{a1} S^2_{a2}

g. s_{a1} s_{a2}

h. VAR_{a1} VAR_{a2}

i. SD_{a1} SD_{a2}

j. Using the data, and M and s for each group, convert each score to a z-score.

k. Using the z-scores you just calculated, determine the percentile for each score.

Appendix B

Statistical Tables

TABLE B.1: PROPORTIONS OF THE AREA UNDER THE NORMAL CURVE

Using Table B.1

To use Table B.1, convert the raw score to a *z*-score using the equation below (also discussed in Appendix A), where *X* is the observed score, *M* is the mean of the data, and *SD* is the standard deviation of the data:

Understanding and Conducting Research in the Health Sciences, First Edition. Christopher J. L. Cunningham, Bart L. Weathington, and David J. Pittenger.

Table B.1

A	B	C	A	B	C	A	B	C
z	Area between M and z	Area beyond z	z	Area between M and z	Area beyond z	z	Area between M and z	Area beyond z
0.00	0.0000	0.5000	0.40	0.1554	0.3446	0.80	0.2881	0.2119
0.01	0.0040	0.4960	0.41	0.1591	0.3409	0.81	0.2910	0.2090
0.02	0.0080	0.4920	0.42	0.1628	0.3372	0.82	0.2939	0.2061
0.03	0.0120	0.4880	0.43	0.1664	0.3336	0.83	0.2967	0.2033
0.04	0.0160	0.4840	0.44	0.1700	0.3300	0.84	0.2995	0.2005
0.05	0.0199	0.4801	0.45	0.1736	0.3264	0.85	0.3023	0.1977
0.06	0.0239	0.4761	0.46	0.1772	0.3228	0.86	0.3051	0.1949
0.07	0.0279	0.4721	0.47	0.1808	0.3192	0.87	0.3078	0.1922
0.08	0.0319	0.4681	0.48	0.1844	0.3156	0.88	0.3106	0.1894
0.09	0.0359	0.4641	0.49	0.1879	0.3121	0.89	0.3133	0.1867
0.10	0.0398	0.4602	0.50	0.1915	0.3085	0.90	0.3159	0.1841
0.11	0.0438	0.4562	0.51	0.1950	0.3050	0.91	0.3186	0.1814
0.12	0.0478	0.4522	0.52	0.1985	0.3015	0.92	0.3212	0.1788
0.13	0.0517	0.4483	0.53	0.2019	0.2981	0.93	0.3238	0.1762
0.14	0.0557	0.4443	0.54	0.2054	0.2946	0.94	0.3264	0.1736
0.15	0.0596	0.4404	0.55	0.2088	0.2912	0.95	0.3289	0.1711
0.16	0.0636	0.4364	0.56	0.2123	0.2877	0.96	0.3315	0.1685
0.17	0.0675	0.4325	0.57	0.2157	0.2843	0.97	0.3340	0.1660
0.18	0.0714	0.4286	0.58	0.2190	0.2810	0.98	0.3365	0.1635
0.19	0.0753	0.4247	0.59	0.2224	0.2776	0.99	0.3389	0.1611
0.20	0.0793	0.4207	0.60	0.2257	0.2743	.99	0.3413	0.1587
0.21	0.0832	0.4168	0.61	0.2291	0.2709	1.01	0.3438	0.1562
0.22	0.0871	0.4129	0.62	0.2324	0.2676	1.02	0.3461	0.1539
0.23	0.0910	0.4090	0.63	0.2357	0.2643	1.03	0.3485	0.1515
0.24	0.0948	0.4052	0.64	0.2389	0.2611	1.04	0.3508	0.1492
0.25	0.0987	0.4013	0.65	0.2422	0.2578	1.05	0.3531	0.1469
0.26	0.1026	0.3974	0.66	0.2454	0.2546	1.06	0.3554	0.1446
0.27	0.1064	0.3936	0.67	0.2486	0.2514	1.07	0.3577	0.1423
0.28	0.1103	0.3897	0.68	0.2517	0.2483	1.08	0.3599	0.1401
0.29	0.1141	0.3859	0.69	0.2549	0.2451	1.09	0.3621	0.1379
0.30	0.1179	0.3821	0.70	0.2580	0.2420	1.10	0.3643	0.1357
0.31	0.1217	0.3783	0.71	0.2611	0.2389	1.11	0.3665	0.1335
0.32	0.1255	0.3745	0.72	0.2642	0.2358	1.12	0.3686	0.1314
0.33	0.1293	0.3707	0.73	0.2673	0.2327	1.13	0.3708	0.1292
0.34	0.1331	0.3669	0.74	0.2704	0.2296	1.14	0.3729	0.1271
0.35	0.1368	0.3632	0.75	0.2734	0.2266	1.15	0.3749	0.1251
0.36	0.1406	0.3594	0.76	0.2764	0.2236	1.16	0.3770	0.1230
0.37	0.1443	0.3557	0.77	0.2794	0.2206	1.17	0.3790	0.1210
0.38	0.1480	0.3520	0.78	0.2823	0.2177	1.18	0.3810	0.1190
0.39	0.1517	0.3483	0.79	0.2852	0.2148	1.19	0.3830	0.1170
1.20	0.3849	0.1151	1.60	0.4452	0.0548	2.00	0.4772	0.0228
1.21	0.3869	0.1131	1.61	0.4463	0.0537	2.01	0.4778	0.0222

Table B.1 (*Continued*)

A	B	C	A	B	C	A	B	C
z	Area between M and z	Area beyond z	z	Area between M and z	Area beyond z	z	Area between M and z	Area beyond z
1.22	0.3888	0.1112	1.62	0.4474	0.0526	2.02	0.4783	0.0217
1.23	0.3907	0.1093	1.63	0.4484	0.0516	2.03	0.4788	0.0212
1.24	0.3925	0.1075	1.64	0.4495	0.0505	2.04	0.4793	0.0207
1.25	0.3944	0.1056	1.65	0.4505	0.0495	2.05	0.4798	0.0202
1.26	0.3962	0.1038	1.66	0.4515	0.0485	2.06	0.4803	0.0197
1.27	0.3980	0.1020	1.67	0.4525	0.0475	2.07	0.4808	0.0192
1.28	0.3997	0.1003	1.68	0.4535	0.0465	2.08	0.4812	0.0188
1.29	0.4015	0.0985	1.69	0.4545	0.0455	2.09	0.4817	0.0183
1.30	0.4032	0.0968	1.70	0.4554	0.0446	2.10	0.4821	0.0179
1.31	0.4049	0.0951	1.71	0.4564	0.0436	2.11	0.4826	0.0174
1.32	0.4066	0.0934	1.72	0.4573	0.0427	2.12	0.4830	0.0170
1.33	0.4082	0.0918	1.73	0.4582	0.0418	2.13	0.4834	0.0166
1.34	0.4099	0.0901	1.74	0.4591	0.0409	2.14	0.4838	0.0162
1.35	0.4115	0.0885	1.75	0.4599	0.0401	2.15	0.4842	0.0158
1.36	0.4131	0.0869	1.76	0.4608	0.0392	2.16	0.4846	0.0154
1.37	0.4147	0.0853	1.77	0.4616	0.0384	2.17	0.4850	0.0150
1.38	0.4162	0.0838	1.78	0.4625	0.0375	2.18	0.4854	0.0146
1.39	0.4177	0.0823	1.79	0.4633	0.0367	2.19	0.4857	0.0143
1.40	0.4192	0.0808	1.80	0.4641	0.0359	2.20	0.4861	0.0139
1.41	0.4207	0.0793	1.81	0.4649	0.0351	2.21	0.4864	0.0136
1.42	0.4222	0.0778	1.82	0.4656	0.0344	2.22	0.4868	0.0132
1.43	0.4236	0.0764	1.83	0.4664	0.0336	2.23	0.4871	0.0129
1.44	0.4251	0.0749	1.84	0.4671	0.0329	2.24	0.4875	0.0125
1.45	0.4265	0.0735	1.85	0.4678	0.0322	2.25	0.4878	0.0122
1.46	0.4279	0.0721	1.86	0.4686	0.0314	2.26	0.4881	0.0119
1.47	0.4292	0.0708	1.87	0.4693	0.0307	2.27	0.4884	0.0116
1.48	0.4306	0.0694	1.88	0.4699	0.0301	2.28	0.4887	0.0113
1.49	0.4319	0.0681	1.89	0.4706	0.0294	2.29	0.4890	0.0110
1.50	0.4332	0.0668	1.90	0.4713	0.0287	2.30	0.4893	0.0107
1.51	0.4345	0.0655	1.91	0.4719	0.0281	2.31	0.4896	0.0104
1.52	0.4357	0.0643	1.92	0.4726	0.0274	2.32	0.4898	0.0102
1.53	0.4370	0.0630	1.93	0.4732	0.0268	2.33	0.4901	0.0099
1.54	0.4382	0.0618	1.94	0.4738	0.0262	2.34	0.4904	0.0096
1.55	0.4394	0.0606	1.95	0.4744	0.0256	2.35	0.4906	0.0094
1.56	0.4406	0.0594	1.96	0.4750	0.0250	2.36	0.4909	0.0091
1.57	0.4418	0.0582	1.97	0.4756	0.0244	2.37	0.4911	0.0089
1.58	0.4429	0.0571	1.98	0.4761	0.0239	2.38	0.4913	0.0087
1.59	0.4441	0.0559	1.99	0.4767	0.0233	2.39	0.4916	0.0084
2.40	0.4918	0.0082	2.80	0.4974	0.0026	3.20	0.4993	0.0007
2.41	0.4920	0.0080	2.81	0.4975	0.0025	3.22	0.4994	0.0006
2.42	0.4922	0.0078	2.82	0.4976	0.0024	3.24	0.4994	0.0006

(*Continued*)

Table B.1 (*Continued*)

A	B	C	A	B	C	A	B	C
z	Area between *M* and *z*	Area beyond *z*	*z*	Area between *M* and *z*	Area beyond *z*	*z*	Area between *M* and *z*	Area beyond *z*
2.43	0.4925	0.0075	2.83	0.4977	0.0023	3.26	0.4994	0.0006
2.44	0.4927	0.0073	2.84	0.4977	0.0023	3.28	0.4995	0.0005
2.45	0.4929	0.0071	2.85	0.4978	0.0022	3.30	0.4995	0.0005
2.46	0.4931	0.0069	2.86	0.4979	0.0021	3.32	0.4995	0.0005
2.47	0.4932	0.0068	2.87	0.4979	0.0021	3.34	0.4996	0.0004
2.48	0.4934	0.0066	2.88	0.4980	0.0020	3.36	0.4996	0.0004
2.49	0.4936	0.0064	2.89	0.4981	0.0019	3.38	0.4996	0.0004
2.50	0.4938	0.0062	2.90	0.4981	0.0019	3.40	0.4997	0.0003
2.51	0.4940	0.0060	2.91	0.4982	0.0018	3.42	0.4997	0.0003
2.52	0.4941	0.0059	2.92	0.4982	0.0018	3.44	0.4997	0.0003
2.53	0.4943	0.0057	2.93	0.4983	0.0017	3.46	0.4997	0.0003
2.54	0.4945	0.0055	2.94	0.4984	0.0016	3.48	0.4997	0.0003
2.55	0.4946	0.0054	2.95	0.4984	0.0016	3.50	0.4998	0.0002
2.56	0.4948	0.0052	2.96	0.4985	0.0015	3.52	0.4998	0.0002
2.57	0.4949	0.0051	2.97	0.4985	0.0015	3.54	0.4998	0.0002
2.58	0.4951	0.0049	2.98	0.4986	0.0014	3.56	0.4998	0.0002
2.59	0.4952	0.0048	2.99	0.4986	0.0014	3.58	0.4998	0.0002
2.60	0.4953	0.0047	3.00	0.4987	0.0013	3.60	0.4998	0.0002
2.61	0.4955	0.0045	3.01	0.4987	0.0013	3.62	0.4999	0.0001
2.62	0.4956	0.0044	3.02	0.4987	0.0013	3.64	0.4999	0.0001
2.63	0.4957	0.0043	3.03	0.4988	0.0012	3.66	0.4999	0.0001
2.64	0.4959	0.0041	3.04	0.4988	0.0012	3.68	0.4999	0.0001
2.65	0.4960	0.0040	3.05	0.4989	0.0011	3.70	0.4999	0.0001
2.66	0.4961	0.0039	3.06	0.4989	0.0011	3.72	0.4999	0.0001
2.67	0.4962	0.0038	3.07	0.4989	0.0011	3.74	0.4999	0.0001
2.68	0.4963	0.0037	3.08	0.4990	0.0010	3.76	0.4999	0.0001
2.69	0.4964	0.0036	3.09	0.4990	0.0010	3.78	0.4999	0.0001
2.70	0.4965	0.0035	3.10	0.4990	0.0010	3.80	0.4999	0.0001
2.71	0.4966	0.0034	3.11	0.4991	0.0009	3.82	0.4999	0.0001
2.72	0.4967	0.0033	3.12	0.4991	0.0009	3.84	0.4999	0.0001
2.73	0.4968	0.0032	3.13	0.4991	0.0009	3.86	0.4999	0.0001
2.74	0.4969	0.0031	3.14	0.4992	0.0008	3.88	0.4999	0.0001
2.75	0.4970	0.0030	3.15	0.4992	0.0008	3.90	0.5000	0.0000
2.76	0.4971	0.0029	3.16	0.4992	0.0008	3.92	0.5000	0.0000
2.77	0.4972	0.0028	3.17	0.4992	0.0008	3.94	0.5000	0.0000
2.78	0.4973	0.0027	3.18	0.4993	0.0007	3.96	0.5000	0.0000
2.79	0.4974	0.0026	3.19	0.4993	0.0007	3.98	0.5000	0.0000

$$z = \frac{(X - M)}{SD}$$

The z-score is a standard deviate that allows you to use the standard normal distribution. The normal distribution has a mean of 0.0 and a standard deviation of 1.0. The normal distribution is symmetrical. The values in Table B.1 represent the proportion of area in the standard normal curve that occurs between specific points. The table contains z-scores between 0.00 and 3.98. Because the normal distribution is symmetrical, the table represents z-scores ranging between −3.98 and 3.98.

Column A of the table represents the z-score. Column B represents the proportion of the curve between the mean and the z-score. Column C represents the proportion of the curve that extends from the z-score to ∞.

Example

Negative z-score:	Positive z-score:
z-score = −1.30	z-score = +1.30

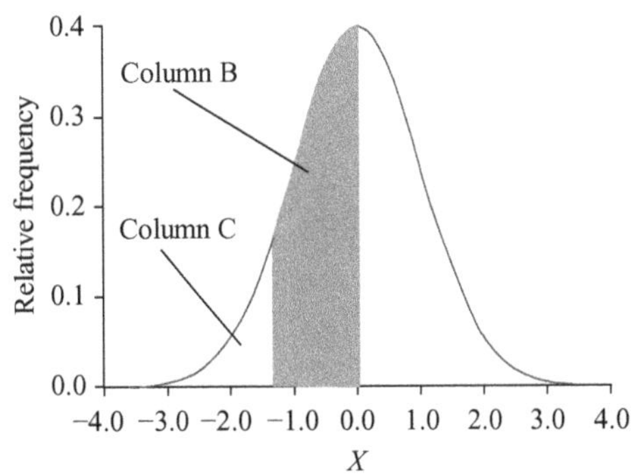

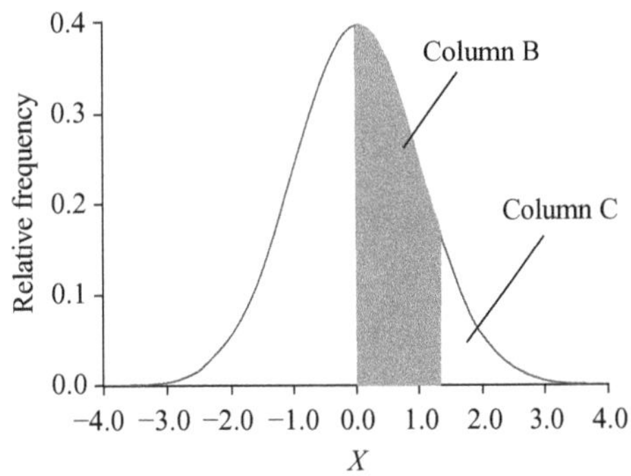

	Column B	Column C	
Negative z-scores			
Area between mean and $-z$	.4032		40.32% of curve
Area less than $-z$		.0968	9.68% of curve
Positive z-scores			
Area between mean and $+z$	.4032		40.32% of curve
Area greater than $+z$		.0968	9.68% of curve
Area **between** $-z$ and $+z$	.4032 + .4032 = .8064 or 80.64% of curve		
Area **below** $-z$ and **above** $+z$	.0968 + .0968 = .1936 or 19.36% of curve		

In the following examples, we add .5000 to the area between the mean and z-score. The .5000 represents the proportion of the curve on the complementary half of the normal curve.

Area at and **below** $+z$ = +1.30	.5000 + .4032 = .9032 or 90.32% of curve
Area at and **above** $-z$ = −1.30	.4032 + .5000 = .9032 or 90.32% of curve

TABLE B.2: 1200 TWO-DIGIT RANDOM NUMBERS

Using Table B.2

This table consists of two-digit random numbers that can range between 00 and 99 inclusive. To select a series of random numbers, select a column and row at random, and then record the numbers. You may move in any direction to generate the sequence of numbers.

Example

A researcher wished to randomly assign participants to one of five treatment conditions. Recognizing that the numbers in Table B.2 range between 00 and 99, the researcher decided to use the following table to convert the random numbers to the five treatment conditions:

Range of random numbers	Treatment condition
00–20	1
21–40	2
41–60	3
61–80	4
81–99	5

Table B.2

	1	2	3	4	5	6	7	8	9	10	11	12	13	14	15	16	17	18	19	20	21	22	23	24
1	43	41	16	31	22	44	10	41	45	00	47	19	43	67	83	02	79	05	98	92	64	82	06	89
2	26	44	01	04	28	85	11	91	23	02	39	79	44	45	93	20	17	91	35	15	25	82	18	41
3	83	39	26	84	04	16	89	79	68	85	61	63	03	20	17	76	95	80	27	39	35	82	10	86
4	65	94	48	27	77	65	34	95	04	51	78	90	14	76	90	83	17	76	69	50	34	01	25	08
5	89	38	32	05	09	49	87	93	21	24	88	74	30	94	26	19	23	72	94	80	90	24	55	44
6	77	80	30	43	26	01	43	46	66	40	52	00	44	69	84	10	48	96	49	85	49	84	97	41
7	43	42	26	74	51	05	56	43	06	80	58	22	57	02	11	95	00	91	88	17	71	98	32	56
8	76	76	61	17	69	06	73	37	77	06	36	28	05	73	31	04	44	33	40	74	46	26	02	99
9	42	05	88	83	15	05	28	52	88	78	88	66	50	80	24	38	31	20	48	73	18	85	18	90
10	46	74	76	34	97	40	59	34	86	11	50	98	69	59	46	74	59	60	98	76	96	42	34	83
11	67	15	82	94	59	55	27	99	02	34	47	34	88	98	72	15	38	73	57	42	56	09	85	83
12	03	58	51	69	14	89	24	06	35	31	16	65	71	76	04	80	01	36	00	67	78	73	07	37
13	79	98	19	32	25	95	89	54	20	78	29	81	96	34	62	53	26	09	02	04	63	95	03	53
14	56	12	61	36	21	69	96	06	22	06	01	80	57	72	23	55	05	74	42	55	91	45	60	91
15	58	80	33	35	75	33	35	42	06	79	73	29	89	73	99	07	05	54	42	77	78	99	33	92
16	31	51	77	53	92	51	35	71	34	46	79	43	76	15	76	46	40	04	36	84	83	64	56	73
17	25	77	95	61	71	10	82	51	57	88	29	59	55	84	71	89	64	34	38	33	11	45	47	19
18	02	12	81	84	23	80	58	65	74	13	46	09	33	66	86	74	94	96	07	22	52	39	31	36

Table B.2 (*Continued*)

	1	2	3	4	5	6	7	8	9	10	11	12	13	14	15	16	17	18	19	20	21	22	23	24
19	18	38	40	30	34	27	70	62	35	71	48	96	73	74	28	61	15	37	23	16	91	29	03	06
20	31	76	47	77	59	14	66	85	27	10	63	58	48	66	66	17	91	16	55	70	30	53	05	94
21	50	93	33	61	20	55	10	61	08	76	62	14	22	65	44	95	75	68	94	76	51	21	22	12
22	45	75	89	11	64	06	22	39	20	04	91	47	16	48	19	93	12	02	17	15	94	74	77	37
23	17	97	59	42	77	26	29	88	66	62	53	28	95	01	10	85	31	10	25	75	10	35	99	60
24	23	25	86	94	12	75	66	93	87	95	09	48	85	43	20	94	00	38	53	45	11	77	01	66
25	63	17	05	28	67	39	72	85	02	34	69	56	53	66	09	38	72	31	85	29	62	18	29	37
26	99	81	28	63	05	26	66	16	66	69	18	56	26	53	29	38	08	04	27	93	54	83	53	15
27	86	72	54	89	57	45	05	82	32	64	93	24	83	44	56	65	29	68	69	14	70	79	92	39
28	42	50	86	19	08	81	57	09	69	35	29	06	52	43	53	99	57	55	30	63	63	67	94	94
29	42	80	75	06	05	62	69	04	90	49	10	48	34	21	63	94	19	99	96	79	83	41	86	38
30	82	48	69	65	59	74	64	25	66	93	32	56	14	57	80	10	36	17	39	48	46	94	88	43
31	24	81	98	33	40	89	60	97	28	64	78	93	07	84	07	02	63	35	64	30	29	49	37	00
32	15	84	59	73	01	21	67	43	43	74	00	28	64	66	03	80	60	08	51	67	51	89	00	46
33	92	31	60	34	23	72	00	19	78	73	80	36	51	54	45	76	17	34	35	74	78	20	49	95
34	05	80	10	40	30	63	25	78	91	13	77	39	90	78	89	17	45	76	28	64	12	37	60	34
35	67	51	92	66	84	33	15	34	42	73	54	93	02	01	19	87	36	58	08	11	58	38	88	98
36	71	44	83	33	92	84	96	76	87	24	59	41	71	36	86	14	54	31	41	25	15	59	74	52
37	43	13	62	58	75	90	94	10	65	16	51	90	01	40	18	21	51	82	69	91	65	91	22	32
38	97	55	94	52	18	65	73	90	55	80	51	05	60	53	01	52	46	57	21	05	76	61	05	23
39	32	75	70	24	04	98	03	79	84	34	50	06	25	00	05	00	04	25	68	58	99	48	06	80
40	23	87	76	65	51	19	93	54	81	09	71	83	97	24	90	01	81	14	70	16	07	16	05	93
41	21	77	33	17	02	64	55	23	21	84	80	02	79	30	61	46	33	94	28	92	44	27	76	20
42	90	11	17	05	24	52	08	39	94	07	43	58	33	72	04	51	81	79	63	70	94	71	71	68
43	89	00	39	09	55	13	96	24	47	81	18	37	82	37	37	01	95	82	38	57	20	20	35	83
44	58	65	18	34	73	85	20	47	04	68	77	28	80	14	37	24	97	62	87	38	09	09	08	50
45	80	35	64	10	03	18	24	41	54	12	99	97	50	14	15	80	71	87	47	79	50	62	87	42
46	87	26	52	18	56	47	76	29	40	08	12	07	40	49	29	70	60	74	20	50	51	00	17	42
47	54	23	81	36	70	93	10	05	39	54	20	49	10	70	49	13	37	59	44	52	98	13	64	48
48	72	08	17	30	70	44	08	10	25	81	53	39	81	67	13	80	74	09	71	06	95	05	17	00
49	34	59	02	12	20	31	15	96	18	12	37	32	25	96	71	52	78	01	77	18	63	66	96	09
50	97	89	00	94	82	17	49	92	29	73	30	17	78	53	45	29	39	24	95	61	63	76	90	86

TABLE B.3: CRITICAL VALUES FOR STUDENT'S *t*-TEST

Using Table B.3

For any given *df*, the table shows the values of $t_{critical}$ corresponding to various levels of probability. The $t_{observed}$ is statistically significant at a given level when it is equal to or greater than the value shown in the table.

For the single-sample *t*-ratio, $df = N - 1$.

For the two-sample *t*-ratio, $df = (n_1 - 1) + (n_2 - 1)$.

Examples

Nondirectional Hypothesis:

H_0: $\mu - \mu = 0$ H_1: $\mu - \mu \neq 0$ $\alpha = .05$ $df = 30$

$t_{critical} = \pm 2.042$ If $|t_{observed}| \geq |t_{critical}|$, then reject H_0.

Directional Hypothesis:

H_0: $\mu - \mu \leq 0$ H_1: $\mu - \mu > 0$ $\alpha = .05$ $df = 30$

$t_{critical} = +1.697$ If $t_{observed} \geq t_{critical}$, then reject H_0.

H_0: $\mu - \mu \geq 0$ H_1: $\mu - \mu < 0$ $\alpha = .05$ $df = 30$

$t_{critical} = -1.697$ If $t_{observed} \leq t_{critical}$, then reject H_0.

Table B.3

	Level of significance of a one-tailed or directional test					
	H_0: $\mu - \mu \geq 0$ or H_0: $\mu - \mu \leq 0$					
	$\alpha = 0.10$	$\alpha = .05$	$\alpha = .025$	$\alpha = .01$	$\alpha = .005$	$\alpha = .0005$
	$1 - \alpha = .90$	$1 - \alpha = .95$	$1 - \alpha = .975$	$1 - \alpha = .99$	$1 - \alpha = .995$	$1 - \alpha = .9995$
	Level of significance of a two-tailed or nondirectional test					
	H_0: $\mu - \mu = 0$					
	$\alpha = .20$	$\alpha = .10$	$\alpha = .05$	$\alpha = .02$	$\alpha = .01$	$\alpha = .001$
df	$1 - \alpha = .80$	$1 - \alpha = .90$	$1 - \alpha = .95$	$1 - \alpha = .98$	$1 - \alpha = .99$	$1 - \alpha = .999$
1	3.078	6.314	12.706	31.821	63.656	636.578
2	1.886	2.920	4.303	6.965	9.925	31.600
3	1.638	2.353	3.182	4.541	5.841	12.924
4	1.533	2.132	2.776	3.747	4.604	8.610
5	1.476	2.015	2.571	3.365	4.032	6.869
6	1.440	1.943	2.447	3.143	3.707	5.959
7	1.415	1.895	2.365	2.998	3.499	5.408
8	1.397	1.860	2.306	2.896	3.355	5.041
9	1.383	1.833	2.262	2.821	3.250	4.781
10	1.372	1.812	2.228	2.764	3.169	4.587
11	1.363	1.796	2.201	2.718	3.106	4.437
12	1.356	1.782	2.179	2.681	3.055	4.318
13	1.350	1.771	2.160	2.650	3.012	4.221

Table B.3 (*Continued*)

	Level of significance of a one-tailed or directional test					
	H_0: $\mu - \mu \geq 0$ or H_0: $\mu - \mu \leq 0$					
	$\alpha = 0.10$	$\alpha = .05$	$\alpha = .025$	$\alpha = .01$	$\alpha = .005$	$\alpha = .0005$
	$1 - \alpha = .90$	$1 - \alpha = .95$	$1 - \alpha = .975$	$1 - \alpha = .99$	$1 - \alpha = .995$	$1 - \alpha = .9995$
	Level of significance of a two-tailed or nondirectional test					
	H_0: $\mu - \mu = 0$					
	$\alpha = .20$	$\alpha = .10$	$\alpha = .05$	$\alpha = .02$	$\alpha = .01$	$\alpha = .001$
df	$1 - \alpha = .80$	$1 - \alpha = .90$	$1 - \alpha = .95$	$1 - \alpha = .98$	$1 - \alpha = .99$	$1 - \alpha = .999$
14	1.345	1.761	2.145	2.624	2.977	4.140
15	1.341	1.753	2.131	2.602	2.947	4.073
16	1.337	1.746	2.120	2.583	2.921	4.015
17	1.333	1.740	2.110	2.567	2.898	3.965
18	1.330	1.734	2.101	2.552	2.878	3.922
19	1.328	1.729	2.093	2.539	2.861	3.883
20	1.325	1.725	2.086	2.528	2.845	3.850
21	1.323	1.721	2.080	2.518	2.831	3.819
22	1.321	1.717	2.074	2.508	2.819	3.792
23	1.319	1.714	2.069	2.500	2.807	3.768
24	1.318	1.711	2.064	2.492	2.797	3.745
25	1.316	1.708	2.060	2.485	2.787	3.725
26	1.315	1.706	2.056	2.479	2.779	3.707
27	1.314	1.703	2.052	2.473	2.771	3.689
28	1.313	1.701	2.048	2.467	2.763	3.674
29	1.311	1.699	2.045	2.462	2.756	3.660
30	1.310	1.697	2.042	2.457	2.750	3.646
40	1.303	1.684	2.021	2.423	2.704	3.551
50	1.299	1.676	2.009	2.403	2.678	3.496
60	1.296	1.671	2.000	2.390	2.660	3.460
70	1.294	1.667	1.994	2.381	2.648	3.435
80	1.292	1.664	1.990	2.374	2.639	3.416
90	1.291	1.662	1.987	2.368	2.632	3.402
100	1.290	1.660	1.984	2.364	2.626	3.390
150	1.287	1.655	1.976	2.351	2.609	3.357
200	1.286	1.653	1.972	2.345	2.601	3.340
500	1.283	1.648	1.965	2.334	2.586	3.310
1000	1.282	1.646	1.962	2.330	2.581	3.300
∞	1.282	1.645	1.960	2.326	2.576	3.290

TABLE B.4: POWER OF STUDENT'S SINGLE-SAMPLE *t*-RATIO

Using Table B.4

This table provides the power $(1 - \beta)$ of the single-sample t-ratio given the effect size, the sample size (n), α, and the directionality of the test.

Example

A researcher plans to conduct a study for which H_0 is $\mu = 12.0$ using a two-tailed t-ratio. The researcher believes that $\alpha = .05$ and that the effect size is .20. Approximately how many participants should be in the sample of power to be approximately .80? According to Table B.4, if the researcher uses 200 participants, the power will be $1 - \beta = .83$.

Note that for Cohen's d, an estimate of effect size

$$d = .20 = \text{"small"} \quad d = .50 = \text{"medium"} \quad d = .80 = \text{"large"}$$

TABLE B.5.1: POWER OF STUDENT'S TWO-SAMPLE *t*-RATIO, ONE-TAILED TESTS

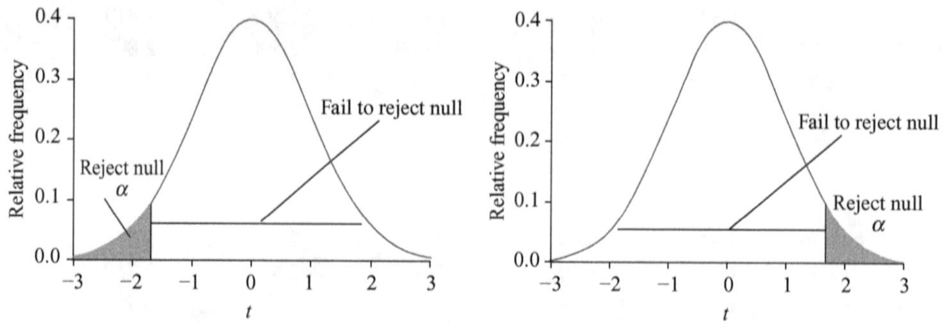

Using Table B.5.1

This table provides the power $(1 - \beta)$ of the two-sample t-ratio given the effect size, the sample size (n), and α when the researcher uses a directional test.

Example

A researcher plans to conduct a study for which H_0 is $\mu_1 \leq \mu_2$ using a one-tailed t-ratio. The researcher believes that $\alpha = .05$ and that the effect size is .20. Approximately how many participants should be in the sample of power to be approximately .80? According to Table B.5.1, if the researcher uses 300 participants in each sample, the power will be $1 - \beta = .81$.

Note that for Cohen's d, an estimate of effect size

$$d = .20 = \text{"small"} \quad d = .50 = \text{"medium"} \quad d = .80 = \text{"large"}$$

Table B.4

Power table: single-sample t-ratio

	$\alpha = .05$, two-tailed					$\alpha = .01$, two-tailed				
n	t_c	.10	.20	.50	.80	t_c	.10	.20	.50	.80
5	2.306	.07	.09	.19	.37	3.355	.02	.03	.07	.16
6	2.228	.07	.09	.22	.44	3.169	.02	.03	.08	.21
7	2.179	.07	.09	.24	.50	3.055	.02	.03	.10	.25
8	2.145	.07	.10	.27	.57	2.977	.02	.03	.11	.30
9	2.120	.07	.10	.30	.62	2.921	.02	.03	.13	.35
10	2.101	.07	.10	.33	.67	2.878	.02	.03	.14	.40
11	2.086	.07	.11	.35	.72	2.845	.02	.03	.16	.45
12	2.074	.07	.11	.38	.76	2.819	.02	.03	.18	.50
13	2.064	.07	.11	.41	.80	2.797	.02	.04	.19	.54
14	2.056	.07	.12	.44	.83	2.779	.02	.04	.21	.59
15	2.048	.07	.12	.46	.86	2.763	.02	.04	.23	.63
16	2.042	.07	.12	.49	.88	2.750	.02	.04	.25	.67
17	2.037	.07	.13	.51	.90	2.738	.02	.04	.27	.71
18	2.032	.07	.13	.54	.92	2.728	.02	.04	.29	.74
19	2.028	.07	.14	.56	.94	2.719	.02	.05	.31	.78
20	2.024	.08	.14	.59	.95	2.712	.02	.05	.33	.80
21	2.021	.08	.14	.61	.96	2.704	.02	.05	.35	.83
22	2.018	.08	.15	.63	.97	2.698	.02	.05	.37	.85
23	2.015	.08	.15	.65	.97	2.692	.02	.05	.39	.87
24	2.013	.08	.16	.67	.98	2.687	.02	.05	.41	.89
25	2.011	.08	.16	.69	.98	2.682	.02	.06	.43	.91
30	2.002	.08	.18	.78	.99	2.663	.02	.07	.53	.96
40	1.991	.09	.23	.89	.99	2.640	.03	.09	.70	.99
50	1.984	.10	.28	.95	.99	2.627	.03	.11	.82	.99
60	1.980	.11	.32	.98	.99	2.618	.04	.14	.90	.99
70	1.977	.13	.37	.99	.99	2.612	.04	.17	.95	.99
80	1.975	.14	.42	.99	.99	2.607	.04	.20	.98	.99
90	1.973	.15	.46	.99	.99	2.604	.05	.23	.99	.99
100	1.972	.16	.50	.99	.99	2.601	.05	.26	.99	.99
150	1.968	.22	.69	.99	.99	2.592	.08	.43	.99	.99
200	1.966	.28	.82	.99	.99	2.588	.11	.59	.99	.99
250	1.965	.34	.90	.99	.99	2.586	.15	.72	.99	.99
300	1.964	.39	.95	.99	.99	2.584	.18	.82	.99	.99
350	1.963	.45	.98	.99	.99	2.583	.22	.89	.99	.99
400	1.963	.51	.99	.99	.99	2.582	.26	.94	.99	.99
500	1.962	.61	.99	.99	.99	2.581	.35	.98	.99	.99
600	1.962	.69	.99	.99	.99	2.580	.43	.99	.99	.99
700	1.962	.76	.99	.99	.99	2.579	.51	.99	.99	.99
800	1.961	.82	.99	.99	.99	2.579	.59	.99	.99	.99
900	1.961	.87	.99	.99	.99	2.579	.66	.99	.99	.99
1000	1.961	.90	.99	.99	.99	2.578	.72	.99	.99	.99

Table B.5.1

Power table: two-sample t-ratio, one-tailed tests

	$\alpha = .05$, one-tailed					$\alpha = .01$, one-tailed				
n	t_c	.10	.20	.50	.80	t_c	.10	.20	.50	.80
5	1.860	.12	.13	.21	.33	2.896	.04	.04	.07	.13
6	1.812	.12	.14	.22	.38	2.764	.03	.04	.08	.15
7	1.782	.12	.14	.24	.42	2.681	.03	.04	.08	.18
8	1.761	.12	.14	.26	.46	2.624	.03	.04	.09	.21
9	1.746	.12	.14	.28	.50	2.583	.03	.04	.10	.23
10	1.734	.12	.14	.29	.54	2.552	.03	.04	.11	.26
11	1.725	.12	.14	.31	.57	2.528	.03	.04	.12	.29
12	1.717	.12	.15	.33	.61	2.508	.03	.04	.13	.32
13	1.711	.12	.15	.35	.64	2.492	.03	.04	.14	.35
14	1.706	.12	.15	.36	.67	2.479	.03	.04	.15	.37
15	1.701	.12	.15	.38	.70	2.467	.03	.04	.16	.40
16	1.697	.12	.16	.40	.73	2.457	.03	.04	.17	.43
17	1.694	.12	.16	.41	.75	2.449	.03	.05	.18	.46
18	1.691	.12	.16	.43	.78	2.441	.03	.05	.19	.49
19	1.688	.12	.16	.45	.80	2.434	.03	.05	.20	.52
20	1.686	.12	.17	.46	.82	2.429	.03	.05	.21	.54
21	1.684	.12	.17	.48	.84	2.423	.03	.05	.22	.57
22	1.682	.12	.17	.50	.85	2.418	.03	.05	.23	.59
23	1.680	.12	.17	.51	.87	2.414	.03	.05	.24	.62
24	1.679	.12	.18	.53	.88	2.410	.03	.05	.25	.64
25	1.677	.12	.18	.54	.89	2.407	.03	.05	.26	.66
30	1.672	.13	.19	.61	.94	2.392	.03	.06	.32	.76
40	1.665	.13	.22	.73	.98	2.375	.03	.07	.44	.89
50	1.661	.14	.25	.82	.99	2.365	.04	.09	.55	.96
60	1.658	.15	.28	.88	.99	2.358	.04	.10	.65	.99
70	1.656	.15	.31	.92	.99	2.354	.04	.12	.73	.99
80	1.655	.16	.34	.95	.99	2.350	.04	.13	.80	.99
90	1.653	.17	.37	.97	.99	2.347	.05	.15	.85	.99
100	1.653	.18	.40	.98	.99	2.345	.05	.17	.90	.99
150	1.650	.21	.53	.99	.99	2.339	.07	.26	.99	.99
200	1.649	.25	.64	.99	.99	2.336	.09	.35	.99	.99
250	1.648	.29	.74	.99	.99	2.334	.10	.45	.99	.99
300	1.647	.33	.81	.99	.99	2.333	.12	.54	.99	.99
350	1.647	.36	.86	.99	.99	2.332	.14	.62	.99	.99
400	1.647	.40	.90	.99	.99	2.331	.17	.69	.99	.99
500	1.646	.47	.96	.99	.99	2.330	.21	.81	.99	.99
600	1.646	.53	.98	.99	.99	2.329	.26	.89	.99	.99
700	1.646	.59	.99	.99	.99	2.329	.30	.94	.99	.99
800	1.646	.64	.99	.99	.99	2.329	.35	.97	.99	.99
900	1.646	.69	.99	.99	.99	2.328	.40	.98	.99	.99
1000	1.646	.74	.99	.99	.99	2.328	.45	.99	.99	.99

TABLE B.5.2: POWER OF STUDENT'S TWO-SAMPLE *t*-RATIO, TWO-TAILED TESTS

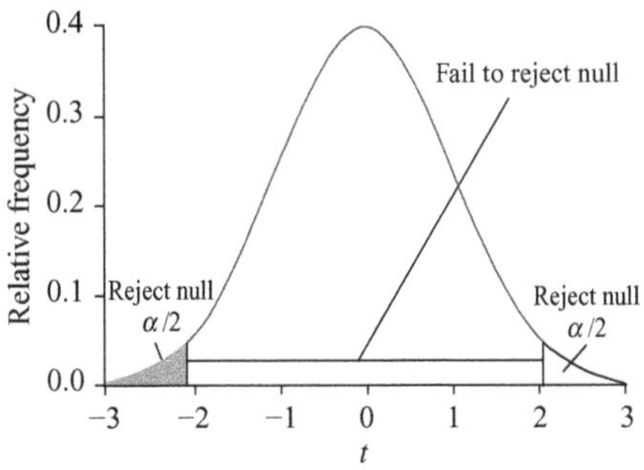

Using Table B.5.2

This table provides the power $(1 - \beta)$ of the two-sample *t*-ratio given the effect size, the sample size (n), and α when the researcher uses a nondirectional test.

Example

A researcher plans to conduct a study for which H_0 is $\mu_1 = \mu_2$ using a two-tailed *t*-ratio. The researcher believes that $\alpha = .05$ and that the effect size is .20. Approximately how many participants should be in the sample of power to be approximately .80? According to Table B.5.2, if the researcher uses 400 participants in each group, the power will be $1 - \beta = .82$.

Note that for Cohen's d, an estimate of effect size

$d = .20 =$ "small" $\quad d = .50 =$ "medium" $\quad d = .80 =$ "large"

TABLE B.6: CRITICAL VALUES FOR PEARSON'S CORRELATION COEFFICIENT

Using Table B.6

For any given *df*, this table shows the values of r corresponding to various levels of probability. The $r_{observed}$ is statistically significant at a given level when it is equal to or greater than the value shown in the table.

Examples

Nondirectional Hypothesis:

H_0: $\rho = 0$ $\qquad$ H_1: $\rho \neq 0$ $\qquad \alpha = .05$, $df = 30$

$r_{critical} = \pm.3494$ $\qquad$ If $|r_{observed}| \geq |r_{critical}|$, then reject H_0.

Table B.5.2

Power table: two-sample t-ratio, two-tailed tests

	$\alpha = .05$, two-tailed					$\alpha = .01$, two-tailed				
n	t_c	.10	.20	.50	.80	t_c	.10	.20	.50	.80
5	2.306	.07	.08	.13	.22	3.355	.02	.02	.04	.08
6	2.228	.07	.08	.14	.26	3.169	.02	.02	.05	.10
7	2.179	.07	.08	.15	.29	3.055	.02	.02	.05	.12
8	2.145	.07	.08	.17	.33	2.977	.02	.02	.06	.14
9	2.120	.07	.08	.18	.36	2.921	.02	.02	.07	.16
10	2.101	.07	.08	.19	.40	2.878	.02	.02	.07	.19
11	2.086	.06	.08	.21	.43	2.845	.02	.02	.08	.21
12	2.074	.06	.08	.22	.47	2.819	.02	.02	.09	.23
13	2.064	.06	.08	.23	.50	2.797	.02	.03	.09	.26
14	2.056	.06	.09	.25	.53	2.779	.02	.03	.10	.28
15	2.048	.06	.09	.26	.56	2.763	.02	.03	.11	.31
16	2.042	.06	.09	.28	.59	2.750	.02	.03	.11	.33
17	2.037	.06	.09	.29	.62	2.738	.02	.03	.12	.36
18	2.032	.06	.09	.30	.65	2.728	.02	.03	.13	.38
19	2.028	.06	.10	.32	.68	2.719	.02	.03	.14	.41
20	2.024	.06	.10	.33	.70	2.712	.02	.03	.15	.43
21	2.021	.07	.10	.35	.72	2.704	.02	.03	.15	.46
22	2.018	.07	.10	.36	.75	2.698	.02	.03	.16	.48
23	2.015	.07	.10	.37	.77	2.692	.02	.03	.17	.51
24	2.013	.07	.10	.39	.79	2.687	.02	.03	.18	.53
25	2.011	.07	.11	.40	.80	2.682	.02	.03	.19	.56
30	2.002	.07	.12	.47	.88	2.663	.02	.04	.24	.67
40	1.991	.07	.14	.60	.96	2.640	.02	.05	.34	.83
50	1.984	.08	.16	.70	.99	2.627	.02	.06	.44	.92
60	1.980	.08	.18	.79	.99	2.618	.02	.07	.54	.97
70	1.977	.09	.21	.85	.99	2.612	.02	.08	.63	.99
80	1.975	.09	.23	.90	.99	2.607	.03	.09	.71	.99
90	1.973	.10	.25	.93	.99	2.604	.03	.10	.78	.99
100	1.972	.10	.28	.96	.99	2.601	.03	.11	.83	.99
150	1.968	.13	.39	.99	.99	2.592	.04	.18	.97	.99
200	1.966	.16	.50	.99	.99	2.588	.05	.26	.99	.99
250	1.965	.19	.60	.99	.99	2.586	.07	.35	.99	.99
300	1.964	.22	.69	.99	.99	2.584	.08	.43	.99	.99
350	1.963	.25	.76	.99	.99	2.583	.10	.51	.99	.99
400	1.963	.28	.82	.99	.99	2.582	.11	.59	.99	.99
500	1.962	.34	.90	.99	.99	2.581	.15	.72	.99	.99
600	1.962	.39	.95	.99	.99	2.580	.18	.82	.99	.99
700	1.962	.45	.98	.99	.99	2.579	.22	.89	.99	.99
800	1.961	.51	.99	.99	.99	2.579	.26	.94	.99	.99
900	1.961	.56	.99	.99	.99	2.579	.31	.96	.99	.99
1000	1.961	.61	.99	.99	.99	2.578	.35	.98	.99	.99

Table B.6

	Level of significance of a one-tailed or directional test					
	H_0: $\rho \le 0$ or H_0: $\rho \ge 0$					
	$\alpha = 0.1$	$\alpha = 0.05$	$\alpha = 0.025$	$\alpha = 0.01$	$\alpha = 0.005$	$\alpha = 0.0005$
	Level of significance of a two-tailed or nondirectional test					
	H_0: $\rho = 0$					
df	$\alpha = 0.2$	$\alpha = 0.1$	$\alpha = 0.05$	$\alpha = 0.02$	$\alpha = 0.01$	$\alpha = 0.001$
1	.9511	.9877	.9969	.9995	.9999	.9999
2	.8000	.9000	.9500	.9800	.9900	.9990
3	.6870	.8054	.8783	.9343	.9587	.9911
4	.6084	.7293	.8114	.8822	.9172	.9741
5	.5509	.6694	.7545	.8329	.8745	.9509
6	.5067	.6215	.7067	.7887	.8343	.9249
7	.4716	.5822	.6664	.7498	.7977	.8983
8	.4428	.5494	.6319	.7155	.7646	.8721
9	.4187	.5214	.6021	.6851	.7348	.8470
10	.3981	.4973	.5760	.6581	.7079	.8233
11	.3802	.4762	.5529	.6339	.6835	.8010
12	.3646	.4575	.5324	.6120	.6614	.7800
13	.3507	.4409	.5140	.5923	.6411	.7604
14	.3383	.4259	.4973	.5742	.6226	.7419
15	.3271	.4124	.4821	.5577	.6055	.7247
16	.3170	.4000	.4683	.5425	.5897	.7084
17	.3077	.3887	.4555	.5285	.5751	.6932
18	.2992	.3783	.4438	.5155	.5614	.6788
19	.2914	.3687	.4329	.5034	.5487	.6652
20	.2841	.3598	.4227	.4921	.5368	.6524
21	.2774	.3515	.4132	.4815	.5256	.6402
22	.2711	.3438	.4044	.4716	.5151	.6287
23	.2653	.3365	.3961	.4622	.5052	.6178
24	.2598	.3297	.3882	.4534	.4958	.6074
25	.2546	.3233	.3809	.4451	.4869	.5974
30	.2327	.2960	.3494	.4093	.4487	.5541
35	.2156	.2746	.3246	.3810	.4182	.5189
40	.2018	.2573	.3044	.3578	.3932	.4896
50	.1806	.2306	.2732	.3218	.3542	.4432
60	.1650	.2108	.2500	.2948	.3248	.4079
70	.1528	.1954	.2319	.2737	.3017	.3798
80	.1430	.1829	.2172	.2565	.2830	.3568
90	.1348	.1726	.2050	.2422	.2673	.3375
100	.1279	.1638	.1946	.2301	.2540	.3211
150	.1045	.1339	.1593	.1886	.2084	.2643
300	.0740	.0948	.1129	.1338	.1480	.1884
500	.0573	.0735	.0875	.1038	.1149	.1464
1000	.0405	.0520	.0619	.0735	.0813	.1038

Directional Hypothesis:

$H_0: \rho \leq 0$	$H_1: \rho > 0$ $\alpha = .05$, $df = 30$
$r_{critical} = +.2960$	If $r_{observed} \geq r_{critical}$, then reject H_0.
$H_0: \rho \geq 0$	$H_1: \rho < 0$ $\alpha = .05$, $df = 30$
$r_{critical} = -.2960$	If $r_{observed} \leq r_{critical}$, then reject H_0.

Note that the relation between the correlation coefficient and the t-ratio is

$$r_c = \frac{t_c}{\sqrt{(n-2)+t_c^2}}$$

TABLE B.7: CRITICAL VALUES FOR SPEARMAN'S RANK-ORDER CORRELATION COEFFICIENT

Using Table B.7

For any given df, the table shows the values of r_S corresponding to various levels of probability. The $r_{S\text{-observed}}$ is statistically significant at a given level when it is equal to or greater than the value shown in the table.

Examples

Nondirectional Hypothesis:

$H_0: \rho_S = 0$	$H_1: \rho_S \neq 0$ $\alpha = .05$ $df = 30$
$r_{critical} = \pm.350$	If $\lvert r_{observed}\rvert \geq \lvert r_{critical}\rvert$, then reject H_0.

Directional Hypothesis:

$H_0: \rho_S \leq 0$	$H_1: \rho_S > 0$ $\alpha = .05$ $df = 30$
$r_{critical} = +.296$	If $r_{observed} \geq r_{critical}$, then reject H_0.
$H_0: \rho_S \geq 0$	$H_1: \rho_S < 0$ $\alpha = .05$ $df = 30$
$r_{critical} = -.296$	If $r_{observed} \leq r_{critical}$, then reject H_0.

When $df > 28$, we can convert the r_S to a t-ratio and then use Table C for hypothesis testing:

$$t = r_S\sqrt{\frac{N-2}{1-r_S^2}}$$

For example, $r_S = .60$, $N = 42$:

$$t = .60\sqrt{\frac{42-2}{1-.60^2}} \quad t = .60\sqrt{\frac{40}{.64}} \quad t = .60\sqrt{62.5}$$

$$t = 4.74,\ df = 40$$

Table B.7

	Level of significance of a one-tailed or directional test					
	H_0: $\rho_S \leq 0$ or H_0: $\rho_S \geq 0$					
	$\alpha = 0.1$	$\alpha = 0.05$	$\alpha = 0.025$	$\alpha = 0.01$	$\alpha = 0.005$	$\alpha = 0.0005$
	Level of significance of a two-tailed or nondirectional test					
	H_0: $\rho_S = 0$					
df	$\alpha = 0.2$	$\alpha = 0.1$	$\alpha = 0.05$	$\alpha = 0.02$	$\alpha = 0.01$	$\alpha = 0.001$
2	1.000	1.000				
3	.800	.900	1.000	1.000		
4	.657	.829	.886	.943	1.000	
5	.571	.714	.786	.893	.929	1.000
6	.524	.643	.738	.833	.881	.976
7	.483	.600	.700	.783	.833	.933
8	.455	.564	.648	.745	.794	.903
9	.427	.536	.618	.709	.755	.873
10	.406	.503	.587	.678	.727	.846
11	.385	.484	.560	.648	.703	.824
12	.367	.464	.538	.626	.679	.802
13	.354	.446	.521	.604	.654	.779
14	.341	.429	.503	.582	.635	.762
15	.328	.414	.485	.566	.615	.748
16	.317	.401	.472	.550	.600	.728
17	.309	.391	.460	.535	.584	.712
18	.299	.380	.447	.520	.570	.696
19	.292	.370	.435	.508	.556	.681
20	.284	.361	.425	.496	.544	.667
21	.278	.353	.415	.486	.532	.654
22	.271	.344	.406	.476	.521	.642
23	.265	.337	.398	.466	.511	.630
24	.259	.331	.390	.457	.501	.619
25	.255	.324	.382	.448	.491	.608
26	.250	.317	.375	.440	.483	.598
27	.245	.312	.368	.433	.475	.589
28	.240	.306	.362	.425	.467	.580
29	.236	.301	.356	.418	.459	.571
30	.232	.296	.350	.412	.452	.563

If $\alpha = .05$, two-tailed,

$$t_{\text{critical}} = 1.684 \quad \text{Reject } H_0\text{: } \rho_S = 0$$

TABLE B.8: *r* TO *Z* TRANSFORMATION

Using Table B.8

This table provides the Fisher *r* to *z* transformation. Both positive and negative values of *r* may be used. For specific transformations, use the following equation:

$$z_r = \frac{1}{2}\log_e\left(\frac{1+r}{1-r}\right)$$

Example

$$r = .25 \rightarrow z_r = .255$$

TABLE B.9: POWER OF PEARSON'S CORRELATION COEFFICIENT

Using Table B.9

This table provides estimates of the power $(1 - \beta)$ of the Pearson correlation coefficient (r) given the effect size, the sample size (n), α, and the directionality of the test.

Example

A researcher plans to conduct a study for which H_0 is $\rho = 0.0$ using a two-tailed test. The researcher believes that $\alpha = .05$ and that the effect size is .30. Approximately how many participants should be in the sample of power to be approximately .80? According to Table B.9, if the researcher uses 90 participants, the power will be $1 - \beta = .82$.

Note that for effect sizes,

$$r = .10 = \text{"small"} \quad r = .30 = \text{"medium"} \quad r = .50 = \text{"large"}$$

TABLE B.10: CRITICAL VALUES FOR THE *F*-RATIO

Using Table B.10

This table provides the critical values required to reject the null hypothesis for the analysis of variance (ANOVA). Note that the bold text represents $\alpha = .01$, whereas the regular text represents $\alpha = .05$. To use the table, you will need to identify the

Table B.8

r	z_r	r	z_r	r	z_r	r	z_r
0.00	0.000	0.25	0.255	0.50	0.549	0.75	0.973
0.01	0.010	0.26	0.266	0.51	0.563	0.76	0.996
0.02	0.020	0.27	0.277	0.52	0.576	0.77	1.020
0.03	0.030	0.28	0.288	0.53	0.590	0.78	1.045
0.04	0.040	0.29	0.299	0.54	0.604	0.79	1.071
0.05	0.050	0.30	0.310	0.55	0.618	0.80	1.099
0.06	0.060	0.31	0.321	0.56	0.633	0.81	1.127
0.07	0.070	0.32	0.332	0.57	0.648	0.82	1.157
0.08	0.080	0.33	0.343	0.58	0.662	0.83	1.188
0.09	0.090	0.34	0.354	0.59	0.678	0.84	1.221
0.10	0.100	0.35	0.365	0.60	0.693	0.85	1.256
0.11	0.110	0.36	0.377	0.61	0.709	0.86	1.293
0.12	0.121	0.37	0.388	0.62	0.725	0.87	1.333
0.13	0.131	0.38	0.400	0.63	0.741	0.88	1.376
0.14	0.141	0.39	0.412	0.64	0.758	0.89	1.422
0.15	0.151	0.40	0.424	0.65	0.775	0.90	1.472
0.16	0.161	0.41	0.436	0.66	0.793	0.91	1.528
0.17	0.172	0.42	0.448	0.67	0.811	0.92	1.589
0.18	0.182	0.43	0.460	0.68	0.829	0.93	1.658
0.19	0.192	0.44	0.472	0.69	0.848	0.94	1.738
0.20	0.203	0.45	0.485	0.70	0.867	0.95	1.832
0.21	0.213	0.46	0.497	0.71	0.887	0.96	1.946
0.22	0.224	0.47	0.510	0.72	0.908	0.97	2.092
0.23	0.234	0.48	0.523	0.73	0.929	0.98	2.298
0.24	0.245	0.49	0.536	0.74	0.950	0.99	2.647

degrees of freedom for the numerator and denominator. The numerator degrees of freedom are those used to determine the mean square for the treatment effect or interaction. The denominator degrees of freedom are those used to determine the mean square for the within-groups or error variance.

Example: One-Factor ANOVA

A researcher conducts a study that produces the following ANOVA summary table:

Source	SS	df	MS	F
Between groups	28.00	2	14.00	3.50
Within groups	156.00	39	4.00	
Total	184.00	41		

Table B.9

α = .05, one tailed						α = .05, two tailed				
Effect size: r						Effect size: r				
n	0.10	0.30	0.50	0.70	0.95	n	0.10	0.30	0.50	0.70	0.95
10	.07	.19	.42	.75	.98	10	.03	.11	.29	.63	.99
11	.07	.21	.46	.80	.99	11	.03	.12	.33	.69	.99
12	.08	.23	.50	.83	.99	12	.04	.14	.37	.74	.99
13	.08	.24	.54	.87	.99	13	.04	.15	.40	.78	.99
14	.08	.26	.57	.89	.99	14	.04	.16	.44	.82	.99
15	.09	.27	.60	.91	.99	15	.04	.17	.47	.85	.99
16	.09	.29	.63	.93	.99	16	.04	.19	.50	.88	.99
17	.09	.31	.66	.94	.99	17	.05	.20	.53	.90	.99
18	.09	.32	.69	.96	.99	18	.05	.21	.56	.92	.99
19	.10	.33	.71	.96	.99	19	.05	.22	.59	.93	.99
20	.10	.35	.73	.97	.99	20	.05	.24	.61	.94	.99
21	.10	.36	.75	.98	.99	21	.05	.25	.64	.95	.99
22	.10	.38	.77	.98	.99	22	.05	.26	.66	.96	.99
23	.11	.39	.79	.98	.99	23	.06	.27	.69	.97	.99
24	.11	.40	.81	.99	.99	24	.06	.28	.71	.97	.99
25	.11	.42	.82	.99	.99	25	.06	.30	.73	.98	.99
26	.11	.43	.84	.99	.99	26	.06	.31	.75	.98	.99
27	.12	.44	.85	.99	.99	27	.06	.32	.76	.98	.99
28	.12	.46	.86	.99	.99	28	.06	.33	.78	.99	.99
29	.12	.47	.88	.99	.99	29	.06	.34	.80	.99	.99
30	.12	.48	.89	.99	.99	30	.07	.35	.81	.99	.99
31	.12	.49	.89	.99	.99	31	.07	.37	.83	.99	.99
32	.13	.50	.90	.99	.99	32	.07	.38	.84	.99	.99
33	.13	.52	.91	.99	.99	33	.07	.39	.85	.99	.99
34	.13	.53	.92	.99	.99	34	.07	.40	.86	.99	.99
35	.13	.54	.93	.99	.99	35	.07	.41	.87	.99	.99
36	.13	.55	.93	.99	.99	36	.07	.42	.88	.99	.99
37	.14	.56	.94	.99	.99	37	.08	.43	.89	.99	.99
38	.14	.57	.94	.99	.99	38	.08	.44	.90	.99	.99
39	.14	.58	.95	.99	.99	39	.08	.45	.91	.99	.99
40	.14	.59	.95	.99	.99	40	.08	.46	.91	.99	.99
50	.17	.69	.98	.99	.99	50	.09	.56	.96	.99	.99
60	.18	.75	.99	.99	.99	60	.11	.64	.98	.99	.99
70	.20	.81	.99	.99	.99	70	.12	.71	.99	.99	.99
80	.22	.85	.99	.99	.99	80	.13	.77	.99	.99	.99
90	.23	.89	.99	.99	.99	90	.15	.82	.99	.99	.99
100	.25	.92	.99	.99	.99	100	.16	.86	.99	.99	.99
200	.40	.99	.99	.99	.99	200	.28	.99	.99	.99	.99
300	.53	.99	.99	.99	.99	300	.40	.99	.99	.99	.99
400	.63	.99	.99	.99	.99	400	.51	.99	.99	.99	.99
500	.72	.99	.99	.99	.99	500	.60	.99	.99	.99	.99

Table B.10

		Degrees of freedom for numerator																		
		α	1	2	3	4	5	6	7	8	9	10	11	12	13	14	15	30	50	100	1000
Degrees of freedom for denominator	1	0.05	161	199	216	225	230	234	237	239	241	242	243	244	245	245	246	250	252	253	254
		0.01	**4052**	**4999**	**5404**	**5624**	**5764**	**5859**	**5928**	**5981**	**6022**	**6056**	**6083**	**6107**	**6126**	**6143**	**6157**	**6260**	**6302**	**6334**	**6363**
	2	0.05	18.51	19.00	19.16	19.25	19.30	19.33	19.35	19.37	19.38	19.40	19.40	19.41	19.42	19.42	19.43	19.46	19.48	19.49	19.49
		0.01	**98.50**	**99.00**	**99.17**	**99.25**	**99.30**	**99.33**	**99.36**	**99.37**	**99.39**	**99.40**	**99.41**	**99.42**	**99.42**	**99.43**	**99.43**	**99.47**	**99.48**	**99.49**	**99.50**
	3	0.05	10.13	9.55	9.28	9.12	9.01	8.94	8.89	8.85	8.81	8.79	8.76	8.74	8.73	8.71	8.70	8.62	8.58	8.55	8.53
		0.01	**34.12**	**30.82**	**29.46**	**28.71**	**28.24**	**27.91**	**27.67**	**27.49**	**27.34**	**27.23**	**27.13**	**27.05**	**26.98**	**26.92**	**26.87**	**26.50**	**26.35**	**26.24**	**26.14**
	4	0.05	7.71	6.94	6.59	6.39	6.26	6.16	6.09	6.04	6.00	5.96	5.94	5.91	5.89	5.87	5.86	5.75	5.70	5.66	5.63
		0.01	**21.20**	**18.00**	**16.69**	**15.98**	**15.52**	**15.21**	**14.98**	**14.80**	**14.66**	**14.55**	**14.45**	**14.37**	**14.31**	**14.25**	**14.20**	**13.84**	**13.69**	**13.58**	**13.47**
	5	0.05	6.61	5.79	5.41	5.19	5.05	4.95	4.88	4.82	4.77	4.74	4.70	4.68	4.66	4.64	4.62	4.50	4.44	4.41	4.37
		0.01	**16.26**	**13.27**	**12.06**	**11.39**	**10.97**	**10.67**	**10.46**	**10.29**	**10.16**	**10.05**	**9.96**	**9.89**	**9.82**	**9.77**	**9.72**	**9.38**	**9.24**	**9.13**	**9.03**
	6	0.05	5.99	5.14	4.76	4.53	4.39	4.28	4.21	4.15	4.10	4.06	4.03	4.00	3.98	3.96	3.94	3.81	3.75	3.71	3.67
		0.01	**13.75**	**10.92**	**9.78**	**9.15**	**8.75**	**8.47**	**8.26**	**8.10**	**7.98**	**7.87**	**7.79**	**7.72**	**7.66**	**7.60**	**7.56**	**7.23**	**7.09**	**6.99**	**6.89**
	7	0.05	5.59	4.74	4.35	4.12	3.97	3.87	3.79	3.73	3.68	3.64	3.60	3.57	3.55	3.53	3.51	3.38	3.32	3.27	3.23
		0.01	**12.25**	**9.55**	**8.45**	**7.85**	**7.46**	**7.19**	**6.99**	**6.84**	**6.72**	**6.62**	**6.54**	**6.47**	**6.41**	**6.36**	**6.31**	**5.99**	**5.86**	**5.75**	**5.66**
	8	0.05	5.32	4.46	4.07	3.84	3.69	3.58	3.50	3.44	3.39	3.35	3.31	3.28	3.26	3.24	3.22	3.08	3.02	2.97	2.93
		0.01	**11.26**	**8.65**	**7.59**	**7.01**	**6.63**	**6.37**	**6.18**	**6.03**	**5.91**	**5.81**	**5.73**	**5.67**	**5.61**	**5.56**	**5.52**	**5.20**	**5.07**	**4.96**	**4.87**
	9	0.05	5.12	4.26	3.86	3.63	3.48	3.37	3.29	3.23	3.18	3.14	3.10	3.07	3.05	3.03	3.01	2.86	2.80	2.76	2.71
		0.01	**10.56**	**8.02**	**6.99**	**6.42**	**6.06**	**5.80**	**5.61**	**5.47**	**5.35**	**5.26**	**5.18**	**5.11**	**5.05**	**5.01**	**4.96**	**4.65**	**4.52**	**4.41**	**4.32**
	10	0.05	4.96	4.10	3.71	3.48	3.33	3.22	3.14	3.07	3.02	2.98	2.94	2.91	2.89	2.86	2.85	2.70	2.64	2.59	2.54
		0.01	**10.04**	**7.56**	**6.55**	**5.99**	**5.64**	**5.39**	**5.20**	**5.06**	**4.94**	**4.85**	**4.77**	**4.71**	**4.65**	**4.60**	**4.56**	**4.25**	**4.12**	**4.01**	**3.92**
	11	0.05	4.84	3.98	3.59	3.36	3.20	3.09	3.01	2.95	2.90	2.85	2.82	2.79	2.76	2.74	2.72	2.57	2.51	2.46	2.41
		0.01	**9.65**	**7.21**	**6.22**	**5.67**	**5.32**	**5.07**	**4.89**	**4.74**	**4.63**	**4.54**	**4.46**	**4.40**	**4.34**	**4.29**	**4.25**	**3.94**	**3.81**	**3.71**	**3.61**
	12	0.05	4.75	3.89	3.49	3.26	3.11	3.00	2.91	2.85	2.80	2.75	2.72	2.69	2.66	2.64	2.62	2.47	2.40	2.35	2.30
		0.01	**9.33**	**6.93**	**5.95**	**5.41**	**5.06**	**4.82**	**4.64**	**4.50**	**4.39**	**4.30**	**4.22**	**4.16**	**4.10**	**4.05**	**4.01**	**3.70**	**3.57**	**3.47**	**3.37**
	13	0.05	4.67	3.81	3.41	3.18	3.03	2.92	2.83	2.77	2.71	2.67	2.63	2.60	2.58	2.55	2.53	2.38	2.31	2.26	2.21
		0.01	**9.07**	**6.70**	**5.74**	**5.21**	**4.86**	**4.62**	**4.44**	**4.30**	**4.19**	**4.10**	**4.02**	**3.96**	**3.91**	**3.86**	**3.82**	**3.51**	**3.38**	**3.27**	**3.18**

(Continued)

Table B.10 (*Continued*)

		Degrees of freedom for numerator																		
	α	1	2	3	4	5	6	7	8	9	10	11	12	13	14	15	30	50	100	1000
14	0.05	4.60	3.74	3.34	3.11	2.96	2.85	2.76	2.70	2.65	2.60	2.57	2.53	2.51	2.48	2.46	2.31	2.24	2.19	2.14
	0.01	**8.86**	**6.51**	**5.56**	**5.04**	**4.69**	**4.46**	**4.28**	**4.14**	**4.03**	**3.94**	**3.86**	**3.80**	**3.75**	**3.70**	**3.66**	**3.35**	**3.22**	**3.11**	**3.02**
15	0.05	4.54	3.68	3.29	3.06	2.90	2.79	2.71	2.64	2.59	2.54	2.51	2.48	2.45	2.42	2.40	2.25	2.18	2.12	2.07
	0.01	**8.68**	**6.36**	**5.42**	**4.89**	**4.56**	**4.32**	**4.14**	**4.00**	**3.89**	**3.80**	**3.73**	**3.67**	**3.61**	**3.56**	**3.52**	**3.21**	**3.08**	**2.98**	**2.88**
16	0.05	4.49	3.63	3.24	3.01	2.85	2.74	2.66	2.59	2.54	2.49	2.46	2.42	2.40	2.37	2.35	2.19	2.12	2.07	2.02
	0.01	**8.53**	**6.23**	**5.29**	**4.77**	**4.44**	**4.20**	**4.03**	**3.89**	**3.78**	**3.69**	**3.62**	**3.55**	**3.50**	**3.45**	**3.41**	**3.10**	**2.97**	**2.86**	**2.76**
17	0.05	4.45	3.59	3.20	2.96	2.81	2.70	2.61	2.55	2.49	2.45	2.41	2.38	2.35	2.33	2.31	2.15	2.08	2.02	1.97
	0.01	**8.40**	**6.11**	**5.18**	**4.67**	**4.34**	**4.10**	**3.93**	**3.79**	**3.68**	**3.59**	**3.52**	**3.46**	**3.40**	**3.35**	**3.31**	**3.00**	**2.87**	**2.76**	**2.66**
18	0.05	4.41	3.55	3.16	2.93	2.77	2.66	2.58	2.51	2.46	2.41	2.37	2.34	2.31	2.29	2.27	2.11	2.04	1.98	1.92
	0.01	**8.29**	**6.01**	**5.09**	**4.58**	**4.25**	**4.01**	**3.84**	**3.71**	**3.60**	**3.51**	**3.43**	**3.37**	**3.32**	**3.27**	**3.23**	**2.92**	**2.78**	**2.68**	**2.58**
19	0.05	4.38	3.52	3.13	2.90	2.74	2.63	2.54	2.48	2.42	2.38	2.34	2.31	2.28	2.26	2.23	2.07	2.00	1.94	1.88
	0.01	**8.18**	**5.93**	**5.01**	**4.50**	**4.17**	**3.94**	**3.77**	**3.63**	**3.52**	**3.43**	**3.36**	**3.30**	**3.24**	**3.19**	**3.15**	**2.84**	**2.71**	**2.60**	**2.50**
20	0.05	4.35	3.49	3.10	2.87	2.71	2.60	2.51	2.45	2.39	2.35	2.31	2.28	2.25	2.22	2.20	2.04	1.97	1.91	1.85
	0.01	**8.10**	**5.85**	**4.94**	**4.43**	**4.10**	**3.87**	**3.70**	**3.56**	**3.46**	**3.37**	**3.29**	**3.23**	**3.18**	**3.13**	**3.09**	**2.78**	**2.64**	**2.54**	**2.43**
21	0.05	4.32	3.47	3.07	2.84	2.68	2.57	2.49	2.42	2.37	2.32	2.28	2.25	2.22	2.20	2.18	2.01	1.94	1.88	1.82
	0.01	**8.02**	**5.78**	**4.87**	**4.37**	**4.04**	**3.81**	**3.64**	**3.51**	**3.40**	**3.31**	**3.24**	**3.17**	**3.12**	**3.07**	**3.03**	**2.72**	**2.58**	**2.48**	**2.37**
22	0.05	4.30	3.44	3.05	2.82	2.66	2.55	2.46	2.40	2.34	2.30	2.26	2.23	2.20	2.17	2.15	1.98	1.91	1.85	1.79
	0.01	**7.95**	**5.72**	**4.82**	**4.31**	**3.99**	**3.76**	**3.59**	**3.45**	**3.35**	**3.26**	**3.18**	**3.12**	**3.07**	**3.02**	**2.98**	**2.67**	**2.53**	**2.42**	**2.32**
23	0.05	4.28	3.42	3.03	2.80	2.64	2.53	2.44	2.37	2.32	2.27	2.24	2.20	2.18	2.15	2.13	1.96	1.88	1.82	1.76
	0.01	**7.88**	**5.66**	**4.76**	**4.26**	**3.94**	**3.71**	**3.54**	**3.41**	**3.30**	**3.21**	**3.14**	**3.07**	**3.02**	**2.97**	**2.93**	**2.62**	**2.48**	**2.37**	**2.27**
24	0.05	4.26	3.40	3.01	2.78	2.62	2.51	2.42	2.36	2.30	2.25	2.22	2.18	2.15	2.13	2.11	1.94	1.86	1.80	1.74
	0.01	**7.82**	**5.61**	**4.72**	**4.22**	**3.90**	**3.67**	**3.50**	**3.36**	**3.26**	**3.17**	**3.09**	**3.03**	**2.98**	**2.93**	**2.89**	**2.58**	**2.44**	**2.33**	**2.22**
25	0.05	4.24	3.39	2.99	2.76	2.60	2.49	2.40	2.34	2.28	2.24	2.20	2.16	2.14	2.11	2.09	1.92	1.84	1.78	1.72
	0.01	**7.77**	**5.57**	**4.68**	**4.18**	**3.85**	**3.63**	**3.46**	**3.32**	**3.22**	**3.13**	**3.06**	**2.99**	**2.94**	**2.89**	**2.85**	**2.54**	**2.40**	**2.29**	**2.18**
26	0.05	4.23	3.37	2.98	2.74	2.59	2.47	2.39	2.32	2.27	2.22	2.18	2.15	2.12	2.09	2.07	1.90	1.82	1.76	1.70
	0.01	**7.72**	**5.53**	**4.64**	**4.14**	**3.82**	**3.59**	**3.42**	**3.29**	**3.18**	**3.09**	**3.02**	**2.96**	**2.90**	**2.86**	**2.81**	**2.50**	**2.36**	**2.25**	**2.14**

27	0.05	4.21	3.35	2.96	2.73	2.57	2.46	2.37	2.31	2.25	2.20	2.17	2.13	2.10	2.08	2.06	1.88	1.81	1.74	1.68
	0.01	**7.68**	**5.49**	**4.60**	**4.11**	**3.78**	**3.56**	**3.39**	**3.26**	**3.15**	**3.06**	**2.99**	**2.93**	**2.87**	**2.82**	**2.78**	**2.47**	**2.33**	**2.22**	**2.11**
28	0.05	4.20	3.34	2.95	2.71	2.56	2.45	2.36	2.29	2.24	2.19	2.15	2.12	2.09	2.06	2.04	1.87	1.79	1.73	1.66
	0.01	**7.64**	**5.45**	**4.57**	**4.07**	**3.75**	**3.53**	**3.36**	**3.23**	**3.12**	**3.03**	**2.96**	**2.90**	**2.84**	**2.79**	**2.75**	**2.44**	**2.30**	**2.19**	**2.08**
29	0.05	4.18	3.33	2.93	2.70	2.55	2.43	2.35	2.28	2.22	2.18	2.14	2.10	2.08	2.05	2.03	1.85	1.77	1.71	1.65
	0.01	**7.60**	**5.42**	**4.54**	**4.04**	**3.73**	**3.50**	**3.33**	**3.20**	**3.09**	**3.00**	**2.93**	**2.87**	**2.81**	**2.77**	**2.73**	**2.41**	**2.27**	**2.16**	**2.05**
30	0.05	4.17	3.32	2.92	2.69	2.53	2.42	2.33	2.27	2.21	2.16	2.13	2.09	2.06	2.04	2.01	1.84	1.76	1.70	1.63
	0.01	**7.56**	**5.39**	**4.51**	**4.02**	**3.70**	**3.47**	**3.30**	**3.17**	**3.07**	**2.98**	**2.91**	**2.84**	**2.79**	**2.74**	**2.70**	**2.39**	**2.25**	**2.13**	**2.02**
31	0.05	4.16	3.30	2.91	2.68	2.52	2.41	2.32	2.25	2.20	2.15	2.11	2.08	2.05	2.03	2.00	1.83	1.75	1.68	1.62
	0.01	**7.53**	**5.36**	**4.48**	**3.99**	**3.67**	**3.45**	**3.28**	**3.15**	**3.04**	**2.96**	**2.88**	**2.82**	**2.77**	**2.72**	**2.68**	**2.36**	**2.22**	**2.11**	**1.99**
32	0.05	4.15	3.29	2.90	2.67	2.51	2.40	2.31	2.24	2.19	2.14	2.10	2.07	2.04	2.01	1.99	1.82	1.74	1.67	1.60
	0.01	**7.50**	**5.34**	**4.46**	**3.97**	**3.65**	**3.43**	**3.26**	**3.13**	**3.02**	**2.93**	**2.86**	**2.80**	**2.74**	**2.70**	**2.65**	**2.34**	**2.20**	**2.08**	**1.97**
33	0.05	4.14	3.28	2.89	2.66	2.50	2.39	2.30	2.23	2.18	2.13	2.09	2.06	2.03	2.00	1.98	1.81	1.72	1.66	1.59
	0.01	**7.47**	**5.31**	**4.44**	**3.95**	**3.63**	**3.41**	**3.24**	**3.11**	**3.00**	**2.91**	**2.84**	**2.78**	**2.72**	**2.68**	**2.63**	**2.32**	**2.18**	**2.06**	**1.95**
34	0.05	4.13	3.28	2.88	2.65	2.49	2.38	2.29	2.23	2.17	2.12	2.08	2.05	2.02	1.99	1.97	1.80	1.71	1.65	1.58
	0.01	**7.44**	**5.29**	**4.42**	**3.93**	**3.61**	**3.39**	**3.22**	**3.09**	**2.98**	**2.89**	**2.82**	**2.76**	**2.70**	**2.66**	**2.61**	**2.30**	**2.16**	**2.04**	**1.92**
35	0.05	4.12	3.27	2.87	2.64	2.49	2.37	2.29	2.22	2.16	2.11	2.07	2.04	2.01	1.99	1.96	1.79	1.70	1.63	1.57
	0.01	**7.42**	**5.27**	**4.40**	**3.91**	**3.59**	**3.37**	**3.20**	**3.07**	**2.96**	**2.88**	**2.80**	**2.74**	**2.69**	**2.64**	**2.60**	**2.28**	**2.14**	**2.02**	**1.90**
36	0.05	4.11	3.26	2.87	2.63	2.48	2.36	2.28	2.21	2.15	2.11	2.07	2.03	2.00	1.98	1.95	1.78	1.69	1.62	1.56
	0.01	**7.40**	**5.25**	**4.38**	**3.89**	**3.57**	**3.35**	**3.18**	**3.05**	**2.95**	**2.86**	**2.79**	**2.72**	**2.67**	**2.62**	**2.58**	**2.26**	**2.12**	**2.00**	**1.89**
38	0.05	4.10	3.24	2.85	2.62	2.46	2.35	2.26	2.19	2.14	2.09	2.05	2.02	1.99	1.96	1.94	1.76	1.68	1.61	1.54
	0.01	**7.35**	**5.21**	**4.34**	**3.86**	**3.54**	**3.32**	**3.15**	**3.02**	**2.92**	**2.83**	**2.75**	**2.69**	**2.64**	**2.59**	**2.55**	**2.23**	**2.09**	**1.97**	**1.85**
40	0.05	4.08	3.23	2.84	2.61	2.45	2.34	2.25	2.18	2.12	2.08	2.04	2.00	1.97	1.95	1.92	1.74	1.66	1.59	1.52
	0.01	**7.31**	**5.18**	**4.31**	**3.83**	**3.51**	**3.29**	**3.12**	**2.99**	**2.89**	**2.80**	**2.73**	**2.66**	**2.61**	**2.56**	**2.52**	**2.20**	**2.06**	**1.94**	**1.82**
42	0.05	4.07	3.22	2.83	2.59	2.44	2.32	2.24	2.17	2.11	2.06	2.03	1.99	1.96	1.94	1.91	1.73	1.65	1.57	1.50
	0.01	**7.28**	**5.15**	**4.29**	**3.80**	**3.49**	**3.27**	**3.10**	**2.97**	**2.86**	**2.78**	**2.70**	**2.64**	**2.59**	**2.54**	**2.50**	**2.18**	**2.03**	**1.91**	**1.79**
44	0.05	4.06	3.21	2.82	2.58	2.43	2.31	2.23	2.16	2.10	2.05	2.01	1.98	1.95	1.92	1.90	1.72	1.63	1.56	1.49
	0.01	**7.25**	**5.12**	**4.26**	**3.78**	**3.47**	**3.24**	**3.08**	**2.95**	**2.84**	**2.75**	**2.68**	**2.62**	**2.56**	**2.52**	**2.47**	**2.15**	**2.01**	**1.89**	**1.76**
46	0.05	4.05	3.20	2.81	2.57	2.42	2.30	2.22	2.15	2.09	2.04	2.00	1.97	1.94	1.91	1.89	1.71	1.62	1.55	1.47
	0.01	**7.22**	**5.10**	**4.24**	**3.76**	**3.44**	**3.22**	**3.06**	**2.93**	**2.82**	**2.73**	**2.66**	**2.60**	**2.54**	**2.50**	**2.45**	**2.13**	**1.99**	**1.86**	**1.74**
48	0.05	4.04	3.19	2.80	2.57	2.41	2.29	2.21	2.14	2.08	2.03	1.99	1.96	1.93	1.90	1.88	1.70	1.61	1.54	1.46
	0.01	**7.19**	**5.08**	**4.22**	**3.74**	**3.43**	**3.20**	**3.04**	**2.91**	**2.80**	**2.71**	**2.64**	**2.58**	**2.53**	**2.48**	**2.44**	**2.12**	**1.97**	**1.84**	**1.72**
50	0.05	4.03	3.18	2.79	2.56	2.40	2.29	2.20	2.13	2.07	2.03	1.99	1.95	1.92	1.89	1.87	1.69	1.60	1.52	1.45
	0.01	**7.17**	**5.06**	**4.20**	**3.72**	**3.41**	**3.19**	**3.02**	**2.89**	**2.78**	**2.70**	**2.63**	**2.56**	**2.51**	**2.46**	**2.42**	**2.10**	**1.95**	**1.82**	**1.70**

(Continued)

Table B.10 (*Continued*)

		Degrees of freedom for numerator																		
	α	1	2	3	4	5	6	7	8	9	10	11	12	13	14	15	30	50	100	1000
55	0.05	4.02	3.16	2.77	2.54	2.38	2.27	2.18	2.11	2.06	2.01	1.97	1.93	1.90	1.88	1.85	1.67	1.58	1.50	1.42
	0.01	**7.12**	**5.01**	**4.16**	**3.68**	**3.37**	**3.15**	**2.98**	**2.85**	**2.75**	**2.66**	**2.59**	**2.53**	**2.47**	**2.42**	**2.38**	**2.06**	**1.91**	**1.78**	**1.65**
60	0.05	4.00	3.15	2.76	2.53	2.37	2.25	2.17	2.10	2.04	1.99	1.95	1.92	1.89	1.86	1.84	1.65	1.56	1.48	1.40
	0.01	**7.08**	**4.98**	**4.13**	**3.65**	**3.34**	**3.12**	**2.95**	**2.82**	**2.72**	**2.63**	**2.56**	**2.50**	**2.44**	**2.39**	**2.35**	**2.03**	**1.88**	**1.75**	**1.62**
65	0.05	3.99	3.14	2.75	2.51	2.36	2.24	2.15	2.08	2.03	1.98	1.94	1.90	1.87	1.85	1.82	1.63	1.54	1.46	1.38
	0.01	**7.04**	**4.95**	**4.10**	**3.62**	**3.31**	**3.09**	**2.93**	**2.80**	**2.69**	**2.61**	**2.53**	**2.47**	**2.42**	**2.37**	**2.33**	**2.00**	**1.85**	**1.72**	**1.59**
70	0.05	3.98	3.13	2.74	2.50	2.35	2.23	2.14	2.07	2.02	1.97	1.93	1.89	1.86	1.84	1.81	1.62	1.53	1.45	1.36
	0.01	**7.01**	**4.92**	**4.07**	**3.60**	**3.29**	**3.07**	**2.91**	**2.78**	**2.67**	**2.59**	**2.51**	**2.45**	**2.40**	**2.35**	**2.31**	**1.98**	**1.83**	**1.70**	**1.56**
80	0.05	3.96	3.11	2.72	2.49	2.33	2.21	2.13	2.06	2.00	1.95	1.91	1.88	1.84	1.82	1.79	1.60	1.51	1.43	1.34
	0.01	**6.96**	**4.88**	**4.04**	**3.56**	**3.26**	**3.04**	**2.87**	**2.74**	**2.64**	**2.55**	**2.48**	**2.42**	**2.36**	**2.31**	**2.27**	**1.94**	**1.79**	**1.65**	**1.51**
100	0.05	3.94	3.09	2.70	2.46	2.31	2.19	2.10	2.03	1.97	1.93	1.89	1.85	1.82	1.79	1.77	1.57	1.48	1.39	1.30
	0.01	**6.90**	**4.82**	**3.98**	**3.51**	**3.21**	**2.99**	**2.82**	**2.69**	**2.59**	**2.50**	**2.43**	**2.37**	**2.31**	**2.27**	**2.22**	**1.89**	**1.74**	**1.60**	**1.45**
125	0.05	3.92	3.07	2.68	2.44	2.29	2.17	2.08	2.01	1.96	1.91	1.87	1.83	1.80	1.77	1.75	1.55	1.45	1.36	1.26
	0.01	**6.84**	**4.78**	**3.94**	**3.47**	**3.17**	**2.95**	**2.79**	**2.66**	**2.55**	**2.47**	**2.39**	**2.33**	**2.28**	**2.23**	**2.19**	**1.85**	**1.69**	**1.55**	**1.39**
150	0.05	3.90	3.06	2.66	2.43	2.27	2.16	2.07	2.00	1.94	1.89	1.85	1.82	1.79	1.76	1.73	1.54	1.44	1.34	1.24
	0.01	**6.81**	**4.75**	**3.91**	**3.45**	**3.14**	**2.92**	**2.76**	**2.63**	**2.53**	**2.44**	**2.37**	**2.31**	**2.25**	**2.20**	**2.16**	**1.83**	**1.66**	**1.52**	**1.35**
200	0.05	3.89	3.04	2.65	2.42	2.26	2.14	2.06	1.98	1.93	1.88	1.84	1.80	1.77	1.74	1.72	1.52	1.41	1.32	1.21
	0.01	**6.76**	**4.71**	**3.88**	**3.41**	**3.11**	**2.89**	**2.73**	**2.60**	**2.50**	**2.41**	**2.34**	**2.27**	**2.22**	**2.17**	**2.13**	**1.79**	**1.63**	**1.48**	**1.30**
400	0.05	3.86	3.02	2.63	2.39	2.24	2.12	2.03	1.96	1.90	1.85	1.81	1.78	1.74	1.72	1.69	1.49	1.38	1.28	1.15
	0.01	**6.70**	**4.66**	**3.83**	**3.37**	**3.06**	**2.85**	**2.68**	**2.56**	**2.45**	**2.37**	**2.29**	**2.23**	**2.17**	**2.13**	**2.08**	**1.75**	**1.58**	**1.42**	**1.22**
1000	0.05	3.85	3.00	2.61	2.38	2.22	2.11	2.02	1.95	1.89	1.84	1.80	1.76	1.73	1.70	1.68	1.47	1.36	1.26	1.11
	0.01	**6.66**	**4.63**	**3.80**	**3.34**	**3.04**	**2.82**	**2.66**	**2.53**	**2.43**	**2.34**	**2.27**	**2.20**	**2.15**	**2.10**	**2.06**	**1.72**	**1.54**	**1.38**	**1.16**

From the summary table:

Degrees of freedom, numeration: $df_N = 2$
Degrees of freedom, denominator: $df_d = 39$
$F_{\text{observed}} = 3.50$

From Table B.10:

Because the exact values of the degrees of freedom for the denominator are not listed, you must interpolate between the two adjacent numbers:

$F_{\text{critical}}\ (2, 38) = 3.24, \alpha = .05$ $F_{\text{critical}}\ (2, 38) = 5.21, \alpha = .01$
$F_{\text{critical}}\ (2, 40) = 3.23, \alpha = .05$ $F_{\text{critical}}\ (2, 40) = 5.15, \alpha = .01.$

Therefore,

$F_{\text{critical}}\ (2, 39) = 3.235, \alpha = .05$ $F_{\text{critical}}\ (2, 39) = 5.18, \alpha = .01$
$F_{\text{observed}} = 3.50 > F_{\text{critical}} = 3.235$ $F_{\text{observed}} = 3.50 < F_{\text{critical}} = 5.18.$
Reject H_0. Do not reject H_0.

Example: Two-Factor ANOVA

Source	*SS*	*df*	*MS*	*F*
Variable *A*	0.067	1	0.067	0.01
Variable *B*	80.433	2	40.217	6.859
AB	58.233	2	29.117	4.966
Within groups	316.600	54	5.863	
Total	455.333	59		

From the summary table:

Critical values:

$\alpha = .05$	$\alpha = .01$
$F_{\text{critical}}\ (1, 54) = 4.02$	$F_{\text{critical}}\ (1, 54) = 7.12$
$F_{\text{critical}}\ (2, 54) = 3.16$	$F_{\text{critical}}\ (2, 54) = 5.01$

	Statistical Decision	
Result	**($\alpha = .05$)**	**($\alpha = .01$)**
Variable *A*: $df_N = 1$, $df_d = 54 \rightarrow F_{\text{observed}} = 0.01$	Do not reject H_0.	Do not reject H_0.
Variable *B*: $df_N = 2$, $df_d = 54 \rightarrow F_{\text{observed}} = 6.86$	Reject H_0.	Reject H_0.
Variable *AB*: $df_N = 2$, $df_d = 54 \rightarrow F_{\text{observed}} = 4.97$	Reject H_0.	Do not reject H_0.

TABLE B.11: CRITICAL VALUES FOR THE F_{MAX} TEST

Using Table B.11

To use this table, divide the largest variance by the smallest variance to create F_{max}. The column labeled n represents the number of subjects in each group. If the sample sizes for the two groups are not equal, determine the average n and round up. The other columns of numbers represent the number of treatment conditions in the study. If the observed value of F_{max} is less than the tabled value, then you may assume that the variances are homogeneous, $\sigma_{smallest} = \sigma_{largest}$.

Example

A researcher conducted a study with six groups. The largest variance as 20 and the smallest variance was 10, with 15 participants in each group. $F_{max} = 2.00$. The critical value of $F_{max} = 4.70$, $\alpha = .05$. Therefore, we do NOT reject the hypothesis that the variances are equivalent. The data do not appear to violate the requirement that there be homogeneity of variance for the ANOVA.

TABLE B.12: CRITICAL VALUES FOR THE STUDENTIZED RANGE TEST

Using Table B.12

This table contains the critical values developed by Tukey for his honestly significant difference (HSD) test. To use the table, you will need the degrees of freedom for the within-groups term in the ANOVA summary table and the number of means to be compared by the HSD test.

Example

A researcher conducted a study with four groups. The degrees of freedom denominator (*df* for the within-groups factor) are 12. Using Table B.12,

$$q_{critical} = 3.62, \alpha = .10$$

$$q_{critical} = 4.20, \alpha = .05$$

$$q_{critical} = 5.50, \alpha = .01$$

TABLE B.13: POWER OF ANOVA

Using Table B.13

The values in this table help you determine the optimal sample size for an ANOVA given the anticipated effect size and α-level.

Table B.11

Number of variances in study										
n	α	2	3	4	5	6	7	8	9	10
4	.05	9.60	15.5	20.6	25.2	29.5	33.6	37.5	41.4	44.6
	.01	**23.2**	**37.0**	**49.0**	**59.0**	**69.0**	**79.0**	**89.0**	**97.0**	**106.0**
5	.05	7.2	10.8	13.7	16.3	18.7	20.8	22.9	24.7	26.5
	.01	**14.9**	**22.0**	**28.0**	**33.0**	**38.0**	**42.0**	**46.0**	**50.0**	**54.0**
6	.05	5.8	8.4	10.4	12.1	13.7	15.0	16.3	17.5	18.6
	.01	**11.1**	**15.5**	**19.1**	**22.0**	**25.0**	**27.0**	**30.0**	**32.0**	**34.0**
7	.05	5.0	6.9	8.4	9.7	10.8	11.8	12.7	13.5	14.3
	.01	**8.9**	**12.1**	**14.5**	**16.5**	**18.4**	**20.0**	**22.0**	**23.0**	**24.0**
8	.05	4.4	6.0	7.2	8.1	9.0	9.8	10.5	11.1	11.7
	.01	**7.5**	**9.9**	**11.7**	**13.2**	**14.5**	**15.8**	**16.9**	**17.9**	**18.9**
9	.05	4.0	5.3	6.3	7.1	7.8	8.4	8.9	9.5	9.9
	.01	**6.5**	**8.5**	**9.9**	**11.1**	**12.1**	**13.1**	**13.9**	**14.7**	**15.3**
10	.05	3.7	4.9	5.7	6.3	6.9	7.4	7.9	8.3	8.7
	.01	**5.9**	**7.4**	**8.6**	**9.6**	**10.4**	**11.1**	**11.8**	**12.4**	**12.9**
12	.05	3.3	4.2	4.8	5.3	5.7	6.1	6.4	6.7	7.0
	.01	**4.9**	**6.1**	**6.9**	**7.6**	**8.2**	**8.7**	**9.1**	**9.5**	**9.9**
15	.05	2.7	3.5	4.0	4.4	4.7	4.9	5.2	5.4	5.6
	.01	**4.1**	**4.9**	**5.5**	**6.0**	**6.4**	**6.7**	**7.1**	**7.3**	**7.5**
20	.05	2.5	2.9	3.3	3.5	3.7	3.9	4.1	4.2	4.4
	.01	**3.3**	**3.8**	**4.3**	**4.6**	**4.9**	**5.1**	**5.3**	**5.5**	**5.6**
30	.05	2.1	2.4	2.6	2.8	2.9	3.0	3.1	3.2	3.3
	.01	**2.6**	**3.0**	**3.3**	**3.4**	**3.6**	**3.7**	**3.8**	**3.9**	**4.0**
60	.05	1.7	1.9	1.9	2.0	2.1	2.2	2.2	2.3	2.3
	.01	**2.0**	**2.2**	**2.3**	**2.4**	**2.4**	**2.5**	**2.5**	**2.6**	**2.6**
∞	.05	1.0	1.0	1.0	1.0	1.0	1.0	1.0	1.0	1.0
	.01	**1.0**	**1.0**	**1.0**	**1.0**	**1.0**	**1.0**	**1.0**	**1.0**	**1.0**

Example: Single-Factor Design

A researcher wishes to conduct a single-factor design with three levels of the independent variable. How many participants will the researcher require in each treatment condition to have power equal to $1 - \beta = .80$ when the effect size is moderate, $f = .25$ and $\alpha = .05$? In this example, $df_N = 2$. According to this table, $1 - \beta = .83$ when there are 55 participants in each treatment condition.

Example: Factorial Design

A researcher designed a 3×4 factorial study. How many participants should the researcher use in each treatment condition to have power equal to $1 - \beta = .80$? Also assume that the effect size is moderate, $f = .25$.

Table B.12

Degrees of freedom for denominator		α	2	3	4	5	6	7	8	9	10	11	12	13	14	15	16	17	18
			Number of means in set																
	1	.10	8.93	13.40	16.40	18.50	20.20	21.50	22.60	23.60	24.50	25.20	25.90	26.50	27.10	27.60	28.10	28.50	29.00
		.05	18.00	27.00	32.80	37.10	40.40	43.10	45.40	47.40	49.10	50.60	52.00	53.20	54.30	55.40	56.30	57.20	58.00
		.01	90.00	13.50	164.00	186.00	202.00	216.00	227.00	237.00	246.00	253.00	260.00	266.00	272.00	277.00	282.00	286.00	290.00
	2	.10	4.13	5.73	6.78	7.54	8.14	8.63	9.05	9.41	9.73	10.00	10.30	10.50	10.70	10.90	11.10	11.20	11.40
		.05	6.09	8.30	9.80	10.90	11.70	12.40	13.00	13.50	14.00	14.40	14.70	15.10	15.40	15.70	15.90	16.10	16.40
		.01	14.00	19.00	22.30	24.70	26.60	28.20	29.50	30.70	31.70	32.60	33.40	34.10	34.80	35.40	36.00	36.50	37.00
	3	.10	3.33	4.47	5.20	5.74	6.16	6.51	6.81	7.06	7.29	7.49	7.67	7.83	7.98	8.12	8.25	8.37	8.78
		.05	4.50	5.91	6.82	7.50	8.04	8.48	8.85	9.18	9.46	9.72	9.95	10.20	10.40	10.50	10.70	10.80	11.00
		.01	8.26	10.60	12.20	13.30	14.20	15.00	15.60	16.20	16.70	17.10	17.50	17.90	18.20	18.50	18.80	19.10	19.30
	4	.10	3.01	3.98	4.59	5.04	5.39	5.69	5.93	6.14	6.33	6.50	6.65	6.78	6.91	7.03	7.13	7.23	7.33
		.05	3.93	5.04	5.76	6.29	6.71	7.05	7.35	7.60	7.83	8.03	8.21	8.37	8.52	8.66	8.79	8.91	9.03
		.01	6.51	8.12	9.17	9.96	10.60	11.10	11.50	11.90	12.30	12.60	12.80	13.10	13.30	13.50	13.70	13.90	14.10
	5	.10	2.85	3.72	4.26	4.66	4.98	5.24	5.44	5.65	5.82	5.97	6.10	6.22	6.34	6.44	6.54	6.63	6.71
		.05	3.64	4.60	5.22	5.67	6.03	6.33	6.58	6.80	6.99	7.17	7.32	7.47	7.60	7.72	7.83	7.93	8.03
		.01	5.70	6.97	7.80	8.42	8.91	9.32	9.67	9.97	10.20	10.50	10.70	10.90	11.10	11.20	11.40	11.60	11.70
	6	.10	2.75	3.56	4.07	4.44	4.73	4.97	5.17	5.34	5.50	5.64	5.76	5.88	5.98	6.08	6.16	6.25	6.33
		.05	3.46	4.34	4.90	5.31	5.63	5.89	6.12	6.32	6.49	6.65	6.79	6.92	7.03	7.14	7.24	7.34	7.43
		.01	5.24	6.33	7.03	7.56	7.97	8.32	8.61	8.87	9.10	9.30	9.49	9.65	9.81	9.95	10.10	10.20	10.30
	7	.10	2.68	3.45	3.93	4.28	4.56	4.78	4.97	5.14	5.28	5.41	5.53	5.64	5.74	5.83	5.91	5.99	6.06
		.05	3.34	4.16	4.69	5.06	5.36	5.61	5.82	6.00	6.16	6.30	6.43	6.55	6.66	6.76	6.85	6.94	7.02
		.01	4.95	5.92	6.54	7.01	7.37	7.68	7.94	8.17	8.37	8.55	8.71	8.86	9.00	9.12	9.24	9.35	9.46
	8	.10	2.63	3.37	3.83	4.17	4.43	4.65	4.83	4.99	5.13	5.25	5.36	5.46	5.56	5.64	5.74	5.83	5.87
		.05	3.26	4.04	4.53	4.89	5.17	5.40	5.60	5.77	5.92	6.05	6.18	6.29	6.39	6.48	6.57	6.65	6.73
		.01	4.74	5.63	6.20	6.63	6.96	7.24	7.47	7.68	7.78	8.03	8.18	8.31	8.44	8.55	8.66	8.76	8.85
	9	.10	2.59	3.32	3.76	4.08	4.34	4.55	4.72	4.87	5.01	5.13	5.23	5.33	5.42	5.51	5.58	5.66	5.72
		.05	3.20	3.95	4.42	4.76	5.02	5.24	5.43	5.60	5.74	5.87	5.98	6.09	6.19	6.28	6.36	6.44	6.51
		.01	4.60	5.43	5.96	6.35	6.66	6.91	7.13	7.32	7.49	7.65	7.78	7.91	8.03	8.13	8.23	8.33	8.41
	10	.10	2.56	3.28	3.70	4.02	4.26	4.47	4.64	4.78	4.91	5.03	5.13	5.23	5.32	5.40	5.47	5.54	5.61
		.05	3.15	3.88	4.33	4.65	4.91	5.12	5.30	5.46	5.60	5.72	5.83	5.93	6.03	6.11	6.19	6.27	6.34
		.01	4.48	5.27	5.77	6.14	6.43	6.67	6.87	7.05	7.21	7.36	7.48	7.60	7.71	7.81	7.91	8.00	8.08
	11	.10	2.54	3.23	3.66	3.97	4.21	4.40	4.57	4.71	4.84	4.95	5.05	5.15	5.23	5.31	5.38	5.45	5.51
		.05	3.11	3.82	4.26	4.57	4.82	5.03	5.20	5.35	5.49	5.61	5.71	5.81	5.90	5.99	6.06	6.18	6.20
		.01	4.39	5.14	5.62	5.97	6.25	6.48	6.67	6.84	6.99	7.13	7.26	7.36	7.46	7.56	7.65	7.73	7.81

12	.10	2.52	3.20	3.62	3.92	4.16	4.35	4.51	4.65	4.78	4.89	4.99	5.08	5.16	5.24	5.31	5.37	5.44
	.05	3.08	3.77	4.20	4.51	4.75	4.95	5.12	5.27	5.40	5.51	5.62	5.71	5.80	5.88	5.95	6.02	6.09
	.01	4.32	5.04	5.50	5.84	6.10	6.32	6.51	6.67	6.81	6.94	7.06	7.17	7.26	7.36	7.44	7.52	7.50
13	.10	2.51	3.18	3.59	3.89	4.12	4.31	4.46	4.60	4.72	4.83	4.93	5.02	5.10	5.18	5.25	5.31	5.37
	.05	3.06	3.73	4.15	4.45	4.69	4.88	5.05	5.19	5.32	5.43	5.53	5.63	5.71	5.79	5.86	5.93	6.00
	.01	4.26	4.96	5.40	5.73	5.98	6.19	6.37	6.53	6.67	6.79	6.90	7.01	7.10	7.19	7.27	7.37	7.42
14	.10	2.99	3.16	3.56	3.83	4.08	4.27	4.42	4.56	4.68	4.79	4.88	4.97	5.05	5.12	5.19	5.26	5.32
	.05	3.03	3.70	4.11	4.41	4.64	4.83	4.99	5.13	5.25	5.36	5.46	5.55	5.64	5.72	5.79	5.85	5.92
	.01	4.21	4.89	5.32	5.63	5.88	6.08	6.26	6.41	6.54	6.66	6.77	6.87	6.96	7.05	7.13	7.20	7.27
16	.10	2.47	3.12	3.52	3.80	4.03	4.21	4.36	4.49	4.61	4.71	4.81	4.89	4.97	5.04	5.11	5.17	5.23
	.05	3.00	3.65	4.05	4.33	4.56	4.74	4.90	5.03	5.15	5.26	5.35	5.44	5.52	5.59	5.66	5.73	5.79
	.01	4.13	4.78	5.19	5.49	5.72	5.92	6.08	6.22	6.35	6.46	6.56	6.66	6.74	6.82	6.90	6.97	7.03
18	.10	2.45	3.10	3.49	3.77	3.98	4.16	4.31	4.44	4.55	4.66	4.75	4.83	4.91	4.98	5.04	5.10	5.16
	.05	2.97	3.61	4.00	4.28	4.49	4.67	4.82	4.96	5.07	5.17	5.27	5.35	5.43	5.50	5.57	5.63	5.69
	.01	4.07	4.70	5.09	5.38	5.60	5.79	5.94	6.08	6.20	6.31	6.41	6.50	6.58	6.65	6.73	6.79	6.85
20	.10	2.44	3.08	3.46	3.74	3.95	4.12	4.27	4.40	4.51	4.61	4.70	4.78	4.86	4.92	4.99	5.05	5.10
	.05	2.95	3.58	3.96	4.23	4.45	4.62	4.77	4.90	5.01	5.11	5.20	5.28	5.36	5.43	5.49	5.55	5.61
	.01	4.02	4.64	5.02	5.29	5.51	5.69	5.84	5.97	6.09	6.19	6.29	6.37	6.45	6.52	6.59	6.65	6.71
24	.10	2.42	3.05	3.42	3.69	3.9	4.07	4.21	4.34	4.45	4.54	4.63	4.71	4.78	4.85	4.91	4.97	5.02
	.05	2.92	3.53	3.9	4.17	4.37	4.54	4.68	4.81	4.92	5.01	5.1	5.18	5.25	5.32	5.38	5.44	5.49
	.01	3.96	4.54	4.91	5.17	5.37	5.54	5.69	5.81	5.92	6.02	6.11	6.19	6.26	6.33	6.39	6.45	6.51
30	.10	2.4	3.02	3.39	3.65	3.85	4.02	4.16	4.28	4.38	4.47	4.56	4.64	4.71	4.77	4.83	4.89	4.94
	.05	2.89	3.49	3.84	4.1	4.3	4.46	4.6	4.72	4.83	4.92	5	5.08	5.15	5.21	5.27	5.33	5.38
	.01	3.89	4.45	4.8	5.05	5.24	5.4	5.54	5.56	5.76	5.85	5.93	6.01	6.08	6.14	6.2	6.26	6.31
40	.10	2.38	2.99	3.35	3.61	3.8	3.96	4.1	4.22	4.32	4.41	4.49	4.56	4.63	4.7	4.75	4.81	4.86
	.05	2.86	3.44	3.79	4.04	4.23	4.39	4.52	4.63	4.74	4.82	4.91	4.98	5.05	5.11	5.16	5.22	5.27
	.01	3.82	4.37	4.7	4.93	5.11	5.27	5.39	5.5	5.6	5.69	5.77	5.84	5.9	5.96	6.02	6.07	6.11
60	.10	2.36	2.96	3.31	3.56	3.76	3.91	4.04	4.16	4.26	4.34	4.42	4.49	4.56	4.62	4.68	4.73	4.78
	.05	2.83	3.4	3.74	3.98	4.16	4.31	4.44	4.55	4.65	4.73	4.81	4.88	4.94	5	5.06	5.11	5.15
	.01	3.76	4.28	4.6	4.82	4.99	5.13	5.25	5.36	5.45	5.53	5.6	5.67	5.73	5.79	5.84	5.89	5.93
120	.10	2.34	2.93	3.28	3.52	3.71	3.86	3.99	4.1	4.19	4.28	4.35	4.42	4.49	4.59	4.6	4.65	4.69
	.05	2.8	3.36	3.69	3.92	4.1	4.24	4.36	4.48	4.56	4.64	4.72	4.78	4.84	4.9	4.95	5	5.04
	.01	3.7	4.2	4.5	4.71	4.87	5.01	5.12	5.21	5.3	5.38	5.44	5.51	5.56	5.61	5.66	5.71	5.75

Table B.13

		$\alpha = .05$		$df_N = 1$			$\alpha = .05$		$df_N = 2$			$\alpha = .05$		$df_N = 3$	
		Effect size, f					Effect size, f					Effect size, f			
n	F_c	.10	.25	.40	.55	F_c	.10	.25	.40	.55	F_c	.10	.25	.40	.55
10	4.414	.08	.19	.40	.65	3.354	.10	.22	.46	.72	2.866	.12	.26	.52	.79
11	4.351	.08	.21	.43	.70	3.316	.10	.24	.50	.77	2.839	.11	.27	.56	.83
12	4.301	.08	.22	.47	.74	3.285	.10	.25	.53	.81	2.816	.11	.29	.60	.87
13	4.260	.08	.23	.50	.78	3.259	.10	.27	.57	.85	2.798	.11	.31	.64	.90
14	4.225	.09	.25	.53	.81	3.238	.10	.28	.60	.88	2.783	.12	.32	.67	.92
15	4.196	.09	.26	.56	.84	3.220	.10	.30	.64	.90	2.769	.12	.34	.71	.94
16	4.171	.09	.28	.59	.87	3.204	.10	.32	.67	.92	2.758	.12	.36	.74	.96
17	4.149	.09	.29	.62	.89	3.191	.10	.33	.70	.94	2.748	.12	.38	.77	.97
18	4.130	.09	.30	.65	.91	3.179	.11	.35	.73	.95	2.739	.12	.39	.79	.98
19	4.113	.10	.32	.68	.92	3.168	.11	.36	.75	.96	2.732	.12	.41	.82	.98
20	4.098	.10	.33	.70	.94	3.159	.11	.38	.78	.97	2.725	.12	.43	.84	.99
21	4.085	.10	.35	.72	.95	3.150	.11	.40	.80	.98	2.719	.12	.45	.86	.99
22	4.073	.10	.36	.75	.96	3.143	.11	.41	.82	.98	2.713	.13	.47	.88	.99
23	4.062	.10	.37	.77	.97	3.136	.12	.43	.84	.99	2.708	.13	.49	.90	.99
24	4.052	.10	.39	.79	.97	3.130	.12	.44	.86	.99	2.704	.13	.50	.91	.99
25	4.043	.11	.40	.80	.98	3.124	.12	.46	.87	.99	2.699	.13	.52	.92	.99
26	4.034	.11	.42	.82	.98	3.119	.12	.48	.89	.99	2.696	.13	.54	.93	.99
27	4.027	.11	.43	.84	.99	3.114	.12	.49	.90	.99	2.692	.14	.56	.94	.99
28	4.020	.11	.44	.85	.99	3.109	.13	.51	.91	.99	2.689	.14	.57	.95	.99
29	4.013	.11	.46	.86	.99	3.105	.13	.52	.92	.99	2.686	.14	.59	.96	.99
30	4.007	.12	.47	.88	.99	3.101	.13	.54	.93	.99	2.683	.14	.61	.97	.99
31	4.001	.12	.48	.89	.99	3.098	.13	.55	.94	.99	2.680	.15	.62	.97	.99
32	3.996	.12	.50	.90	.99	3.094	.13	.57	.95	.99	2.678	.15	.64	.98	.99
33	3.991	.12	.51	.91	.99	3.091	.14	.58	.96	.99	2.675	.15	.65	.98	.99
34	3.986	.13	.52	.92	.99	3.088	.14	.60	.96	.99	2.673	.15	.67	.98	.99
35	3.982	.13	.54	.93	.99	3.085	.14	.61	.97	.99	2.671	.15	.68	.99	.99
36	3.978	.13	.55	.93	.99	3.083	.14	.62	.97	.99	2.669	.16	.70	.99	.99
37	3.974	.13	.56	.94	.99	3.080	.14	.64	.98	.99	2.667	.16	.71	.99	.99
38	3.970	.13	.57	.95	.99	3.078	.15	.65	.98	.99	2.666	.16	.72	.99	.99
39	3.967	.14	.59	.95	.99	3.076	.15	.66	.98	.99	2.664	.17	.74	.99	.99
40	3.963	.14	.60	.96	.99	3.074	.15	.68	.98	.99	2.663	.17	.75	.99	.99
45	3.949	.15	.65	.98	.99	3.065	.16	.73	.99	.99	2.656	.18	.81	.99	.99
50	3.938	.16	.70	.99	.99	3.058	.18	.78	.99	.99	2.651	.20	.85	.99	.99
55	3.929	.17	.75	.99	.99	3.052	.19	.83	.99	.99	2.646	.21	.89	.99	.99
60	3.921	.18	.79	.99	.99	3.047	.20	.86	.99	.99	2.643	.22	.92	.99	.99
70	3.910	.21	.85	.99	.99	3.040	.23	.92	.99	.99	2.637	.26	.96	.99	.99
80	3.901	.23	.90	.99	.99	3.034	.26	.95	.99	.99	2.633	.29	.98	.99	.99
90	3.894	.25	.93	.99	.99	3.030	.28	.97	.99	.99	2.630	.32	.99	.99	.99
100	3.889	.28	.96	.99	.99	3.026	.31	.99	.99	.99	2.627	.35	.99	.99	.99
110	3.884	.30	.97	.99	.99	3.023	.34	.99	.99	.99	2.625	.38	.99	.99	.99
120	3.881	.32	.98	.99	.99	3.021	.37	.99	.99	.99	2.624	.41	.99	.99	.99
130	3.878	.35	.99	.99	.99	3.019	.39	.99	.99	.99	2.622	.45	.99	.99	.99
140	3.875	.37	.99	.99	.99	3.017	.42	.99	.99	.99	2.621	.48	.99	.99	.99

Table B.13 (*Continued*)

		$\alpha = .05$		$df_N = 1$			$\alpha = .05$		$df_N = 2$			$\alpha = .05$		$df_N = 3$	
		Effect size, *f*					Effect size, *f*					Effect size, *f*			
n	F_c	.10	.25	.40	.55	F_c	.10	.25	.40	.55	F_c	.10	.25	.40	.55
150	3.873	.39	.99	.99	.99	3.016	.45	.99	.99	.99	2.620	.51	.99	.99	.99
160	3.871	.42	.99	.99	.99	3.015	.47	.99	.99	.99	2.619	.54	.99	.99	.99
170	3.869	.44	.99	.99	.99	3.014	.50	.99	.99	.99	2.618	.57	.99	.99	.99
180	3.868	.46	.99	.99	.99	3.013	.53	.99	.99	.99	2.617	.60	.99	.99	.99
190	3.866	.48	.99	.99	.99	3.012	.55	.99	.99	.99	2.617	.62	.99	.99	.99
200	3.865	.50	.99	.99	.99	3.011	.58	.99	.99	.99	2.616	.65	.99	.99	.99
300	3.857	.69	.99	.99	.99	3.006	.78	.99	.99	.99	2.612	.85	.99	.99	.99
10	2.579	.13	.29	.57	.83	2.386	.14	.32	.61	.87	2.246	.16	.34	.65	.89
11	2.557	.13	.31	.61	.87	2.368	.14	.33	.66	.90	2.231	.15	.36	.69	.93
12	2.540	.13	.32	.65	.91	2.354	.14	.35	.70	.93	2.219	.15	.38	.74	.95
13	2.525	.13	.34	.69	.93	2.342	.14	.37	.74	.95	2.209	.15	.40	.77	.97
14	2.513	.13	.36	.73	.95	2.332	.14	.39	.77	.97	2.200	.15	.42	.81	.98
15	2.503	.13	.38	.76	.96	2.323	.14	.42	.80	.98	2.193	.15	.45	.84	.99
16	2.494	.13	.40	.79	.97	2.316	.14	.44	.83	.99	2.186	.15	.47	.87	.99
17	2.486	.13	.42	.82	.98	2.309	.14	.46	.86	.99	2.181	.15	.49	.89	.99
18	2.479	.13	.44	.84	.99	2.303	.14	.48	.88	.99	2.176	.15	.51	.91	.99
19	2.473	.13	.46	.87	.99	2.298	.14	.50	.90	.99	2.171	.16	.54	.93	.99
20	2.467	.13	.48	.89	.99	2.294	.15	.52	.92	.99	2.167	.16	.56	.94	.99
21	2.463	.14	.50	.90	.99	2.290	.15	.54	.93	.99	2.164	.16	.58	.95	.99
22	2.458	.14	.52	.92	.99	2.286	.15	.56	.95	.99	2.161	.16	.60	.96	.99
23	2.454	.14	.54	.93	.99	2.283	.15	.58	.96	.99	2.158	.16	.62	.97	.99
24	2.451	.14	.56	.94	.99	2.280	.15	.60	.96	.99	2.155	.17	.65	.98	.99
25	2.447	.14	.58	.95	.99	2.277	.16	.62	.97	.99	2.153	.17	.67	.98	.99
26	2.444	.15	.59	.96	.99	2.274	.16	.64	.98	.99	2.151	.17	.69	.99	.99
27	2.441	.15	.61	.97	.99	2.272	.16	.66	.98	.99	2.149	.17	.70	.99	.99
28	2.439	.15	.63	.97	.99	2.270	.16	.68	.99	.99	2.147	.18	.72	.99	.99
29	2.436	.15	.65	.98	.99	2.268	.17	.70	.99	.99	2.145	.18	.74	.99	.99
30	2.434	.16	.66	.98	.99	2.266	.17	.72	.99	.99	2.143	.18	.76	.99	.99
31	2.432	.16	.68	.99	.99	2.264	.17	.73	.99	.99	2.142	.18	.77	.99	.99
32	2.430	.16	.70	.99	.99	2.263	.17	.75	.99	.99	2.141	.19	.79	.99	.99
33	2.428	.16	.71	.99	.99	2.261	.18	.76	.99	.99	2.139	.19	.80	.99	.99
34	2.426	.17	.73	.99	.99	2.260	.18	.78	.99	.99	2.138	.19	.82	.99	.99
35	2.425	.17	.74	.99	.99	2.258	.18	.79	.99	.99	2.137	.20	.83	.99	.99
36	2.423	.17	.76	.99	.99	2.257	.19	.80	.99	.99	2.136	.20	.84	.99	.99
37	2.422	.17	.77	.99	.99	2.256	.19	.82	.99	.99	2.135	.20	.85	.99	.99
38	2.420	.18	.78	.99	.99	2.255	.19	.83	.99	.99	2.134	.21	.87	.99	.99
39	2.419	.18	.80	.99	.99	2.254	.20	.84	.99	.99	2.133	.21	.88	.99	.99
40	2.418	.18	.81	.99	.99	2.253	.20	.85	.99	.99	2.132	.21	.89	.99	.99
45	2.413	.20	.86	.99	.99	2.248	.22	.90	.99	.99	2.128	.23	.93	.99	.99
50	2.408	.21	.90	.99	.99	2.245	.23	.93	.99	.99	2.125	.25	.95	.99	.99
55	2.405	.23	.93	.99	.99	2.242	.25	.96	.99	.99	2.123	.27	.97	.99	.99

(*Continued*)

Table B.13 (*Continued*)

		$\alpha = .05$		$df_N = 1$			$\alpha = .05$		$df_N = 2$			$\alpha = .05$		$df_N = 3$	
		Effect size, f					Effect size, f					Effect size, f			
n	F_c	.10	.25	.40	.55	F_c	.10	.25	.40	.55	F_c	.10	.25	.40	.55
60	2.402	.25	.95	.99	.99	2.239	.27	.97	.99	.99	2.121	.29	.98	.99	.99
70	2.398	.28	.98	.99	.99	2.236	.31	.99	.99	.99	2.117	.33	.99	.99	.99
80	2.395	.32	.99	.99	.99	2.233	.35	.99	.99	.99	2.115	.38	.99	.99	.99
90	2.392	.35	.99	.99	.99	2.231	.39	.99	.99	.99	2.113	.42	.99	.99	.99
100	2.390	.39	.99	.99	.99	2.229	.43	.99	.99	.99	2.112	.46	.99	.99	.99
110	2.388	.43	.99	.99	.99	2.228	.47	.99	.99	.99	2.110	.51	.99	.99	.99
120	2.387	.46	.99	.99	.99	2.227	.51	.99	.99	.99	2.109	.55	.99	.99	.99
130	2.386	.50	.99	.99	.99	2.226	.54	.99	.99	.99	2.109	.59	.99	.99	.99
140	2.385	.53	.99	.99	.99	2.225	.58	.99	.99	.99	2.108	.63	.99	.99	.99
150	2.384	.57	.99	.99	.99	2.224	.62	.99	.99	.99	2.107	.66	.99	.99	.99
160	2.383	.60	.99	.99	.99	2.223	.65	.99	.99	.99	2.107	.70	.99	.99	.99
170	2.382	.63	.99	.99	.99	2.223	.68	.99	.99	.99	2.106	.73	.99	.99	.99
180	2.382	.66	.99	.99	.99	2.222	.71	.99	.99	.99	2.106	.76	.99	.99	.99
190	2.381	.69	.99	.99	.99	2.222	.74	.99	.99	.99	2.105	.79	.99	.99	.99
200	2.381	.71	.99	.99	.99	2.222	.77	.99	.99	.99	2.105	.81	.99	.99	.99
300	2.378	.90	.99	.99	.99	2.219	.93	.99	.99	.99	2.103	.96	.99	.99	.99
10	8.285	.02	.07	.19	.38	5.488	.03	.10	.25	.48	4.377	.04	.12	.30	.57
11	8.096	.02	.08	.21	.42	5.390	.03	.10	.27	.53	4.313	.04	.13	.34	.63
12	7.945	.02	.09	.23	.47	5.312	.03	.11	.30	.59	4.261	.04	.14	.37	.68
13	7.823	.03	.09	.26	.52	5.248	.03	.12	.33	.64	4.218	.04	.15	.41	.73
14	7.721	.03	.10	.28	.56	5.194	.03	.13	.36	.68	4.182	.04	.16	.44	.78
15	7.636	.03	.11	.31	.60	5.149	.03	.14	.39	.72	4.152	.04	.17	.48	.81
16	7.562	.03	.11	.33	.64	5.110	.03	.15	.42	.76	4.126	.04	.18	.51	.85
17	7.499	.03	.12	.36	.68	5.077	.03	.16	.45	.80	4.103	.04	.19	.55	.88
18	7.444	.03	.13	.38	.71	5.047	.04	.16	.49	.83	4.083	.04	.20	.58	.90
19	7.396	.03	.14	.41	.75	5.021	.04	.18	.52	.86	4.066	.04	.22	.61	.92
20	7.353	.03	.15	.43	.78	4.998	.04	.19	.54	.88	4.050	.04	.23	.64	.94
21	7.314	.03	.15	.46	.80	4.977	.04	.20	.57	.90	4.036	.04	.24	.67	.95
22	7.280	.03	.16	.48	.83	4.959	.04	.21	.60	.92	4.024	.04	.25	.70	.96
23	7.248	.03	.17	.51	.85	4.942	.04	.22	.63	.93	4.012	.05	.27	.73	.97
24	7.220	.03	.18	.53	.87	4.927	.04	.23	.66	.95	4.002	.05	.28	.75	.98
25	7.194	.03	.19	.56	.89	4.913	.04	.24	.68	.96	3.992	.05	.30	.78	.98
26	7.171	.03	.20	.58	.90	4.900	.04	.25	.70	.96	3.984	.05	.31	.80	.99
27	7.149	.03	.21	.60	.92	4.888	.04	.26	.73	.97	3.976	.05	.33	.82	.99
28	7.129	.04	.22	.62	.93	4.877	.04	.28	.75	.98	3.968	.05	.34	.84	.99
29	7.110	.04	.23	.65	.94	4.867	.04	.29	.77	.98	3.961	.05	.35	.86	1.00
30	7.093	.04	.24	.67	.95	4.858	.04	.30	.79	.99	3.955	.05	.37	.87	1.00
31	7.077	.04	.25	.69	.96	4.849	.04	.31	.81	.99	3.949	.05	.38	.89	1.00
32	7.062	.04	.26	.71	.96	4.841	.05	.33	.82	.99	3.944	.05	.40	.90	1.00
33	7.048	.04	.27	.72	.97	4.833	.05	.34	.84	.99	3.938	.05	.41	.91	1.00
34	7.035	.04	.28	.74	.97	4.826	.05	.35	.86	1.00	3.934	.05	.43	.92	1.00
35	7.023	.04	.29	.76	.98	4.819	.05	.36	.87	1.00	3.929	.06	.44	.93	1.00

Table B.13 (*Continued*)

		$\alpha = .05$	$df_N = 1$				$\alpha = .05$	$df_N = 2$				$\alpha = .05$	$df_N = 3$		
		Effect size, *f*					Effect size, *f*					Effect size, *f*			
n	F_c	.10	.25	.40	.55	F_c	.10	.25	.40	.55	F_c	.10	.25	.40	.55
36	7.011	.04	.30	.77	.98	4.813	.05	.38	.88	1.00	3.925	.06	.46	.94	1.00
37	7.000	.04	.31	.79	.98	4.807	.05	.39	.89	1.00	3.921	.06	.47	.95	1.00
38	6.990	.04	.32	.80	.99	4.802	.05	.40	.90	1.00	3.917	.06	.49	.96	1.00
39	6.981	.04	.33	.82	.99	4.796	.05	.42	.91	1.00	3.913	.06	.50	.96	1.00
40	6.971	.05	.34	.83	.99	4.791	.05	.43	.92	1.00	3.910	.06	.52	.97	1.00
45	6.932	.05	.39	.88	1.00	4.770	.06	.49	.96	1.00	3.895	.07	.59	.99	1.00
50	6.901	.06	.44	.92	1.00	4.753	.06	.55	.98	1.00	3.883	.07	.66	.99	1.00
55	6.876	.06	.49	.95	1.00	4.739	.07	.61	.99	1.00	3.874	.08	.72	1.00	1.00
60	6.855	.07	.54	.97	1.00	4.727	.08	.67	.99	1.00	3.866	.09	.77	1.00	1.00
70	6.822	.08	.63	.99	1.00	4.709	.09	.76	1.00	1.00	3.853	.11	.85	1.00	1.00
80	6.798	.09	.71	1.00	1.00	4.696	.10	.83	1.00	1.00	3.844	.12	.91	1.00	1.00
90	6.779	.10	.78	1.00	1.00	4.686	.12	.89	1.00	1.00	3.837	.14	.95	1.00	1.00
100	6.765	.11	.83	1.00	1.00	4.677	.14	.93	1.00	1.00	3.831	.16	.97	1.00	1.00
110	6.753	.13	.88	1.00	1.00	4.671	.15	.95	1.00	1.00	3.827	.18	.99	1.00	1.00
120	6.743	.14	.91	1.00	1.00	4.665	.17	.97	1.00	1.00	3.823	.21	.99	1.00	1.00
130	6.734	.15	.94	1.00	1.00	4.660	.19	.98	1.00	1.00	3.820	.23	1.00	1.00	1.00
140	6.727	.17	.96	1.00	1.00	4.656	.21	.99	1.00	1.00	3.817	.25	1.00	1.00	1.00
150	6.721	.18	.97	1.00	1.00	4.653	.23	.99	1.00	1.00	3.815	.28	1.00	1.00	1.00
160	6.715	.20	.98	1.00	1.00	4.650	.25	1.00	1.00	1.00	3.813	.30	1.00	1.00	1.00
170	6.710	.21	.99	1.00	1.00	4.647	.27	1.00	1.00	1.00	3.811	.33	1.00	1.00	1.00
180	6.706	.23	.99	1.00	1.00	4.645	.29	1.00	1.00	1.00	3.809	.35	1.00	1.00	1.00
190	6.702	.25	.99	1.00	1.00	4.643	.31	1.00	1.00	1.00	3.808	.38	1.00	1.00	1.00
200	6.699	.26	1.00	1.00	1.00	4.641	.33	1.00	1.00	1.00	3.806	.40	1.00	1.00	1.00
300	6.677	.43	1.00	1.00	1.00	4.629	.54	1.00	1.00	1.00	3.798	.65	1.00	1.00	1.00
10	3.767	.05	.15	.36	.65	3.377	.06	.17	.41	.71	3.103	.07	.19	.45	.76
11	3.720	.05	.16	.40	.71	3.339	.06	.18	.45	.76	3.071	.07	.20	.50	.81
12	3.681	.05	.17	.43	.76	3.308	.06	.19	.49	.81	3.046	.07	.22	.54	.85
13	3.649	.05	.18	.47	.80	3.283	.06	.20	.53	.85	3.024	.07	.23	.58	.89
14	3.622	.05	.19	.51	.84	3.261	.06	.22	.57	.89	3.007	.07	.25	.62	.92
15	3.600	.05	.20	.55	.87	3.243	.06	.23	.61	.91	2.992	.07	.26	.67	.94
16	3.580	.05	.21	.59	.90	3.228	.06	.25	.65	.94	2.979	.07	.28	.70	.96
17	3.563	.05	.23	.62	.93	3.214	.06	.26	.69	.95	2.967	.06	.30	.74	.97
18	3.548	.05	.24	.66	.94	3.202	.06	.28	.72	.97	2.957	.06	.31	.77	.98
19	3.535	.05	.26	.69	.96	3.191	.06	.29	.75	.98	2.948	.06	.33	.80	.99
20	3.523	.05	.27	.72	.97	3.182	.06	.31	.78	.98	2.940	.07	.35	.83	.99
21	3.513	.05	.29	.75	.98	3.174	.06	.33	.81	.99	2.933	.07	.37	.86	.99
22	3.503	.05	.30	.78	.98	3.166	.06	.35	.84	.99	2.927	.07	.39	.88	.99
23	3.495	.05	.32	.80	.99	3.159	.06	.36	.86	.99	2.921	.07	.41	.90	.99
24	3.487	.05	.33	.83	.99	3.153	.06	.38	.88	.99	2.916	.07	.43	.91	.99
25	3.480	.05	.35	.85	.99	3.147	.06	.40	.90	.99	2.911	.07	.45	.93	.99
26	3.473	.05	.37	.87	.99	3.142	.06	.42	.91	.99	2.907	.07	.47	.94	.99

(*Continued*)

Table B.13 (*Continued*)

		$\alpha = .05$		$df_N = 1$			$\alpha = .05$		$df_N = 2$			$\alpha = .05$		$df_N = 3$	
		Effect size, f					Effect size, f					Effect size, f			
n	F_c	.10	.25	.40	.55	F_c	.10	.25	.40	.55	F_c	.10	.25	.40	.55
27	3.467	.06	.38	.88	.99	3.137	.06	.44	.93	.99	2.902	.07	.49	.95	.99
28	3.461	.06	.40	.90	.99	3.132	.06	.46	.94	.99	2.899	.07	.51	.96	.99
29	3.456	.06	.42	.91	.99	3.128	.06	.48	.95	.99	2.895	.07	.53	.97	.99
30	3.451	.06	.43	.93	.99	3.124	.07	.49	.96	.99	2.892	.07	.55	.97	.99
31	3.447	.06	.45	.94	.99	3.120	.07	.51	.96	.99	2.889	.07	.57	.98	.99
32	3.443	.06	.47	.95	.99	3.117	.07	.53	.97	.99	2.886	.08	.59	.98	.99
33	3.439	.06	.48	.95	.99	3.114	.07	.55	.98	.99	2.883	.08	.61	.99	.99
34	3.435	.06	.50	.96	.99	3.111	.07	.57	.98	.99	2.881	.08	.62	.99	.99
35	3.431	.06	.52	.97	.99	3.108	.07	.58	.98	.99	2.878	.08	.64	.99	.99
36	3.428	.07	.53	.97	.99	3.105	.07	.60	.99	.99	2.876	.08	.66	.99	.99
37	3.425	.07	.55	.98	.99	3.103	.07	.62	.99	.99	2.874	.08	.68	.99	.99
38	3.422	.07	.57	.98	.99	3.101	.08	.64	.99	.99	2.872	.08	.69	.99	.99
39	3.419	.07	.58	.98	.99	3.098	.08	.65	.99	.99	2.870	.09	.71	.99	.99
40	3.417	.07	.60	.99	.99	3.096	.08	.67	.99	.99	2.869	.09	.73	.99	.99
45	3.406	.08	.67	.99	.99	3.087	.09	.74	.99	.99	2.861	.10	.80	.99	.99
50	3.397	.09	.74	.99	.99	3.080	.10	.80	.99	.99	2.855	.11	.85	.99	.99
55	3.389	.09	.80	.99	.99	3.074	.11	.86	.99	.99	2.850	.12	.90	.99	.99
60	3.383	.10	.84	.99	.99	3.069	.12	.90	.99	.99	2.846	.13	.93	.99	.99
70	3.374	.12	.91	.99	.99	3.062	.14	.95	.99	.99	2.839	.16	.97	.99	.99
80	3.367	.14	.95	.99	.99	3.056	.16	.98	.99	.99	2.835	.18	.99	.99	.99
90	3.362	.17	.98	.99	.99	3.052	.19	.99	.99	.99	2.831	.21	.99	.99	.99
100	3.357	.19	.99	.99	.99	3.048	.22	.99	.99	.99	2.828	.25	.99	.99	.99
110	3.354	.22	.99	.99	.99	3.045	.25	.99	.99	.99	2.826	.28	.99	.99	.99
120	3.351	.24	.99	.99	.99	3.043	.28	.99	.99	.99	2.824	.32	.99	.99	.99
130	3.348	.27	.99	.99	.99	3.041	.31	.99	.99	.99	2.822	.35	.99	.99	.99
140	3.346	.30	.99	.99	.99	3.039	.34	.99	.99	.99	2.820	.39	.99	.99	.99
150	3.344	.33	.99	.99	.99	3.038	.38	.99	.99	.99	2.819	.43	.99	.99	.99
160	3.343	.36	.99	.99	.99	3.036	.41	.99	.99	.99	2.818	.46	.99	.99	.99
170	3.341	.39	.99	.99	.99	3.035	.44	.99	.99	.99	2.817	.50	.99	.99	.99
180	3.340	.42	.99	.99	.99	3.034	.48	.99	.99	.99	2.816	.53	.99	.99	.99
190	3.339	.45	.99	.99	.99	3.033	.51	.99	.99	.99	2.816	.57	.99	.99	.99
200	3.338	.48	.99	.99	.99	3.033	.54	.99	.99	.99	2.815	.60	.99	.99	.99
300	3.332	.73	.99	.99	.99	3.027	.80	.99	.99	.99	2.811	.86	.99	.99	.99

First, determine the degrees of freedom for each effect in the ANOVA:

$$df_A = 2 = (3-1) \quad j = \text{levels of factor A}$$

$$df_B = 3 = (4-1) \quad k = \text{levels of factor B}$$

$$df_{AB} = 6 = (3-1)(4-1)$$

Next, adjust the degrees of freedom using the following equation. For this example, assume that the sample size is 10:

$$n'_{\text{effect}} = \frac{jk(n_{ij}-1)}{df_{\text{effect}}+1}+1$$

	df_N		Adjusted sample size	Rounded* sample size	Estimated power
Factor A	2	$n' = \frac{12(10-1)}{2+1}+1$	$n' = 37$	$n' = 40$	$1-\beta \approx .68$
Factor B	3	$n' = \frac{12(10-1)}{3+1}+1$	$n' = 28$	$n' = 30$	$1-\beta \approx .61$
Factor AB	6	$n' = \frac{12(10-1)}{6+1}+1$	$n' = 16.429$	$n' = 16$	$1-\beta \approx .45$

*The adjusted sample size has been rounded to match the closest values in the power tables.

Note that for effect sizes in this type of analysis,

$$f = .10 = \text{“small”} \quad f = .25 = \text{“medium”} \quad f = .40 = \text{“large”}$$

TABLE B.14: CRITICAL VALUES FOR THE CHI-SQUARE TEST

Using Table B.14

For any given *df*, the table shows the values of χ^2_{critical} corresponding to various levels of probability. The χ^2_{observed} is statistically significant at a given level when it is equal to or greater than the value shown in the table.

The following table lists methods for determining the degrees of freedom for different types of the χ^2 test.

Goodness-of-fit test	$df = k - 1$	where k represents the number of categories.
Test of independence	$df = (r - 1)(c - 1)$	where r and c represent the number of rows and columns.

Table B.14

df	$\alpha = 0.995$	$\alpha = 0.99$	$\alpha = 0.975$	$\alpha = 0.95$	$\alpha = 0.05$	$\alpha = 0.025$	$\alpha = 0.01$	$\alpha = 0.005$
1	0.000	0.000	0.001	0.004	3.841	5.024	6.635	7.879
2	0.010	0.020	0.051	0.103	5.991	7.378	9.210	10.597
3	0.072	0.115	0.216	0.352	7.815	9.348	11.345	12.838
4	0.207	0.297	0.484	0.711	9.488	11.143	13.277	14.860
5	0.412	0.554	0.831	1.145	11.070	12.832	15.086	16.750
6	0.676	0.872	1.237	1.635	12.592	14.449	16.812	18.548
7	0.989	1.239	1.690	2.167	14.067	16.013	18.475	20.278
8	1.344	1.647	2.180	2.733	15.507	17.535	20.090	21.955
9	1.735	2.088	2.700	3.325	16.919	19.023	21.666	23.589
10	2.156	2.558	3.247	3.940	18.307	20.483	23.209	25.188
11	2.603	3.053	3.816	4.575	19.675	21.920	24.725	26.757
12	3.074	3.571	4.404	5.226	21.026	23.337	26.217	28.300
13	3.565	4.107	5.009	5.892	22.362	24.736	27.688	29.819
14	4.075	4.660	5.629	6.571	23.685	26.119	29.141	31.319
15	4.601	5.229	6.262	7.261	24.996	27.488	30.578	32.801
16	5.142	5.812	6.908	7.962	26.296	28.845	32.000	34.267
17	5.697	6.408	7.564	8.672	27.587	30.191	33.409	35.718
18	6.265	7.015	8.231	9.390	28.869	31.526	34.805	37.156
19	6.844	7.633	8.907	10.117	30.144	32.852	36.191	38.582
20	7.434	8.260	9.591	10.851	31.410	34.170	37.566	39.997
21	8.034	8.897	10.283	11.591	32.671	35.479	38.932	41.401
22	8.643	9.542	10.982	12.338	33.924	36.781	40.289	42.796
23	9.260	10.196	11.689	13.091	35.172	38.076	41.638	44.181
24	9.886	10.856	12.401	13.848	36.415	39.364	42.980	45.558
25	10.520	11.524	13.120	14.611	37.652	40.646	44.314	46.928
26	11.160	12.198	13.844	15.379	38.885	41.923	45.642	48.290
27	11.808	12.878	14.573	16.151	40.113	43.195	46.963	49.645
28	12.461	13.565	15.308	16.928	41.337	44.461	48.278	50.994
29	13.121	14.256	16.047	17.708	42.557	45.722	49.588	52.335
30	13.787	14.953	16.791	18.493	43.773	46.979	50.892	53.672
31	14.458	15.655	17.539	19.281	44.985	48.232	52.191	55.002
32	15.134	16.362	18.291	20.072	46.194	49.480	53.486	56.328
33	15.815	17.073	19.047	20.867	47.400	50.725	54.775	57.648
34	16.501	17.789	19.806	21.664	48.602	51.966	56.061	58.964
35	17.192	18.509	20.569	22.465	49.802	53.203	57.342	60.275
36	17.887	19.233	21.336	23.269	50.998	54.437	58.619	61.581
37	18.586	19.960	22.106	24.075	52.192	55.668	59.893	62.883
38	19.289	20.691	22.878	24.884	53.384	56.895	61.162	64.181
39	19.996	21.426	23.654	25.695	54.572	58.120	62.428	65.475
40	20.707	22.164	24.433	26.509	55.758	59.342	63.691	66.766
50	27.991	29.707	32.357	34.764	67.505	71.420	76.154	79.490
60	35.534	37.485	40.482	43.188	79.082	83.298	88.379	91.952
70	43.275	45.442	48.758	51.739	90.531	95.023	100.425	104.215
80	51.172	53.540	57.153	60.391	101.879	106.629	112.329	116.321
90	59.196	61.754	65.647	69.126	113.145	118.136	124.116	128.299
100	67.328	70.065	74.222	77.929	124.342	129.561	135.807	140.170

Examples

$\alpha = .05$ $df = 30$

$\chi^2_{\text{critical}} = 43.773$ If $\chi^2_{\text{observed}} \leq \chi^2_{\text{critical}}$, then reject H_0.

TABLE B.15: CRITICAL VALUES FOR THE MANN–WHITNEY *U*-TEST

Using Table B.15

This table provides the critical values for the Mann–Whitney *U*-test. Note that when calculating this statistic, you can determine the value of U and U'. When calculating U, its value must be less than or equal to the tabled value to be considered statistically significant at the level of α selected. When calculating U', its value must be greater than or equal to the tabled value to be considered statistically significant at the level of α selected.

Table B.15

Critical values for U and U' for a directional test at $\alpha = .005$ or a nondirectional test at $\alpha = .01$

To reject the null hypothesis for the two sample sizes, U must be equal to or **less than** the smaller of the tabled values and U' must be equal to **or greater than** the larger of the tabled values.

$N2$ \ $N1$	1	2	3	4	5	6	7	8	9	10	11	12	13	14	15	16	17	18	19	20
1	–	–	–	–	–	–	–	–	–	–	–	–	–	–	–	–	–	–	–	–
2	–	–	–	–	–	–	–	–	–	–	–	–	–	–	–	–	–	–	0	0
																			38	40
3	–	–	–	–	–	–	–	–	0	0	0	1	1	1	2	2	2	2	3	3
									27	30	33	35	38	41	43	46	49	52	54	57
4	–	–	–	–	–	0	0	1	1	2	2	3	3	4	5	5	6	6	7	8
						24	28	31	35	38	42	45	49	52	55	59	62	66	69	72
5	–	–	–	–	0	1	1	2	3	4	5	6	7	7	8	9	10	11	12	13
					25	29	34	38	42	46	50	54	58	63	67	71	75	79	83	87
6	–	–	–	0	1	2	3	4	5	6	7	9	10	11	12	13	15	16	17	18
				24	29	34	39	44	49	54	59	63	68	73	78	83	87	92	97	102
7	–	–	–	0	1	3	4	6	7	9	10	12	13	15	16	18	19	21	22	24
				28	34	39	45	50	56	61	67	72	78	83	89	94	100	105	111	116
8	–	–	–	1	2	4	6	7	9	11	13	15	17	18	20	22	24	26	28	30
				31	38	44	50	57	63	69	75	81	87	94	100	106	112	118	124	130
9	–	–	0	1	3	5	7	9	11	13	16	18	20	22	24	27	29	31	33	36
			27	35	42	49	56	63	70	77	83	90	97	104	111	117	124	131	138	144
10	–	–	0	2	4	6	9	11	13	16	18	21	24	26	29	31	34	37	39	42
			30	38	46	54	61	69	77	84	92	99	106	114	121	129	136	143	151	158
11	–	–	0	2	5	7	10	13	16	18	21	24	27	30	33	36	39	42	45	48
			33	42	50	59	67	75	83	92	100	108	116	124	132	140	148	156	164	172
12	–	–	1	3	6	9	12	15	18	21	24	27	31	34	37	41	44	47	51	54
			35	45	54	63	72	81	90	99	108	117	125	134	143	151	160	169	177	186
13	–	–	1	3	7	10	13	17	20	24	27	31	34	38	42	45	49	53	56	60
			38	49	58	68	78	87	97	106	116	125	125	144	153	163	172	181	191	200
14	–	–	1	4	7	11	15	18	22	26	30	34	38	42	46	50	54	58	63	67
			41	52	63	73	83	94	104	114	124	134	144	154	164	174	184	194	203	213
15	–	–	2	5	8	12	16	20	24	29	33	37	42	46	51	55	60	64	69	73
			43	55	67	78	89	100	111	121	132	143	153	164	174	185	195	206	216	227
16	–	–	2	5	9	13	18	22	27	31	36	41	45	50	55	60	65	70	74	79
			46	59	71	83	94	106	117	129	140	151	163	174	185	196	207	218	230	241
17	–	–	2	6	10	15	19	24	29	34	39	44	49	54	60	65	70	75	81	86
			49	62	75	87	100	112	124	148	148	160	172	184	195	207	219	231	242	254
18	–	–	2	6	11	16	21	26	31	37	42	47	53	58	64	70	75	81	87	92
			52	66	79	92	105	118	131	143	156	169	181	194	206	218	231	243	255	268
19	–	0	3	7	12	17	22	28	33	39	45	51	56	63	69	74	81	87	93	99
		38	54	69	83	97	111	124	138	151	164	177	191	203	216	230	242	255	268	281
20	–	0	3	8	13	18	24	30	36	42	48	54	60	67	73	79	86	92	99	105
		40	57	72	87	102	116	130	144	158	172	186	200	213	227	241	254	268	281	295

Table B.15 (*Continued*)

Critical values for *U* and *U*′ for a directional test at $\alpha = .01$ or a nondirectional test at $\alpha = .02$

To reject the null hypothesis for the two sample sizes, *U* must be equal to or **less than** the smaller of the tabled values and *U*′ must be equal to **or greater than** the larger of the tabled values.

*N*2 \ *N*1	1	2	3	4	5	6	7	8	9	10	11	12	13	14	15	16	17	18	19	20
1	–	–	–	–	–	–	–	–	–	–	–	–	–	–	–	–	–	–	–	–
2	–	–	–	–	–	–	–	–	–	–	–	–	0	0	0	0	0	0	1	1
													26	28	30	32	34	36	37	39
3	–	–	–	–	–	–	0	0	1	1	1	2	2	2	3	3	4	4	4	5
							21	24	26	29	32	34	37	40	42	45	47	50	52	55
4	–	–	–	–	0	1	1	2	3	3	4	5	5	6	7	7	8	9	9	10
					20	23	27	30	33	37	40	43	47	50	53	57	60	63	67	70
5	–	–	–	0	1	2	3	4	5	6	7	8	9	10	11	12	13	14	15	16
				20	24	28	32	36	40	44	48	52	56	60	64	68	72	76	80	84
6	–	–	–	1	2	3	4	6	7	8	9	11	12	13	15	16	18	19	20	22
				23	28	33	38	42	47	52	57	61	66	71	75	80	84	89	94	98
7	–	–	0	1	3	4	6	7	9	11	12	14	16	17	19	21	23	24	26	28
			21	27	32	38	43	49	54	59	65	70	75	81	86	91	96	102	107	112
8	–	–	0	2	4	6	7	9	11	13	15	17	20	22	24	26	28	30	32	34
			24	30	36	42	49	55	61	67	73	79	84	90	96	102	108	114	120	126
9	–	–	1	3	5	7	9	11	14	16	18	21	23	26	28	31	33	36	38	40
			26	33	40	47	54	61	67	74	81	87	94	100	107	113	120	126	133	140
10	–	–	1	3	6	8	11	13	16	19	22	24	27	30	33	36	38	41	44	47
			29	37	44	52	59	67	74	81	88	96	103	110	117	124	132	139	146	153
11	–	–	1	4	7	9	12	15	18	22	25	28	31	34	37	41	44	47	50	53
			32	40	48	57	65	73	81	88	96	104	112	120	128	135	143	151	159	167
12	–	–	2	5	8	11	14	17	21	24	28	31	35	38	42	46	49	53	56	60
			34	43	52	61	70	79	87	96	104	113	121	130	138	146	155	163	172	180
13	–	0	2	5	9	12	16	20	23	27	31	35	39	43	47	51	55	59	63	67
		26	37	47	56	66	75	84	94	103	112	121	130	139	148	157	166	175	184	193
14	–	0	2	6	10	13	17	22	26	30	34	38	43	47	51	56	60	65	69	73
		28	40	50	60	71	81	90	100	110	120	130	139	149	159	168	178	187	197	207
15	–	0	3	7	11	15	19	24	28	33	37	42	47	51	56	61	66	70	75	80
		30	42	53	64	75	86	96	107	117	128	138	148	159	169	179	189	200	210	220
16	–	0	3	7	12	16	21	26	31	36	41	46	51	56	61	66	71	76	82	87
		32	45	57	68	80	91	102	113	124	135	146	157	168	179	190	201	212	222	233
17	–	0	4	8	13	18	23	28	33	38	44	49	55	60	66	71	77	82	88	93
		34	47	60	72	84	96	108	120	132	143	155	166	178	189	201	212	224	234	247
18	–	0	4	9	14	19	24	30	36	41	47	53	59	65	70	76	82	88	94	100
		36	50	63	76	89	102	114	126	139	151	163	175	187	200	212	224	236	248	260
19	–	1	4	9	15	20	26	32	38	44	50	56	63	69	75	82	88	94	101	107
		37	53	67	80	94	107	120	133	146	159	172	184	197	210	222	235	248	260	273
20	–	1	5	10	16	22	28	34	40	47	53	60	67	73	80	87	93	100	107	114
		39	55	70	84	98	112	126	140	153	167	180	193	207	220	233	247	260	273	286

(*Continued*)

Table B.15 (*Continued*)

Critical values for U and U' for a directional test at $\alpha = .025$ or a nondirectional test at $\alpha = .05$

To reject the null hypothesis for the two sample sizes, U must be equal to or **less than** the smaller of the tabled values and U' must be equal to **or greater than** the larger of the tabled values.

		*N*1																			
		1	2	3	4	5	6	7	8	9	10	11	12	13	14	15	16	17	18	19	20
*N*2	1	–	–	–	–	–	–	–	–	–	–	–	–	–	–	–	–	–	–	–	–
	2	–	–	–	–	–	–	–	0	0	0	0	1	1	1	1	1	2	2	2	2
									16	18	20	22	23	25	27	29	31	32	34	36	38
	3	–	–	–	–	0	1	1	2	2	3	3	4	4	5	5	6	6	7	7	8
						15	17	20	22	25	27	30	32	35	37	40	42	45	47	50	52
	4	–	–	–	0	1	2	3	4	4	5	6	7	8	9	10	11	11	12	13	13
					16	19	22	25	28	32	35	38	41	44	47	50	53	57	60	63	67
	5	–	–	0	1	2	3	5	6	7	8	9	11	12	13	14	15	17	18	19	20
				15	19	23	27	30	34	38	42	46	49	53	57	61	65	68	72	76	80
	6	–	–	1	2	3	5	6	8	10	11	13	14	16	17	19	21	22	24	25	27
				17	22	27	31	36	40	44	49	53	58	62	67	71	75	80	84	89	93
	7	–	–	1	3	5	6	8	10	12	14	16	18	20	22	24	26	28	30	32	34
				20	25	30	36	41	46	51	56	61	66	71	76	81	86	91	96	101	106
	8	–	0	2	4	6	8	10	13	15	17	19	22	24	26	29	31	34	36	38	41
			16	22	28	34	40	46	51	57	63	69	74	80	86	91	97	102	108	111	119
	9	–	0	2	4	7	10	12	15	17	20	23	26	28	31	34	37	39	42	45	48
			18	25	32	38	44	51	57	64	70	76	82	89	95	101	107	114	120	126	132
	10	–	0	3	5	8	11	14	17	20	23	26	29	33	36	39	42	45	48	52	55
			20	27	35	42	49	56	63	70	77	84	91	97	104	111	118	125	132	138	145
	11	–	0	3	6	9	13	16	19	23	26	30	33	37	40	44	47	51	55	58	62
			22	30	38	46	53	61	69	76	84	91	99	106	114	121	129	136	143	151	158
	12	–	1	4	7	11	14	18	22	26	29	33	37	41	45	49	53	57	61	65	69
			23	32	41	49	58	66	74	82	91	99	107	115	123	131	139	147	155	163	171
	13	–	1	4	8	12	16	20	24	28	33	37	41	45	50	54	59	63	67	72	76
			25	35	44	53	62	71	80	89	97	106	115	124	132	141	149	158	167	175	184
	14	–	1	5	9	13	17	22	26	31	36	40	45	50	55	59	64	67	74	78	83
			27	37	47	51	67	76	86	95	104	114	123	132	141	151	160	171	178	188	197
	15	–	1	5	10	14	19	24	29	34	39	44	49	54	59	64	70	75	80	85	90
			29	40	50	61	71	81	91	101	111	121	131	141	151	161	170	180	190	200	210
	16	–	1	6	11	15	21	26	31	37	42	47	53	59	64	70	75	81	86	92	98
			31	42	53	65	75	86	97	107	118	129	139	149	160	170	181	191	202	212	222
	17	–	2	6	11	17	22	28	34	39	45	51	57	63	67	75	81	87	93	99	105
			32	45	57	68	80	91	102	114	125	136	147	158	171	180	191	202	213	224	235
	18	–	2	7	12	18	24	30	36	42	48	55	61	67	74	80	86	93	99	106	112
			34	47	60	72	84	96	108	120	132	143	155	167	178	190	202	213	225	236	248
	19	–	2	7	13	19	25	32	38	45	52	58	65	72	78	85	92	99	106	113	119
			36	50	63	76	89	101	114	126	138	151	163	175	188	200	212	224	236	248	261
	20	–	2	8	13	20	27	34	41	48	55	62	69	76	83	90	98	105	112	119	127
			38	52	67	80	93	106	119	132	145	158	171	184	197	210	222	235	248	261	273

Table B.15 (*Continued*)

Critical values for U and U' for a directional test at $\alpha = .05$ or a nondirectional test at $\alpha = .10$

To reject the null hypothesis for the two sample sizes, U must be equal to or **less than** the smaller of the tabled values and U' must be equal to **or greater than** the larger of the tabled values.

*N*2	*N*1																			
	1	2	3	4	5	6	7	8	9	10	11	12	13	14	15	16	17	18	19	20
1	–	–	–	–	–	–	–	–	–	–	–	–	–	–	–	–	–	–	0	0
																			19	20
2	–	–	–	–	0	0	0	1	1	1	1	2	2	2	3	3	3	4	4	4
					10	12	14	15	17	19	21	22	24	26	27	29	31	32	34	36
3	–	–	0	0	1	2	2	3	3	4	5	5	6	7	7	8	9	9	10	11
			9	12	14	16	19	21	24	26	28	31	33	35	38	40	42	45	47	49
4	–	–	0	1	2	3	4	5	6	7	8	9	10	11	12	14	15	16	17	18
			12	15	18	21	24	27	30	33	36	39	42	45	48	50	53	56	59	62
5	–	0	1	2	4	5	6	8	9	11	12	13	15	16	18	19	20	22	23	25
		10	14	18	21	25	29	32	36	39	43	47	50	54	57	61	65	68	72	75
6	–	0	2	3	5	7	8	10	12	14	16	17	19	21	23	25	26	28	30	32
		12	16	21	25	29	34	38	42	46	50	55	59	63	67	71	76	80	84	88
7	–	0	2	4	6	8	11	13	15	17	19	21	24	26	28	30	33	35	37	39
		14	19	24	29	34	38	43	48	53	58	63	67	72	77	82	86	91	96	101
8	–	1	3	5	8	10	13	15	18	20	23	26	28	31	33	36	39	41	44	47
		15	21	27	32	38	43	49	54	60	65	70	76	81	87	92	97	103	108	113
9	–	1	3	6	9	12	15	18	21	24	27	30	33	36	39	42	45	48	51	54
		17	24	30	36	42	48	54	60	66	72	78	84	90	96	102	108	114	120	126
10	–	1	4	7	11	14	17	20	24	27	31	34	37	41	44	48	51	55	58	62
		19	26	33	39	46	53	60	66	73	79	86	93	99	106	112	119	125	132	138
11	–	1	5	8	12	16	19	23	27	31	34	38	42	46	50	54	57	61	65	69
		21	28	36	43	50	58	65	72	79	87	94	101	108	115	122	130	137	144	151
12	–	2	5	9	13	17	21	26	30	34	38	42	47	51	55	60	64	68	72	77
		22	31	39	47	55	63	70	78	86	94	102	109	117	125	132	140	148	156	163
13	–	2	6	10	15	19	24	28	33	37	42	47	51	56	61	65	70	75	80	84
		24	33	42	50	59	67	76	84	93	101	109	118	126	134	143	151	159	167	176
14	–	2	7	11	16	21	26	31	36	41	46	51	56	61	66	71	77	82	87	92
		26	35	45	54	63	72	81	90	99	108	117	126	135	144	153	161	170	179	188
15	–	3	7	12	18	23	28	33	39	44	50	55	61	66	72	77	83	88	94	100
		27	38	48	57	67	77	87	96	106	115	125	134	144	153	163	172	182	191	200
16	–	3	8	14	19	25	30	36	42	48	54	60	65	71	77	83	89	95	101	107
		29	40	50	61	71	82	92	102	112	122	132	143	153	163	173	183	193	203	213
17	–	3	9	15	20	26	33	39	45	51	57	64	70	77	83	89	96	102	109	115
		31	42	53	65	76	86	97	108	119	130	140	151	161	172	183	193	204	214	225
18	–	4	9	16	22	28	35	41	48	55	61	68	75	82	88	95	102	109	116	123
		32	45	56	68	80	91	103	114	123	137	148	159	170	182	193	204	215	226	237
19	–	4	10	17	23	30	37	44	51	58	65	72	80	87	94	101	109	116	123	130
		34	47	59	72	84	96	108	120	132	144	156	167	179	191	203	214	226	238	250
20	0	4	11	18	25	32	39	47	54	62	69	77	84	92	100	107	115	123	130	138
	20	36	49	62	75	88	101	113	126	138	151	163	176	188	200	213	225	237	250	262

Index

Understanding and Conducting Research in the Health Sciences, First Edition. Christopher J. L. Cunningham, Bart L. Weathington, and David J. Pittenger.

www.ingramcontent.com/pod-product-compliance
Lightning Source LLC
LaVergne TN
LVHW080032301125
826312LV00003B/54
* 9 7 8 1 1 1 8 1 3 5 4 0 2 *